Nadia Nedjah, Ajith Abraham and Luiza de Macedo Mourelle (Eds.)

Computational Intelligence in Information Assurance and Security

Studies in Computational Intelligence, Volume 57

Editor-in-chief
Prof. Janusz Kacprzyk
Systems Research Institute
Polish Academy of Sciences
ul. Newelska 6
01-447 Warsaw
Poland
E-mail: kacprzyk@ibspan.waw.pl

Further volumes of this series
can be found on our homepage:
springer.com

Vol. 33. Martin Pelikan, Kumara Sastry, Erick Cantú-Paz (Eds.)
Scalable Optimization via Probabilistic Modeling, 2006
ISBN 978-3-540-34953-2

Vol. 34. Ajith Abraham, Crina Grosan, Vitorino Ramos (Eds.)
Swarm Intelligence in Data Mining, 2006
ISBN 978-3-540-34955-6

Vol. 35. Ke Chen, Lipo Wang (Eds.)
Trends in Neural Computation, 2007
ISBN 978-3-540-36121-3

Vol. 36. Ildar Batyrshin, Janusz Kacprzyk, Leonid Sheremetor, Lotfi A. Zadeh (Eds.)
Preception-based Data Mining and Decision Making in Economics and Finance, 2006
ISBN 978-3-540-36244-9

Vol. 37. Jie Lu, Da Ruan, Guangquan Zhang (Eds.)
E-Service Intelligence, 2007
ISBN 978-3-540-37015-4

Vol. 38. Art Lew, Holger Mauch
Dynamic Programming, 2007
ISBN 978-3-540-37013-0

Vol. 39. Gregory Levitin (Ed.)
Computational Intelligence in Reliability Engineering, 2007
ISBN 978-3-540-37367-4

Vol. 40. Gregory Levitin (Ed.)
Computational Intelligence in Reliability Engineering, 2007
ISBN 978-3-540-37371-1

Vol. 41. Mukesh Khare, S.M. Shiva Nagendra (Eds.)
Artificial Neural Networks in Vehicular Pollution Modelling, 2007
ISBN 978-3-540-37417-6

Vol. 42. Bernd J. Krämer, Wolfgang A. Halang (Eds.)
Contributions to Ubiquitous Computing, 2007
ISBN 978-3-540-44909-6

Vol. 43. Fabrice Guillet, Howard J. Hamilton (Eds.)
Quality Measures in Data Mining, 2007
ISBN 978-3-540-44911-9

Vol. 44. Nadia Nedjah, Luiza de Macedo Mourelle, Mario Neto Borges, Nival Nunes de Almeida (Eds.)
Intelligent Educational Machines, 2007
ISBN 978-3-540-44920-1

Vol. 45. Vladimir G. Ivancevic, Tijana T. Ivancevic
Neuro-Fuzzy Associative Machinery for Comprehensive Brain and Cognition Modeling, 2007
ISBN 978-3-540-47463-0

Vol. 46. Valentina Zharkova, Lakhmi C. Jain
Artificial Intelligence in Recognition and Classification of Astrophysical and Medical Images, 2007
ISBN 978-3-540-47511-8

Vol. 47. S. Sumathi, S. Esakkirajan
Fundamentals of Relational Database Management Systems, 2007
ISBN 978-3-540-48397-7

Vol. 48. H. Yoshida (Ed.)
Advanced Computational Intelligence Paradigms in Healthcare, 2007
ISBN 978-3-540-47523-1

Vol. 49. Keshav P. Dahal, Kay Chen Tan, Peter I. Cowling (Eds.)
Evolutionary Scheduling, 2007
ISBN 978-3-540-48582-7

Vol. 50. Nadia Nedjah, Leandro dos Santos Coelho, Luiza de Macedo Mourelle (Eds.)
Mobile Robots: The Evolutionary Approach, 2007
ISBN 978-3-540-49719-6

Vol. 51. Shengxiang Yang, Yew Soon Ong, Yaochu Jin Honda (Eds.)
Evolutionary Computation in Dynamic and Uncertain Environment, 2007
ISBN 978-3-540-49772-1

Vol. 52. Abraham Kandel, Horst Bunke, Mark Last (Eds.)
Applied Graph Theory in Computer Vision and Pattern Recognition, 2007
ISBN 978-3-540-68019-2

Vol. 53. Huajin Tang, Kay Chen Tan, Zhang Yi
Neural Networks: Computational Models and Applications, 2007
ISBN 978-3-540-69225-6

Vol. 54. Fernando G. Lobo, Cláudio F. Lima and Zbigniew Michalewicz (Eds.)
Parameter Setting in Evolutionary Algorithms, 2007
ISBN 978-3-540-69431-1

Vol. 55. Xianyi Zeng, Yi Li, Da Ruan and Ludovic Koehl (Eds.)
Computational Textile, 2007
ISBN 978-3-540-70656-4

Vol. 56. Akira Namatame, Satoshi Kurihara and Hideyuki Nakashima (Eds.)
Emergent Intelligence of Networked Agents
ISBN 978-3-540-71073-8

Vol. 57. Nadia Nedjah, Ajith Abraham and Luiza de Macedo Mourella (Eds.)
Computational Intelligence in Information Assurance and Security
ISBN 978-3-540-71077-6

Nadia Nedjah
Ajith Abraham
Luiza de Macedo Mourelle
(Eds.)

Computational Intelligence in Information Assurance and Security

With 100 Figures and 28 Tables

Springer

Nadia Nedjah
Universidade do Estado do Rio de Janeiro
Faculdade de Engenharia
Sala 5022-D
Rua São Francisco Xavier 524
20550-900, MARACANÃ-RJ
Brazil
E-mail: nadia@eng.uerj.br

Luiza de Macedo Mourelle
Universidade do Estado do Rio de Janeiro
Faculdade de Engenharia
Sala 5022-D
Rua São Francisco Xavier 524
20550-900, MARACANÃ-RJ
Brazil
E-mail: ldmm@eng.uerj.br

Ajith Abraham
Centre for Quantifiable Quality of Service
in Communication Systems (Q2S)
Centre of Excellence
Norwegian University of Science
and Technology
O.S. Bragstads plass 2E
N-7491 Trondheim
Norway
E-mail: ajith.abraham@ieee.org

ISSN print edition: 1860-949X

ISBN 978-3-642-09016-5 e-ISBN 978-3-540-71078-3

Springer is a part of Springer Science+Business Media
springer.com

Softcover reprint of the hardcover 1st edition 2007

Cover design: deblik, Berlin

Preface

The global economic infrastructure is becoming increasingly dependent upon information technology, with computer and communication technology being essential and vital components of Government facilities, power plant systems, medical infrastructures, financial centers and military installations to name a few. Finding effective ways to protect information systems, networks and sensitive data within the critical information infrastructure is challenging even with the most advanced technology and trained professionals.

This volume provides the academic and industrial community with a medium for presenting original research and applications related to information assurance and security using computational intelligence techniques. The included chapters communicate current research on information assurance and security regarding both the theoretical and methodological aspects, as well as various applications in solving real world information security problems using computational intelligence.

In Chapter 1, which is entitled *Cryptography and Cryptanalysis Through Computational Intelligence*, the authors give a brief introduction to cryptography and Computational Intelligence methods, followed by a short survey of the applications of Computational Intelligence to cryptographic problems. Their contribution in this field is presented. Specifically, some cryptographic problems are viewed as discrete optimization tasks and Evolutionary Computation methods are utilized to address them. The authors show that Artificial Neural Networks are effective in approximating some cryptographic functions. They also present some theoretical issues of Ridge Polynomial Networks and cryptography.

In Chapter 2, which is entitled *Multimedia Content Protection Based on Chaotic Neural Networks*, the author gives a brief introduction to chaotic neural network based data encryption, including stream ciphers, block ciphers and hash functions. Then, he analyzes chaotic neural networks' properties that are suitable for data encryption, such as parameter sensitivity, random similarity, diffusion property, confusion property, one-way property, etc. The author also gives some proposals to design chaotic neural network based cipher or hash function, and uses these ciphers to construct media encryption and authentication methods.

In Chapter 3, which is entitled *Evolutionary Regular Substitution Boxes for Secure Cryptography Using Nash equilibrium*, the authors focus on engineering regular Substitution Boxes or S-boxes that present high non-linearity and low auto-correlation properties using evolutionary computation. There are three properties that need to be optimised: regularity, non-linearity and auto-correlation. The authors exploit the Nash equilibrium-based multi-objective evolutionary algorithm to engineer resilient substitution boxes.

In Chapter 4, which is entitled *Industrial Applications Using Wavelet Packets for Gross Error Detection*, the authors address the Gross Error Detection using uni-variate signal-based approaches and propose an algorithm for the peak noise level determination in measured signals. More specifically, they present developed algorithms and results using two uni-variate, signal-based approaches regarding performance, parameterization, commissioning, and on-line applicability. They base their approach on two algorithms: the Median Absolute Deviation (MAD) a nd the wavelet-based one. Many findings are drawn form the comparison.

In Chapter 5, which is entitled *Immune-inspired Algorithm for Anomaly Detection*, the author presents a new theory: the Danger theory and Dendritic cells, and explores the relevance of those to the application domain of security. He introduces an immune based anomaly detection approach from the abstraction of Danger theory. He also presents the derivation of bio-inspired anomaly detection from the DC functionality with Danger theory, and depicts two examples of how the proposed approach can be applied for computer and network security issues with preliminary results.

In Chapter 6, which is entitled *How to Efficiently Process Uncertainty within a Cyberinfrastructure*, the authors propose a simple solution to the problem of estimating uncertainty of the results of applying a black-box algorithm – without sacrificing privacy and confidentiality of the algorithm.

In Chapter 7, which is entitled *Fingerprint Recognition Using a Hierarchical Approach*, the authors introduce a topology-based approach to fingerprint recognition utilizing both global and local fingerprint features. They also describe a new hierarchical approach to fingerprint matching which can be used to accelerate the speed of fingerprint identification in large databases. Furthermore, the authors propose to apply Radial Basis Functions to model fingerprint's elastic deformation, which greatly increases the system's tolerance to distorted images. They claim that experimental results confirm that the proposed hierarchical matching algorithm achieves very good performance with respect to both speed and accuracy.

In Chapter 8, which is entitled *Smart Card Security*, the authors describe the various attacks that can be applied to smart cards, and the subsequent countermeasures required in software to achieve a secure solution. A case study on the various generations of the European mobile telephone networks is given as an example of how the deployment of countermeasures has changed due to the described attacks.

In Chapter 9, which is entitled *Governance of Information Security: New Paradigm of Security Management*, the author provides a structured approach of security governance to corporate executives. He summarises previous studies on the governance and security management to be able to explain the components and requirements of a governance framework for corporate security. The author provides a governance framework for corporate security, which consists of four domains and two relationship categories.

We are very much grateful to the authors of this volume and to the reviewers for their tremendous service by critically reviewing the chapters. The editors would like also to thank Prof. Janusz Kacprzyk, the editor-in-chief of the Studies in Computational Intelligence Book Series and Dr. Thomas Ditzinger, Springer Verlag, Germany for the editorial assistance and excellent cooperative collaboration to produce this important scientific work. We hope that the reader will share our excitement to present this volume on **Computational Intelligence in Information Assurance and Security** and will find it useful.

November 2006

Nadia Nedjah[1], **Ajith Abraham**[2] and **Luiza M. Mourelle**[1] (Eds.)

[1]State University of Rio de Janeiro, Brazil
[2]IITA Professorship Program, Chung-Ang University, Korea

Contents

List of Figures

List of Tables

1

Cryptography and Cryptanalysis Through Computational Intelligence

E.C. Laskari[1,4], G.C. Meletiou[2,4], Y.C. Stamatiou[3,4], and M.N. Vrahatis[1,4]

[1] Computational Intelligence Laboratory, Department of Mathematics, University of Patras, GR–26110 Patras, Greece
elena@math.upatras.gr, vrahatis@math.upatras.gr
[2] A.T.E.I. of Epirus, Arta, Greece, P.O. Box 110, GR–47100 Arta, Greece
gmelet@teiep.gr
[3] University of Ioannina, Department of Mathematics, GR–45110 Ioannina, Greece istamat@uoi.gr
[4] University of Patras Artificial Intelligence Research Center (UPAIRC), University of Patras, GR–26110 Patras, Greece

The past decade has witnessed an increasing interest in the application of Computational Intelligence methods to problems derived from the field of cryptography and cryptanalysis. This phenomenon can be attributed both to the effectiveness of these methods to handle hard problems, and to the major importance of automated techniques in the design and cryptanalysis of cryptosystems.

This chapter begins with a brief introduction to cryptography and Computational Intelligence methods. A short survey of the applications of Computational Intelligence to cryptographic problems follows, and our contribution in this field is presented. Specifically, some cryptographic problems are viewed as discrete optimization tasks and Evolutionary Computation methods are utilized to address them. Furthermore, the effectiveness of Artificial Neural Networks to approximate some cryptographic functions is studied. Finally, theoretical issues of Ridge Polynomial Networks and cryptography are presented.

The experimental results reported suggest that problem formulation and representation are critical determinants of the performance of Computational Intelligence methods in cryptography. Moreover, since strong cryptosystems should not reveal any patterns of the encrypted messages or their inner structure, it appears that Computational Intelligence methods can constitute a first measure of the cryptosystems' security.

E.C. Laskari et al.: *Cryptography and Cryptanalysis Through Computational Intelligence*, Studies in Computational Intelligence (SCI) **57**, 1–49 (2007)
www.springerlink.com

1.1 Introduction

A basic task of cryptography is the transformation, or *encryption*, of a given message into another message which appears meaningful only to the intended recipient through the process of *decryption*. The message that undergoes encryption is called the *plaintext* (or cleartext), while the transformed message is called *ciphertext*. *Cryptanalysis* refers to the process of discovering the plaintext from the ciphertext without knowing the decryption key. A cryptographic algorithm, *cipher*, is a mathematical function employed for the encryption and decryption of messages. Ciphers can be divided into two categories, the *symmetric-key* and the *public-key* ciphers. In symmetric-key ciphers the sender and the receiver of the message secretly choose the key that will be used for encryption and decryption. A drawback of this type of cryptosystems is that it requires prior communication of the key between the sender and the receiver, through a secure channel, before any message is sent.

Public-key ciphers are designed in such a way that the key used for encryption is publicly available and differs from the key used in decryption, which is secret. Although these two keys are functionally interrelated, the computation of the secret key from the public key is computationally intractable. Thus, using the public key anyone can send an encrypted message, but only the owner of the secret key can perform the decryption. Next, we briefly present the cryptographic algorithms that are used in the reported cryptanalysis experiments.

1.1.1 Block ciphers

A block cipher is a function which maps n-bit plaintext blocks to n-bit ciphertext blocks, where n is a chosen blocklength. The function is parameterized by a k-bit key K, which takes values from a subset $\mathcal{K}$, the *key space*, of the set of all k-bit vectors. The function must be invertible to allow unique decryption. Block ciphers can be either symmetric-key or public-key [85].

A *Feistel cipher* is an iterated block cipher based on the repetitive computation of simple functions, called *round functions*, on the input data, for a predetermined number of rounds. The resulting function maps an n–bit plaintext P, to a ciphertext C. In a Feistel cipher the currently computed (by the round function) n–bit word is divided into $(n/2)$–bit parts, the left part L_i and the right part R_i [32]. Then, the ith round, $1 \leqslant i \leqslant r$, has the following effect:

$$L_i = R_{i-1}, \quad R_i = L_{i-1} \oplus F_i(R_{i-1}, K_i), \tag{1.1}$$

where K_i is the subkey used in the ith round (derived from the cipher key K), and F_i is an arbitrary round function for the ith round. After the last round function has been performed, the two halves are swapped and the outcome is the ciphertext C of the Feistel cipher, i.e. $C = (R_r, L_r)$. The encryption procedure of Feistel ciphers is illustrated in Fig. 1.1.

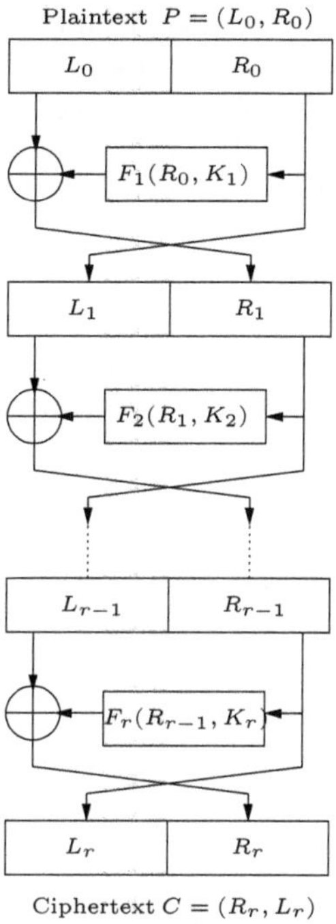

Fig. 1.1. The encryption procedure of Feistel ciphers

On Feistel based cryptosystems the decryption function is simply derived from the encryption function by applying the subkeys, K_i, and the round functions, F_i, in reverse order. This renders the Feistel structure an attractive choice for software and hardware implementations.

One of the most widely used (for non-critical applications) Feistel ciphers is the Data Encryption Standard (DES) [94]. DES is a symmetric-key cryptosystem, meaning that the parties exchanging information possess the same key. It processes plaintext blocks of $n = 64$ bits, producing 64–bit ciphertext blocks. The effective key size is $k = 64$ bits, 8 of which can be used as parity bits. The main part of the round function is the F function, which works on the right half of the data, using a subkey of 48 bits and eight S-boxes. *S-boxes* are mappings that transform 6 bits into 4 bits in a nonlinear manner and constitute the only nonlinear component of DES. The 32 output bits of the F function are XORed with the left half of the data and the two halves are subsequently exchanged. A detailed description of the DES algorithm can be found in [85, 125].

Two of the most powerful cryptanalytic attacks for Feistel based ciphers, rely on the exploitation of specific weaknesses of the S-boxes of the target cryptoalgorithm. These attacks are the *Linear Cryptanalysis* [78, 79] and the *Differential Cryptanalysis* [6, 7], which were successfully applied first to the cryptanalysis of DES. Differential Cryptanalysis (DC) is a *chosen plaintext* attack. In chosen plaintext attacks the opponent has temporary access to the encryption function and thus he/she can choose some plaintexts and construct the corresponding ciphertexts. DC analyzes the effect of particular differences in plaintext pairs on the differences of the resulting ciphertext pairs. These differences can be used to assign probabilities to the possible keys and to identify bits of the key that was used in the encryption process. This method usually works on a number of pairs of plaintexts having a specific difference and relies on the resulting ciphertext pairs only. For cryptosystems similar to DES, the difference is chosen to be a fixed XORed value of the two plaintexts.

To locate the most probable key, DC employs *characteristics*. Note that any pair of encrypted plaintexts is associated with the XOR value of its two plaintexts, the XOR value of its ciphertexts, the XOR values of the inputs of each round in the two encryption executions and the XOR values of the outputs of each round in the two encryption executions. These XOR values form an r-round characteristic [6]. More formally, an *r-round characteristic* is a tuple $\Omega = (\Omega_P, \Omega_\Lambda, \Omega_C)$, where Ω_P and Ω_C are n bit numbers and Ω_Λ is a list of r elements $\Omega_\Lambda = (\Lambda_1, \Lambda_2, \ldots, \Lambda_r)$, each of which is a pair of the form $\Lambda_i = (\lambda_I^i, \lambda_O^i)$, where λ_I^i and λ_O^i are $n/2$ bit numbers and n is the block size of the cryptosystem [6]. A characteristic satisfies the following requirements:

(a) λ_I^1 is the right half of Ω_P,
(b) λ_I^2 is the left half of $\Omega_P \oplus \lambda_O^1$,
(c) λ_I^r is the right half of Ω_C,
(d) λ_I^{r-1} is the left half of $\Omega_C \oplus \lambda_O^r$, and
(e) for every i, $2 \leqslant i \leqslant r-1$, it holds that $\lambda_O^i = \lambda_I^{i-1} \oplus \lambda_I^{i+1}$.

To each characteristic is assigned the probability of a random pair with the chosen plaintext XOR, Ω_P, having the round XOR values, Λ_i, and the ciphertext XOR, Ω_C, specified in the characteristic. Each characteristic allows the search for a particular set of bits in the subkey of the last round: the bits that enter particular S-boxes, depending on the chosen characteristic. The characteristics that are most useful are those that have a maximal probability and a maximal number of subkey bits whose occurrences can be counted.

DC is a statistical method that rarely fails. A more extended analysis on DC and its results on DES for different numbers of rounds is provided in [6]. DC was the first theoretical cryptanalysis for DES requiring (on average) less steps than the brute force attack, i.e., testing all $2^{56} = 72\,057\,594\,037\,927\,936$ possible keys. Although this number appears prohibitive, a brute-force attack on 56-bit DES, using technology standards of previous decades, has been successfully launched. A specially designed hardware with appropriate software, designed and built by the Cryptography Research, the Advanced Wireless Technologies, and the EFF (Electronic Frontier Foundation) reached a rate of key searches of about 90 billion keys per second. Their prototype, called Deep Crack, contains 29 boards each containing 64 specially designed chips. The achieved key search led to the determination of the key in the RSA DES challenge ($10.000 worth) after an, approximately, 56 hours search effort on July 15 in 1998. Moreover, total cost remained at a relatively low level, below $250.000 (much more lower today) which renders their achievement even more important and worrying as far as the security of 56-bit DES is concerned. However, as if anticipating this attack, the National Institute of Standards and Technology (NIST) had already initiated in 1997 an international contest, accepting proposals for what will become the new standard to replace DES. The contest winner was called *Advanced Encryption Standard* (AES) and it is expected to withstand attacks for a period of at least 30 years, as Miles Smid, the manager of the security technology division of NIST, stated. AES became the government standard and it is also used by private companies (on a royalty-free basis). In 1998, NIST announced the acceptance of fifteen candidate algorithms (in the first round of the process) and resorted to the cryptography community to investigate their security and efficiency. After reviewing the studies and the relevant reports, NIST selected five finalists (Round 2). Among these Rijndael, proposed by Daemen and Rijmen was selected as the new DES. The other four finalists were: MARS (proposed by IBM), RC6 (proposed by RSA Laboratories), Serpent by Anderson, Biham, and Knudsen and Twofish by Schneier, Kelsey, Whiting, Wagner, Hall, and Ferguson. In a third round, NIST concluded that the *Rijndael* cipher should become the Advanced Encryption Standard. Since then, various attacks have been proposed on this cipher but none with devastating effects.

In Sect. 1.4.2 the problem of finding some missing bits of the key that is used in a simple Feistel cipher, namely the Data Encryption Standard with four and six rounds, respectively is studied.

1.1.2 Public key cryptographic schemes

Public key cryptography is intimately related to a number of hard and complex mathematical problems from the fields of computational algebra, number theory, probability theory, mathematical logic, Diophantine's complexity and algebraic geometry. Such problems are the factorization [112], the discrete logarithm [1, 96, 104] and others [86]. Cryptosystems rely on the assumption that these problems are computationally intractable, in the sense that their computation cannot be completed in polynomial time.

Discrete Logarithm Problem (DLP): DLP amounts to the development of an efficient algorithm for the computation of an integer x that satisfies the relation $\alpha^x = \beta$, where α is a fixed primitive element of a finite field $\mathbb{F}_q$ (i.e., α is a generator of the multiplicative group $\mathbb{F}_q^*$ of $\mathbb{F}_q$) and β is a non–zero element of the field. We assume that x is the smallest nonnegative integer with $\alpha^x = \beta$. Then, x is called the *index*, or the *discrete logarithm*, of β. In the special case of a finite field $\mathbb{Z}_p$ of prime order p, a primitive root g modulo p is selected. If u is the smallest nonnegative integer with

$$g^u \equiv h \pmod{p}, \tag{1.2}$$

then u is called the *index*, or the *discrete logarithm*, of h [1, 96, 104].

The security of various public and symmetric key cryptosystems [1, 22, 29, 82, 83, 93, 95, 96, 104, 133], namely the *Diffie–Hellman exchange protocol* [25], the *El Gamal public key cryptosystem*, as well as, *the El Gamal digital signature scheme* [29], relies on the assumption that DLP is computationally intractable.

Diffie–Hellman key Problem (DHP): DHP is defined as follows [22, 80, 133]. Let α be a fixed primitive element of a finite field $\mathbb{F}_q$; x, y, satisfying, $0 \leqslant x, y \leqslant q-2$, denote the private keys of two users; and $\beta = \alpha^x$, $\gamma = \alpha^y$ represent the corresponding public keys. Then, the problem amounts to computing α^{xy} from β and γ, where α^{xy} is the symmetric key for secret communication between the two users. Consider the special case of the DHP, where $\beta = \gamma$. The term *Diffie–Hellman Mapping* refers to the mapping,

$$\beta = \alpha^x \longmapsto \alpha^{x^2}. \tag{1.3}$$

Diffie–Hellman Mapping Problem (DHMP): The definition of the DHMP follows naturally from the aforementioned definition of DHP. The two problems, DHP and DHMP, are computationally equivalent, as the following relation holds $\alpha^{x^2}\alpha^{y^2}\alpha^{2xy} = \alpha^{(x+y)^2}$, and the computation of α^{xy} from α^{2xy} is feasible (square roots over finite fields).

For the discrete logarithm problem and the Diffie–Hellman key problem, the following theorem holds:

Theorem 1. *Let G be a cyclic group of order m and $G = \langle\alpha\rangle$, where $\alpha^m = e$, and e is the neutral element of the group. Then the well–known cryptosystems of DLP and DHP based on the group G can be represented by matrices of the form*

$$\mathbf{x}^{\top} W \mathbf{y}, \tag{1.4}$$

where $\mathbf{x}, \mathbf{y}$ are vectors and $W = \{w_{ij}\}_{i,j=1}^{m} = \alpha^{-ij}$.

Proof. The proof follows by taking into consideration that there exists a prime p such that $m|p-1$ (or $m|p^n - 1 = q - 1$). Then G can be considered as a subgroup of the multiplicative group of $\mathbb{Z}_p^*$ (or $\mathrm{GF}^*(p,n)$) and according to [22, 68, 81, 93], such a representation exists.

Factorization problem: The factorization problem on the other hand, is related to the RSA cryptosystem and its variants [112]. The security of this cryptosystem relies on the computational intractability of the factorization of a positive integer $N = p \times q$, where p and q are distinct odd primes [85]. The factorization of N is equivalent to determining $\phi(N)$ from N, where $\phi(N) = (p-1) \times (q-1)$ [112]. Numerous techniques, including algebraic, number theoretic, soft computing and interpolation methods, have been proposed to tackle the aforementioned problems [1, 22, 62, 96, 120].

In Sects. 1.4.1 and 1.4.3, the DLP, the DHP, the DHMP and the factorization problem are studied in different settings utilizing Evolutionary Computation methods and Artificial Neural Networks, respectively.

1.1.3 Elliptic Curve based cryptosystems

Cryptographic systems based on elliptic curves were proposed in [57,90] as an alternative to conventional public key cryptosystems. Their main advantage is the use smaller parameters (in terms of bits) compared to the conventional cryptosystems (e.g. RSA). This is due to the apparently increased difficulty of the *Elliptic Curve Discrete Logarithm Problem* (ECDLP), which constitutes the underlying mathematical problem. ECDLP is believed to require more time to solve than the time required for the solution of its finite field analogue, the Discrete Logarithm Problem (DLP). The security of cryptosystems that rely on discrete logarithms, relies on the hypothesis these problems cannot be solved in polynomial time. Numerous techniques that exploit algebraic and number theoretic methods, software oriented methods, as well as, approximation and interpolation techniques [22, 67, 80, 83, 134], have been proposed to speed up the solution of these two types of the discrete logarithm problem.

An *Elliptic Curve* over a prime finite field $\mathbb{F}_p$, where $p > 3$ and prime, is denoted by $E(\mathbb{F}_p)$ and is defined as the set of all pairs $(x, y) \in \mathbb{F}_p$ (points in affine coordinates) that satisfy the equation $y^2 = x^3 + ax + b$, where $a, b \in \mathbb{F}_p$, with the restriction $4a^3 + 27b^2 \neq 0$. These points, together with a special point denoted by $\mathcal{O}$, called *point at infinity*, and an appropriately defined point addition operation form an Abelian group. This is the *Elliptic Curve group* and the point $\mathcal{O}$ is its identity element (see [8, 121] for more details on this group). The *order m of an elliptic curve* is defined as the number of points in $E(\mathbb{F}_p)$. According to Hasse's theorem (see e.g., [8,121]) it holds that $p + 1 - 2\sqrt{p} \leqslant m \leqslant p + 1 + 2\sqrt{p}$. The *order of a point* $P \in E(\mathbb{F}_p)$ is the smallest positive integer, n, for which $nP = \mathcal{O}$. From Lagrange's theorem, it holds that the order of a point is a divisor of the order of the elliptic curve.

DLP can be described as follows. Let G be any group and h one of its elements. Then, the DLP for G to the base $g \in G$ consists of determining an integer, u, such that $g^u = h$, when the group operation is written as multiplication, or, $ug = h$ when the group operation is written as addition. In groups formed by elliptic curve points the group operation is addition. Therefore, let E be an elliptic curve over a finite field $\mathbb{F}_q$, P a point on $E(\mathbb{F}_q)$ of order n, and Q a point on $E(\mathbb{F}_q)$, such that $Q = tP$, with $0 \leqslant t \leqslant (n-1)$. The ECDLP amounts to determining the value of t. The best algorithms to solve the ECDLP require an exponential number of expected steps, in contrast to the best algorithms known today for the DLP defined over the multiplicative group of $\mathbb{F}_q$, which require sub-exponential time in the size of the used group. In Sect. 1.4.4 the problem of computing the least significant bit of the ECDLP using Artificial Neural Netwroks is studied with interesting results.

1.2 Computational Intelligence Background and Methods

Alan Turing is considered to be the first who conceived the idea of Artificial and Computational Intelligence as early as 1950, when he hypothesized that computers that mimic the processes of the human brain can be developed. This hypothesis implies that any reasoning can be carried out on a large enough deterministic computer. Turing's hypothesis remains a vision, but it has inspired a great amount of research in the effort to embed intelligence to computers.

Although there is no commonly accepted definition, Computational Intelligence (CI) can be considered as the study of adaptive mechanisms that enable intelligent behavior of a system in complex and changing environments [28,30]. These mechanisms exhibit the ability to learn, or adapt to new situations, such that one or more attributes of reason, such as generalization, discovery, association and abstraction, are perceived to be possessed by the system. To enable intelligent behavior, CI systems are often designed to model aspects of biological and natural intelligence. Thus, CI systems are usually hybrids of paradigms such as Evolutionary Computation systems, Artificial Neural Networks and Fuzzy systems, supplemented with elements of reasoning.

1.2.1 Evolutionary Computation

Evolutionary Computation (EC) is a branch of CI that draws its inspiration from evolutionary mechanisms such as natural selection and adaptive behavior, to design optimization and classification methods. Natural selection refers to the survival of the fittest through reproduction. An offspring must retain those characteristics of its parents that are best suited to survive in a given environment. Offsprings that are weak lose the battle of survival. The EC paradigms that form this class are *Genetic Algorithms* (GA), *Genetic Programming* (GP), *Evolutionary Programming* (EP), *Evolution Strategies* (ES) and *Differential Evolution* (DE). The social and adaptive behavior of animals organized in groups inspired the development of another class of EC methods, namely *Swarm Intelligence* (SI). These methods model the social procedures of living organisms that are organized into groups and act for a common cause. Typical examples are the search for food mechanisms of bird flocks, fish schools and ant colonies. The study of many biological processes of social and adaptive behavior led to the opinion that social sharing of information among the individuals of a population can generate an evolutionary advantage [28]. Paradigms of EC that belong to this class of methods are the *Particle Swarm Optimization* (PSO) method and the *Ant Colony Optimization* (ACO) method. In the following sections a brief description of each paradigm of EC is given. Since, the PSO and DE methods will be used in our experiments in Sect. 1.4, they are more thoroughly described.

Genetic Algorithms

The experimentation of biologists in simulating natural genetic systems using computers gave rise to Genetic Algorithms (GA). John Holland is regarded as the creator of the field of GAs. He studied machine intelligence and machine learning and developed the abilities of GAs to artificial systems [45]. These systems had the ability to adapt to changes of the environment and also exhibited self–adaptation in the sense that they could adjust their operations according to their interaction with the environment. Among the innovations of Holland was the use of a population of individuals for the search procedure instead of a single search point.

The basic concepts of GAs is natural evolution and genetic inheritance. In natural evolution each biological specie has to search for the most appropriate adaptations to a complex and changing environment to ensure its survival. GAs are based on the idea that the knowledge and experience that a specie gains passes in the chromosomes of its members. For this reason the vocabulary used for GAs is that of genetics. Thus, the individuals of the population are called *chromosomes* or *genotypes*. Each chromosome consists of parts called genes and each of them is responsible for the inheritance of one or more characteristics. The evolution procedure of a population of chromosomes corresponds to a search on the space of possible problem solutions and has to balance between two different scopes, the *exploitation* of the best solutions and the *exploration* of the search space. The evolution procedure of GAs is implemented using two operators, *crossover* and *mutation*. These operators alter chromosomes to produce better ones. The selection of the new population is completed using as a criterion a fitness measure. Regarding the representation of the chromosomes, GAs usually employ binary representation, but GA methods that use other arithmetic systems, including floating point numbers, have also been developed. GAs have been successfully applied to optimization problems arising in different fields such as applied mechanics and design, time–scheduling, the traveling salesman's problem, optimal control and robotics, and economics among others [3, 24, 34, 38, 87].

Evolutionary Programming

The field of Evolutionary Programming (EP) was developed by Larry Fogel [35] parallel to that of GAs. The aim of EP was the evolution of Artificial Intelligence by predicting the changes of the environment. The environment in EP is described as a sequence of symbols from a finite set and the evolution algorithm provides as output a new symbol. This symbol has to maximize the fitness function that is used as a measure for the accuracy of the prediction. For the representation of each individual of the population finite state machines were chosen. Evolutionary Programming, just like GAs, uses the principle of the selection of the fittest for the new population, but only the mutation operator is used for altering the individuals of the population. To this initial version of EP two more basic concepts have been added. The first regards the ability of handling continuous parameters in addition to the discrete ones, and the second is the ability of self–adaptation. Using these new advances, EP can address optimization and classification problems with applications in several scientific fields as for example economics [33, 34].

Evolution Strategies

In the 1970s, Ingo Rechenberg and Hans-Paul Schwefel used the idea of mutation trying to obtain the optimal design for a sequence of joints in a liquid transition pipe. The classical optimization techniques that make use of the gradient of the fitness function were unable to handle the problem and the only solution was the experimentation with mutation. Using mutation they caused a small perturbation to the best existing problem solutions in order to explore in a stochastic manner the neighborhoods in the search space of the problem. This experimentation was the beginning of the development of Evolution Strategies, which were established in 1973 [109]. Evolution Strategies can be considered as evolutionary programs that use floating point representation and employ a *recombination* and a mutation operator. They have been used for the solution of several optimization problems with continuously changing parameters and they have been recently extended for discrete problems [42].

Genetic Programming

Genetic Programming (GP) was developed more recently by Koza [61]. The idea behind GP is the following: instead of constructing an evolutionary program to solve the problem, to locate in the space of computational programs the most proper one for the specific case. GP provides means to achieve this goal. A population of executable computational programs is created and every individual program competes with the rest. The non efficient programs become idle while the best ones reproduce by means of operators such as crossover and mutation. The evaluation of the programs is done using a fitness on a predefined set of problems.

Differential Evolution

The Differential Evolution algorithm (DE) [126] is a parallel direct numerical search method, that utilizes N, D–dimensional parameter vectors $x_{i,G}$, $i = 1, 2, \ldots, N$, as a population for each iteration (generation) of the algorithm. At each generation, the *mutation* and *crossover* (*recombination* [103, 127]) operators are applied on the individuals, to produce a new population, which is subsequently subjected to the selection phase.

For each vector $x_{i,G}$, $i = 1, 2, \ldots, N$, a *mutant vector* is generated through the following equation:

$$v_{i,G+1} = x_{r_1,G} + F\left(x_{r_2,G} - x_{r_3,G}\right), \tag{1.5}$$

where $r_1, r_2, r_3 \in \{1, 2, \ldots, N\}$, are random indexes, mutually different and different from i, and $F \in (0, 2]$. Consequently, N must be greater than, or equal to, 4. Following the mutation phase, the crossover operator is applied on the mutant vector yielding the *trial vector*, $u_{i,G+1} = (u_{1i,G+1}, u_{2i,G+1}, \ldots, u_{Di,G+1})$, where,

$$u_{ji,G+1} = \begin{cases} v_{ji,G+1}, & \text{if}\,(\text{randb}(j) \leqslant CR) \text{ or } j = \text{rnbr}(i), \\ x_{ji,G}, & \text{if}\,(\text{randb}(j) > CR) \text{ and } j \neq \text{rnbr}(i), \end{cases} \tag{1.6}$$

for $j = 1, 2, \ldots, D$; where randb(j), is the jth evaluation of a uniform random number generator in the range $[0, 1]$; CR is the (user specified) crossover constant in the range $[0, 1]$; and rnbr(i) is a randomly chosen index from the set $\{1, 2, \ldots, D\}$. To decide whether or not the vector $u_{i,G+1}$ will be a member of the population of the next generation, it is compared to the initial vector $x_{i,G}$. Thus,

$$x_{i,G+1} = \begin{cases} u_{i,G+1}, & \text{if } \; f(u_{i,G+1}) < f(x_{i,G}), \\ x_{i,G}, & \text{otherwise}. \end{cases} \tag{1.7}$$

The DE algorithm that utilizes the mutation operator of (1.5) is called the standard variant of the DE algorithm. Different mutation operators define the other variants of the DE algortihm. The mutation operators that have been applied with promising results [126], are the following:

$$v_{i,G+1} = x_{\text{best},G} + F(x_{r_1,G} - x_{r_2,G}), \tag{1.8}$$
$$v_{i,G+1} = x_{i,G} + F(x_{\text{best},G} - x_{i,G}) + F(x_{r_1,G} - x_{r_2,G}), \tag{1.9}$$
$$v_{i,G+1} = x_{\text{best},G} + F(x_{r_1,G} + x_{r_2,G} - x_{r_3,G} - x_{r_4,G}), \tag{1.10}$$
$$v_{i,G+1} = x_{r_1,G} + F(x_{r_2,G} + x_{r_3,G} - x_{r_4,G} - x_{r_5,G}), \tag{1.11}$$

where, $x_{\text{best},G}$, corresponds to the best individual of the Gth generation, r_1, r_2, r_3, r_4, $r_5 \in \{1, 2, \ldots, N\}$, are mutually different random indexes and $x_{i,G}$ is the current individual of generation G.

Particle Swarm Optimization

Particle Swarm Optimization (PSO) method is a population–based algorithm that exploits a population of individuals, to identify promising regions of the search space. In this context, the population is called *swarm* and the individuals are called *particles*. Each particle moves with an adaptable velocity within the search space, and retains in its memory the best position it ever encountered. In the *global* variant of the PSO the best position ever attained by all individuals of the swarm is communicated to all the particles. In the *local* variant, each particle is assigned to a neighborhood consisting of a prespecified number of particles. In this case, the best position ever attained by the particles that comprise the neighborhood is communicated among them [28].

Assume a D–dimensional search space, $\mathbf{S} \subset \mathbb{R}^D$, and a swarm of N particles. The ith particle is in effect a D–dimensional vector $X_i = (x_{i1}, x_{i2}, \ldots, x_{iD})^\top$. The *velocity* of this particle is also a D–dimensional vector, $V_i = (v_{i1}, v_{i2}, \ldots, v_{iD})^\top$. The *best previous position* ever encountered by the i–th particle is a point in $\mathbf{S}$, denoted by $P_i = (p_{i1}, p_{i2}, \ldots, p_{iD})^\top$. Assume g, to be the index of the particle that attained the best previous position among all the individuals of the swarm (global variant of PSO) or among all individuals of the neighborhood of the i-th particle (local variant of PSO).

Then, according to the *constriction factor* version of PSO the swarm is manipulated using the following equations [21]:

$$V_i^{(t+1)} = \chi \left(V_i^{(t)} + c_1 r_1 \left(P_i^{(t)} - X_i^{(t)} \right) + c_2 r_2 \left(P_g^{(t)} - X_i^{(t)} \right) \right), \quad (1.12)$$

$$X_i^{(t+1)} = X_i^{(t)} + V_i^{(t+1)}, \quad (1.13)$$

where $i = 1, 2, \ldots, N$; χ is the constriction factor; c_1 and c_2 denote the *cognitive* and *social* parameters respectively; r_1, r_2 are random numbers uniformly distributed in the range $[0, 1]$; and t, stands for the counter of iterations. The value of the constriction factor is typically obtained through the formula $\chi = 2\kappa / |2 - \varphi - \sqrt{\varphi^2 - 4\varphi}|$, for $\varphi > 4$, where $\varphi = c_1 + c_2$, and $\kappa = 1$. The default parameter values found in the literature [21] are $\chi = 0.729$ and $c_1 = c_2 = 2.05$. Different configurations of χ as well as a theoretical analysis of the derivation of the above formula can be found in [21].

In a different version of PSO a parameter called *inertia weight* is used, and the swarm is manipulated according to the formulae [28, 52, 118]:

$$V_i^{(t+1)} = w V_i^{(t)} + c_1 r_1 \left(P_i^{(t)} - X_i^{(t)} \right) + c_2 r_2 \left(P_g^{(t)} - X_i^{(t)} \right), \quad (1.14)$$

$$X_i^{(t+1)} = X_i^{(t)} + V_i^{(t+1)}, \quad (1.15)$$

where $i = 1, 2, \ldots, N$; and w is the inertia weight, while all other variables are the same as in the constriction factor version. There is no explicit formula for the determination of the factor w, which controls the impact of the previous history of velocities on the current one. However, since a large inertia weight facilitates global exploration (searching new areas), while a small one tends to facilitate local exploration (fine–tuning the current search area), it appears intuitively appealing to initially set it to a large value and gradually decrease it to obtain more refined solutions. The superiority of this approach against the selection of a constant inertia weight, has been experimentally verified [118]. An initial value around 1.2 and a gradual decline toward 0.1 is considered a good choice for w. Proper fine–tuning of the parameters c_1 and c_2, results in faster convergence and alleviation of local minima. As default values, $c_1 = c_2 = 2$ have been proposed, but experimental results indicate that alternative configurations, depending on the problem at hand, can produce superior performance [52, 98].

In order to avoid velocities from assuming large values that lead to fluctuation of the particles over the search space, and destroy the dynamic of the method, a maximum value, $V_{\max}$, is set for each coordinate of the velocity.

Typically, the swarm and the velocities, are initialized randomly in the search space. For more sophisticated techniques, see [97,99]. The performance of the PSO method for the Integer Programming problem and the Minimax problem was studied in [71,72], respectively, with very promising results.

Ant Colony Optimization

The Ant Colony Optimization (ACO) algorithm is a Swarm Intelligence method for tackling, in general, Combinatorial Optimization problems, like the traveling salesman problem and telecommunications scheduling. It exploits a population of members called *artificial ants* and it has been inspired from experiments with real ant colonies. In these experiments it was discovered that after a small time interval groups of ants choose the shortest between two routes to transfer food to their nest. This ability becomes possible by a chemical substance, called *pheromone*, which ants leave in the environment, that serves as an indirect communication mechanism. Thus, at the beginning the route chosen by ants appears to be random but with the progress of time the possibility of choosing the shortest path becomes higher as the quantity of pheromone on this path increases faster compared to the quantity of pheromone on longer paths. This simple idea is implemented by the ACO methods to locate solutions and address hard optimization problems [10,27].

1.2.2 Artificial Neural Networks

The complex and parallel functionality of the human brain has motivated the design of *Artificial Neural Networks* (ANNs). An ANN can be considered as a massively parallel distributed processor, comprised of simple units called *neurons*, and characterized by an inherent ability to acquire knowledge from data through a learning process. Knowledge is stored at the interneuron connection strengths, called *weights*, making it thus available for use [41]. Each artificial neuron implements a local computation. The output of this computation is determined by the neuron's input and its activation function. The overall functionality of a network is determined by its *topology* (architecture), i.e. the number of neurons and their interconnection pattern, the training algorithm applied, and its neuron characteristics [46,102].

ANNs can be categorized based on their topology, their functionality, their training methods, and other characteristics. Regarding their topology, the most simple ANNs have only one layer of neurons and are called single-layer ANNs, while the ones with more than one layers of neurons are called multilayer ANNs. Furthermore, ANNs with acyclic interneuron connections are called Feedforward Neural Networks (FNNs), while those with feedback loops are called Recurrent Neural Networks (RNNs). The most commonly used

ANNs are FNNs. A Feedforward Neural Network is a network with acyclic and one-way directed interneuron connections, where neurons can be grouped into layers. Thus, the network's topology can be described by a series of integers each representing the number of units that belong to the corresponding layer.

The functionality of ANNs is based on the type of neurons they consist of, and their activation function. In general, the are two types of neurons, summing and product neurons. Summing neurons apply their activation function over the sum of the weighted inputs, while product neurons apply their activation function over the product of the weighted inputs (see Sect. 1.5). The activation function determines the output of the neuron, and several types of activation functions can be used. The most commonly encountered ones are the *linear function* (1.16), the *threshold function* (1.17), the *sigmoid function* (1.18), the *hyperbolic tangent function* (1.19) and the *Gaussian function* (1.20).

$$f_1(x) = \alpha x, \tag{1.16}$$

$$f_2(x) = \begin{cases} \alpha_1, & \text{if } x \geqslant \theta, \\ \alpha_2, & \text{if } x < \theta, \end{cases} \tag{1.17}$$

$$f_3(x) = \frac{1}{1 + e^{-\lambda_1 x}}, \tag{1.18}$$

$$f_4(x) = \tanh(\lambda_2 x), \tag{1.19}$$

$$f_5(x) = e^{-x^2/\sigma^2}, \tag{1.20}$$

where $\alpha, \alpha_1, \alpha_2, \theta, \lambda_1, \lambda_2$ are constants and σ^2 is the variance of the Gaussian distribution. The training methods for ANNs can be divided into three categories, *supervised learning* methods in which case the ANNs must adapt to given data so as to produce a specific output; *unsupervised learning* where ANNs have to discover patterns on the input data; and *reinforcement learning* that aims at rewarding ANNs for good performance and penalize them otherwise [30, 39, 41, 58].

In the case of supervised learning, the goal of training is to assign to the weights (free parameters) of the network, W, values such that the difference between the desired output (target) and the actual output of the network is minimized. The adaptation process starts by presenting to the network a series of patterns for which the desired outputs are a priori known, and computing a total error function $E = \sum_{k=1}^{P} E_k$. In this equation, P is the number of patterns and E_k is the partial network error with respect to the kth pattern. For the computation of the partial network error a variety of error (distance) functions can be used [74, 132]. Usually, it is computed by summing the squared difference between the actual network outputs and the desired outputs for the corresponding pattern. The training patterns can be presented numerous times to the network. Each pass of all the patterns that belong to the training set, T, is called a *training epoch*. The total number of epochs required can be considered as the speed of the training algorithm. Several training algorithms can be found in [41, 73, 75, 76, 103, 110, 130].

The computational power of neural networks derives from their parallel and distributed structure and their inherent ability to adapt to specific problems, learn, and generalize. These characteristics allow ANNs to solve complex problems. In [46,131] the following statement has been proved: *"Standard feedforward networks with only a single hidden layer can approximate any continuous function uniformly on any compact set and any measurable function to any desired degree of accuracy"*. It has also been proved [102] that *"a single hidden layer feedforward network with a fixed number of units in the hidden layer, has a lower bound on the degree of the approximation of any function"*. The lower bound obstacle can be alleviated if more than one hidden layers are used. Mairov and Pincus in [102] have proved that, *"on the unit cube in $\mathbb{R}^n$ any continuous function can be uniformly approximated, to within any error by using a two hidden layer network having $2n + 1$ units in the first layer and $4n+3$ units in the second"*. Furthermore, Anthony in [2] has proved that *"there is a 2-layer threshold network capable of computing any Boolean function"*. These results imply that any lack of success in applications can be attributed to inadequate training, an insufficient number of hidden units, or the lack of a deterministic relationship between input and target.

ANNs have been applied in several scientific fields and addressed efficiently and effectively a number of hard and complex problems. Some classes of ANN applications are *function approximation*, aiming at learning the functional relationship between the inputs and the desired output, *optimization*, i.e., finding the optimal parameter values in an optimization problem, *data mining*, aiming at discovering hidden patterns in data, *classification*, i.e. prediction of the class of an input vector, *pattern matching*, where the scope is to produce a pattern that is best associated with a given input vector, *pattern completion*, where the scope is to complete missing parts of a given input vector, and *control*, where, given an input vector, an appropriate action is suggested [30].

1.2.3 Fuzzy systems

Traditional set theory and binary-valued logic both require two values of parameters, be part of a set or not, and 0 or 1, respectively. Human reasoning, however, includes a measure of uncertainty, and hence is not exact. Fuzzy sets and fuzzy logic allow what is referred to as approximate reasoning. With *fuzzy sets*, an element belongs to a set with a certain degree of certainty. *Fuzzy logic* allows reasoning with these uncertain facts to infer new facts, with a degree of certainty associated with each fact. In a sense, fuzzy sets and fuzzy logic allow the modeling of common sense [30]. The uncertainty in fuzzy systems is referred to as non statistical uncertainty, which should not be confused with statistical uncertainty. Statistical uncertainty is based on the laws of probability, whereas non statistical uncertainty is based on vagueness, imprecision and/or ambiguity. Statistical uncertainty is resolved through observations. For example, when a coin is tossed we are certain what the outcome is, while before tossing the coin, we know that the probability of each outcome is 50%. Nonstatistical uncertainty, or fuzziness, is an inherent property of a system and cannot be altered or resolved by observation. Fuzzy systems have been applied to control systems, gear transmission and braking systems in vehicles, controlling lifts, home appliances, and controlling traffic signals, among others [30, 128].

1.3 Review of Cryptography and Cryptanalysis Through Computational Intelligence

Computational Intelligence methods have been successfully applied in numerous scientific fields. Evolutionary Computation (EC) algorithms share a common characteristic, namely that they do not require good mathematical properties, such as continuity or differentiability, for the objective function of the underlying problem. Therefore, they are applicable to hard real–world optimization problems that involve discontinuous objective functions and/or disjoint search spaces [34, 52, 117]. Artificial Neural Networks (ANNs) have also been applied to many scientific fields and problem classes and provided very promising results, due to their parallel and distributed structure and their inherent ability to adapt, learn and generalize. The use of automated techniques in the design and cryptanalysis of cryptosystems is desirable as it minimizes the need for time-consuming human interaction with the search process [17]. However, due to its nature the field of cryptography and cryptanalysis is quite demanding and complex. Thus, the application of an efficient and effective tool such as Computational Intelligence (CI) to the field of cryptology comes naturally. A brief survey of the research relating the two fields follows.

The works of Peleg and Rosenfeld [100] in 1979, Hunter and McKenzie [47] in 1983, Carrol and Martin [13] in 1986 and King and Bahler [53] in 1992, that used relaxation algorithms for breaking simple substitution ciphers, can be considered as predecessors to the application of EC methods in cryptanalysis. In 1993 Spillman et al. [123, 124] and Mathews [77] introduced the use of genetic algorithms for addressing simple substitution, transposition and knapsack ciphers, while later in the same year Forsyth and Safavi-Naini [36] proposed the simulated annealing method for attacking a simple substitution algorithm. In 1995, Jakobsen [49] proposed some simplified hill-climbing techniques for addressing the problems of [36, 124] and in 1996 Vertan and Geangala [129] used genetic algorithms for breaking the Merkle–Hellman cryptosystem. Also, in 1997 Bagnall et al. [4] presented a ciphertext-only attack for a simplified version of an Enigma rotor machine using genetic algorithms.

In 1998 A. Clark proposed in his Ph.D. thesis [17] the tabu search algorithm for cryptanalysis and compared several heuristic techniques, including genetic algorithms, for breaking classical cryptosystems. In his thesis, it was also proved that the knapsack cipher attack of [123] was flawed and, furthermore, Millan, A. Clark and Dawson proposed the construction of Boolean functions with good cryptographic properties utilizing smart hill-climbing techniques and genetic algorithms [88, 89]. Continuing the work of Millan, A. Clark and Dawson, in [18] J. Clark and Jacob presented a two stage optimization for the design of Boolean functions and more recently they proposed new attacking techniques of cryptographic primitives based on fault injection and timing analysis which are effective in breaking a specific kind of identification schemes using simulated annealing [19]. In [12] Burnett et al. designed the S-boxes of MARS, one of the five AES finalists, using hill-climbing and genetic algorithms. Also, J. Clark et al. in [20] provided an improvement for the design of S-boxes using simulated annealing. In 2002 Hernández et al. proposed a new cryptanalytic technique for TEA with reduced number of rounds, which also proved to be useful in distinguishing a block cipher from a random permutation, by applying genetic algorithms [43, 44]. Finally, in 2004 Barbieri et al. [5] described a method for generating good linear block error-correcting codes that uses genetic algorithms, following the idea of genetic approach to code generation of Dontas and Jong [26].

Over the last fifteen years, just a few research studies have appeared to relate ANNs with cryptography and cryptanalysis. In 1991 Rivest wrote a survey article about the relationship between cryptography and machine learning [111], emphasizing on how these fields can contribute ideas to each other. Blum, Furst, Kearns and Lipton in [9] presented how to construct good cryptographic primitives based on problems in learning that are believed to be intractable. Working on the same concept, Pointcheval in [105, 106] used an NP-Complete problem based on ANNs for the design of certain type of secure identification problems but later Knudsen and Meier in [56] demonstrated that this scheme is less secure than what was previously believed. In 1998, Ramzan in his Ph.D. thesis [107] broke the Unix Crypt cryptosystem, a simlified variant of the Enigma cryptosystem, using ANNs. In 2001, an application of cryptology to the field of ANNs was proposed by Chang and Lu [14]. Specifically, they proposed oblivious polynomial evaluation protocols that can operate directly to floating point numbers and gave as example the oblivious learning of an ANN. Also, in [135] a general paradigm for building ANNs for visual cryptography is presented. In [50, 51] Karras and Zorkadis used Feedforward and Hopfield neural networks to improve and strengthen traditional pseudorandom stream generators for the secure management of communication systems. ANNs have also been used for the development of a new key exchange system which is based on a new phenomenon, the synchronization of ANNs [54, 55, 113]. The synchronization of ANNs is a kind of mutual training of ANNs on common inputs. However, in [55] it was shown that this key exchange protocol can be broken in three different ways, using genetic algorithms, genetic attack and probabilistic attack, respectively. Lately, in [91,115,116] some techniques for the improvement of the previously proposed protocol were presented. Finally, the idea of applying ANNs for the construction of S-boxes was presented in [60].

1.4 Applying Computational Intelligence in Cryptanalysis

In the following sections our results obtained from the application of CI methods in the cryptanalysis of known cryptosystems are presented. Specifically, in the first section cryptographic problems derived from classical public key cryptosystems are formulated as discrete optimization tasks and EC methods are applied to address them. In the next section, EC methods are considered for the partial cryptanalysis of Feistel ciphers. The effectiveness of ANNs for classical cryptographic problems and problems of elliptic curve cryptography, follow. Lastly, the relationship between a specific class of ANNs, namely the Ridge Polynomial Networks, and theoretical results of cryptography is presented.

1.4.1 Cryptanalysis as Discrete Optimization Task

In this section three problems encountered in the field of cryptology are formulated as discrete optimization tasks and two EC algorithms, namely PSO method and DE algorithm, are applied for their cryptanalysis. The reported results suggest that the formulation of the problems as discrete optimization tasks preserves their complexity which makes it difficult for the methods to extract pieces of information [64, 70]. This fact suggests that the main issue when using EC methods in cryptanalysis is the proper definition of the fitness function, i.e., avoiding the deceptive landscapes that lead in results not better than random search, which was also later mentioned in [48]. Thus, the first conclusion derived by these experiments is that, due to the proven complexity of the cryptographic problems, when EC methods are applied to cryptanalysis special attention must be paid to the design of the fitness function so as to include as much information as possible for the target problem. The second conclusion is that EC methods (and CI methods in general) can be used as a quick practical assessment for the efficiency and the effectiveness of proposed cryptographic systems. Specifically, since strong cryptosystems must not reveal any patterns of the encrypted messages or their inner structure (as this could lead to their cryptanalysis), CI methods can be used as a first measure for the evaluation of new cryptographic schemes before more formal methods (which may be complex to apply) are employed for their analysis.

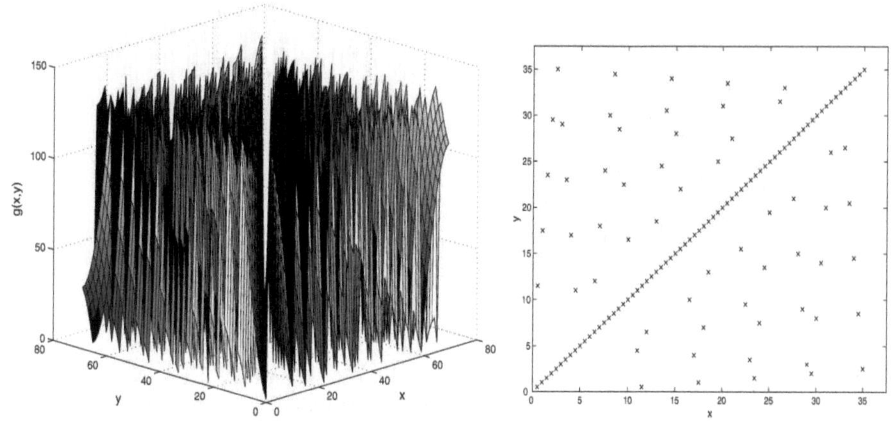

Fig. 1.2. (a) Plot of function $g(x,y) = x^2 - y^2 \pmod N$, for $N = 143$ and (b) contour plot of function $g(x,y) = x^2 - y^2 \pmod N$, for $N = 143$ at value $g = 0$

Problem Formulation

All three problems considered below are derived from the factorization problem described in Sect. 1.1.2. The first problem is defined as follows: given a composite integer N, find pairs of x, $y \in \mathbb{Z}_N^*$, such that $x^2 \equiv y^2 \pmod N$, with $x \not\equiv \pm y \pmod N$. This problem is equivalent to finding non-trivial factors of N, as N divides $x^2 - y^2 = (x-y)(x+y)$, but N does not divide either $x - y$ or $x + y$. Thus, the $\gcd(x-y, N)$ is a non-trivial factor of N [85].

We formulate the problem as a discrete optimization task by defining the minimization function $f : \{1, 2, \ldots, N-1\} \times \{1, 2, \ldots, N-1\} \mapsto \{0, 1, \ldots, N-1\}$, with

$$f(x,y) = x^2 - y^2 \pmod N, \tag{1.21}$$

subject to the constraints $x \not\equiv \pm y \pmod N$. The constraint $x \equiv -y \pmod N$ can be incorporated in the problem by changing the function domain. In this case, the problem reduces to minimizing the function $g : \{2, 3, \ldots, (N-1)/2\} \times \{2, 3, \ldots, (N-1)/2\} \mapsto \{0, 1, \ldots, N-1\}$, with

$$g(x,y) = x^2 - y^2 \pmod N, \tag{1.22}$$

subject to the constraint $x \not\equiv y \pmod N$. This is a 2-dimensional minimization problem and the global minimum of the function g is zero. For simplicity, we will call the minimization problem of function g, i.e., finding a global minimizer (x^*, y^*) of the function g, subject to the constraint $x \not\equiv y \pmod N$, as **Problem 1**. An illustration of the function $g(x,y)$ for $N = 11 \times 13 = 143$ is depicted in Fig. 1.2 (a), and the contour plot of function g at the global minimum $g(x,y) = 0$ is shown in Fig. 1.2 (b).

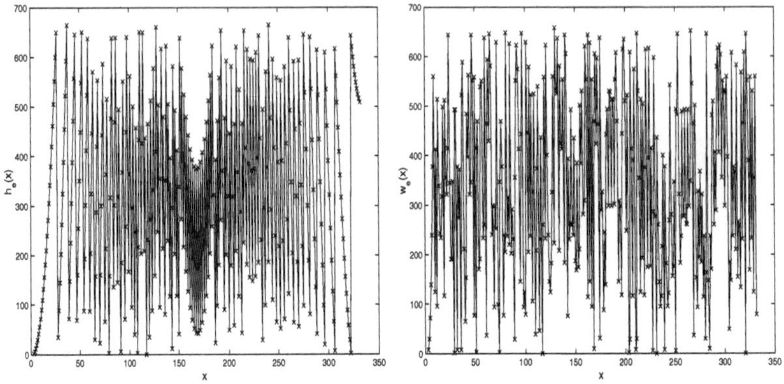

Fig. 1.3. (a) Plot of function $h_e(x) = (x-1)(x-2)(\bmod N)$, for $N = 667$ and (b) plot of function $w_e(x) = (x+1)(x-1)(x-2)(\bmod N)$, for $N = 667$

We also consider the following minimization problems. Let us define the minimization function $h : \{1, 2, \ldots, N{-}1\} \mapsto \{0, 1, \ldots, N{-}1\}$, with

$$h(x) = (x-a)(x-b) \pmod N, \tag{1.23}$$

where a, b non–zero integers and $x \not\equiv a(\bmod N)$, $x \not\equiv b(\bmod N)$. A test case of this problem is the function

$$h_e(x) = (x-1)(x-2) \pmod N, \tag{1.24}$$

where $x \not\equiv 1(\bmod N)$ and $x \not\equiv 2(\bmod N)$. This is 1-dimensional minimization problem with global minimum zero. We will refer to the minimization of function $h_e(x)$, subject to the constraints $x \not\equiv 1(\bmod N)$ and $x \not\equiv 2(\bmod N)$, as **Problem 2**. Figure 1.3 (a) depicts the function $h_e(x)$ for the small value of $N = 23 \times 29 = 667$.

In a more general setting, we can consider the minimization of the function

$$w(x) = (x-a)(x-b)\cdots(x-m) \pmod N, \tag{1.25}$$

where $x \in \{0, 1, \ldots, N{-}1\}$ and $x \not\equiv \{a, b, \ldots, m\}(\bmod N)$. We study the case function

$$w_e(x) = (x+1)(x-1)(x-2) \pmod N, \tag{1.26}$$

with $x \not\equiv \{-1, 1, 2\}(\bmod N)$. We will refer to the 1-dimensional minimization of function $w_e(x)$, subject to the constraints with $x \not\equiv \{-1, 1, 2\}(\bmod N)$, as **Problem 3.** In Fig. 1.3 (b) an illustration of the function $w_e(x)$ for the small value $N = 23 \times 29$ is shown.

Experimental Setup and Results

Both Particle Swarm Optimization (PSO) [21] and Differential Evolution (DE) [126] methods are applied on the problems formulated in the previous section (Problems 1,2,3) and compared with the simple random search technique. The global and local PSO variants of both the inertia weight and the constriction factor versions, as well as the DE variants with the mutation operators of (1.5) and (1.8), are used. The typical parameter values for the PSO variants are used (see Sect. 1.2), and the local variant of the PSO is tested for neighborhood size equal to 1. For the PSO, preliminary experiments on the specific problems indicated that the value of maximum velocity $V_{\max}$ of the particles affects its performance significantly. The most promising results were produced using the values $V_{\max} = ((N-7)/10, (N-7)/10)$ for Problem 1, and the value $V_{\max} = (N-4)/5$ for Problems 2 and 3, and therefore they are adopted in all the experiments. The parameters of the DE algorithm are set at the values $F = 0.5$ and $CR = 0.5$. In all cases, the populations are constrained to lie within the feasible region of the corresponding problem.

For the minimization of function g (Problem 1), the performance of the methods is investigated for several values of N, in the range $N = 199 \times 211 = 41\ 989$ up to $N = 691 \times 701 = 484\ 391$. For each value of N considered, 100 independent runs are performed. The corresponding results are shown in Table 1.1. In this table, PSOGW corresponds to the global variant of PSO method with inertia weight; PSOGC is the global variant of PSO with constriction factor; PSOLW is PSO's local variant with inertia weight; PSOLC is PSO's local variant with constriction factor, DE1 corresponds to the DE algorithm with the mutation operator of (1.5) and DE2 to the DE algorithm with the mutation operator of (1.8). Random search results are denoted as RS. A run is considered to be successful if the algorithm identifies a global minimizer within a prespecified number of function evaluations. The function evaluations threshold is taken equal to the cardinal of integers in the domain of the target function. The success rates of each algorithm, i.e. the fraction of times it achieved a global minimizer within the prespecified threshold, the minimum number, the median, the mean value and the standard deviation of function evaluations (F.E.) needed for success, are also reported in the table.

The experimental results of Table 1.1 indicate that the variants of the PSO method outperform in success rates the variants of the DE method over these problem instances (i.e., different values of the parameter N) and with the parameter setup used. Moreover, the performance of the DE method decreases as the value of N increases while PSO appears to be more stable with respect to this parameter. However, in contrast to the known behavior of EC methods, the best success rates achieved are relatively low (around 50%) and the random search technique (RS) outperforms both EC methods and their variants. This fact suggests that the almost random behavior of the specific kind of problem makes it quite difficult for the methods to extract knowledge about their dynamics. In the cases where EC methods located a global minimizer they

Table 1.1. Results for the minimization of function g (see (1.22))

N	Method	Suc.Rate	mean F.E.	St.D. F.E.	median F.E.	min F.E.
	PSOGW	56%	8844.643	5992.515	8325.000	660
	PSOGC	48%	7149.375	5272.590	5355.000	330
	PSOLW	51%	8329.412	6223.142	7050.000	270
$N = 199 \times 211$	PSOLC	51%	7160.588	6001.276	5940.000	420
	DE1	4%	517.500	115.866	465.000	450
	DE2	9%	5476.667	6455.651	1830.000	60
	RS	66%	9104.015	5862.358	8700.500	22
	PSOGW	41%	16210.244	11193.375	15090.000	120
	PSOGC	45%	16818.667	12664.632	13800.000	630
	PSOLW	58%	18455.690	12870.897	14520.000	270
$N = 293 \times 307$	PSOLC	50%	16374.000	13597.782	13365.000	120
	DE1	7%	1598.571	1115.488	1470.000	120
	DE2	19%	17815.263	12484.580	16290.000	2730
	RS	64%	21548.531	13926.751	20852.500	57
	PSOGW	53%	31965.849	24423.975	27570.000	780
	PSOGC	45%	32532.667	22652.983	33210.000	1740
	PSOLW	55%	31472.182	23394.791	22620.000	720
$N = 397 \times 401$	PSOLC	54%	38156.111	22925.970	37665.000	750
	DE1	1%	1680.000	0.000	1680.000	1680
	DE2	12%	27722.500	17498.736	28620.000	180
	RS	60%	27302.567	21307.031	23607.500	145
	PSOGW	56%	49893.750	37515.327	44640.000	930
	PSOGC	55%	49975.636	36727.380	41760.000	300
	PSOLW	55%	49207.091	34053.904	50430.000	2010
$N = 499 \times 503$	PSOLC	46%	48443.478	34677.039	43470.000	1920
	DE1	1%	2480.000	0.000	2480.000	2480
	DE2	8%	67245.000	35114.316	64770.000	14730
	RS	61%	54139.443	38642.970	48743.000	140
	PSOGW	52%	72175.000	48653.823	71550.000	600
	PSOGC	51%	81476.471	53666.543	75100.000	5000
	PSOLW	49%	78651.020	48197.105	67400.000	11200
$N = 599 \times 601$	PSOLC	52%	69542.308	48837.949	53050.000	2500
	DE1	2%	4700.000	4808.326	4700.000	1300
	DE2	5%	8620.000	8078.180	9300.000	800
	RS	64%	86123.656	47504.284	89392.500	904
	PSOGW	46%	207443.478	163585.340	214800.000	800
	PSOGC	46%	175426.086	138118.794	149200.000	800
	PSOLW	60%	196993.334	146204.518	144500.000	9200
$N = 691 \times 701$	PSOLC	52%	209307.692	163833.606	200100.000	1800
	DE1	2%	23800.000	25000.000	23800.000	21000
	DE2	10%	71000.000	95357.642	15200.000	1600
	RS	60%	185932.334	126355.926	154999.000	2828

required a quite small number of function evaluations with respect to the cardinal of the domain of the function. Finally, it is important to note that in the experiments where the EC methods failed in obtaining a global minimizer, they located a local minimizer with value close to the global minimum.

Similar results are obtained for the minimization of the functions h_e (Problem 2) and w_e (Problem 3), and for $N = 103 \times 107$ they are reported in Table 1.2. In the case of Problem 3, the success rates of PSO method and its variants are high (around 80%) while the performance of DE variants remains low. However, random search again outperforms the EC methods in terms of success rates.

Table 1.2. Results for functions h_e (see (1.24)) and w_e (see (1.26)), for $N = 103 \times 107$

Function	Method	Suc.Rate	mean F.E.	St.D. F.E.	median F.E.	min F.E.
	PSOGW	51%	2013.333	1483.535	1500.000	100
	PSOGC	57%	1974.035	1609.228	1420.000	60
	PSOLW	59%	1677.288	1254.688	1420.000	60
h_e	PSOLC	58%	2385.862	1676.898	2040.000	120
	DE1	1%	100.000	0.000	100.000	100
	DE2	1%	80.000	0.000	80.000	80
	RS	65%	2099.646	1448.007	2056.000	6
	PSOGW	79%	1382.785	1265.927	820.000	40
	PSOGC	84%	1402.857	1442.194	930.000	40
	PSOLW	80%	1757.750	1544.267	1110.000	40
w_e	PSOLC	85%	1416.000	1329.034	880.000	40
	DE1	1%	60.000	0.000	60.000	60
	DE2	1%	80.000	0.000	80.000	80
	RS	96%	1507.969	1328.913	1104.000	7

1.4.2 Cryptanalysis of Feistel Ciphers through Evolutionary Computation Methods

In this section two different instances of a problem introduced by the Differential Cryptanalysis of a Feistel cryptosystem are considered and formulated as optimization tasks. Specifically, the problem of finding some missing bits of the key that is used in a simple Feistel cipher, namely the Data Encryption Standard with four and six rounds, respectively is studied [65, 66]. The two instances are complementary, since every problem of missing bits of keys in Differential Cryptanalysis of Feistel ciphers can be categorized into one of the two cases.

The performance of PSO and the DE on this problem is studied. Experimental results for DES reduced to four rounds show that the optimization methods considered, located the solution efficiently, as they required a smaller number of function evaluations compared to the brute force approach. For DES reduced to six rounds the effectiveness of the proposed algorithms depends on the construction of the objective function.

Problem Formulation

DES reduced to four rounds

For DES reduced to four rounds Differential Cryptanalysis (DC) uses a one–round characteristic occurring with probability 1, recovering at the first step of the cryptanalysis 42 bits of the subkey of the last round. Considering the case where the subkeys are calculated using the DES key scheduling algorithm, the 42 bits given by DC are actual key bits of the key and there are 14 key bits still missing for the completion of the key. The brute force attack (i.e. search among all 14-bit keys) requires testing 2^{14} trials. The right key should satisfy the known plaintext XOR value for all the pairs that are used by DC. An alternative approach is to use a second characteristic that corresponds to the missing bits and attempt a more careful counting on the key bits of the last two rounds, which is however more complicated.

Instead of using the aforementioned approaches to find the missing key bits, we formulate the problem of computing the missing bits as an integer optimization problem [71]. Since the right key should satisfy the known plaintext XOR value for all the pairs that are used by DC, these ciphertexts can be used for the evaluation of possible solutions provided by optimization methods. Thus, let $\mathbf{X}$ be a 14–dimensional vector, where each of its components corresponds to one of the 14 unknown key bits. Such a vector represents a possible solution of the optimization problem. Also, let np be the number of ciphertext pairs used by DC to obtain the right 42 key bits. Then we can construct the 56 bits of the key, using the 42 bits which are recovered by DC and the 14 components of $\mathbf{X}$ in the proper order. With the resulting key, decrypt the np ciphertext pairs and count the number of decrypted pairs that satisfy the known plaintext XOR value, denoted as $cnp_{\mathbf{X}}$. Thus, the objective function f is the difference between the desired output np and the actual output $cnp_{\mathbf{X}}$, i.e. $f(\mathbf{X}) = np - cnp_{\mathbf{X}}$. The global minimum of the function f is zero and the global minimizer is with high probability the actual key. A first study of this approach is given in [64].

DES reduced to six rounds

The cryptanalysis of DES reduced to six rounds is, as expected, more complicated than that of the four round version, since the best characteristic that can be used has probability less than 1. In particular, DC uses two characteristics of probability $p_{sr} = 1/16$ to provide 42 bits of the right key. Again, there are 14 bits of the key missing. However, in this case the right key may not be suggested by all ciphertext pairs. This happens because not all the corresponding plaintexts pairs are *right pairs.* A pair is called *right pair with respect to an r-round characteristic $\Omega = (\Omega_P, \Omega_\Lambda, \Omega_C)$ and an independent key K*, if it holds that $P' = \Omega_P$, where P' is the pair's XOR value, and for the first r rounds of the encryption of the pair using the independent key K the input and output XORs of the ith round are equal to λ_I^i and λ_O^i, respectively [6].

The probability that a pair with plaintext XOR equal to Ω_P of the characteristic is a right pair using a fixed key is approximately equal to the probability of the characteristic. A pair which is not a right pair is called *wrong pair* and it does not necessarily suggest the right key as a possible value. The study of right and wrong pairs has shown that the right key appears with the probability of the characteristic from the right pairs and some other random occurrences from wrong pairs. In conclusion, if all the pairs of DC (right and wrong) are used in the predefined objective function f, the function's minimum value will change depending on the specific pairs used. On the other hand, if the right pairs are filtered and are solely used in the objective function f, the function's global minimum will be constant with value equal to 0, as in the case of missing bits of DES reduced to four rounds. As the filtering of the right pairs is not always possible and easy, we study the behavior of the proposed approach using the objective function f with all the pairs of DC.

Experimental Setup and Results

Both the PSO and DE methods were applied considering each component of the possible solution as a real number in the range $[0, 1]$ and all populations were constrained to lie within the feasible region of the problem. For the evaluation of the suggested solutions, the technique of rounding off the real values of the solution to the nearest integer [71, 108] was applied. For the PSO method we have considered both the global and local variants, and for the DE algorithm all five variants described in Sect. 1.2. A maximum value for the velocity, $V_{\max} = 0.5$, of the PSO method was set in order to avoid the swarm's explosion, i.e. avoid velocities from assuming large values that lead to fluctuation of the particles over the search space, and thus destroy the dynamic of the method. The parameters of PSO were set at the default values, i.e. $\chi = 0.729$ and $c_1 = c_2 = 2.05$, found in the literature [21], and the parameters of DE were set at equal values $CR = F = 0.5$.

The proposed approach was tested for several different initial keys and number of pairs, np. For each setting, the size of each population was equal to 100 and the performance of the methods was investigated on 100 independent runs. A run is considered to be successful if the algorithm identifies the global minimizer within a prespecified number of function evaluations. The function evaluations threshold for both problems was taken equal to 2^{14}. For the missing bits of the key of DES reduced to four rounds, the results for six different keys, k_i, $i = 1, 2, \ldots, 6$, and for test pairs, np, equal to 20 and 50 are reported in Tables 1.3, 1.4, respectively. In the tables, PSOCG denotes the global variant of the constriction factor version of PSO, PSOCL1 denotes the PSO's local variant with neighborhood size equal to 1, PSOCL2 corresponds to PSO's local variant with neighborhood size equal to 2 (see Sect. 1.2), and DE1, DE2, DE3, DE4, DE5 denote the five DE variants of (1.5),(1.8),(1.9),(1.10),(1.11), respectively. Each table reports the success rate (Suc.Rate) of each algorithm, that is the proportion of the times it achieved the global minimizer within the prespecified threshold, and the mean value of function evaluations (Mean F.E.) over the successful experiments.

The results for the problem of recovering the missing (after the application of DC) bits of DES reduced to four rounds suggest that the proposed approach is able to locate the global minimizer i.e. the 14 missing bits of the key, with relatively low computational cost compared to the brute force attack. The success rates of all versions of the two methods are high. For np equal to 20 (Table 1.3) success rates range from 93% to 100%, with an average of 99.3%. For np equal to 50 (Table 1.4) the success rates lie in the region from 90% to 100%, with mean 99.4%. The improvement of the success rate as the number of ciphertext pairs, np, increases is expected as the larger number of ciphertext pairs used for the evaluation of the possible solutions reduces the possibility of a wrong 14-tuple being suggested as the right one. The mean number of function evaluations required to locate the global minimizer (over all variants) is 1309 in the case of $np = 20$ and 982 in the case of $np = 50$. This implies that, as more ciphertext pairs are incorporated in the objective function, not only the evaluation becomes more accurate, but also the global minimizer becomes easier to locate. However, the number of ciphertext pairs used by the proposed approach should not exceed the number of the ciphertext pairs that are used by DC for the initial problem, as this would increase the total cost of the cryptanalysis in terms of encryptions and decryptions.

With respect to the different variants of the PSO method, the local variant with neighborhood size 2 (PSOLC2) accomplished success rates close to 100%, in all instances of the first problem, with an average of 1489 function evaluations. The global variant of PSO (PSOGC) achieved success rates from 93% to 100% in different instances of the problem, but with an average of 898 function evaluations. This means that, although the global variant of PSO exhibits overall lower success rates, in the cases where both local and global variants of PSO are able to locate the minimizer, the global variant requires less function evaluations than the local variant.

The DE variants exhibited a stable and similar to each other behavior, with mean success rates 100% in all cases. One minor exception was DE4 that achieved mean success rates of 99% on two instances of the problem. DE1 required the lowest mean number of function evaluations (576) among all the considered methods and their variants.

For the missing bits of the key of DES reduced to six rounds, where both right and wrong pairs are used in construction of the objective function, the

Table 1.3. Results for DES reduced to four rounds for six different keys using $np = 20$ test pairs

key	Method	Suc.Rate	Mean F.E.
k_1	PSOGC	99%	742.42
	PSOLC1	100%	1773.00
	PSOLC2	100%	1255.00
	DE1	100%	614.00
	DE2	100%	1406.00
	DE3	100%	780.00
	DE4	100%	588.00
	DE5	100%	1425.00
k_2	PSOGC	99%	911.11
	PSOLC1	100%	2665.00
	PSOLC2	100%	1650.00
	DE1	100%	603.00
	DE2	100%	1518.00
	DE3	100%	879.00
	DE4	100%	615.00
	DE5	100%	1649.00
k_3	PSOGC	94%	1117.02
	PSOLC1	99%	2447.48
	PSOLC2	100%	1688.00
	DE1	99%	693.94
	DE2	100%	1497.00
	DE3	100%	805.00
	DE4	100%	690.00
	DE5	100%	1427.00
k_4	PSOGC	96%	876.04
	PSOLC1	100%	2089.00
	PSOLC2	100%	1418.00
	DE1	99%	701.01
	DE2	100%	1378.00
	DE3	100%	843.00
	DE4	100%	568.00
	DE5	100%	1362.00
k_5	PSOGC	97%	900.00
	PSOLC1	99%	1979.80
	PSOLC2	100%	1496.00
	DE1	100%	662.00
	DE2	100%	1493.00
	DE3	100%	848.00
	DE4	100%	662.00
	DE5	100%	1542.00
k_6	PSOGC	93%	1457.00
	PSOLC1	95%	4475.79
	PSOLC2	99%	2913.13
	DE1	100%	651.00
	DE2	100%	1717.00
	DE3	100%	1063.00
	DE4	99%	725.25
	DE5	100%	1583.00

Table 1.4. Results for DES reduced to four rounds for six different keys using $np = 50$ test pairs

key	Method	Suc.Rate	Mean F.E.
k_1	PSOGC	99%	860.61
	PSOLC1	100%	1698.00
	PSOLC2	100%	1141.00
	DE1	100%	485.00
	DE2	100%	1215.00
	DE3	100%	785.00
	DE4	100%	553.00
	DE5	100%	1382.00
k_2	PSOGC	94%	741.49
	PSOLC1	100%	1367.00
	PSOLC2	100%	1100.00
	DE1	99%	490.91
	DE2	100%	1081.00
	DE3	100%	669.00
	DE4	100%	521.00
	DE5	100%	1128.00
k_3	PSOGC	99%	631.31
	PSOLC1	100%	1217.00
	PSOLC2	100%	1035.00
	DE1	100%	385.00
	DE2	100%	1006.00
	DE3	100%	546.00
	DE4	100%	409.00
	DE5	100%	1016.00
k_4	PSOGC	90%	947.78
	PSOLC1	98%	2292.88
	PSOLC2	100%	1588.00
	DE1	98%	666.33
	DE2	100%	1342.00
	DE3	100%	838.00
	DE4	99%	649.50
	DE5	100%	1294.00
k_5	PSOGC	100%	707.00
	PSOLC1	100%	1763.00
	PSOLC2	100%	1193.00
	DE1	100%	445.00
	DE2	100%	1127.00
	DE3	100%	684.00
	DE4	100%	465.00
	DE5	100%	1131.00
k_6	PSOGC	96%	880.21
	PSOLC1	100%	2009.00
	PSOLC2	100%	1390.00
	DE1	100%	507.00
	DE2	100%	1250.00
	DE3	100%	692.00
	DE4	100%	563.00
	DE5	100%	1230.00

Table 1.5. Results for DES reduced to six rounds for six different keys using $np = 200$ test pairs

key	Method	Suc.Rate	Mean F.E.
	PSOGC	26%	7038.46
	PSOLC1	9%	2188.89
	PSOLC2	8%	3862.50
	DE1	36%	5191.67
k_1	DE2	52%	5515.39
	DE3	41%	5807.32
	DE4	51%	6364.71
	DE5	59%	6855.93
	PSOGC	24%	5037.50
	PSOLC1	3%	1500.00
	PSOLC2	7%	2357.14
	DE1	34%	6535.29
k_2	DE2	58%	6968.97
	DE3	40%	5945.00
	DE4	39%	6897.44
	DE5	61%	6932.79
	PSOGC	41%	4902.44
	PSOLC1	6%	4533.33
	PSOLC2	5%	7340.00
	DE1	48%	5070.83
k_3	DE2	61%	6967.21
	DE3	53%	6698.11
	DE4	48%	5889.58
	DE5	56%	7926.79
	PSOGC	47%	4912.77
	PSOLC1	13%	4407.69
	PSOLC2	23%	4134.78
	DE1	57%	6491.23
k_4	DE2	76%	7594.74
	DE3	66%	6418.18
	DE4	72%	5741.67
	DE5	76%	7001.32
	PSOGC	36%	5575.00
	PSOLC1	4%	1950.00
	PSOLC2	5%	4700.00
	DE1	51%	5688.24
k_5	DE2	62%	7803.23
	DE3	57%	5229.83
	DE4	53%	5377.36
	DE5	64%	6387.50
	PSOGC	37%	5624.32
	PSOLC1	5%	2920.00
	PSOLC2	9%	3377.78
	DE1	49%	5681.63
k_6	DE2	63%	7380.95
	DE3	50%	7048.00
	DE4	51%	5621.57
	DE5	64%	7679.69

results for the same six different keys tested for DES reduced to four rounds, and for test pairs, np, equal to 200, are reported in Table 1.5.

From Tables 1.3,1.4,1.5, we observe that there is a considerable difference between the success rates for the case of four rounds and the case of six rounds. This can be attributed to the fact that in the former case we work with a characteristic that occurs with probability 1 while in the latter case we work with a characteristic with smaller probability (1/16). This means that in the set of 200 ciphertext pairs used by the objective function, approximately 12 pairs are right and suggest the right tuple, while the remaining 188 pairs suggest tuples at random, thus decreasing, the possibility of suggestion of the right tuple. Consequently, since the objective function becomes more effective when more right pairs are available or equivalently, when the probability of the utilized characteristic is large, it is expected that in the four round case the performance of the methods should be better than in the six round case. Although, the wrong pairs used in the objective function of DES for six rounds are misleading for the evaluation of the right tuple of missing bits, the global variant of PSO (PSOGC) and all DE variants (DE1–DE5) were able to locate the missing bits on an average of 35% of independent runs for PSOGC and 55% for the DE variants over all six different keys tested (Table 1.5). The function evaluations required for the location of the right 14–tuple of missing bits in this case are on average 5600 for all methods.

Finally, an interesting observation from the results of the proposed approach is that in the case of DES reduced to four rounds all methods in independent runs were able to locate four different 14–tuples satisfying the condition criterion of the objective function. These four solutions of the problem differed in two fixed positions, the 10th and the 36th, of the DES key. In the case of DES reduced to six rounds just one solution, the right one, was located by all methods.

The results indicate that the proposed methodology is efficient in handling this type of problems, since on DES reduced to four rounds it managed to address the problem at hand using an average of 576 function evaluations in contrast with the brute force approach that requires $2^{14} = 16384$ evaluations. Furthermore, the results of DES reduced to six rounds suggest that the effectiveness of the proposed approach depends mainly on the construction of the objective function. This approach is also applicable to all Feistel cryptosystems that are amenable to differential cryptanalysis, thus motivating its use for other Feistel cryptosystems. Finally, as a future direction, we are interested in studying the effectiveness of the proposed approach not just for missing bits of the key produced by Differential Cryptanalysis but also for all the bits of the key of Feistel ciphers.

1.4.3 Utilizing Artificial Neural Networks to Address Cryptographic Problems

In this section we consider the Artificial Neural Networks approach and study its performance on some cryptographic problems [69]. Specifically, we study the approximation of the *Discrete Logarithm Problem* (DLP) and the *Diffie Hellman key–exchange protocol Problem* (DHP) over the finite field $\mathbb{Z}_p$, where p is prime, and the *factorization problem related to the RSA cryptosystem* [112] (all three problems are presented in Sect. 1.1.2).

Experimental Setup and Results

Training algorithms: In this study the ANN training algorithms considered were the Standard Back Propagation (BP) [114], the Back Propagation with Variable Stepsize (BPVS) [75], the Resilient Back Propagation (RPROP) [110], the On-Line Adaptive Back Propagation (OABP) [73] and the Scaled Conjugate Gradient (SCG) method [92]. All methods were extensively tested with a wide range of parameters. In most of the testing cases, the training methods did not exhibit significantly different performance, except from BP, which encountered difficulties in training most of the times.

Network architecture: Since the definition of an "optimal" network architecture for any particular problem is quite difficult and remains an open problem, we tested a variety of topologies with different numbers of hidden layers and with various numbers of neurons at each layer. The results reported are the best results obtained for each problem. The architecture used is described with a series of integers denoting the number of neurons at each layer.

Data normalization: To make the adaptation of the network easier, the data are transformed through the normalization procedure, that takes place before training. Assuming that the data presented to the network are in $\mathbb{Z}_p$, where p is a prime number, the space $S = [-1, 1]$, is split in p sub-spaces. Thus, numbers in the data set are transformed to analogous ones in the space S. At the same time, the network output is transformed to a number within $\mathbb{Z}_p$ using the inverse operation.

Network evaluation: For the evaluation of the network performance we first measured the percentage of the training data, for which the network was able to compute the exact target value. This measure is denoted by μ_0. However, as network output was restricted within the range $[-1, 1]$, very small differences in output, rendered the network unable to compute the exact target but rather to be very close to it. This fact resulted in the insufficiency of the μ_0 measure as a performance indicator. Thus we employed the $\mu_{\pm v}$ measure. This measure represents the percentage of the data for which the difference between desired and actual output does not exceed $\pm v$ of the real target.

We note that the "near" measure, $\mu_{\pm v}$, has different meaning for the DLP and the DHMP. The "near" $\mu_{\pm v}$ measure is very important in the case of the DLP. If the size of the $(\pm v)$ interval is $O(\log(p))$ then the "near" measure can replace the "complete" μ_0 one. In general, for some small values of v the "near" measure is acceptable since the discrete logarithm computation can be verified i.e. computation of exponents over finite fields [104]. However, the verification of the Diffie–Hellman Mapping is an open problem. Sets of possible values for the Diffie–Hellman Mapping can be used to compute sets of possible values for the Diffie–Hellman key. The values of the Diffie–Hellman key can be tested in practice; they are symmetric keys of communication between the two users. The percentage of success for the "near" measure for DHMP can be compared with the corresponding percentage for the DLP. The results of the comparison can be related to the conjecture that the two problems are computationally equivalent.

In [84], both DLP and DHMP for several small prime numbers p have been tested. The input patterns were different values of the input variable of the discrete logarithm function and the Diffie–Hellman Mapping respectively, and the target patterns were the values of the corresponding function, for fixed chosen values of generators g and primes p. The ANNs in this case succeeded in training and generalizing, reaching up to 100%. Next, larger primes were tested rendering the task of training networks harder. Having so many numbers normalized in the range $[-1, 1]$ posed problems for the adaptation process. Thus, small changes in the network output caused complete failure, requiring the use of larger architectures, i.e., more nodes and layers. In cases with very large primes, the network performance on training was very poor. Some indicative results on training are reported in Table 1.6.

Table 1.6. Results for networks trained on the DLP and DHMP

p	Topology	Epochs	μ_0	$\mu_{\pm 2}$	$\mu_{\pm 5}$	$\mu_{\pm 10}$	Problem
83	$1-5-5-1$	20000	20%	30%	48%	70%	DLP
	$1-5-5-1$	20000	20%	35%	51%	70%	DHMP
97	$1-5-5-1$	25000	20%	30%	48%	70%	DLP
	$1-5-5-1$	20000	20%	35%	51%	70%	DHMP

The DLP is also studied in a different setting. More specifically, we have studied the case where, for several values of the prime p and the primitive root g, the value of $h = g^u (\bmod p)$, remains fixed. The input patterns consisted of pairs of primes p and the corresponding primitive roots g and the target patterns were the corresponding values of u, such that $\log_g h \equiv u (\bmod p)$, for a chosen fixed value h. We have tested the DLP in this setting for p assuming values between 101 and 2003, with several network topologies and training methods. In this case, there was a differentiation among the results obtained by different methods. For small values of p, i.e. from 101 to 199, the best results on the approximation of u, were obtained by the AOBP method. For larger values of p, the best results were given by the SCG method. Results on this new setting are reported in Table 1.7. All these results refer to training the ANNs on the approximation of the value of discrete logarithm u. Comparing the results exhibited in Tables 1.6 and 1.7, it seems that for the DLP problem, the approximation capability of the FNNs is better in the new setting.

Table 1.7. Results for networks trained on the second setting of the DLP

Range of p	Topology	Epochs	μ_0	$\mu_{\pm 15}$	$\mu_{\pm 20}$	$\mu_{\pm 30}$	$\mu_{\pm 40}$
101 – 199	2 – 15 – 1	600000	100%	100%	100%	100%	100%
503 – 1009	2 – 25 – 1	600000	82%	93%	96%	96%	98%
1009 – 2003	2 – 30 – 1	600000	17%	40%	46.7%	51.8%	54.1%
1009 – 2003	2 – 3 – 3 – 3 – 1	20000	7.5%	34.3%	44.8%	64.2%	71.6%

The ability of neural networks to address the RSA cryptosystem has also been investigated. In a previous work of ours, we have tried to approximate the $\phi(N)$ mapping, $N \mapsto \phi(N)$, with input patterns being numbers $N = p \times q$, where p and q are primes, and as target patterns the $\phi(N) = (p-1) \times (q-1)$ numbers. In this case the normalization problem was no more an obstacle. What is really interesting in this case is the generalization performance of the networks. Clearly, the networks were able not only to adapt to the training data, but also to achieve very good results with respect to the test sets [84]. Indicative results on networks trained for the $\phi(N)$ mapping are exhibited in Table 1.8.

Table 1.8. Results for networks trained for the $\phi(N)$ mapping with $N = p \times q \leqslant 10^4$

Topology	Epochs	μ_0	$\mu_{\pm 2}$	$\mu_{\pm 5}$	$\mu_{\pm 10}$	$\mu_{\pm 20}$
1 – 5 – 5 – 1	80000	3%	15%	35%	65%	90%
1 – 7 – 8 – 1	50000	6%	20%	50%	70%	100%

The factorization problem is also viewed in a different setting. More specifically, approximating the value of the function $p^2 + q^2$, given the value of N, leads directly to the factorization of N to its factors p and q. Thus, we tested the ANNs on the approximation of the aforementioned function for several instances of N. The results for this problem are reported in Table 1.9.

Table 1.9. Results for the second setting of the factorization problem for N ranging from 143 to 1003

Topology	Epochs	μ_0	$\mu_{\pm 15}$	$\mu_{\pm 20}$	$\mu_{\pm 30}$	$\mu_{\pm 40}$
$1-15-1$	200000	35.1%	36.8%	42.1%	43.8%	45.6%
$1-20-1$	600000	35.1%	43.8%	45.6%	52.6%	56.2%

Although the two settings of the factorization problem are computationally equivalent the approximation capabilities of the FNNs for this problem seem to be better for the first setting.

It is known that if a method for computing indices over finite fields is available, then the RSA cryptosystem breaks. In other words, the DLP is no easier than the factorization problem related to the RSA, which is confirmed by our experimental results.

In this study we consider only FNNs. In a future correspondence we intend to apply various other networks and learning techniques including non-monotone neural networks [11], probabilistic neural networks [122], self-organized maps [58], recurrent networks and radial basis function networks [41] among others. All data sets used in our experiments are available upon request.

1.4.4 Artificial Neural Networks Applied on Problems Related to Elliptic Curve Cryptography

In this section we study the performance of ANNs on the problem of computing the least significant bit of the discrete logarithm of a point over elliptic curves. The computation of the least significant bit of the discrete logarithm over elliptic curves with known odd order is important for cryptographic applications as it leads to the computation of all bits of the discrete logarithm. The results of this first attempt to address the specific problem using ANNs indicate that ANNs are able to adapt to the data presented with high accuracy, while the response of ANNs to unknown data is slightly higher than random selection. Another important finding is that ANNs require a small amount of storage for the known patterns in contrast to the storage needed for the data set itself [63].

Problem Formulation

For the discrete logarithm problem over elliptic curves, the following proposition is derived from the bit security of discrete logarithms over any cyclic group [101, 125].

Proposition 1. *Given an elliptic curve, E, over a finite field, $\mathbb{F}_q$, with known order n, and an oracle for a bit of the discrete logarithm that does not correspond to any power of 2 that divides the order n, then all the bits of the discrete logarithm can be computed in polynomial time.*

Remark 1. Currently, there is no polynomial algorithm for finding the order of an elliptic curve. Furthermore, the complexity for the computation of the discrete logarithm problem over elliptic curves with no knowledge of its order is exponential and, hence, it remains a computationally difficult task.

From Proposition 1 it is derived that in the case of an elliptic curve with odd order n, an oracle that gives the least significant bit of the discrete logarithm of a point over the elliptic curve leads to the computation of all bits of the discrete logarithm in polynomial time. Furthermore, prime order elliptic curves are considered more secure. Thus, our focus is on the computation of the least significant bit of the discrete logarithm of a point over elliptic curves of odd order. Complexity estimates for the computation of bits of the discrete logarithm over different fields can be found in [22, 120].

In relation to our problem, the considered Boolean function is defined as follows. Assume an elliptic curve $E(\mathbb{F}_p)$ and let $P = (x_P, y_P)$, $Q = (x_Q, y_Q)$ be two points of $E(\mathbb{F}_p)$, such that $Q = tP$, with $0 \leqslant t \leqslant (n-1)$. Define the Boolean function $f : \{0,1\}^{4\lceil \log p \rceil} \mapsto \{0,1\}$, with

$$f(x_P, y_P, x_Q, y_Q) = \mathrm{lsb}(t), \tag{1.27}$$

which has inputs the coordinates x_P, y_P, x_Q, y_Q, in binary representation, and outputs the least significant bit (lsb) of t, i.e., 1 if the least significant bit of t is 1, and 0 otherwise. In general, a Boolean circuit that computes this function can be exponentially large in $\lceil \log p \rceil$ [22]. For the computation of this Boolean function we employ Artificial Neural Networks. Here, we focus on FNNs for the approximation of the Boolean function derived by elliptic curve cryptography defined in (1.27). For the general problem of the computation of a Boolean function by a FNNs, the following theorem proved in [2], supports the effectiveness of the proposed approach.

Theorem 2. *There is a threshold network with one hidden layer capable of computing any Boolean function.*

Experimental Setup and Results

Training ANNs with threshold units requires the use of training methods that do not employ information about the derivatives of the error function. Furthermore, as shown in [23], analog neural networks can be more powerful than neural networks using thresholds, even for the computation of Boolean functions. Thus, we study the performance of ANNs using the hyperbolic tangent activation function of (1.19), which approximates a threshold function as λ_2 tends to infinity. In all experiments the output layer consists of two neurons, and the neuron with the highest output value determines the class in which the computed bit is classified. Thus, if the first neuron's output value is smaller than the value of the second neuron, the bit is considered to belong to Class 0, which corresponds to a "0" value of the bit, and vice versa. This setting enables us to use training methods that employ derivatives of the error function. In particular, we have studied the performance of three training algorithms each from a different category of training algorithms, namely the Resilient Back Propagation method (RPROP) [110], the Adaptive On–line Back Propagation method (AOBP) [73] and the Differential Evolution algorithm (DE) [126]. Regarding the topology of the networks, we have tested a variety of topologies with various numbers of neurons at each layer. We report only the best results obtained for each problem.

For the construction of the datasets the ECC_LIB library for elliptic curve cryptography [59] was used. The performance of ANNs was tested for three different datasets of the considered Boolean function that correspond to randomly chosen p's of bit length 14, 20, and 32, respectively, where $\mathbb{F}_p$ is the finite field over which the elliptic curve is constructed. All data sets used are available upon request.

At each experiment the dataset was randomly partitioned into a training set and a test set. Two thirds of the dataset were assigned to the training set and the remaining one third comprised the test set. To evaluate the network performance, first we measure the average of the percentage of the training set over all 10 experiments, for which the network was able to correctly predict the least significant bit. Then, the network's performance is evaluated by measuring the average percentage of the test set over all experiments.

The best results, for the prescribed setting and $\lambda_2 = 1$, were obtained using the AOBP method and are reported in Tables 1.10, 1.11 and 1.12, respectively. The results indicate that for three bit lengths, ANNs are able to adapt to the training data with an average accuracy of 90%. With respect to the test sets, ANNs achieved for all three bit lengths an average accuracy of 57%, i.e. a slightly higher than random selection. Regarding the training epochs required in each case, as the bit length of p increases, more epochs are needed to achieve the same accuracy.

Another interesting finding regarding the training set, is that the network is able to learn the training patterns and respond correctly about the least significant bit of the discrete logarithm, using less storage than that required by the corresponding dataset. The results for the data compression are reported in Table 1.13. In Table 1.13, "BL(p)" denotes the bit length of p, "Data Stor." denotes the storage bits required for the dataset, "ANN Stor." denotes the storage bits required for the network weights and "Accuracy" corresponds to the accuracy of the network to identify the desired value for both classes.

An interesting line of further research is to study the performance of ANNs for larger values of p and elliptic curves of different order, as well as, in other related problems such as the computation of the order of elliptic curves.

1.5 Ridge Polynomial Networks for Cryptography

Ridge Polynomial Networks (RPNs) belong to the class of ANNs that are based on product type of neurons, i.e., neurons apply their activation function over the product of the weighted inputs. RPNs exhibit several advantages compared to ANNs based on summing units. The computation of the least significant bit of the discrete logarithm is important for cryptographic applications as it is related to the computation of all bits of the discrete logarithm [40]. For this reason, in this section we relate these two aspects providing some theoretical conclusions and insights for future research.

For completeness purposes let us first introduce the Pi-Sigma networks that are the building components of the RPNs, along with some theoretical background of RPNs. A Pi-Sigma network (PSN) is a feedforward network

Table 1.10. Results for p of bit length 14, using $56-3-2$ topology

Epochs		Train			Test		
		Class 0	Class 1	Accuracy	Class 0	Class 1	Accuracy
500	Class 0	168	33	83.58%	30	24	55.56%
	Class 1	48	151	75.88%	23	23	50.00%
650	Class 0	184	17	91.54%	33	21	61.11%
	Class 1	32	167	83.92%	23	23	50.00%
700	Class 0	183	18	91.04%	33	21	61.11%
	Class 1	30	169	84.92%	21	25	54.35%
1000	Class 0	186	15	92.54%	33	21	61.11%
	Class 1	25	174	87.44%	18	28	60.87%

Table 1.11. Results for p of bit length 20, using $80-3-2$ topology

Epochs		Train			Test		
		Class 0	Class 1	Accuracy	Class 0	Class 1	Accuracy
2000	Class 0	186	14	93.0%	32	26	55.17%
	Class 1	23	177	88.5%	17	25	59.52%
3000	Class 0	191	9	95.5%	30	28	51.72%
	Class 1	19	181	90.5%	21	21	50.00%
4000	Class 0	194	6	98.0%	32	26	55.17%
	Class 1	18	182	91.0%	19	23	54.76%
6000	Class 0	196	4	98.0%	33	25	56.90%
	Class 1	17	183	91.5%	20	22	52.38%

Table 1.12. Results for p of bit length 32, using $128-3-2$ topology

Epochs		Train			Test		
		Class 0	Class 1	Accuracy	Class 0	Class 1	Accuracy
4000	Class 0	193	5	97.47%	36	21	63.16%
	Class 1	16	186	92.08%	20	23	53.49%
5000	Class 0	193	5	97.47%	36	21	63.16%
	Class 1	15	187	92.57%	19	24	55.81%
8000	Class 0	193	5	97.47%	35	22	61.40%
	Class 1	14	188	93.07%	18	25	58.14%
9000	Class 0	193	5	97.47%	35	22	61.40%
	Class 1	14	188	93.07%	16	27	62.79%

Table 1.13. Data compression results

BL(p)	Data Stor.	ANN Stor.	Accuracy
14	23200	8400	89.99%
20	32800	11856	94.75%
32	52000	18768	95.27%

with a single "hidden" layer of linear units that uses product units in the output layer, i.e., it uses *products of sums* of input components. The presence of only one layer of adaptive weights at PSNs results in fast training. There are two types of PSNs, the *Analog Pi-Sigma Networks* (APSNs) and the *Binary Pi-Sigma Networks* (BPSNs). A generalization of APSN, the Ridge polynomial networks (RPNs) is proved to have universal approximation capability [37]. BPSNs, on the other hand, are capable of realizing any Boolean function [119].

In Fig. 1.4 a PSN with a single output is illustrated. This network is a fully connected two-layered feedforward network. However, the summing layer is not "hidden" as in the case of the Multilayer Perceptron (MLP), since the weights from this layer to the output layer are fixed at the value 1. This property contributes to reducing the required training time.

Let $\mathbf{x} = (1, x_1, \ldots, x_N)^\top$ be an $N+1$-dimensional augmented input column vector, where x_k denotes the k-th component of $\mathbf{x}$. The inputs are weighted by K $(N+1)$-dimensional weight vectors $\mathbf{w}_j = (w_{0j}, w_{1j}, \ldots, w_{Nj})^\top$, $j = 1, 2, \ldots, K$ and summed by a layer of K linear summing units, where k is the desired order of the network. The output of the jth summing unit, h_j, is given as follows:

$$h_j = \mathbf{w}_j^\top \mathbf{x} = \sum_{k=1}^{N} w_{kj} x_k + w_{0j}, \quad j = 1, 2, \ldots, K. \tag{1.28}$$

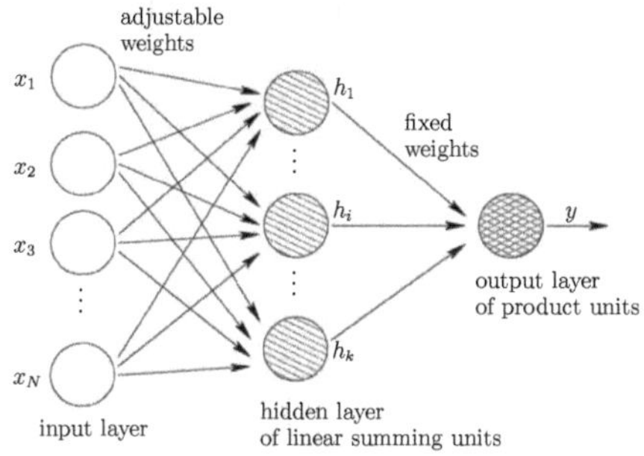

Fig. 1.4. A Pi-Sigma network (PSN) with one output unit

The output y of the network is given by:

$$y = \sigma(\prod_{j=1}^{K} h_j) = \sigma(net), \tag{1.29}$$

where $\sigma(\cdot)$ is a suitable activation function and $net = \prod_{j=1}^{K} h_j$. In the above, w_{kj} is an adjustable weight from input x_k to the jth summing unit and w_{0j} is the threshold of the jth summing unit. Weights can assume arbitrary real values.

The network shown in Fig. 1.4 is called a K-th order PSN since K summing units are incorporated. The total number of adjustable weight connections, including the adjustable thresholds, for a K-th order PSN with N-dimensional inputs is $(N+1)K$. If multiple outputs are required, an independent summing layer is needed for each output. Thus, for an M-dimensional output vector $\mathbf{y}$, a total of $\sum_{i=1}^{M}(N+1)K_i$ adjustable weight connections are present, where K_i is the number of summing units for the ith output. This enables the network to be incrementally expandable, since the order can be increased by adding another summing unit and associated weights, without disturbing any connection previously established. PSNs can handle both analog and binary input/output by using a suitable nonlinear activation function $\sigma(\cdot)$.

Regarding the approximation capabilities of PSNs, although the activation function is applied on a K-th order polynomial when K summing units are used and the exponents i_j sum up to K, this does not mean that only K-th order terms can be used, since by letting an extra input (the bias) be fixed at value 1, terms of order less than K are also realized. This K-th order polynomial, however, does not have full degrees of freedom since the coefficients are composed of sums and products of w_{kj}s and thus are not independent. Thus, a PSN cannot uniformly approximate all continuous multivariate functions that can be defined on a compact set. However, the theory of ridge polynomials can be used to show that universal approximation capability can be achieved simply by summing the outputs of APSNs of different orders. The resulting network is a generalization of PSN, which is called *Ridge Polynomial Network* (RPN), and is developed as follows.

For $\mathbf{x} = (x_1, \ldots, x_N)^\top$ and $\mathbf{w} = (w_1, \ldots, w_N)^\top \in \mathbb{R}^N$, we denote as $\langle \mathbf{x}, \mathbf{w} \rangle$ their inner product, i.e., $\langle \mathbf{x}, \mathbf{w} \rangle = \sum_{i=1}^{N} x_i w_i$. For a given compact set $C \subset \mathbb{R}^N$, all functions defined on C in the form of $f(\langle \mathbf{x}, \mathbf{w} \rangle)$, where f is a continuous function in one variable, are called ridge functions. A *ridge polynomial* is a ridge function that can be represented as

$$\sum_{i=0}^{n} \sum_{j=1}^{m} a_{ij} \langle \mathbf{x}, \mathbf{w} \rangle^i, \tag{1.30}$$

for some $a_{ij} \in \mathbb{R}$ and $\mathbf{w}_{ij} \in \mathbb{R}^N$.

It was proved in [15,16] that any polynomial in $\mathbb{R}^N$ with degree less than or equal to k can be represented by a ridge polynomial and, furthermore, it can be realized by an RPN [37]. From these results and the Stone-Weierstrass theorem the uniform approximation capability of the ridge polynomials of (1.30) is implied [37]. For applications of the RPNs in approximation and root finding see [31].

The ridge polynomial network (RPN), is defined as a feedforward network based on the generalized form of ridge polynomials [37]:

$$p(\mathbf{x}) = \sum_{j=1}^{n_{\text{total}}} \prod_{i=1}^{j} (\langle \mathbf{x}, \mathbf{w}_{ji} \rangle + w_{ij}), \tag{1.31}$$

where $n_{\text{total}} = \sum_{l=0}^{k} n_l$, and approximates an unknown function f on a compact set $C \subset \mathbb{R}^N$ as

$$\begin{aligned} f(\mathbf{x}) \approx (\langle \mathbf{x}, \mathbf{w}_{11} \rangle + w_{11}) + (\langle \mathbf{x}, \mathbf{w}_{21} \rangle + w_{21})(\langle \mathbf{x}, \mathbf{w}_{22} \rangle + w_{22}) + \cdots \\ +(\langle \mathbf{x}, \mathbf{w}_{N1} \rangle + w_{N1}) \cdots (\langle \mathbf{x}, \mathbf{w}_{NN} \rangle + w_{NN}). \end{aligned} \tag{1.32}$$

Each product term in (1.32) can be obtained as the output of a PSN with linear output units. Thus, the formulation of RPNs can be considered as a generalization of PSNs. Figure 1.5 represents a generic network architecture of the RPN using PSNs as building blocks. The RPN has only a single layer of adjustable weights which is beneficial in terms of training speed. Note that (1.32) serves as the basis for an incremental learning algorithm where PSNs of successively higher orders can be added until a desirable level of accuracy is obtained.

The uniform approximation capability of RPNs, their faster training compared to other kinds of ANNs, and the ability to perform incremental training, renders them a promising methodology for application to cryptological problems. Some conclusions derived by relating these two fields follow, providing an insight for future research.

Considering the computation of the rightmost bit of the discrete logarithm by real polynomials, the following theorem holds:

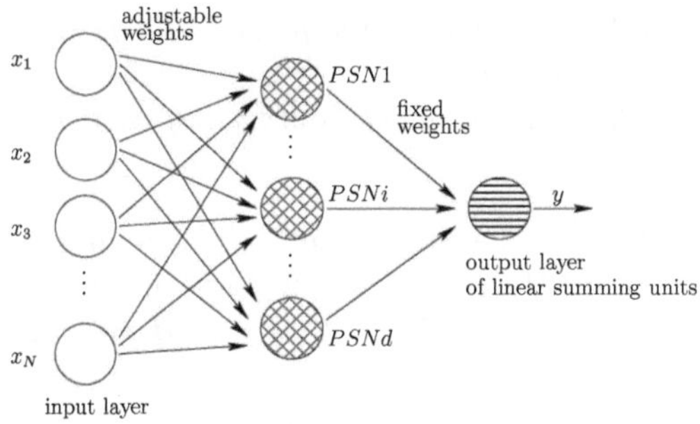

Fig. 1.5. A ridge polynomial network (RPN) with one linear output unit

Theorem 3 ([120]). *Let $0 \leqslant M < M+H \leqslant p-1$. Assume that a polynomial $f(X) \in \mathbb{R}[X]$ is such that $f(x) \geqslant 0$ if x is a quadratic residue modulo p and $f(x) < 0$ otherwise, for every element $x \in S$ from some set $S \subseteq \{M+1,\dots,M+H\}$ of cardinality $|S| \geqslant H-s$. Then for any $\varepsilon > 0$ the bound*

$$\deg f \geqslant \begin{cases} H/2 - 2s - 1 - p^{1/2}\log p, & \text{for any } H, \\ \mathcal{C}(\varepsilon)\min\{H, H^2p^{-1/2}\} - 2s - 1, & \text{if } p^{1/4+\varepsilon} \leqslant H \leqslant p^{1/2+\varepsilon}, \\ (p-1)/2 - 2s - 1, & \text{if } N = 0, H = p-1, \end{cases} \quad (1.33)$$

holds, where $\mathcal{C}(\varepsilon) > 0$ depends only on ε.

Moreover, regarding real multivariate polynomials that compute the rightmost bit of the discrete logarithm, the following theorem holds:

Theorem 4 ([120]). *Let α_0, α_1 be two distinct real numbers, $r = \lfloor \log p \rfloor$, and let a polynomial $f(X_1,\dots,X_r) \in \mathbb{R}[X_1,\dots,X_r]$ be such that $f(\alpha_{u_1},\dots,\alpha_{u_r}) \geqslant 0$, if x is a quadratic residue modulo p and $f(\alpha_{u_1},\dots,\alpha_{u_r}) < 0$, otherwise, where $x = u_1 \dots u_r$ is the bit representation of x, $1 \leqslant x \leqslant 2^r - 1$. Then f is of degree $\deg f \geqslant \log r + o(\log r)$ and contains at least $\operatorname{spr} f \geqslant 0.25r + o(r)$ distinct monomials.*

Since, the universal approximation capability of RPNs is achieved by summing the outputs of PSNs of different orders, for the approximation of the real polynomials that compute the rightmost bit of the discrete logarithm the following corollary is derived.

Corollary 1. *Assume the polynomial $f(X) \in \mathbb{R}[X]$ that satisfies the conditions of Theorem 3, i.e., it is such that $f(x) \geqslant 0$ if x is a quadratic residue modulo p and $f(x) < 0$ otherwise, for every element $x \in S$ from some set $S \subseteq \{M+1,\dots,M+H\}$ of cardinality $|S| \geqslant H-s$, and the polynomial $f(X_1,\dots,X_r)$ that satisfies the conditions of Theorem 4, i.e., it is such that $f(\alpha_{u_1},\dots,\alpha_{u_r}) \geqslant 0$, if x is a quadratic residue modulo p and $f(\alpha_{u_1},\dots,\alpha_{u_r}) < 0$, otherwise, where $x = u_1 \dots u_r$ is the bit representation of x, $1 \leqslant x \leqslant 2^r - 1$. Then, the following conclusions hold.*

(a) There exist two RPNs that realize the polynomial $f(X)$ and the polynomial $f(X_1,\ldots,X_r)$, respectively, and
(b) The maximum orders of the nodes comprising each RPN, are required to be for the first case equal to

$$\deg f(X) \geqslant \begin{cases} H/2 - 2s - 1 - p^{1/2}\log p, & \textit{for any} \quad H, \\ C(\varepsilon)\min\{H, H^2 p^{-1/2}\} - 2s - 1, & \textit{if} \quad p^{1/4+\varepsilon} \leqslant H \leqslant p^{1/2+\varepsilon}, \\ (p-1)/2 - 2s - 1, & \textit{if} \quad N = 0, H = p-1, \end{cases} \tag{1.34}$$

where $C(\varepsilon) > 0$ depends only on ε, and equal to

$$\deg f(X_1,\ldots,X_r) \geqslant \log r + o(\log r), \tag{1.35}$$

for the second case.

Additionally, the inherent ability of RPNs to expand the final polynomial that they realize after training, as an ordinary multivariate polynomial, may prove of major importance in the field of cryptography. Finally, relating the fact that many cryptographic results consider the use of Boolean functions with the capability of BPSNs to realize any Boolean function, can provide theoretical conclusions for this class of functions too.

1.6 Summary

The past decade has witnessed an increasing interest in the application of Computational Intelligence (CI) methods to problems derived from the field of cryptography and cryptanalysis. This is not only due to the effectiveness of these methods observed in many other scientific fields but also due to the need for automated techniques in the design and cryptanalysis of cryptosystems.

In this contribution, a brief review to cryptography and CI methods is initially provided. Then, a short survey of the applications of CI to cryptographic problems follows, and our contribution in this field is analytically presented. More specifically, at first three cryptographic problems derived from classical public key cryptosystems are formulated as discrete optimization tasks and Evolutionary Computation (EC) methods are applied to address them. Next, EC methods for the partial cryptanalysis of a Feistel cipher, the Data Encryption Standard reduced to four and six rounds, respectively, are considered. The effectiveness of Artificial Neural Networks (ANNs) for classical cryptographic problems and problems related to elliptic curve cryptography, follow. Lastly, some theoretical results are derived based on the composition of a specific class of ANNs, namely the Ridge Polynomial Networks, with theoretical issues of cryptography.

The experimental results presented for each considered problem suggest that problem formulation and representation are critical determinants of the performance of CI methods in cryptography. Regarding the application of EC methods in cryptanalysis, the proper definition of the fitness function such that no deceptive landscapes are created, is of major importance. Furthermore, the performance of ANNs in cryptographic problems depends on the problem formulation and data representation. A second conclusion derived is that EC methods (and CI methods in general) can be used as a practical assessment for the efficiency and the effectiveness of proposed cryptographic systems, meaning that they can "sense" flawed cryptographic schemes by finding patterns before more complex methods are employed for their analysis.

References

1. Adleman L (1979) A subexponential algorithm for discrete logarithm problem with applications to cryptography. In: Proceedings of the 20th FOCS, pp. 55–60
2. Anthony M (2003) Boolean functions and artificial neural networks. Technical report, CDAM, The London School of Economics and Political Science. CDAM Research Report LSE-CDAM-2003-01
3. Bäck T (1996) Evolutionary Algorithms in Theory and Practice : Evolution Strategies, Evolutionary Programming, Genetic Algorithms. Oxford University Press
4. Bagnall T, McKeown G. P, Rayward-Smith V. J (1997) The cryptanalysis of a three rotor machine using a genetic algorithm. In: Bäck T (ed) Proceedings of the 7th International Conference on Genetic Algorithms (ICGA97), San Francisco, CA, Morgan Kaufmann
5. Barbieri A, Cagnoni S, Colavolpe G (2004) A genetic approach for generating good linear block error-correcting codes. Lecture Notes in Computer Science 3103:1301–1302
6. Biham E, Shamir A (1991) Differential cryptanalysis of DES–like cryptosystems. Journal of Cryptology
7. Biham E, Shamir A (1993) Differential Cryptanalysis of the Data Encryption Standard. Springer–Verlag
8. Blake I (1999) Elliptic Curves in Cryptography. London Mathematical Society Lecture Notes Series vol. 265. Cambridge University Press
9. Blum A, Furst M, Kearns M, Lipton R. J (1994) Cryptographic primitives based on hard learning problems. Lecture Notes in Computer Science 773:278–291
10. Bonabeau E, Dorigo M, Théraulaz G (1999) From Natural to Artificial Swarm Intelligence. Oxford University Press, New York
11. Boutsinas B, Vrahatis M. N (2001) Artificial nonmonotonic neural networks. Artificial Intelligence 132:1–38
12. Burnett L, Carter G, Dawson E, Millan W (2001) Efficient methods for generating Mars-like S-boxes. Lecture Notes in Computer Science 1978(4):300–313
13. Carrol J, Martin S (1986) The automated cryptanalysis of substitution ciphers. Cryptologia 10(4):193–209
14. Chang Y.-C, Lu C.-J (2001) Oblivious polynomial evaluation and oblivious neural learning. Lecture Notes in Computer Science 2248:369–384

15. Chui C.K, Li X (1991) Realization of neural networks with one hidden layer. Technical report, Center for Approximation Theory, Dept. of Mathematics, Texas A&M University
16. Chui C.K, and Li X (1992) Approximation by Ridge functions and neural networks with one hidden layer. Journal of Approximation Theory 70:131–141
17. Clark A (1998) Optimisation Heuristics for Cryptography. PhD Thesis, Queensland University of Technology, Australia
18. Clark J.A, Jacob J.L (2000) Two-stage optimisation in the design of Boolean functions. Lecture Notes in Computer Science 1841:242–254
19. Clark J.A, Jacob J.L (2002) Fault injection and a timing channel on an analysis technique. Lecture Notes in Computer Science 2332:181–196
20. Clark J.A, Jacob J.L, Stepney S (2004) The design of S-boxes by Simulated Annealing. In: CEC 2004: International Conference on Evolutionary Computation, Portland OR, USA, June 2004, pp. 1517–1524. IEEE
21. Clerc M, Kennedy J (2002) The particle swarm–explosion, stability, and convergence in a multidimensional complex space. IEEE Transactions on Evolutionary Computation 6(1):58–73
22. Coppersmith D, Shparlinski I (2000) On polynomial approximation of the discrete logarithm and the Diffie–Hellman mapping. Journal of Cryptology 13:339–360
23. DasGupta B, Schnitger G (1996) Analog versus discrete neural networks. Neural Computation 8(4):805–818
24. De Jong K.A (1985) Genetic algorithms: A 10 year perspective. In: Proceedings of the First International Conference on Genetic Algorithms pp. 169–177. Lawrence Erlbaum Associates.
25. Diffie W, Hellman M.E (1976) New directions in cryptography. IEEE Transactions on Information Theory IT-22(6):644–654
26. Dontas K, Jong K (1990) Discovery of maximal distance codes using genetic algorithms. In: Proceedings of the Second International IEEE Conference on Tools for Artificial Intelligence pp. 805–811
27. Dorigo M, Gambardella L.M (1997) Ant colonies for the traveling salesman problem. BioSystems 43:73–81
28. Eberhart R.C, Simpson P, Dobbins R (1996) Computational Intelligence PC Tools. Academic Press
29. ElGamal T (1985) A public key cryptosystem and a signature scheme based on discrete logarithms. IEEE Transactions on Information Theory 31(4):469–472
30. Engelbrecht A (2002) Computational Intelligence: An Introduction. John Wiley & Sons
31. Epitropakis M.G, Vrahatis M.N (2005) Root finding and approximation approaches through neural networks. SIGSAM Bulletin: Communications in Computer Algebra, ACM Press 39(4):118–121
32. Feistel H (1973) Cryptography and computer privacy. Scientific American
33. Fogel D.B (1993) Evolving behaviours in the iterated prisoner's dilemma. Evolutionary Computation 1(1):77–97
34. Fogel D.B (1995) Evolutionary Computation: Towards a New Philosophy of Machine Intelligence. IEEE Press, Piscataway, NJ
35. Fogel D.B, Owens A.J, Walsh M.J (1966) Artificial Intelligence Through Simulated Evolution. John Wiley, Chichester, UK
36. Forsyth W.S, Safavi-Naini R (1993) Automated cryptanalysis of substitution ciphers. Cryptologia 17(4):407–418

37. Ghosh J, Shin Y (1992) Efficient higher-order neural networks for classification and function approximation. International Journal of Neural Systems 3:323–350
38. Goldberg D.E (1989) Genetic Algorithms in Search, Optimization, and Machine Learning. Addison Wesley, Reading, MA
39. Hassoun M.H (1995) Foundamentals of Artificial Neural Networks. MIT Press, Cambridge, MA
40. Håstad J, Näslund M (2004) The security of all RSA and discrete log bits. Journal of the ACM 51(2):187–230
41. Haykin S (1999) Neural Networks, A Comprehensive Foundation. Prentice Hall, New Jersey, USA, 2nd edition edition
42. Herdy M (1991) Application of the evolution strategy to discrete optimization problems. Lecture Notes in Computer Science pp. 188–192
43. Hernández J, Isasi P, Ribagorda A (2002) An application of genetic algorithms to the cryptoanalysis of one round TEA. In: Proc. of the 2002 Symposium on Artificial Intelligence and its Application
44. Hernández J, Sierra J, Isasi P, Ribagorda A (2002) Genetic cryptoanalysis of two rounds TEA. Lecture Notes in Computer Science 2331:1024–1031
45. Holland J.H (1975) Adaptation in Natural and Artificial Systems. University of Michigan Press, Ann Arbor
46. Hornik K (1989) Multilayer feedforward networks are universal approximators. Neural Networks 2:359–366
47. Hunter D, McKenzie A (1983) Experiments with relaxation algorithms for breaking simple substitution ciphers. The Computer Journal 26(1):68–71
48. Isasi P, Hernández J (2004) Introduction to the applications of evolutionary computation in computer security and cryptography. Computational Intelligence 20(3):445–449
49. Jakobsen T (1995) A fast method for cryptanalysis of substitution ciphers. Cryptologia 19(3):265–274
50. Karras D, Zorkadis V (2002) Strong pseudorandom bit sequence generators using neural network techniques and their evaluation for secure communications. Lecture Notes in Artificial Intelligence 2557:615–626
51. Karras D, Zorkadis V (2003) On neural network techniques in the secure management of communication systems through improving and quality assessing pseudorandom stream generators. Neural Networks 16:899–905
52. Kennedy J, Eberhart R.C (2001) Swarm Intelligence. Morgan Kaufmann Publishers
53. King J, Bahler D (1992) An implementation of probabilistic relaxation in the cryptanalysis of simple substitution ciphers. Cryptologia 16(3):215–225
54. Kinzel W, Kanter I (2002) Interacting neural networks and cryptography. In: Kramer B (ed) Advances in Solid State Physics. vol. 42, pp. 383–391. Springer–Verlag
55. Klimov A, Mityagin A, Shamir A (2002) Analysis of neural cryptography. Lecture Notes in Computer Science 2501:288–298
56. Knudsen L.R, Meier W (1999) A new identification scheme based on the perceptrons problem. Lecture Notes in Computer Science 1592:363–374
57. Koblitz N (1987) Elliptic curve cryptosystems. Mathematics of Computation 48:203–209
58. Kohonen T (2000) Self-Organizing Maps. Springer-Verlag, Berlin, 3rd edition

59. Konstantinou E, Stamatiou Y, Zaroliagis C (2002) A software library for elliptic curve cryptography. Lecture Notes in Computer Science 2461:625–637
60. Kotlarz P, Kotulski Z (2005) On application of neural networks for s-boxes design. Lecture Notes in Artificial Intelligence 3528:243–248
61. Koza J.R (1992) Genetic Programming: On the Programming of Computers by Means of Natural Selection. MIT Press, Cambridge, MA
62. Lange T, Winterhof A (2002) Incomplete character sums over finite fields and their application to the interpolation of the discrete logarithm by boolean functions. Acta Arithmetica 101(3):223–229
63. Laskari E.C, Meletiou G.C, Stamatiou Y.C, Tasoulis D.K, Vrahatis M.N (2006) Assessing the effectiveness of artificial neural networks on problems related to elliptic curve cryptography. Mathematical and Computer Modelling. to appear
64. Laskari E.C, Meletiou G.C, Stamatiou Y.C, Vrahatis M.N (2005) Evolutionary computation based cryptanalysis: A first study. Nonlinear Analysis: Theory, Methods and Applications 63:e823–e830
65. Laskari E.C, Meletiou G.C, Vrahatis M.N (2005) Problems of Cryptography as Discrete Optimization Tasks. Nonlinear Analysis: Theory, Methods and Applications 63:e831–e837
66. Laskari E.C, Meletiou G.C, Stamatiou Y.C, Vrahatis M.N (2006) Applying evolutionary computation methods for the cryptanalysis of Feistel ciphers. Applied Mathematics and Computation. to appear
67. Laskari E.C, Meletiou G.C, Tasoulis D.K, Vrahatis M.N (2005) Aitken and Neville inverse interpolation methods over finite fields. Applied Numerical Analysis and Computational Mathematics 2(1):100–107
68. Laskari E.C, Meletiou G.C, Tasoulis D.K, Vrahatis M.N (2005) Transformations of two cryptographic problems in terms of matrices. SIGSAM Bulletin: Communications in Computer Algebra, ACM Press 39(4):127–130
69. Laskari E.C, Meletiou G.C, Tasoulis D.K, Vrahatis M.N (2006) Studying the performance of artificial neural networks on problems related to cryptography. Nonlinear Analysis Series B: Real World Applications 7(5):937–942
70. Laskari E.C, Meletiou G.C, Vrahatis M.N (2004) The discrete logarithm problem as an optimization task: A first study. In: Proceedings of the IASTED International Conference on Artificial Intelligence and Applications pp. 1–6. ACTA Press
71. Laskari E.C, Parsopoulos K.E, Vrahatis M.N (2002) Particle swarm optimization for integer programming. In: Proceedings of the IEEE 2002 Congress on Evolutionary Computation pp. 1576–15812. IEEE Press.
72. Laskari E.C, Parsopoulos K.E, Vrahatis M.N (2002) Particle swarm optimization for minimax problems. In: Proceedings of the IEEE 2002 Congress on Evolutionary Computation pp. 1576–1581. IEEE Press.
73. Magoulas G.D, Plagianakos V.P, Vrahatis M.N (2001) Adaptive stepsize algorithms for on-line training of neural networks. Nonlinear Analysis T.M.A. 47(5):3425–3430
74. Magoulas G.D, Vrahatis M.N (2006) Adaptive algorithms for neural network supervised learning: a deterministic optimization approach. International Journal of Bifurcation and Chaos 16(7):1929–1950
75. Magoulas G.D, Vrahatis M.N, Androulakis G.S (1997) Effective backpropagation training with variable stepsize. Neural Networks 10(1):69–82

76. Magoulas G.D, Vrahatis M.N, Androulakis G.S (1999) Increasing the convergence rate of the error backpropagation algorithm by learning rate adaptation methods. Neural Computation 11(7):1769–1796
77. Mathews R (1993) The use of genetic algorithms in cryptanalysis. Cryptologia 17(4):187–201
78. Matsui M (1994) Linear cryptanalysis method for DES cipher. Lecture Notes in Computer Science 765:386–397
79. Matsui M, Yamagishi A (1992) new method for known plaintext attack of feal cipher. Lecture Notes in Computer Science pp. 81–91
80. Maurer U, Wolf S (1999) The relationship between breaking the diffie-hellman protocol and computing discrete logarithms. SIAM Journal on Computing 28:1689–1721
81. Meletiou G.C (1992) A polynomial representation for exponents in $\mathbb{Z}_p$. Bulletin of the Greek Mathematical Society 34:59–63
82. Meletiou G.C (1993) Explicit form for the discrete logarithm over the field GF(p, k). Archivum Mathematicum (Brno) 29(1–2):25–28
83. Meletiou G.C, Mullen G.L (1992) A note on discrete logarithms in finite fields. Applicable Algebra in Engineering, Communication and Computing 3(1):75–79
84. Meletiou G.C, Tasoulis D.K, Vrahatis M.N (2003) Cryptography through interpolation approximation and computational inteligence methods. Bulletin of the Greek Mathematical Society 48:61–75
85. Menezes A, van Oorschot P, Vanstone S (1996) Handbook of applied cryptography. CRC Press series on discrete mathematics and its applications. CRC Press
86. Merkle R.C, Hellman M.E (1978) Hiding information and signatures in trapdoor knapsacks. IEEE Transactions on Information Theory 24:525–530
87. Michalewicz Z (1994) Genetic Algorithms + Data Structures = Evolution Programs. Springer, Berlin
88. Millan W, Clark A, Dawson E (1997) Smart hill climbing finds better boolean functions. In: Proceedings of the 4th Workshop on Selected Areas in Cryptography
89. Millan W, Clark A, Dawson E (1999) Boolean function design using hill climbing methods. Lecture Notes in Computer Science 1587:1–11
90. Miller V (1986) Uses of elliptic curves in cryptography. Lecture Notes in Computer Science 218:417–426
91. Mislovaty R, Perchenok Y, Kanter I, Kinzel W (2002) Secure key-exchange protocol with an absence of injective functions. Phys. Rev. E 66(6):066102–1–066102–5
92. Møller M.F (1993) A scaled conjugate gradient algorithm for fast supervised learning. Neural Networks 6:525-533
93. Mullen G.L, White D (1986) A polynomial representation for logarithms in GF(q). Acta Arithmetica 47:255–261
94. National Bureau of Standards, U.S. Department of Commerce, FIPS pub. 46. Data Encryption Standard. January 1977.
95. Niederreiter H (1990) A short proof for explicit formulas for discrete logarithms in finite fields. Applicable Algebra in Engineering, Communication and Computing 1:55–57
96. Odlyzko A (2000) Discrete logarithms: The past and the future. Designs, Codes, and Cryptography 19(2–3):129–145

97. Parsopoulos K.E, Vrahatis M.N (2002) Initializing the particle swarm optimizer using the nonlinear simplex method. In: Grmela A, Mastorakis N.E (eds) Advances in Intelligent Systems, Fuzzy Systems, Evolutionary Computation pp. 216–221
98. Parsopoulos K.E, Vrahatis M.N (2002) Recent approaches to global optimization problems through particle swarm optimization. Natural Computing 1(2–3):235–306
99. Parsopoulos K.E, Vrahatis, M.N (2004) On the computation of all global minimizers through particle swarm optimization. IEEE Transactions on Evolutionary Computation 8(3):211–224
100. Peleg S, Rosenfeld A (1979) Breaking substitution ciphers using a relaxation algorithm. Communications of the ACM 22(11):598–605
101. Peralta R (1986) Simultaneous security of bits in the discrete log. Lecture Notes in Computer Science 219:62–72
102. Pincus A (1999) Approximation theory of the mlp model in neural networks. Acta Numerica pp. 143–195
103. Plagianakos V, Vrahatis M.N (2002) Parallel Evolutionary Training Algorithms for "Hardware-Friendly" Neural Networks. Natural Computing 1:307–322
104. Pohlig S.C, Hellman M (1978) An improved algorithm for computing logarithms over GF(p) and its cryptographic significance. IEEE Transactions on Information Theory 24:106–110
105. Pointcheval D (1994) Neural networks and their cryptographic applications. In: Charpin P (ed) INRIA, Livres de resumes EUROCODE'94
106. Pointcheval D (1995) A new identification scheme based on the perceptrons problem. Lecture Notes in Computer Science 950:318–328
107. Ramzan Z (1998) On Using Neural Networks to Break Cryptosystems. PhD Thesis. Laboratory of Computer Science, MIT
108. Rao S.S (1996) Engineering Optimization–Theory and Practice. Wiley Eastern, New Delhi
109. Rechenberg I (1973) Evolutionsstrategie: Optimierung technischer Systeme nach Prinzipien der biologischen Evolution. Frommann-Holzboog Verlag, Stuttgart, Germany
110. Riedmiller M, Braun H (1993) A direct adaptive method for faster backpropagation learning: The RPROP algorithm. In: Proceedings of the IEEE International Conference on Neural Networks pp. 586–591
111. Rivest R (1991) Cryptography and machine learning. Lecture Notes in Computer Science 739:427–439
112. Rivest R, Shamir A, Adleman L (1978) A method for obtaining digital signatures and public key cryptosystems. Communications of the ACM 21:120–126
113. Rosen-Zvi M, Kanter I, Kinzel W (2002) Cryptography based on neural networks – analytical results. Journal of Physics A: Mathematical and General 35(47):L707–L713
114. Rumelhart D, Hinton G, Williams R (1986) Learning internal representations by error propagation. In: RumelhartD.E, McClelland J.L (eds) Parallel distributed processing: Explorations in the microstructure of cognition. vol. 1 pp. 318–362. MIT Press
115. Ruttor A, Kinzel W, Kanter I (2005) Neural cryptography with queries. Journal of Statistical Mechanics pp. P01009
116. Ruttor A, Kinzel W, Shacham L, Kanter I (2004) Neural cryptography with feedback. Physical Review E 69(4):046110–1–046110–7

117. Schwefel H.-P (1995) Evolution and Optimum Seeking. Wiley, New York
118. Shi Y, Eberhart R.C (1998) A modified particle swarm optimizer. In: Proceedings of the IEEE Conference on Evolutionary Computation. Anchorage, AK
119. Shin Y, Ghosh J (1991) Realization of Boolean functions using binary Pi-Sigma networks. In: Proceedings of the Conference on Artificial Neural Networks in Engineering. St. Louis
120. Shparlinski I (ed) (2003) Cryptographic Applications of Analytic Number Theory. Progress in Computer Science and Applied Logic. Birkhäuser Verlag
121. Silverman J.H (1986) The Arithmetic of Elliptic Curves. Springer-Verlag
122. Specht D.F (1990) Probabilistic neural networks. Neural Networks 3(1):109–118
123. Spillman R (1993) Cryptanalysis of knapsack ciphers using genetic algorithms. Cryptologia 17(4):367–377
124. Spillman R, Janssen M, Nelson B, Kepner M (1993) Use of a genetic algorithm in the cryptanalysis of simple substitution ciphers. Cryptologia 17(1):31–44
125. Stinson D (1995) Cryptography: Theory and Practice (Discrete Mathematics and Its Applications). CRC Press
126. Storn R, Price K (1997) Differential evolution–a simple and efficient heuristic for global optimization over continuous spaces. Journal of Global Optimization 11:341–359
127. Tasoulis D.K, Pavlidis N.G, Plagianakos V.P, Vrahatis M.N (2004) Parallel Differential Evolution. In: Proceedings of the IEEE 2004 Congress on Evolutionary Computation (CEC 2004), Portland
128. Terano T, Asai K, Sugeno M (1992) A Complete Introduction to the Field: Fuzzy Systems Theory and Its Applications. Academic Press
129. Vertan C, Geangala C (1996) Breaking the Merkle-Hellman cryptosystem by genetic algorithms: Locality versus performance. In: Zimmermann H, Negoita M, Dascalu D (eds), Real World Applications of Intelligent Technologies pp. 201–208. Editura Academiei Romanie, Bucharest
130. Vrahatis M.N, Androulakis G.S, Lambrinos J.N, Magoulas G.D (2000) A class of gradient unconstrained minimization algorithms with adaptive stepsize. Journal of Computational and Applied Mathematics 114(2):367–386
131. White H (1990) Connectionist nonparametric regression: Multilayer feedforward networks can learn arbitrary mappings. Neural Networks 3:535–549
132. Wilson D, Martinez T (1997) Improved heterogeneous distance functions. Journal of Artificial Intelligence Research 6:1–34
133. Winterhof A (2001) A note on the interpolation of the Diffie-Hellman mapping. Bulletin of Australian Mathematical Society 64(3):475–477
134. Winterhof A (2002) Polynomial interpolation of the discrete logarithm. Designs, Codes and Cryptography 25(1):63–72
135. Yue T.-W, Chiang S (2001) The general neural-network paradigm for visual cryptography. Lecture Notes in Computer Science 2084:196–206

2

Multimedia Content Protection Based on Chaotic Neural Networks

Shiguo Lian

SAMI Lab,France Telecom R&D Beijing
2 Science South Rd,Haidian District,
Beijing,100080,China
shiguo.lian@orange-ft.com

For chaotic neural network has both chaos' and neural network's properties, it is regarded as the most close one to human's thinking, and has been attracting more and more researchers since the past decade. The properties, such as parameter sensitivity, random similarity, learning ability, etc., make it suitable for information protection, such as data encryption, data authentication, intrusion detection, etc. For its simple structure generates complicated or random sequences, it has been used to design the encryption algorithms that are of lower time-cost than traditional computing-based ones, such as DES, IDEA or AES. These algorithms are suitable for multimedia encryption or authentication.

In this chapter, by investigating chaotic neural networks' properties, the low-cost cipher and hash function based on chaotic neural networks are proposed and used to encrypt and authenticate images or videos. First, this chapter gives a brief introduction to chaotic neural network based data encryption, including stream ciphers, block ciphers and hash functions. Secondly, the chapter analyzes chaotic neural networks' properties that are suitable for data encryption, such as parameter sensitivity, random similarity, diffusion property, confusion property, one-way property, etc. Thirdly, the chapter gives some proposals to design chaotic neural network based cipher or hash function, and uses these ciphers to construct media encryption and authentication methods. In media encryption, the stream cipher and partial encryption method are constructed, which are used to encrypt images or videos. In media authentication, the chaotic neural network based hash function and media authentication method are designed, which can detect the malicious tampering of the media content. The analyses and experiments show the practicability of these encryption algorithms. Finally, conclusions are drawn, and some open issues in this field are presented.

S. Lian: *Multimedia Content Protection Based on Chaotic Neural Networks*, Studies in Computational Intelligence (SCI) **57**, 51–78 (2007)
www.springerlink.com

2.1 Introduction

In recent years, it is founded that there is chaos phenomenon in human's brain [1,2], and chaos theory can be used to explain some abnormal activities in the brain [3]. Thus, chaos dynamics provides a chance for the research in neural networks, such as using neural networks to generate chaos or constructing chaotic neural networks. After knowing the existence of chaos in human's brain, some scientists propose the theory of chaotic neural networks. The earliest work was done by Aibara [4] and Inoue et al [5,6] in 1990 and 1991. From then on, chaotic neural networks attract more and more researchers, and are used in more and more applications, such as algorithm optimization, pattern recognition and some other applications. Among these applications, information security is a challenging and interesting one.

Neural networks are used to design data protection schemes because of their complicated and time-varying structures [7]. For example, the ciphers [8,9] are constructed based on the random sequences generated from the neural networks. For the property of initial-value sensitivity, ergodicity or random similarity, chaos is also introduced to data protection. For example, the block cryptosystem is designed via iterating a chaotic map [10], and the one-way hash is constructed based on the chaotic map with changeable parameters [11]. As a combination of neural networks and chaos, chaotic neural networks (CNN) are expected to be more suitable for data encryption. For example, it is reported [12] that faster synchronization can be obtained by jointing neural network's synchronization and chaos' synchronization.

Due to both of the properties, chaotic neural networks are regarded more suitable for data encryption. Till now, various chaotic neural network based encryption algorithms have been reported, which can be classified into two classes: stream cipher and block cipher. The first one uses neural networks to construct stream ciphers, in which, neural network is used to produce pseudo-random sequences. Cauwenberghs [13] made use of VLSI to construct cell neurons that produce the random vectors suitable for data encryption. Chan et al [14] took advantage of Hopfield neural networks' chaotic property to produce pseudo-random sequence that is used to encrypt plaintexts. Karras et al [15] proposed the method to evaluate the property of the pseudorandom sequence generated from chaotic neural networks, and reported that the sequence generated from chaotic neural networks may be of better performance than the one generated from conventional methods. Caponetto et al [16] designed a secure communication scheme based on cell neural networks. These stream ciphers' security depends on the sequences' randomness.

The second one uses neural network to construct block ciphers, which makes use of chaotic neural networks' properties to encrypt plaintext block by block. Yue et al [17] constructed the image encryption algorithm based on chaotic neural networks, which encrypts several gray images into some binary images. Yen et al [18] used perception neurons to construct a block cipher, in which, the weight matrix and threshold vector are used as keys, and the neurons' input and output act as plaintext and ciphertext, respectively. Lian et al [19] proposed a cipher based on chaotic neural network, which is used to encrypt compressed images partially. Yee et al [20] designed a symmetric block cipher based on the multi-layer perception network. Generally, block ciphers' security depends on their computing security [21–23] (It is not practical from the viewpoint of computing complexity for attackers to break the cryptosystem in condition of not knowing the key) that is determined by the confusion and diffusion criterias.

With the development of computer technology and Internet technology, multimedia data (images, videos or audios) are widely used in human's daily life. Taking videos for example, in order to keep secure, some sensitive videos need to be protected. Till now, various encryption algorithms have been reported, such as DES, RSA, IDEA or AES. However, most of them are used in text/binary data not in video data, for videos are often of large volumes with real-time operations. These operations include browsing, displaying, cutting, copying or bit-rate control, etc. In order to meet these applications, new encryption algorithms should be studied. For chaotic signal's generation is often of lower cost than traditional pseudorandom sequence's generation, chaos based cipher has been used to encrypt such large-voluminal data as image or video. Fridrich [24] constructed the block cipher based on Baker map, Cat map or Standard map and used it to encrypt images, Chen et al [25] constructed the one based on 3-Dimensional Cat map and used it to encrypt real-time images, Yen et al [17] designed the image encryption algorithm based on chaotic neural network, Lian et al [19] proposed the chaotic neural network based cipher to encrypt images or videos, and Li et al [26] proposed a chaos-based video encryption scheme (CVES) and used it to encrypt videos. These chaotic ciphers are more efficient compared with traditional ones.

In this chapter, the chaotic neural network's properties suitable for data encryption are investigated, and the design of efficient stream cipher and hash function based on chaotic neural networks are presented. Based on these ciphers, the secure and efficient media encryption and authentication schemes are proposed. The analyses and experiments show the practicability of the schemes. The rest of the chapter is arranged as follows. In Section 2, the chaotic neural networks' generation is presented, together with their cipher-suitable properties, such as diffusion, one-way, parallelization, parameter sensitivity or random similarity, etc. The media encryption methods based on chaotic neural networks are proposed in Section 3, including a brief introduction to media encryption, the design of chaotic neural networks based stream cipher, the design of the partial encryption methods for compressed images or videos, and some experimental results on the proposed encryption schemes. Similarly, in Section 4, the media authentication scheme based on chaotic neural networks is presented, including a brief introduction to media authentication, the design of hash function based on chaotic neural networks, the design of the authentication method, and the experiments and analyses.

Then, some open issues and future work are discussed in Section 5. Finally, in Section 6, some conclusions are drawn.

2.2 Chaotic neural networks' generation and properties

2.2.1 Chaotic neural network's generation

Generally, three methods are used to construct chaotic neural networks: 1) adopt neural networks' learning ability to approach a chaotic dynamic system; 2) analyze the condition that makes a neural network generate chaos; 3) design the chaotic neuron model and then the corresponding chaotic neural network.

In the first method, the chaos system is known, and the problem is to find a neural network that approaches the chaos system with it confirmed that the performance of the neural network is similar with the one of the chaos system. Thus, the synchronized neural network is a chaotic neural network. For example, Ren et al [27] used Elman network [28] to approach Hennon chaotic system, and the synchronization algorithm is based on feedback chaos control theory [29]. Generally, this method's difficulty is to realize chaos synchronization [30, 31].

In the second method, the neural network is known, and the problem is to find the condition, under which, the neural network generates chaos. Among the existing neural networks, Hopfield network attracts the most researchers. The basic idea is to introduce chaos dynamics to Hopfield network. For example, Zou et al [32] found chaotic attractors in the dynamic activity of two cells' cellular neural networks, and they [33] reported that 3-order autonomous cellular neural network has chaotic attractors. Generally, the nonlinear property of neural network determines that it itself may have chaotic property, while the difficulty is to find the suitable condition.

The third method is to construct the chaotic neural network by designing a chaotic neuron model. The well-known examples include the Global Coupling Map (GCM) model [34], Aihara's chaotic neural network model [35] and Inoue's chaotic neural network model [36]. Among them, the GCM model proposed by Kaneko is composed of many neurons, each of which satisfies chaotic Logistic map [37]. Aihara's model is composed of some neurons, and each neuron includes the feedback coming from the internal neurons. Inoue's model is composed of some neurons, each of which is a coupling chaotic oscillator. There are some other methods to design chaotic neurons, such as the one using network model and Lorenz equation [38] and the one using a four-level feedback neural network [39].

Among the mentioned methods, the third one is more practical than the others. For the first one, the constructed chaotic neural network acts just as the original chaos system. For the second one, the process to find the chaotic dynamic is time cost and it limits the number of the chaotic neural network. The third one gives a general method to construct chaotic neural networks, which is easy to produce the chaotic neural network with both the chaos system's property and the neural network's property. Thus, the third method is considered in the following content.

2.2.2 Chaotic neural network's properties suitable for data encryption

Chaotic neural networks have both chaos' properties and neural networks' properties, such as diffusion, one-way property, parallel implementation, confusion, parameter sensitivity and randomness similarity.

Diffusion property

The neuron layer of a neural network has the property of diffusion. Taking the simple neuron layer in Fig. 2.1 for example, the neuron layer is composed of n neurons, f is a linear function, P is the input, and C is the output. It is defined as

$$C = \begin{bmatrix} c_0 \\ c_1 \\ \vdots \\ c_{n-1} \end{bmatrix} = f\left(\begin{bmatrix} w_{0,0} & w_{0,1} & \cdots & w_{0,n-1} \\ w_{1,0} & w_{1,1} & \cdots & w_{1,n-1} \\ \vdots & \vdots & \vdots & \vdots \\ w_{n-1,0} & w_{n-1,1} & \cdots & w_{n-1,n-1} \end{bmatrix} \begin{bmatrix} p_0 \\ p_1 \\ \vdots \\ p_{n-1} \end{bmatrix} + \begin{bmatrix} b_0 \\ b_1 \\ \vdots \\ b_{n-1} \end{bmatrix} \right)$$

where

$$c_i = f(\sum_{j=0}^{n-1} w_{i,j}p_j + b_i) = \sum_{j=0}^{n-1} w_{i,j}p_j + b_i (i = 0, 1, \cdots, n-1)$$

Thus, each element c_i $(i = 0, 1, \cdots, n-1)$ in the output C is in relation with all the elements p_j $(j = 0, 1, \cdots, n-1)$ in the input P. That is, the change of one element in P may cause the changes of all the elements in C. This property is similar to the diffusion criteria [22] of the cipher with high computational security. In the diffusion criteria, the slight change in the plaintext or the key causes great changes in the ciphertext. The diffusion property of neural network makes it a potential choice for cipher designing.

One-way property

Some neural networks have the one-way property. Fig. 2.2 shows a simple neuron model with one-way property. In this neuron, the input is composed of n elements $p_0, p_1, \cdots, p_{n-1}$, while the output is a unique element c. It is defined as

$$c = f(\sum_{j=0}^{n-1} w_j p_j + b)$$

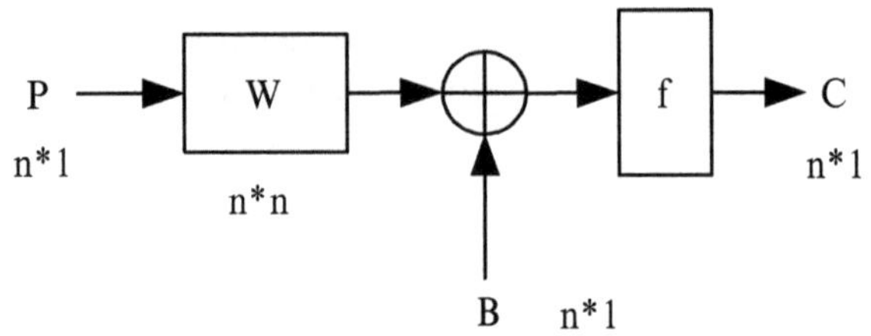

Fig. 2.1. A simple neuron layer with diffusion property

As can be seen, it is easy to compute c from $P = [p_0, p_1, \cdots, p_{n-1}]$, while difficult to compute P from c. The difficulty is equal to solve a singular equation. Thus, it is a one-way process from the input P to the output c. The one-way property makes neural network a suitable chaotic for hash function. As a hash function, it should be easy to compute the hash value from the plaintext, while difficult to compute the plaintext from the hash value. The hash value is often much shorter than the plaintext, which makes it suitable for digital signature or data authentication. According to this case, neural network may be used in hash function.

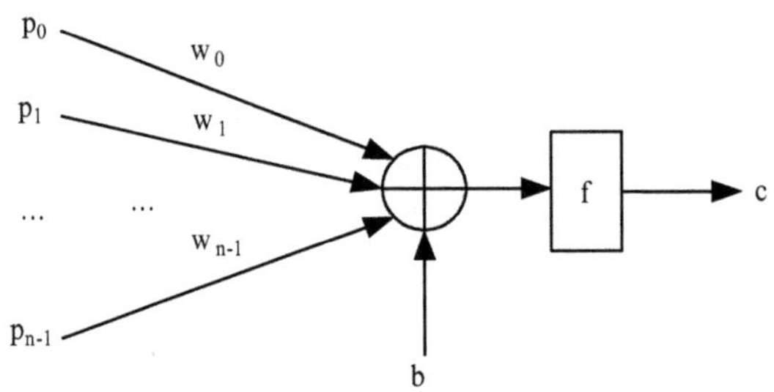

Fig. 2.2. A simple neuron layer with one-way property

Parallel implementation

Parallel implementation is an important property of neural networks, which benefits from neural networks' structure. Generally, in each layer of a neural network, the neurons are paralleled. Thus, each neuron can implement certain functionality independently. This property makes it suitable for parallel data processing. Fig. 2.3 shows the traditional data processing scheme (a) and the neural network based scheme (b). In (a), the data blocks $p_0, p_1, \cdots, p_{n-1}$ are operated one by one in order, which is named waterline scheme. In (b), the data blocks $p_0, p_1, \cdots, p_{n-1}$ are operated in a parallel way, which is named parallel scheme. Generally, the parallel scheme (b) saves much time cost than the waterline scheme (a).

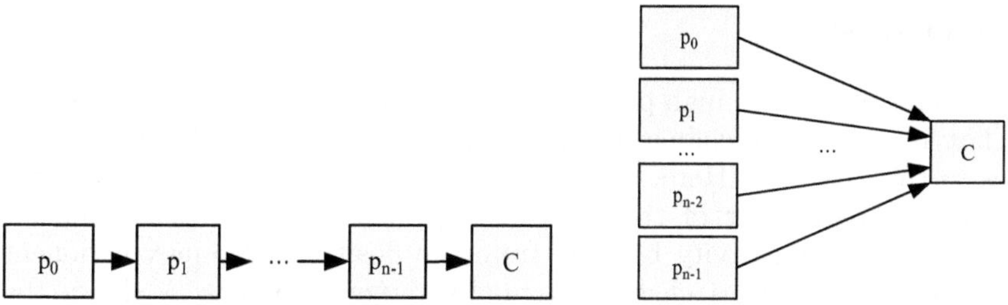

(a) Waterline scheme (b) Neural network based scheme

Fig. 2.3. Comparison of data block processing schemes

Confusion

Confusion is a special property caused by the nonlinear structure of chaos systems or chaotic neural networks. The ergodicity and randomness of chaotic dynamic make the output depend on the input in a nonlinear and complicated manner [37]. Fig. 2.4 shows the piecewise linear chaotic map [37]. Here, x is the input, $g()$ is the chaotic map, and $y = g(x)$ is the output. Thus, for each output y, there exists 4 corresponding inputs x. Furthermore, if the chaotic map is iterated for n times, $y = g^n(x)$, then each output y corresponds to 4^n inputs x. Thus, it is difficult to determine the exact input. This property is similar to the confusion criteria of the block cipher with high computational security [22]. In the confusion criteria, the bit of ciphertext depends on all the bits of the plaintext and the key in a complicated way. The confusion property of chaos or chaotic neural networks makes them a potential choice for cipher designing.

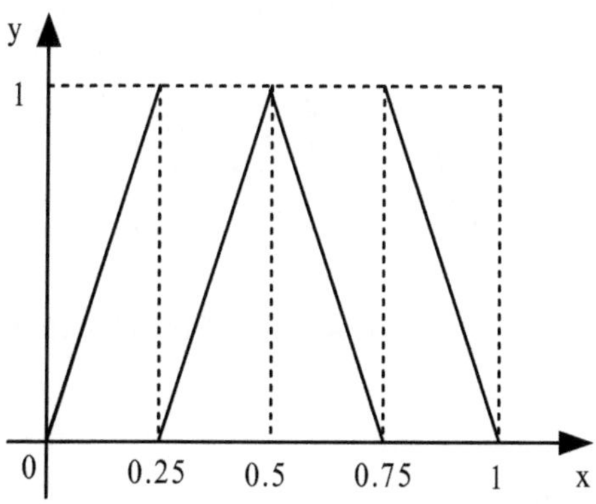

Fig. 2.4. The piecewise linear chaotic map

Parameter sensitivity

Chaos systems often have high parameter sensitivity, known as butterfly effect [37]. That is, the slight change in the original parameter will causes great changes in the final one. Here, the parameter can be either the initial value or the control parameter of the chaotic map. Taking the initial value for example, the high sensitivity benefits from the positive Lyapunov exponent of the chaos system. In a chaos system, Lyapunov exponent λ shows the bifurcation property of the chaotic orbit. It is defined as

$$\Delta Y = e^{\lambda t}|\Delta X| = (e^{\lambda})^{t}|\Delta X|,$$

where ΔX is the slight difference in the initial value, and ΔY is the difference after t times of iteration. For a chaos system, $\lambda > 0$ is confirmed. Thus, the difference ΔY increases with t. Fig. 2.5 shows an example of chaotic Logistic map with $\Delta X = 0.000001$ and $t = 50$. As can be seen, the two sequences are close when $t < 20$, while the iterated difference is $\Delta Y = 0.5975$ that is much larger than ΔX when $t = 50$. The parameter sensitivity is similar to ciphers' key sensitivity. That is, a slight change in the decryption key causes great changes in the decrypted data. Thus, the sensitive parameter of chaotic maps can act as the encryption key in data encryption.

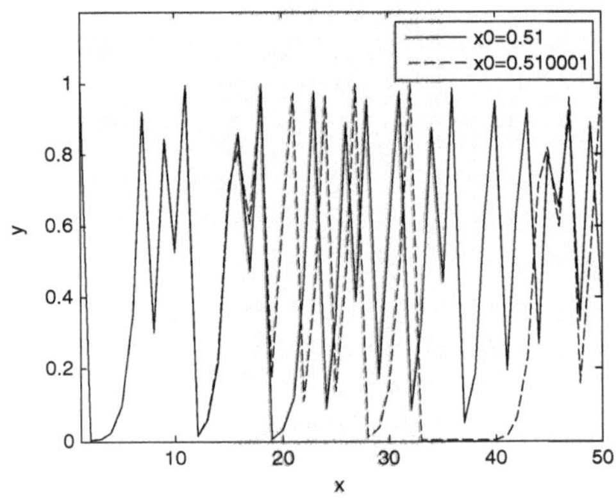

Fig. 2.5. The initial-value sensitivity of the chaotic Logistic map

Randomness Similarity

For chaos systems or chaotic neural networks, the produced sequence is often of random distribution. This property benefits from the property of ergodicity. That is, the chaotic sequence experiences all the states in the chaotic space.

Taking the chaotic Logistic map for example, the chaotic sequence x_i experiences all the value in the section [0,1]. Additionally, the chaotic sequence has some pseudorandom properties in autocorrelation function, cross-correlation function or frequency statistics. Fig. 2.6 shows an example on the chaotic Logistic map with $x_0 = 0.1$ and $x_1 = 0.11$. As can be seen, the x_0-sequence's autocorrelation (a) is similar to impulse function, and the cross-correlation (b) between x_0-sequence and x_1-sequence is near to 0. These properties make the chaos system a choice for pseudorandom sequence generation.

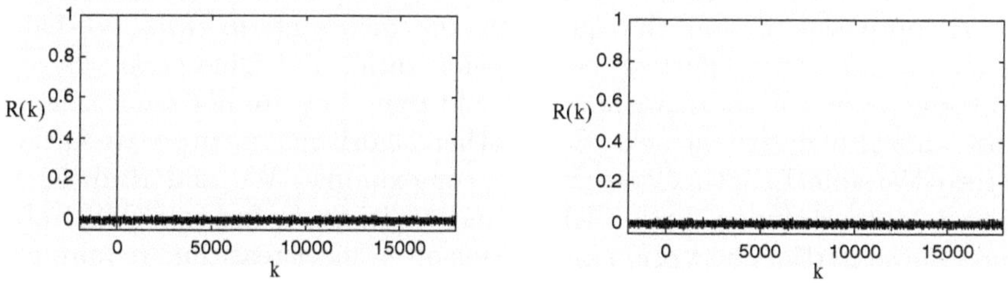

(a) autocorrelation (b) cross-correlation

Fig. 2.6. The statistical results of chaotic sequence

2.3 Multimedia content encryption based on chaotic neural networks

2.3.1 Introduction to multimedia content encryption

In the past decade, some image or video encryption algorithms have been reported, which can be classified into two categories: complete encryption and partial encryption. The former one encrypts the raw data or the compressed data directly with traditional or chaotic ciphers. This category can be further divided into two types. The first one is to permute raw images or videos directly [40–42], which reduce media data's understandability by changing adjacent pixels' relationship. For example, Maniccam et al [42] proposed an algorithm that permutes images or videos with SCAN patterns. These algorithms are often used to encrypt TV signals that are not compressed. The second one is to encrypt the compressed data directly [43, 44]. For example, Qiao et al [43] proposed Video Encryption Algorithm (VEA) that encrypts only half of the plaintext with DES and decreases the time cost to 46%. These algorithms are often of high security, but also of high complexity, and often change the file format. According to these properties, complete encryption algorithm is more suitable for secure video storing than for real-time transmission.

Compared with complete encryption, partial encryption incorporates encryption into compression and encrypts images or videos partially. This is based on the assumption that an encryption scheme is regarded as secure if the cost for breaking it is not smaller than the one paid for the media data. Some algorithms encrypt only the DCT coefficients or motion vectors in the compressed media data [45–47]. For example, Shi et al [45] proposed the algorithms that encrypt only the signs of DCT coefficients, and Yen et al [47] proposed the one that encrypts motion vectors with the Chaotic Pseudo-Random Sequence Generator (CPRSG). Some other algorithms permute DCT coefficients partially or completely during compression [48–51]. For example, Tosum et al [48] proposed the one that permutes 64 coefficients in three sub-bands in order to reduce the effect on compression ratio, and Qiao et al [49] analyzed these permutation algorithms and told that they are not secure enough against known-plaintext attacks. Some other algorithms incorporate encryption into Variable Length Code (VLC). For example, Wu and Kuo [52, 53] proposed Multi-Huffman Trees (MHT) and Multi-State Indexes (MSI) algorithms. These partial encryption algorithms often meet real-time requirement and keep file format unchanged. For these features, they are more suitable for such real-time applications as video transmission or video access.

According to the above analyses, partial encryption is preferred in order to obtain good tradeoff between security, real-time operation and format compliance [19, 54, 55]. Some media encryption algorithms based on chaos or chaotic neural networks have been presented. However, most of them belong to complete encryption. For example, Fridrich [24] and Chen et al [25] proposed the block cipher based on 2-D chaotic map to encrypt uncompressed images, Yen et al [17] proposed the chaotic neural network based scheme to encrypt uncompressed images, and Li et al [26] proposed the chaos based block cipher to encrypt compressed videos. These algorithms cannot keep the synchronization information in the file format. According to this case, the partial encryption algorithm based on chaotic neural networks will be presented in the following content. Considering that the encrypted part is often of variable size, the adopted cipher should support the plaintext with variable size. Thus, the chaotic neural network based cipher will be designed, that is not only efficient but also supports variable-size plaintext.

2.3.2 The cipher based on chaotic neural network

For the random similarity of chaotic neural networks, they are suitable for pseudorandom number generation. Here, we propose a stream cipher based on a chaotic neural network, which is a modification of the one developed in [18]. The new cipher proposed here is described as follows.

The encryption process

The encryption process is synthesised in (2.1). Here, $f(x)$ is a function taking value 1 if $x \geq 0$ and 0 otherwise, P and C are the plaintext and the ciphertext, respectively. The weight ω and θ are determined by encryption/decryption keys, which will be presented in the following content. P is composed of N bits, and $P = [p_0, p_1, \cdots, p_{N-1}]^T (p_i = 0 \text{ or } 1)$. Similarly, $C = [c_0, c_1, \cdots, c_{N-1}]^T (c_i = 0 \text{ or } 1)$, $\theta = [\theta_0, \theta_1, \cdots, \theta_{N-1}]^T (\theta_i = \pm 1/2)$, and

$$\omega = \begin{bmatrix} \omega_{0,0} & 0 & \cdots & 0 \\ 0 & \omega_{1,1} & \ddots & \vdots \\ \vdots & \ddots & \ddots & 0 \\ 0 & \cdots & 0 & \omega_{N-1,N-1} \end{bmatrix} (\omega_{i,i} = \pm 1, i = 0, \cdots, N-1).$$

$$C = f(\omega P + \theta) \tag{2.1}$$

The decryption process is symmetric to the encryption process. In decryption, the plaintext P is replaced by the ciphertext C, and the ciphertext C by the plaintext P. In the proposed cipher, the size of the plaintext, N, is arbitrary.

Generation of the parameters

The parameters ω and θ are generated from a chaotic sequence generator, as shown in Fig. 2.7. Here, the chaotic sequence X is firstly generated, then quantized and finally used to decide the two parameters. The process is described in detail as below.

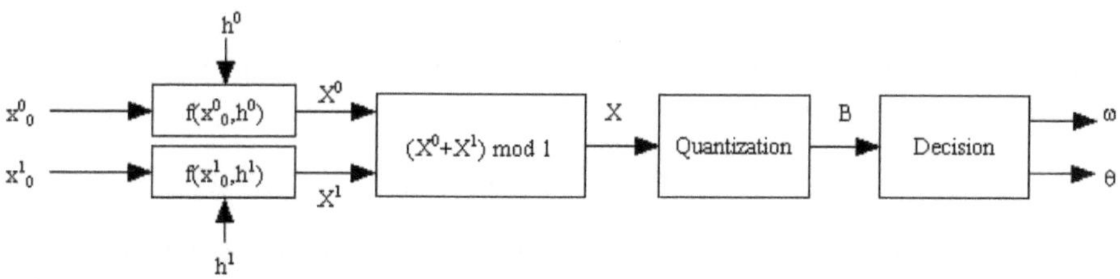

Fig. 2.7. Parameter generation of the chaotic neural network

Firstly, the chaotic skew tent map, shown in Eq. 2.2, is used to generate a chaotic sequence $X = x_0, x_1, \cdots, x_{N-1}$.

$$x_{j+1} = f(x_j, h) = \begin{cases} x_j/h & 0 < x_j \le h \\ (1 - x_j)/(1 - h) & h < x_j \le 1 \end{cases} (j = 0, 1, \cdots, N-1) \tag{2.2}$$

Here, the parameters h and x_0 both range from 0 to 1, which are used for the key. According to the parameter sensitivity, the chaotic map is iterated for more than 50 times, and the 51-th iteration corresponds to x_0, 52-th iteration corresponds to x_1, and so on. Considering that the word length of computer implementation is limited, the computing process may degrade the chaos' properties [56]. In order to reduce the degradation, two sequences X^0 and X^1 and compounded together, which produces X. Among them, X^0 is produced by the parameter pair (x_0^0, h^0), and X^1 by (x_0^1, h^1). And the compounding operation is

$$(x_i^0 + x_i^1) mod 1 = \begin{cases} x_i^0 + x_i^1, & 0 < x_i^0 + x_i^1 \leq 1 \\ x_i^0 + x_i^1 - 1, & 1 < x_i^0 + x_i^1 \leq 2 \end{cases} (i = 0, 1, \cdots, N-1)$$

Then, the chaotic sequence X is quantized, which produces the chaotic binary sequence B=$b_0, b_1, \cdots, b_{N-1}$. A simple example of quantization is

$$b_i = \begin{cases} 1, x_i \leq 0.5 \\ 0, else \end{cases}$$

Finally, the chaotic binary sequence B is used to determine the parameters ω and θ, as shown below. For example, if $b_i = 0$, then $\omega_{i,i} = -1$, otherwise, $\omega_{i,i} = 1$. Similarly, if $b_i = 0$, then $\theta_i = -1/2$, otherwise, $\theta_i = 1/2$.

$$\omega_{i,i} = \begin{cases} -1, & b_i = 0 \\ 1, & b_i = 1 \end{cases}$$

$$\theta_i = \begin{cases} -1/2, & b_i = 0 \\ 1/2, & b_i = 1 \end{cases}$$

The chained encryption mode

A chained encryption mode is used here to enhance the cryptosystem's security. The encryption process is

$$C_i = E(P_{i-1} \oplus P_i, K_i)$$

Here, P_i, P_{i-1}, K_i are i-th plaintext, i-1-th plaintext and i-th key respectively, $\oplus$ means the bitwise operation, and $E()$ is the encryption algorithm. The decryption process is

$$\begin{aligned} P_i &= D(C_i, K_i) \oplus P_{i-1} = D(C_i, K_i) \oplus D(C_{i-1}, K_{i-1}) \oplus P_{i-2} \\ &= D(C_i, K_i) \oplus D(C_{i-1}, K_{i-1}) \oplus \cdots \oplus D(C_1, K_1) \oplus D(C_0, K_0) \oplus P_{-1} \end{aligned}$$

Here, $D()$ is the decryption algorithm. Seen from the equation, the i-th ciphertext depends on all the previous ones, but is not affected by the following ones. Thus, the encrypted data stream can be directly cut off at the end, which makes it suitable for applications with the bit-rate control requirement.

Performances of the proposed stream cipher

In this cipher, for simplicity, the perception neuron is used. Some other networks can also be used to construct the stream cipher, which will cause higher computing cost. Additionally, compared with the ciphers proposed in [13–17], this cipher is composed of fixed computing instead of floating computing, which makes the cryptosystem independent of the computer's precision. Compared with the one proposed in [18], this cipher contains a chaotic pseudorandom sequence generator with good properties, adopts the chained encryption mode that enhances the system security, and supports the plaintext with variable size.

The randomness of the chaotic pseudorandom sequence is tested [57] under the condition of $x_0^0 = 0.1, h^0 = 0.3, x_0^1 = 0.6$ and $h^1 = 0.8$. The results are shown in Table 2.1. As can be seen, the produced binary sequence is of good randomness. In the key of the stream cipher, the four parameters are all of high sensitivity that benefits from the chaotic map's properties mentioned above. Fig. 2.8 gives the test results of the cross-correlation between the plaintext and the ciphertext, and of the one between different ciphertexts. In (a), the key is $K_0 = [0.2, 0.3, 0.4, 0.8]$. In (b), the keys are $K_0 = [0.2, 0.3, 0.4, 0.8]$ and $K_1 = [0.2, 0.3 + 10^{-10}, 0.4, 0.8]$. As can be seen, the ciphertext is nearly independent from the plaintext, and the ciphertexts are nearly independent from each other. Because of high sensitivity, each parameter can be more than 16 bits, and the key space is more than 64 bits.

Table 2.1. Randomness test of the produced binary sequence. The sequence is of 20000 bits with the key $K = [0.1, 0.3, 0.6, 0.8]$.

Monobit	Poker	Run test				Long runs
		Run length	Frequency	Run length	Frequency	
9979	20.38	1	4986	7	97	None
		2	2492	8	58	
		3	1279	9	22	
		4	643	10	12	
		5	336	11	3	
		6	179	12	1	

For this cipher, a kind of known-plaintext attack is to solve Eq. 2.1. Taking the i-th plaintext-ciphertext couple for example, in order to obtain ω_i and θ_i, 4 times of attempts should be carried out. Thus, the attacker should attempt for 2^{2N} times in order to obtain P. If N is bigger than 64, the difficulty is no smaller than the brute-force attack, $2^{2N} \geq 2^{128}$. This property keeps the cryptosystem secure against this kind of known-plaintext attack.

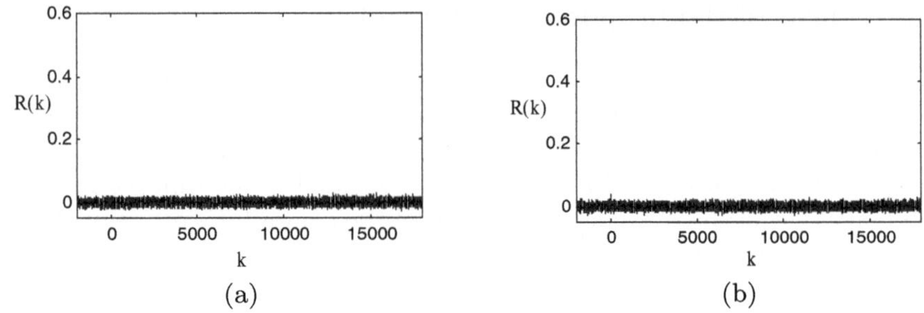

Fig. 2.8. Security test of the stream cipher. (a) Correlation between plaintext and ciphertext, (b) Correlation between different ciphertexts

2.3.3 Selective video encryption based on Advanced Video Coding

The proposed encryption scheme

In Advanced Video Coding (AVC) [58], the produced data stream is composed of intra-prediction mode, motion vector difference (MVD), residue data and some ancillary information. Considering that the residue data are often near to zero because of inter/intra-prediction, encrypting only the residue data is not secure enough. Similarly, encrypting only the intra-prediction mode [59] is not secure enough because it is fragile to replacement attacks [60, 61]. To encrypt the intra-prediction mode, MVD and residue data completely may be secure, while it is often time cost. Thus, it is reasonable to encrypt these parameters partially. In the proposed partial encryption scheme, the intra-prediction mode, the residue data's signs and the MVD's signs are encrypted with the proposed chaotic neural network based stream cipher. Here, the intra-prediction modes are encrypted after the entropy encoding, while residue data and MVDs are encrypted after quantization and before entropy encoding.

Security analysis

Compared with the permutation scheme [59], the proposed partial encryption scheme encrypts both the texture information and the motion information, which makes it difficult to recognize the video's content. Fig. 2.9 shows the encryption results of different sample videos, where (b) and (e) are encrypted with the permutation scheme, (c) and (f) are encrypted with the proposed partial encryption scheme. As can be seen, (c) and (f) are more chaotic than (b) and (e), which shows that the proposed scheme is of higher perception security.

The security against replacement attack is analyzed by comparing the proposed scheme and the permutation scheme. For the permutation scheme, only the intra-prediction mode is permuted. By replacing the encrypted prediction

(a) Original (b) Existing scheme (c) Proposed scheme

(d) Original (e) Existing scheme (f) Proposed scheme

Fig. 2.9. Results of video encryption

modes with certain value, the decoded video Fig. 2.10(a) is intelligible. Differently, besides the prediction mode, the MVD and residue data are encrypted in the proposed scheme, which makes the reconstructed video Fig. 2.10(b) still unintelligible. It proves that the proposed encryption scheme is more secure against replacement attacks.

(a) The existing scheme (b) The proposed scheme

Fig. 2.10. Security against replacement attacks

Time efficiency

The time efficiency of the encryption/decryption process is measured by the time ratio between encryption/decryption and compression/decompression. Seen from Table 2.2, the encryption/decryption operation holds no more than 10% of the compression/decompression operation. It shows that the encryption/decryption process does not affect the compression/decompression process greatly, which makes it practical to combines them together.

Table 2.2. Test of time efficiency

Video	Size	Time ratio	
		Encryption/Compression	Decryption/Decompression
Foreman	QCIF	1.6%	5.4%
Bus	QCIF	2.2%	6.8%
Flower	QCIF	1.2%	6.2%
Akiyo	QCIF	0.9%	5.0%
Mother	QCIF	1.1%	4.7%
Silent	CIF	1.6%	5.7%
Mobile	CIF	1.6%	4.1%
Football	CIF	2.6%	8.2%
Akiyo	CIF	1.9%	6.1%
Foreman	CIF	1.4%	5.3%

2.4 Multimedia content authentication based on chaotic neural networks

2.4.1 Introduction to multimedia content authentication

Different from media encryption that protects media data's confidentiality, media authentication aims to protect media data's integrity. Different from text/binary data authentication, media data (image or video) authentication prefers to content authentication. For example, such operations on an image as recompression or adding noise are often acceptable, while the malicious tempering on an image is not permitted. Thus, media authentication algorithms should be robust to some acceptable operations while sensitive to unacceptable operations. For the special requirement, media authentication technology has been attracting more and more researchers since the past decade.

In image authentication, hash function is often used. Till now, some hash functions have been reported [62–66]. Traditional hash, such as MD5 [66], is sensitive to slight changes. Thus, it is more suitable for the images that permit no modifications. To detect only the tampered regions, the image should be partitioned into regions, and each region produces a hash value with at least 128-bit. The extra channel for these regions' hash is often limited. The hash function proposed by Fridrich et al [64] produces only one bit for each data region, which reduces the length of hash value, but also decreases its security. The one proposed by Lin and Chang [65] uses the traditional hash to authenticate the extracted features between two DCT blocks, which produces at least 128 bits for each DCT block if MD5 is used. This length requires much storage space, and cannot be accepted for some applications. Therefore, contradictions exist between the hash value's length and the security. It is necessary to study new hash that produces suitable length of hash value and also keeps secure. For the one-way property, neural network has been used to construct hash function [67]. In the following content, a hash function based on a chaotic neural network is presented. The hash function can be controlled by the key, and its hash length is also controllable. Thus, the hash function is applied to the features extracted from the image regions, which produces the hash value with a suitable length. The produced hash value is then embedded into the corresponding image region. The scheme can realize self-authentication.

2.4.2 The hash function based on chaotic neural network

The proposed hash function is composed of two parts: data confusion and data diffusion, as shown in Fig. 2.11(a). Here, an $M \times N$ image is partitioned into m data regions, each data region produces a feature vector $P_i(i = 0, 1, \cdots, m-1)$, and each feature vector is composed of n floating feature pixels: $p_0, p_1, \cdots, p_{n-1}(0 \leq p_i \leq 1)$. Generally, each feature vector is composed of nr bits, and each r bits are transformed into a floating pixel ranging in $[0, 1]$. Then each feature vector P_i is confused firstly with a confusion neuron layer under the control of W_c and B_c. The confusion process generates a string of confused feature vectors $P_0', P_1', \cdots, P_{m-1}'$. Then, each of the confused feature vectors is diffused with a diffusion neuron layer under the control of W_d and B_d, which produces a hash value $C_i'(i = 0, 1, \cdots, m-1)$ corresponding to each data region. These regions' hash values are combined to compose a hash vector C that is the produced hash value.

The confusion neuron layer shown in Fig. 2.11(b) is defined as

$$\begin{cases} P_i' = f(W_{ci}P_i + B_{ci}), \\ f(x) = x \end{cases}$$

where '+' denotes module 1 addition, P_i and B_{ci} are composed of floating data ranging in [0,1], and W_{ci} is composed of binary data '0' or '1'. W_{ci} is defined as

$$W_{ci} = \begin{bmatrix} 0\,1 & 0 & \cdots \\ 0 \cdots 1 & 0 \\ \vdots \cdots\cdots \vdots \\ 1\,0 & \cdots 0 \end{bmatrix}_{n \times n}.$$

Here, there is only one '1' in each row or each column of W_{ci}. Thus, if $B_{ci} = 0$, then W_{ci} permutes P_i only. Otherwise, P_i is both confused and diffused. Here, both W_{ci} and B_{ci} are generated under the control of the key.

Similarly, the diffusion neuron layer shown in Fig. 2.11(c) is defined as

$$\begin{cases} C_i = g(W_{di}P_i' + B_{di}) \\ g(x) = l^q(x) \\ l(x) = 4x(1-x) \end{cases},$$

where '+' denotes module 1 addition, $l()$ is the chaotic Logistic map with $q \geq 50$, $l^q()$ means to iterate $l()$ for q times, W_{di} and B_{di} are produced by the

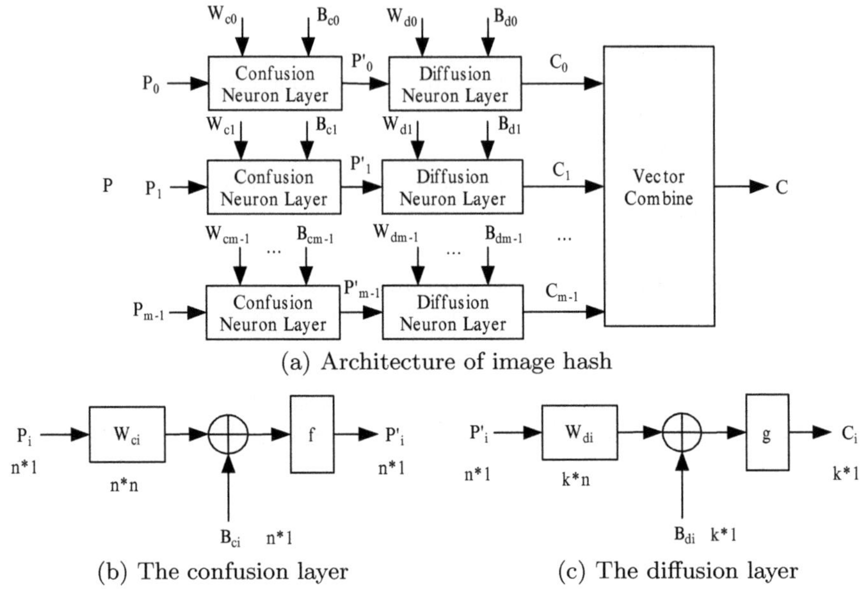

Fig. 2.11. The image hash based on chaotic neural network

key. In the following content of the section, '+' denotes module 1 addition. W_{di} is a matrix of k rows and n columns, which transforms n pixels into k pixels. The final hash value with kr bits is formed by extracting r bits from each of the k pixels. The process is defined as

$$\begin{aligned}
C_i &= \begin{bmatrix} c_{i,0} \\ c_{i,1} \\ \vdots \\ c_{i,k-1} \end{bmatrix} = g(W_{di}P_i' + B_{di}) = g(P_i^{"}) \\
&= g\left(\begin{bmatrix} w_{i,0,0} & w_{i,0,1} & \cdots & w_{i,0,n-1} \\ w_{i,1,0} & w_{i,1,1} & \cdots & w_{i,1,n-1} \\ \vdots & \vdots & \vdots & \vdots \\ w_{i,k-1,0} & w_{i,k-1,1} & \cdots & w_{i,k-1,n-1} \end{bmatrix} \begin{bmatrix} p'_{i,0} \\ p'_{i,1} \\ \vdots \\ p'_{i,n-1} \end{bmatrix} + \begin{bmatrix} b_{i,0} \\ b_{i,1} \\ \vdots \\ b_{i,k-1} \end{bmatrix} \right) \\
&= g\left(\begin{bmatrix} p^{"}_{i,0} \\ p^{"}_{i,1} \\ \vdots \\ p^{"}_{i,k-1} \end{bmatrix} \right) .
\end{aligned}$$

Thus, in order to decrease the length of image hash, k is often a value smaller than n. And $g()$ is defined as

$$\sum_{j=0}^{k-1} c_{i,j} = g(\sum_{j=0}^{k-1} p''_{i,j}) = l^q(\sum_{j=0}^{k-1} p''_{i,j}).$$

For each image, the confusion process and diffusion process are applied to each data region, respectively. Thus, the image hash $H()$ is defined as

$$C = \begin{bmatrix} C_0 \\ C_1 \\ \vdots \\ C_{m-1} \end{bmatrix} = \begin{bmatrix} g(W_{d0}f(W_{c0}P_0 + B_{c0}) + B_{d0}) \\ g(W_{d1}f(W_{c1}P_1 + B_{c1}) + B_{d1}) \\ \vdots \\ g(W_{dm-1}f(W_{cm-1}P_{m-1} + B_{cm-1}) + B_{dm-1}) \end{bmatrix} = H(P, K),$$

where K is the key that produces W_c, B_c, W_d and B_d.

Additionally, the chaotic neural network composed of the confusion neuron and diffusion neuron can be serialized in order to reduce the length of the produced hash value. Thus, if the original image feature is of L_0 bits, the hash value is of L_c bits, and the serial layer is T, then the following condition is satisfied.

$$L_c = L_0 \cdot (k/n)^T$$

Since k is often smaller than n, the image signature with suitable length can be obtained through choosing suitable serial layer T.

2.4.3 The proposed image authentication scheme

Based on the proposed image hash, an image authentication scheme is presented in Fig. 2.12. At the sender end, the feature P is extracted from an image V, from which, the hash value C is computed under the control of K_1. Then, C is embedded into the image V under the control of K_2. The process is described as

$$S = Emb(V, C, K_2) = Emb(V, H(P, K_1), K_2) = Emb(V, H(Fea(V), K_1), K_2),$$

where S is the embedded image, $Fea()$ is the feature extraction process, and $Emb()$ is the embedding algorithm. The extracted feature determines the robustness of the authentication process. Taking wavelet transformation for example, in order to keep robust to recompression and sensitive to the malicious tampering, the high bit-planes of the code blocks in the low frequency-band are preferred. For example, if an image is transformed by 4-level wavelet transformation, then only the 7 highest bit-planes of the code blocks in the 10 lowest frequency-bands are selected as features. The computed hash value should be embedded into the image itself imperceptibly [68], which saves extra space for hash value's storage. To keep imperceptibility, the signature is embedded into the position with little effect on images' quality, for example, the 8-th highest bit-plane. The embedding position is controlled by the key.

Similarly, at the receiver end, the embedded hash value C is extracted from the image, which is compared with the computed one C'. If they are same to

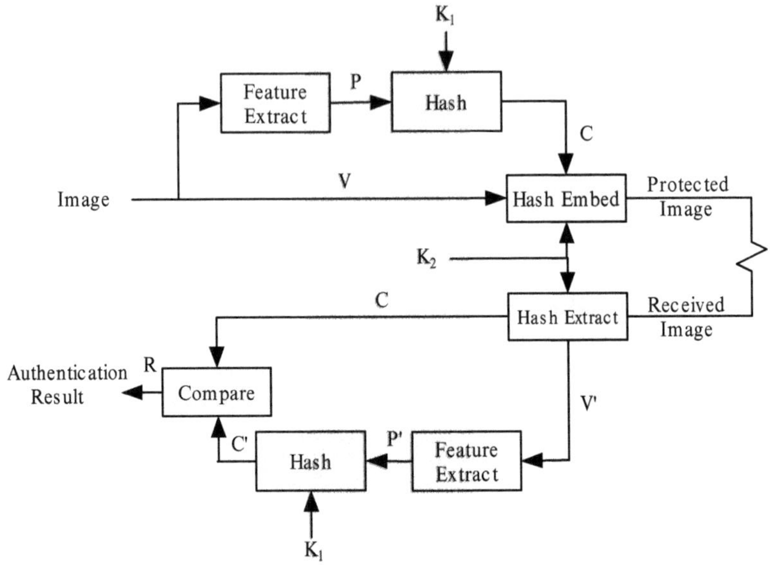

Fig. 2.12. The proposed image authentication scheme

each other, then the received image is authenticated. Otherwise, the image's integrity is destroyed, and the difference between C and C' determines the regions that have been tampered. The authentication process is defined as

$$\begin{cases} (C, V') = Ext(S, K_2) \\ R = Com(C, C') = Com(C, H(P', K_1)), \end{cases}$$

where R is the authentication result, $Ext()$ is the extraction algorithm [68], and $Com()$ is to determine the difference between C and C'.

2.4.4 Performance analysis

Security

In the proposed hash function, the confusion layer and diffusion layer are combined together to obtain high computing security, which obeys the cryptographic design principles [57]. It is different from the function proposed in [67], which depends on the randomness of the sequence produced by a chaotic neural network. Some other neural networks with more confusion layers or diffusion layers can also be used to design a hash function with higher security.

The proposed image hash is nearly a one-way function. That is, it is difficult to recover V from C. The signature process is controlled by the K, which makes brute-force attack difficult. For attackers knowing both P and C, the difficulty to solve a singular equation (the diffusion function) makes the attacks unpractical. For select-image attackers, they have to break the confusion process and to decode P'' from C before to solve the singular equation. But the brute-force space of confusion process is $n \cdot (n-1) \cdot (n-2) \cdot \ldots \cdot 1$, which is bigger than 2^n if $n > 4$. In practice, the feature vector is often of more than 16 pixels, which makes brute-force attacks difficult. What's more, the iterated Logistic map is difficult to solve, which requires 2^q times of brute-force attempts. Thus, the brute-force space of decoding P'' is no smaller than 2^{50} when $q \geq 50$. And the overall space of select-image attack is 2^{n+q} that make it difficult to falsify an image hash.

The hash function is of high plaintext sensitivity. The confusion process and diffusion process make the signature sensitive to the plaintext. From the diffusion function, we get

$$c_{i,j} = g(\sum_{t=0}^{n-1} w_{i,j,t} p'_{i,t} + b_{i,j}),$$

where j varies from 0 to k-1. As can be seen, $c_{i,j}$ is in relation with $p'_{i,t} (t = 0, 1, \cdots, n-1)$, that is, a bit of difference in the plaintext causes great differences in the hash value. Additionally, in the proposed hash function, the control parameters are independent of the plaintext. These properties make attackers difficult to find another plaintext with the same signature.

The image hash is controlled by the key. That is, for the same image, various hash values can be generated, while only the user with the correct key can authenticated it. This property makes it difficult for attackers to falsify hash values. The same advantage can be drawn from the key-based hash embedding process. Only the authenticated user can embed or extract the hash value correctly.

Authentication performance

The proposed image hash can keep robust to acceptable operations, such as recompression or slight noise. Table 2.3 gives the experimental results on various images, where the embedding bit-plane varies from 6 to 8 and (n,k,r)=(64,4,4). The 10 test images include Lena, Airplane, Couple, Girl, Boats, Cameraman, and so on. The hash length of each data region is 16. Here, the JPEG2000 compression ratio changes from 5 to 9, and the zero-mean Gaussian noise's variance varies from 0.1 to 0.0001. As can be seen, when 6 or 7 bit-planes are used as the feature, all the test images are correctly authenticated. Only 7 or 5 images are correctly authenticated when 8 bit-planes are hashed and the compression ratio is no smaller than 8. In practice, $C \leq 8$ and $N \leq 0.01$ are acceptable operations, and now the number of the hashed bit-planes should be no bigger than 7.

Additionally, this scheme is sensitive to malicious tampering, such as cutting, pasting or substitution. Fig. 2.13 gives the experimental results on image

Table 2.3. Robustness of the proposed image hash

Compress, Noise	6 Bit-planes		7 Bit-planes		8 Bit-planes	
	Compress	Noise	Compress	Noise	Compress	Noise
5, 0.0001	10	10	10	10	10	10
6, 0.001	10	10	10	10	10	10
8, 0.01	10	10	10	10	7	7
9, 0.1	10	10	10	10	5	4

Plane, where the image is tampered by cutting and substitution, the number of the hashed bit-planes varies from 6 to 8, and (n,k,r)=(64,4,4). As can be seen, two tampered data regions are not detected if only 6 bit-planes are hashed, while all the tampered regions are detected when 7 or 8 bit-planes are hashed. It shows that the more the hashed bit-planes, the higher the system's sensitivity. Through experiments on various images, the number of the hashed bit-planes is recommended as 7.

(a) Original (b) Watermarked (c) Tampered

(d) Authentication(6b) (e) Authentication(7b) (f) Authentication(8b)

Fig. 2.13. Sensitivity of the authentication scheme. (a) original, (b) embedded, (c) tampered, (d) authentication result of 6 bit-planes, (e) authentication result of 7 bit-planes, and (f) authentication result of 8 bit-planes

Computing complexity

The computing complexity depends on the data volumes to be hashed. Taking the time ratio between image authentication and image compression as a measurement, the results are shown in Table 2.4. Here, the embedding bit-plane varies from 6 to 8, (n,k,r)=(64,4,4), and JPEG2000 is applied under the compression ratio of 8. As can be seen, the time ratio increases with the rise of the number of hashed bit-planes, and it is no bigger than 20% when the number of the hashed bit-plane is not bigger than 7. Thus, the authentication process does not affect compression process greatly, which makes it practical to combine them together to realize real time image compression and authentication.

Table 2.4. Test of computing complexity

Image(Size,Color)	Time ratio of authentication/compression		
	6 Bit-planes	7 Bit-planes	8 Bit-planes
Lena (256×256, Colorful)	9.2%	11.6%	18.6%
Airplane (256×256, Colorful)	10.8%	16.4%	21.9%
Couple (256×256, Colorful)	8.9%	12.7%	16.8%
Girl (128×128, Gray)	9.4%	16.3%	19.9%
Boats (512×512, Gray)	10.5%	15.6%	22.8%
Cameraman (256×256, Gray)	7.8%	14.7%	18.5%

2.5 Future work and discussions

Considering that multimedia data need special encryption or authentication algorithms that are time efficient and support the plaintext with variable size, some new data encryption or authentication algorithms are expected. In this chapter, the cipher and hash function based on chaotic neural networks are investigated, and proved workable. However, for further practicability, some improvements are expected.

Firstly, the implementation in integer form is preferred, which is easier to realize digital information protection. Now, most of the chaotic neural network based algorithm is implemented in floating form, which makes the system rely on the word length of the computer. In integer form, the computing resolution is confirmed, and it is easy to realize synchronization between different computers. The difficulty is to make the integer implementation keep the same performance with the floating implementation.

Secondly, the parallel property is to be used. Compared with traditional ciphers, chaotic neural network's parallel implementation makes the ciphers based on chaotic neural networks easy in efficiency improvement. To adopt the parallel structure to realize lightweight encryption is an attractive and challenging work.

Thirdly, the chaotic neural networks with more input neurons and neural layers are preferred. In the chapter, only some simple chaotic neural networks are considered. In fact, the neural networks with more complicated structure, such as more layers or more input neurons, may bring higher security. Of course, the computing efficiency should be considered in order to obtain suitable tradeoff.

Fourthly, the security of the ciphers based on chaotic neural networks should be analyzed thoroughly. Besides such traditional analysis methods for ciphers as brute-force space, linear analysis or differential analysis, some special attacks should be considered, such as chaos mode analysis [69] or neural network structure analysis. This work should be done to strengthen the ciphers based on chaotic neural networks.

Fifthly, the partial encryption scheme should be carefully designed. According to partial encryption, the encrypted media data should be of high perception security, and keeps independent from the unencrypted data. Additionally, such media data with fixed pattern are not suitable for encryption, such as the file header or synchronization information, because they are fragile to replacement attacks. Thus, the encryption scheme should be considered besides that the cipher is secure.

2.6 Conclusions

In this chapter, the chaotic neural network's application in media encryption and authentication is investigated. By using the chaotic neural network's properties, such as diffusion, confusion, one-way, parameter sensitivity, random similarity, the cipher and hash function suitable for media encryption and authentication are designed. Based on the cipher, the partial encryption scheme for AVC video is proposed and analyzed. The experimental results show the scheme is secure in perception and efficient in practice. Based on the hash function, the image authentication scheme is proposed to authenticate media data's integrity. The analyses and experiments show that the scheme is secure against malicious tampering and robust to such acceptable operations as compression or noise. These attempts prove the practicability of media encryption based on chaotic neural networks.

Although chaotic neural network's properties show its potential value in media data encryption, using chaotic neural network to encrypt media data is just beginning, with much work to be done. The efficiency and computing resolution are to be improved by changing the floating implementation into integer implementation. Additionally, the security should be thoroughly analyzed by considering of both traditional attacks and some special attacks.

However, considering of the urgent need of media protection, it should be confirmed that the future of the research work is still bright.

2.7 Acknowledgements

The author wants to thank Prof. Z. Wang and Mr. J. Wang for their great help and to thank reviewers for their valuable advices.

References

1. Babloyantz A, Salazar J, Nicolis C (1985) Evidence for Chaotic Dynamics of Brain Activity during the Sleep Cycle. Phys Lett 111:152–156
2. Wasserman P (1993) Advanced Methods in Neural Computing. New York: Van Nostrand Reinhold, New York
3. Chay T, Fan Y (1993) Evolution of Periodic States and Chaos in Two Types of Neural Models. In: Ditto W(ed.) Chaos in Biology and Medicine, Proc SPIE, Washington: SPIE, 2036:100–114
4. Aibara K (1990) Chaotic Neural Networks. Physical Letter A 144(6,7):334–340.
5. Inoue M, Nagayoshi A (1991) A Chaos Neuro-computer. Physical Letter A 158(8):373–376
6. Inoue M, Nakamoto K (1993) Epilepsy in a Chaos Neuro-computer Model. In SPIE Proceeding, Chaos in Biology and Medicine, 2306:77–84
7. Chan C, Cheng L (1998) Pseudorandom Generator Based on Clipped Hopfield Neural Network. In: Proceedings of the 1998 IEEE International Symposium on Circuits and Systems, 3:183–186
8. Chan C, Chan C, Cheng L (2001) Software Generation of Random Number by using Neural Network. In Proceeding of 5th International Conference on Artificial Neural Networks and Genetic Algorithms (ICANNGA), Pragus, Czech Republic
9. Guo D, Cheng L, Cheng L (1999) A New Symmetric Probabilistic Encryption Scheme Based on Chaotic Attractors of Neural networks. Applied Intelligence 10(1):71–84
10. Xiang T, Liao X, Tang G, Chen Y, Wong K (2006) A novel block cryptosystem based on iterating a chaotic map. Physics Letters A 349(1-4):109–115
11. Xiao D, Liao X, Deng S (2005) One-way Hash function construction based on the chaotic map with changeable-parameter. Chaos, Solitons & Fractals 24(1):65–71
12. Rachel M, Einat K, Wolfgang K (2003) Public Channel Cryptography by Synchronization of Neural Networks and Chaotic Maps. Physical Review Letters 91(11):118701/1–118701/4
13. Cauwenberghs G (1999) Delta-sigma cellular automata for analog VLSI random vector generation. IEEE Transactions on Circuits and Systems II: Analog and Digital Signal Processing 46(3):240–250
14. Chan C, Cheng L (2001) The convergence properties of a clipped Hopfield network and its application in the design of keystream generator. IEEE Transactions on Neural Networks 12(2):340–348

15. Karras D, Zorkadis V (2003) On neural network techniques in the secure management of communication systems through improving and quality assessing pseudorandom stream generators. Neural Networks 16(5-6):899–905
16. Caponetto R, Lavorgna M, Occhipinti L (1996) Cellular neural networks in secure transmission applications. In Proceedings of the IEEE International Workshop on Cellular Neural Networks and their Applications, 411–416
17. Yue T, Chiang S (2000) A neural network approach for visual cryptography. In Proceedings of the IEEE-INNS-ENNS International Joint Conference on Neural Networks 5:494–499
18. Yen J, Guo J (1999) A Chaotic Neural Network for Signal Encryption/Decryption and Its VLSI Architecture. In: Proc. 10th (Taiwan) VLSI Design/CAD Symposium, 319-322
19. Lian S, Chen G, Cheung A, Wang Z (2004) A Chaotic-Neural-Network-Based Encryption Algorithm for JPEG2000 Encoded Images. In Proceedings of 2004 IEEE Symposium on Neural Networks (ISNN2004),Springer LNCS, 3174:627–632
20. Yee L, Silva D (2002) Application of multilayer perception networks in symmetric block ciphers. In Proceedings of the 2002 International Joint Conference on Neural Networks, 2:1455–1458
21. Vanstone S, Menezes A, Oorschot P (1996) Handbook of Applied Cryptography. CRC Press
22. Shannon C (1949) Communication theory of secrecy systems. Bell System Technical Journal 28:656–715
23. Li C, Li S, Zhang D, Chen G (2004) Cryptanalysis of a Chaotic Neural Network Based Multimedia Encryption Scheme. In Proceedings of the 2004 Pacific-Rim Conference on Multimedia (PCM2004), Springer LNCS, 3333:418–425
24. Fridrich J (1998) Symmetric ciphers based on two-dimensional chaotic maps. Int. J. Bifurcation and Chaos 8(6):1259–1284
25. Chen G, Mao Y, Chui C (2004) A symmetric image encryption scheme based on 3D chaotic cat maps. Chaos, Solitons and Fractals 12:749–761
26. Li S, Zheng X, Mou X, Cai Y (2002) Chaotic Encryption Scheme for Real-time Digital Video. In Proceedings of SPIE, Electronic Imaging, Real-time Imaging V1, 4666:149–160
27. Ren X, Hu G, Tan Z (2000) Synchronization of chaotic neural network and its application in secure communication. Journal of Shanghai Jiaotong University 34(6):744–747
28. Elman J (1990) Finding structure in time. Cognitive Science 14:179-211
29. Pyragas K (1992)Continuous control of chaos by self controlling feedback. Phys Lett A 170:421–428
30. Ushio T (1995) Chaotic synchronization and controlling chaos based on contraction maps. Phys Lett A 198:14–22
31. Kocarev L, Parlitz U (1995) General approach for chaotic synchronization with application to communication. Phys Rev Lett 74(25):5028–5031
32. Zou F, Nossek J (1991) A Chaotic Attractor with Cellular Neural Networks. IEEE Transaction on Circuits and Systems 38(7):811–812
33. Zou F, Joset A (1993) Nossek. Bifurcation and chaos in Cellular Neural Networks. IEEE Transaction on Circuits and Systems-I: Fundamental Theory and Applications 40(3):166–173
34. Kaneko K (1990) Clustering, coding, switching, hierarchical ordering and control in a network of chaotic elements. Physics D 41(22):137–172

35. Aihara K, Takabe T, Toyoda M (1990) Chaotic Neural Networks. Phys Lett A 144(6):333–340
36. Inoue M, Nagayoshi A (1991) A Chaos Neuro-computer. Phys Lett A 158(8):373–376
37. Addison P (1997) Fractals and Chaos - an illustrated course, Institute of physics publishing
38. Fang Y, Yu Y (1993) Dynamic analysis of chaotic neurons. In Proceedings of Chinese Conference on Neural Networks, 199-203
39. Nagayama I (1995) Spectral Analysis of the Chaotic Behavior Obtained by Neural Networks. In Proceedings of 1995 International Conference on Neural Networks, 28–31
40. Kudelski A (1994) Method for scrambling and unscrambling a video signal. United States Patent 5375168
41. Access control system for the MAC/packet family: EUROCRYPT. European Standard EN 50094, CENELEC (1992)
42. Maniccam S, Bourbakis N (2004) Image and video encryption using SCAN patterns. Pattern Recognition 37(4):725–737
43. Qiao L, Nahrstedt K (1997) A new algorithm for MPEG video encryption. In Proceeding of the First International Conference on Imaging Science, Systems and Technology (CISST'97),21–29
44. Romeo R, Romdotti G, Mattavelli M, Mlynek D (1999) Cryptosystem architectures for very high throughput multimedia encryption: the RPK solution. In Proceedings of ICECS'99, 1:261–264
45. Shi C, Wang S, Bhargava, B (1999) MPEG video encryption in real-time using secret key cryptography. In Proc. of International Conference of Parallel and Distributed Processing Techniques and Applications (PDPTA'99), 2822–2828
46. Chiaraluce F, Ciccarelli L, Gambi E, Pierleoni P, Reginelli M (2002) A New Chaotic Algorithm for Video Encryption. IEEE Transactions on Consumer Electronics 48(4):838–844
47. Yen J, Guo J (1999) A new MPEG encryption system and its VLSI architecture. In Proceeding of IEEE Workshop on Signal Processing Systems,430–437
48. Tosum A, Feng W (2000) Efficient multi-layer coding and encryption of MPEG video streams. In Proceedings of IEEE International Conference on Multimedia and Expo (I), 119–122
49. Qiao L, Nahrstedt K, Tam I (1997) Is MPEG encryption by using random list instead of zigzag order secure. In Proceeding of IEEE International Symposium on Consumer Electronics, 226–229
50. Shi C, Bhargava B (1998) A fast MPEG video encryption algorithm. In Proceedings of the 6th ACM International Multimedia Conference, 81–88
51. Zeng W, Lei S (2003) Efficient frequency domain selective scrambling of digital video. IEEE Trans Multimedia 5(1):118–129
52. Wu C, Kuo C (2000) Fast encryption methods for audiovisual data confidentiality. In SPIE International Symposia on Information Technologies 2000, Proceedings of SPIE 4209:284–295
53. Wu C, Kuo C (2001) Efficient multimedia encryption via entropy codec design. In SPIE International Symposium on Electronic Imaging 2001, Proceedings of SPIE 4314
54. Lian S, Sun J, Wang Z (2004) Perceptual Cryptography on SPIHT Compressed Images or Videos. In Proceedings of 2004 IEEE Conference on Multimedia & Expo (ICME2004), 3:2195–2198

55. Lian S, Sun J, Wang Z (2004) A Novel Image Encryption Scheme Based-on JPEG Encoding. In Proceedings of the Eighth International Conference on Information Visualization (IV), 217–220
56. Li S, Chen G, Mou X (2005) On the Dynamical Degradation of Digital Piecewise Linear Chaotic Maps. International Journal of Bifurcation and Chaos 15(10):3119-3151
57. Security Requirements for Cryptographic Modules (Change Notice). Federal Information Processing Standards Publication (FIPS PUB) 140-1 (2001)
58. ITU-T Rec. H.264/ISO/IEC 11496-10. Advanced Video Coding. Final Committee Draft, Document JVT-E022 (2002)
59. Ahn J, Shim H, Jeon B, Choi I (2004) Digital Video Scrambling Method Using Intra Prediction Mode. PCM2004, Springer, LNCS 3333:386–393
60. Lian S, Sun J, Zhang D, Wang Z (2004) A Selective Image Encryption Scheme Based on JPEG2000 Codec. In Proceedings of 2004 Pacific-Rim Conference on Multimedia, Springer LNCS 3332:65–72
61. Podesser M, Schmidt H, Uhl A (2002) Selective Bitplane Encryption for Secure Transmission of Image Data in Mobile Environments. In: CD-ROM Proceedings of the 5th IEEE Nordic Signal Processing Symposium (NORSIG 2002). Tromso-Trondheim, Norway
62. Fridrich J, Goljan M, Du R (2001) Invertible Authentication. In: Proceeding of SPIE, Security and Watermarking of Multimedia Contents, San Jose, California, 3971:197–208
63. Fridrich J (2000) Visual Hash Functions for Obvious Watermarking. In: Proceeding of SPIE Photonic West Electronic Imaging 2000, Security and Watermarking of Multimedia Contents, San Jose, California, 286–294
64. Fridrich J, Goljan M (2000) Robust Hash Functions for Digital Watermarking. The International Conference on Information Technology: Coding and Computing Las Vegas, Nevada, 178–183
65. Lin C, Chang S (1998) A Robust Image Authentication Method Surviving JPEG Lossy Compression. In: Proceedings of SPIE Storage and Retrieval of Image/Video Databases, San Jose, 3312:296–307
66. Rivest R (1992) The MD5 Message Digest Algorithm. RFC 1321
67. Xiao D, Liao X (2004) A Combined Hash and Encryption Scheme by Chaotic Neural Network. In Proceedings of ISNN2004, 2:633–638
68. Yeung M, Mintzer F (1997) An Invisible Watermarking Technique for Image Verification. In: Proceeding of ICIP'97, Santa Barbara, California, 2:680–683
69. Zhusubaliyev Z, Mosekilde E (2003) Bifurcations and Chaos in Piecewise-Smooth Dynamical Systems. World Scientific, Singapore.

3

Evolutionary Regular Substitution Boxes

Nadia Nedjah[1] and Luiza de Macedo Mourelle[2]

[1] Department of Electronics Engineering and Telecommunications,
Engineering Faculty,
State University of Rio de Janeiro,
Rua São Francisco Xavier, 524, Sala 5022-D,
Maracanã, Rio de Janeiro, Brazil
nadia@eng.uerj.br, http://www.detel.eng.uerj.br

[2] Department of System Engineering and Computation,
Engineering Faculty,
State University of Rio de Janeiro,
Rua São Francisco Xavier, 524, Sala 5022-D,
Maracanã, Rio de Janeiro, Brazil
ldmm@eng.uerj.br, http://www.desc.eng.uerj.br

A substitution box (or S-box) is simply a transformation of an input sequence of bits into another. The input and output sequences have not necessarily the same number of bits. In cryptography, an S-box constitutes a cornerstone component of symmetric key algorithms. In block ciphers, they are typically used to obscure the relationship between the plaintext and the ciphertext. Non-linear and non-correlated S-boxes are the most secure linear and differential cryptanalysis. In this chapter, we focus on engineering regular S-boxes, presenting high non-linearity and low auto-correlation properties using evolutionary computation. Hence, there are three properties that need to be optimised: regularity, non-linearity and auto-correlation. We use the Nash equilibrium-based multi-objective evolutionary algorithm to engineer resilient substitution boxes.

3.1 Introduction

In cryptography, confusion and diffusion are two important properties of a secure cipher as identified in [1]. Confusion allows one to make the relationship between the encryption key and ciphertext as complex as possible while diffusion allows one to reduce as much as possible the dependency between the plaintext and the corresponding ciphertext. Substitution (a plaintext symbol is replaced by another) has been identified as a mechanism for primarily

N. Nedjah and L. de Macedo Mourelle: *Evolutionary Regular Substitution Boxes*, Studies in Computational Intelligence (SCI) **57**, 79–88 (2007)
www.springerlink.com

confusion. Conversely transposition (rearranging the order of symbols) is a technique for diffusion. In modern cryptography, other mechanisms are used, such as linear transformations [1]. Product ciphers use alternating substitution and transposition phases to achieve both confusion and diffusion respectively. Here we concentrate on confusion using non-linear and non-correlated substitution boxes or simply S-boxes.

It is well-known that the more linear and the less auto-correlated the S-box is, the more resilient the cryptosystem that uses them. However, engineering a regular S-box that has the highest non-linearity and lowest auto-correlation properties is an *NP*-complete problem. Evolutionary computation is the ideal tool to deal with this type of problems. As there are three objectives that need to be reached, which are maximal regularity, maximal non-linearity yet minimal auto-correlation, we propose to use multi-objective evolutionary optimisation. Therefore, we exploit the game theroy [3] and more specifically the well-known Nash equilibrium strategy [4] to to engineer such *resilient* substitution boxes.

The rest of this chapter is organised in five sections. First, in Section 3.2, we define more formally S-boxes as well as their characteristics. Subsequently, in Section 3.3, we present the multi-objective evolutionary algorithm used to perform the evolution i.e., the Nash equilibrium-based evolutionary algorithm [3]. Thereafter, in Section 3.4, we describe the S-box encoding and the genetic operators used followed by the definition and implementation of an S-box fitness evaluation with respect to all three considered objectives. Then, in Section 3.5, we evaluate the performance of the evolutionary process and assess the quality of the evolved resilient S-boxes. Also, we compare the characteristics of the engineered S-boxes to those of Advanced Encryption Standard (AES) [1] and Data Encryption Standard (DES) [2]. Last but not least, in Section 3.6, we summarise the content of the chapter and draw some useful conclusions.

3.2 Preliminaries for Substitution Boxes

S-Boxes play a basic and fundamental role in many modern block ciphers. In block ciphers, they are typically used to obscure the relationship between the plaintext and the ciphertext. Perhaps the most notorious S-boxes are those used in DES [2]. S-boxes are also used in modern cryptosystems based on AES [1] and Kasumi [5]. All three are called Feistel cryptographic algorithms [6] and have the simplified structure depicted in Fig. 3.1.

An S-box can simply be seen as a Boolean function of n inputs and m outputs, often with $n > m$. Considerable research effort has been invested in designing resilient S-boxes that can resist the continuous cryptanalyst's attacks. In order to resist linear and differential cryptanalysis [7, 8], S-boxes need to be confusing or non-linear and diffusing or non auto-correlated.

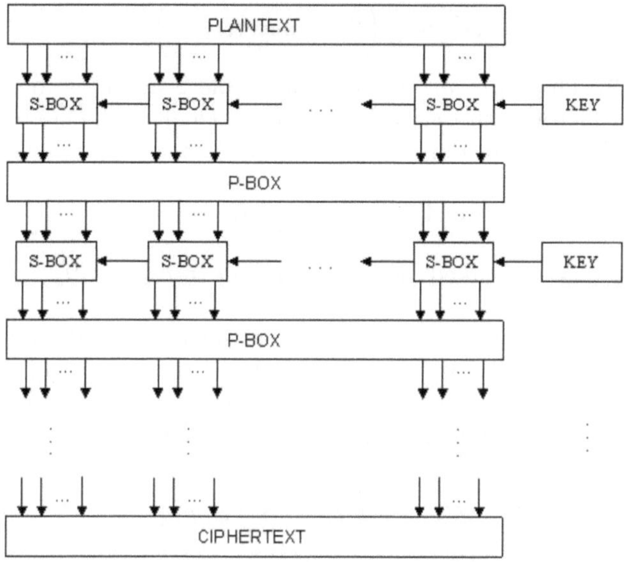

Fig. 3.1. The simplified structure of Feistel cryptographic algorithm

S-boxes also need to be non-regular.

Definition 1. A *simple* S-box $\mathcal{S}$ is a Boolean function defined as $\mathcal{S}$: $\mathcal{B}^n \longmapsto \mathcal{B}$.

Definition 2. A *linear* simple S-boxes $\mathcal{L}$ is defined in (3.1):

$$\mathcal{L}_\beta(x) = \bigoplus_{i=0}^{i=m} \beta_i.\mathcal{L}(x_i) \tag{3.1}$$

Definition 3. The *polarity* of a simple S-box $\mathcal{S}$ is defined in (3.2):

$$\hat{\mathcal{S}}(x) = (-1)^{\mathcal{S}(x)} \tag{3.2}$$

Definition 4. The *uncorrelation factor* of Two simple S-boxes $\mathcal{S}$ and $\mathcal{S}^{'}$ is defined in (3.3):

$$\mathcal{U}_{\mathcal{S},\mathcal{S}'} = \sum_{x \in \mathcal{B}^n} \hat{\mathcal{S}}(x) \times \hat{\mathcal{S}}'(x) \tag{3.3}$$

Definition 5. Two simple S-boxes $\mathcal{S}$ and $\mathcal{S}\prime$ are said to be uncorrelated if and only $\mathcal{U}_{\mathcal{S},\mathcal{S}'} = 0$.

Definition 6. The *non-linearity* of a simple S-box $\mathcal{S}$ is measured by its uncorrelation factor with all possible linear simple S-Boxes defined in (3.4):

$$\mathcal{N}_{\mathcal{S}} = \frac{1}{2}\left(2^n - \max_{\alpha \in \mathcal{B}^n} |\mathcal{U}_{\mathcal{S},\mathcal{L}}|\right) \tag{3.4}$$

Definition 7. The *auto-correlation* of a simple S-box $\mathcal{S}$ is measured by its uncorrelation factor with derivatives S-boxes $\mathcal{D}(x) = \mathcal{S}(x) \oplus \alpha$, for all $\alpha \in \mathcal{B}^n \setminus \{0^n\}$ and defined in (3.5):

$$\mathcal{A}_{\mathcal{S}} = \max_{\alpha \in \mathcal{B}^n \setminus \{0^n\}} |\mathcal{U}_{\mathcal{S},\mathcal{D}}| \tag{3.5}$$

Note that $\mathcal{U}_{\mathcal{S},\mathcal{D}}$ is also called *Walsh Hadamard transform* [9].

Definition 8. A simple S-box is said to be *balanced* if and only the number of combinations $x \in \mathcal{B}^n$ such that $\mathcal{S}(x) = 0$ and the number of combinations $y \in \mathcal{B}^n$ such that $\mathcal{S}(y) = 1$ are the same. The *balance* of a simple S-box is measured using its *Hamming weight*, defined in (3.6).

$$\mathcal{W}_{\mathcal{S}} = \frac{1}{2}\left(2^n - \sum_{x \in \mathcal{B}^n} \hat{\mathcal{S}}(x)\right) \tag{3.6}$$

3.3 Nash Equilibrium-based Evolutionary Algorithms

This approach is inspired by the Nash strategy for economics and game theory [3,4]. The multi- objective optimisation process based on this strategy is non-cooperative in the sense that each objective is optimised separately. The basic idea consists of associating an agent or player to every objective. Each agent attempts to optimise the corresponding objective fixing the other objectives to their best values so far. As proven by Nash in [3], the Nash equilibrium point should be reached when no player can improve further the corresponding objective.

Let m be the number of objectives $f_1, \ldots, f_m$. The multi-objective genetic algorithms based on Nash strategy assigns the optimisation of objective f_i to $player_i$, each of which has its own population. The process is depicted in Figure 3.2. Basically, it is a parallel genetic algorithm [10] with the exception that there are several criteria to be optimised. When a player, say $player_i$, completes an evolution generation, say t, it sends the local best solution reached f_i^t to the all $player_j$, $j \in \{1, \ldots, m\} \setminus \{i\}$, which will then be used to fix objective f_i to f_i^t during the next generation $t+1$. This evolutionary process is repeated iteratively until no player can further improve the associated criteria.

3.4 Evolving Resilient S-Boxes

In general, two main important concepts are crucial to any evolutionary computation: individual encoding and fitness evaluation. One needs to know how

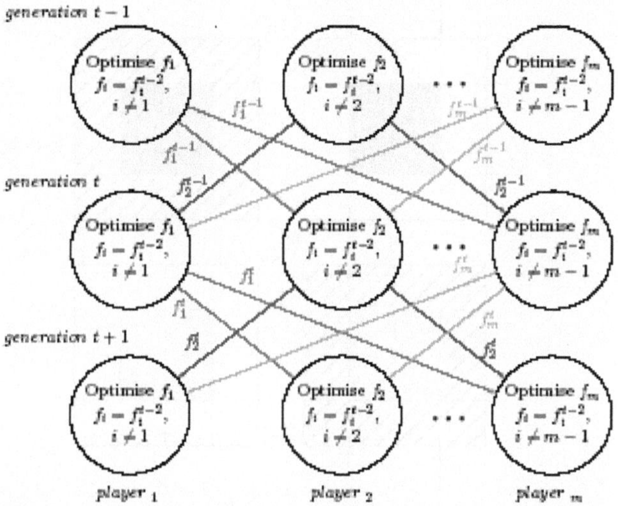

Fig. 3.2. Multi-objective optimisation using Nash strategy

to appreciate the solutions with respect to each one of the multiple objectives. So In the section, we concentrate on these two aspects for evolving optimal S-boxes.

3.4.1 S-Box encoding and genetic operators

We encode an S-box simply using a matrix of bytes. The mutation operator chooses an entry randomly and changes its value, using a randomised byte. Crossover of S-Boxes is implemented using four-point crossover. This is described in Fig. 3.3. The four-point crossover can degenerate to either the triple, double or single-point crossover. Moreover, these can be either *horizontal* or *vertical.*

The four-point crossover degenerates into a triple-point one whenever one of the crossover points (say p) coincides with one of the sides of the S-box. It is a horizontal triple-point when p is one of the vertical crossover points, as in Fig. 3.4–(a) and Fig. 3.4–(b) and horizontal otherwise, as in Fig. 3.4–(c) and Fig. 3.4–(d).

The four-point crossover degenerates into a double-point one whenever either both horizontal points or vertical points coincide with the low and high or left and right limits of the S-box respectively. The horizontal double-point crossover is described in Fig. 3.5–(a) and the horizontal one in Fig. 3.5–(b).

The four-point crossover can also yield a horizontal and vertical single-point crossover as described in Fig. 3.6–(a) and Fig. 3.6–(b) respectively. The vertical single-point crossover happens when the horizontal crossover points

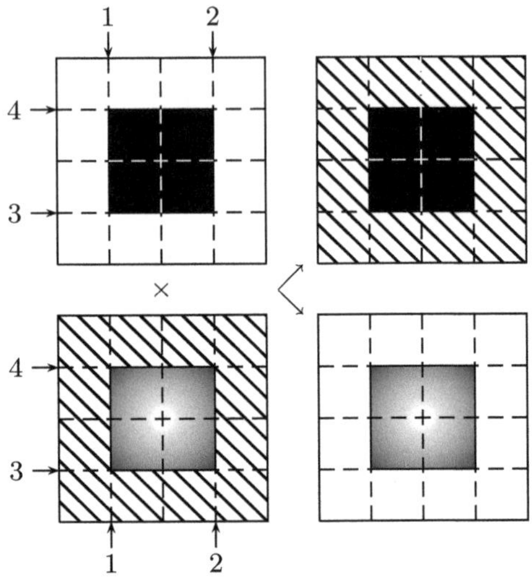

Fig. 3.3. Four-point crossover of S-boxes

coincide with the low and high limits of the S-box and the one of the vertical points coincides with the left or right limit of the S-box. On the other hand, the horizontal single-point crossover happens when the vertical crossover points coincide with the left and right limits of the S-box and one of the horizontal crossover points coincide with low or high limit of the S-box.

3.4.2 S-Box evaluation

Now, let us generalise the definitions of balance, non-linearity and auto-correlation to non-simple S-boxes, i.e. S-boxes defined as $\mathcal{S}$: $\mathcal{B}^n \longmapsto \mathcal{B}^m$.

Definition 9. An S-box $\mathcal{S}$ defined as $\mathcal{S}$: $\mathcal{B}^n \longmapsto \mathcal{B}^m$ is a concatenation of m simple S-boxes $\mathcal{S}_i$ with $1 \leq i \leq m$, such as in (3.7):

$$\mathcal{S}(x) = \mathcal{S}_1(x)\mathcal{S}_2(x)\dots\mathcal{S}_m(x) \tag{3.7}$$

Definition 10. The *non-linearity* of S-box $\mathcal{S}$ is measured by $\mathcal{N}_{\mathcal{S}}^*$ defined in (3.8):

$$\begin{aligned} \mathcal{N}_{\mathcal{S}}^* &= \min_{\beta \in \mathcal{B}^m \setminus \{0^m\}} \mathcal{N}_{\mathcal{S}_\beta(x)}, \text{ wherein} \\ \mathcal{S}_\beta(x) &= \bigoplus_{i=0}^{i=m} \beta_i \mathcal{S}_i(x) \end{aligned} \tag{3.8}$$

Definition 11. The *auto-correlation* of S-box $\mathcal{S}$ is measured by $\mathcal{A}_{\mathcal{S}}^*$ defined in (3.9):

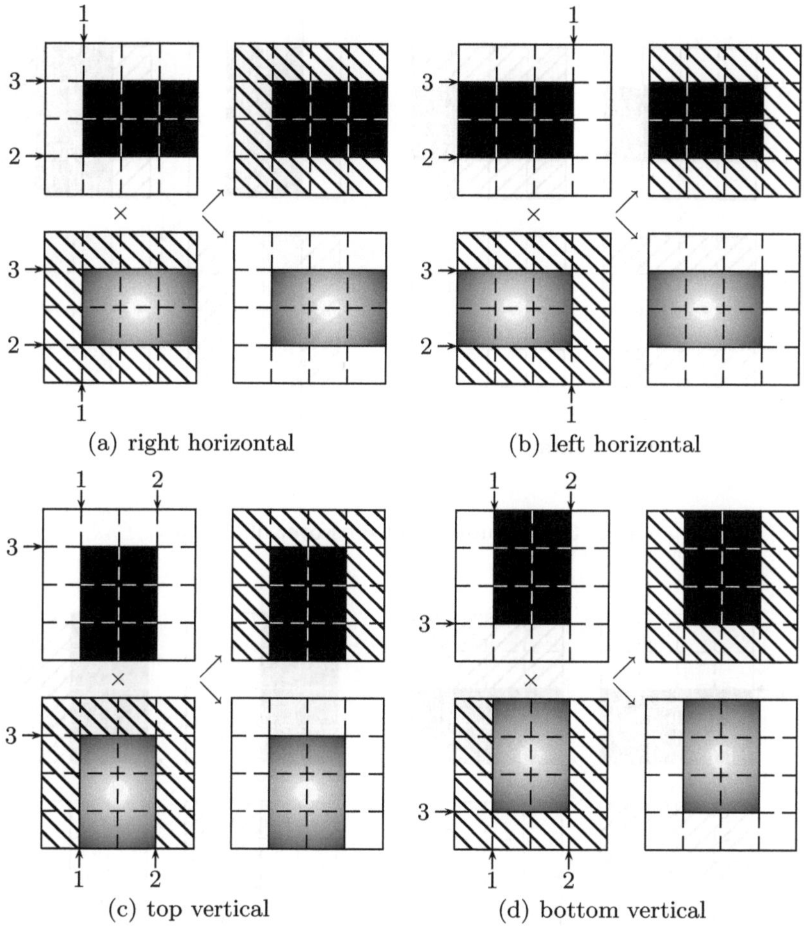

Fig. 3.4. Triple-point crossover of S-boxes

$$\begin{aligned} \mathcal{A}^*_{\mathcal{S}} &= \max_{\beta \in \mathcal{B}^m \setminus \{0^m\}} \mathcal{A}_{\mathcal{S}_\beta(x)}, \text{ wherein} \\ \mathcal{S}_\beta(x) &= \bigoplus_{i=0}^{i=m} \beta_i \mathcal{S}_i(x) \end{aligned} \tag{3.9}$$

Definition 12. An S-box $\mathcal{S}$ is said to be *regular* if and only if for each $\omega \in \mathcal{B}^m$ there exists exactly the same number of $x \in \mathcal{B}^n$ such that $\mathcal{S}(x) = \omega$. The *regularity* of an S-box can be measured by $\mathcal{W}^*_{\mathcal{S}}$ defined in (3.10):

$$\begin{aligned} \mathcal{W}^*_{\mathcal{S}} &= \max_{\beta \in \mathcal{B}^m \setminus \{0^m\}} \left|\mathcal{W}_{\mathcal{S}_\beta(x)}\right|, \text{ wherein} \\ \mathcal{S}_\beta(x) &= \bigoplus_{i=0}^{i=m} \beta_i \mathcal{S}_i(x) \end{aligned} \tag{3.10}$$

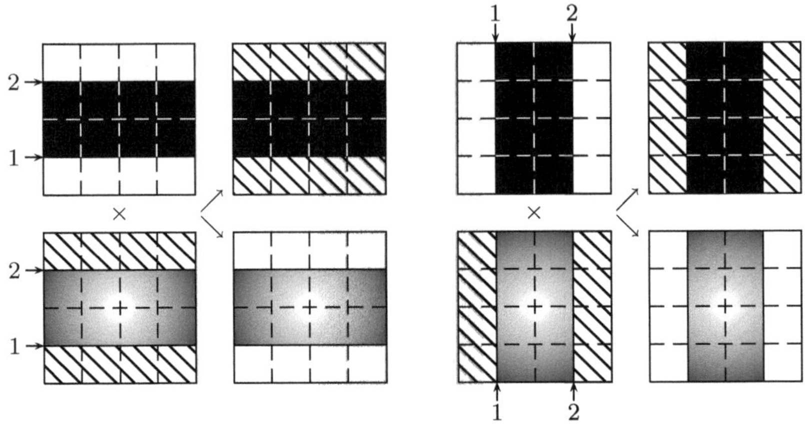

(a) horizontal (b) vertical

Fig. 3.5. Double-point crossover of S-boxes

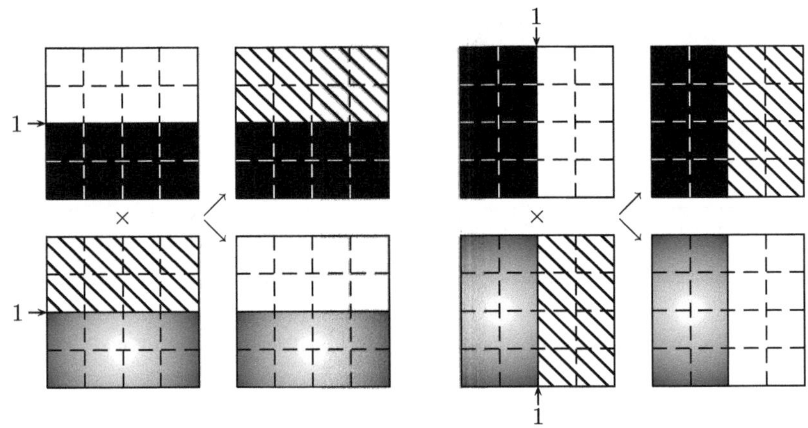

(a) horizontal (b) vertical

Fig. 3.6. Single-point crossover of S-boxes

Note that a regular S-box $\mathcal{S}$ has $\mathcal{W}^*_{\mathcal{S}} = 2^{n-1}$.

The optimisation objectives consist of maximising regularity of the S-box as well as its non-linearity while minimising its auto-correlation. These are stated in (3.11):

$$\begin{cases} \max\limits_{\mathcal{S}} \mathcal{W}^*_{\mathcal{S}} \\ \max\limits_{\mathcal{S}} \mathcal{N}^*_{\mathcal{S}} \\ \min\limits_{\mathcal{S}} \mathcal{A}^*_{\mathcal{S}} \end{cases} \tag{3.11}$$

3.5 Performance Results

The Nash algorithm [3, 4], described in Section 3.3 was implemented using multi-threading available in Java^{TM}. As we have a three-objective optimisation (maximising regularity, maximising non-linearity and minimising auto-correlation), our implementation has three agents, one per objective, all three running in parallel in a computer with a Hyper-Threaded Pentium-IV processor of 3.2GHz. We used S-boxes of different size (i.e. number of input and output bits) to evaluate the performance of our approach. All the yield S-boxes were regular. The non-linearity and the auto-correlation criteria for the best S-boxes yield are given in Table 3.1 together with those obtained in [11] and [12]. For all the benchmarks, our approach performed better producing S-boxes that are *more* non-linear and *less* auto-correlated than those presented in [11] and [12]. The best solutions were always found after at most 500 generations.

Table 3.1. Characteristics of the best S-boxes $\mathcal{S}^+$ by Millan et al. [11], Clarck et al. [12] and our approach

input×output	Millan et al.		Clarck et al.		Our approach	
	$\mathcal{N}^*_{\mathcal{S}+}$	$\mathcal{A}^*_{\mathcal{S}+}$	$\mathcal{N}^*_{\mathcal{S}+}$	$\mathcal{A}^*_{\mathcal{S}+}$	$\mathcal{N}^*_{\mathcal{S}}$	$\mathcal{A}^*_{\mathcal{S}}$
8×2	108	56	114	32	116	34
8×3	106	64	112	40	114	42
8×4	104	72	110	48	110	42
8×5	102	72	108	56	110	56
8×6	100	80	106	64	106	62
8×7	98	80	104	72	102	70

In order to compare the solutions produced by the three compared approaches, we introduce the concept of *dominance* relation between two S-boxes in a multi-objective optimisation.

Definition 13. An S-box $\mathcal{S}_1$ *dominates* another S-box $\mathcal{S}_2$, denoted by $S_1 \succ S_2$ or interchangeably solution $\mathcal{S}_2$ *is dominated by* solution $\mathcal{S}_1$ if and only if $\mathcal{S}_1$ is no worse than $\mathcal{S}_2$ with respect to all objectives and $\mathcal{S}_1$ is strictly better than $\mathcal{S}_2$ in at least one objective. Otherwise, S-box $\mathcal{S}_1$ does not dominate solution $\mathcal{S}_2$ or interchangeably $\mathcal{S}_2$ is not dominated by $\mathcal{S}_1$.

So, at the light of Definition 13, we can see that all the S-boxes evolved by our approach dominate those yield by Burnett et al. [11]. Furthermore, the 8×4, 8×5 and 8×6 S-boxes we produced dominate those generated by Clarck et al. too. Nevertheless, the 8×2, 8×3 and 8×7 S-boxes are non-dominated.

3.6 Conclusion

In this chapter, we used a multi-objective evolutionary algorithm based on the concept of Nash equilibrium to evolve resilient innovative evolutionary S-boxes. The produced S-boxes are regular in the sense of Definition 12. Moreover, considering the dominance relationship between solution in multi-objective optimisation as defined in Definition 13, the generated S-boxes are better than those obtained by both Burnett et al. and Clark et al. This is encouraging to pursue further evolution of more complex S-boxes. Also, as a future work, we intend to apply genetic programming to evolve the resilient S-boxes as minimal Boolean equations so that the hardware of the S-boxes is given for free.

References

1. Daemen, J. and Rijmen, V., The Design of Rijndael: AES – The Advanced Encryption Standard, Springer-Verlag, 2002.
2. National Institute of Standard and Technology, Data Encryption Standard, Federal Information Processing Standards 46, November 1977.
3. Nash, J.F., Equilibrium points in n-person games, Proceedings of the National Academy of Sciences, vol. 36, pp.48–49, 1950.
4. Nash, J.F., Non-cooperative game, Annals of Mathematics, vol. 54, no. 2, pp. 286–295, 1951.
5. Kang, J.S., Yi, O., Hong, D. and Cho, H., Pseudo-randomness of MISTY-type transformations and the block cipher KASUMI, Information Security and Privacy, The 6th Australasian Conference, Lecture Notes in Computer Science, vol. 2119, pp. 60–73, Springer-Verlag, 2001.
6. Menezes A. J., Van Oorschot, P. C. and Vanstone, S. A., Handbook of Applied Cryptography, CRC Press, 1996.
7. Matsui, M., Linear cryptanalysis method for DES cipher, T. Helleseth (Ed.), Advances in Cryptology, vol. 765, Lecture Notes in Computer Science, pp. 386–397, Springer-Verlag, 1994.
8. Biham, E. and Shamir, A., Differential Cryptanalysis of DES-like Cryptosystems, Advances of Cryptology, vol. 537, pp. 2–21, Springer-Verlag, 1990.
9. Beauchamp, K. G., Walsh Functions and Their Applications. New York: Academic, 1975.
10. Dorigo, M. and Maniezzo, M., Parallel Genetic Algorithms: Introduction and Overview of Current Research, in Parallel Genetic Algorithms, J.Stender (Ed.), IOS Press, 1993.
11. Millan, W., Burnett, L, Cater, G., Clark, J.A. and Dawson, E., Evolutionary Heuristics for Finding Cryptographically Strong S-Boxes, Lecture Notes in Computer Science, vol. 1726, pp. 263–274, Springer-Verlag, 1999.
12. Clark, J.A., Jacob, J.L. and Stepney, S., The Design of S-Boxes by Simulated Annealing, New Generation Computing, vol. 23, no. 3, pp. 219–231, Ohmsha and Springer, 2005.

4

Industrial Applications Using Wavelet Packets for Gross Error Detection

Paolo Mercorelli[1] and Alexander Frick[2]

[1] University of Applied Sciences Wolfsburg
Dep. of Vehicles, Production and Control Engineering
Robert-Koch-Platz 12., 38440 Wolfsburg, Germany
Phone: +49-(0)5361-831615 Fax. +49-(0)5361-831602.
p.mercorelli@fh-wolfsburg.de

[2] ABB Utilities GmbH UTD/PAT2
Kallstadter Strasse 1
DE 68309 Mannheim, Germany
Phone: +49 621 381 4539 Fax: +49 621 381 2244.
alexander.frick@de.abb.com

This chapter addresses Gross Error Detection using uni-variate signal-based approaches and an algorithm for the peak noise level determination in measured signals. Gross Error Detection and Replacement (GEDR) may be carried out as a pre-processing step for various model-based or statistical methods. More specifically, this work presents developed algorithms and results using two uni-variate, signal-based approaches regarding performance, parameterization, commissioning, and on-line applicability. One approach is based on the Median Absolute Deviation (MAD) whereas the other algorithm is based on wavelets. In addition, an algorithm, which was developed for the parameterization of the MAD algorithm, is also utilized to determine an initial variance (or peak noise level) estimate of measured variables for other model-based or statistical methods. The MAD algorithm uses a wavelet approach to set the variance of the noise in order to initialize the algorithm. The findings and accomplishments of this investigation are:

1. Both GEDR algorithms, MAD based and wavelet based, show good robustness and sensitivity with respect to one type of Gross Errors (GEs), namely outliers.
2. The MAD based GEDR algorithm, however, performs better with respect to both robustness and sensitivity.
3. The algorithm developed to detect the peak noise level is accurate for a wide range of S/N ratios in the presence of outliers.

P. Mercorelli and A. Frick: *Industrial Applications Using Wavelet Packets for Gross Error Detection*, Studies in Computational Intelligence (SCI) **57**, 89–127 (2007)
www.springerlink.com

4. The two developed algorithms based on wavelets (Algorithms for GEDR and peak noise level estimation) do not require the specification of any additional parameters which makes parameterization fully automated as a result. The MAD algorithm only requires the result of the peak noise level detection algorithm that has been developed and therefore is also fully parameterized.
5. The sensitivity and robustness of the algorithms is demonstrated using computer-generated as well as experimental data.
6. The algorithms work both, for off and on-line cases where in the on-line case computation times are small so that application on standard Distributed Control Systems (DCS) for large plants should be feasible.
7. A translation from Matlab to C (C++) was done using the Matlab Compiler and a COM wrapper was implemented as a functional prototype, which can be directly accessed through an excel spreadsheet.

To conclude, the developed algorithms are totally general and they are present in some industrial software platforms to detect sensor outliers. Furthermore, it is currently integrated in the inferential modelling platform of the unit responsible for Advanced Control and Simulation Solutions within ABB's (Asea Brown Boveri) industry division. Experimental results using sensor measurements of temperature, pressure and Enthalpy in a Distillation Column are presented in the chapter.

4.1 Introduction

Gross error detection and replacement (GEDR) is essential for many model-based or statistical methods in the process control field. An example would be parameter estimation (PE) and data reconciliation (DR). Here, gross error (GE) free data is assumed prior to PE and DR. Currently, only simple GEDR based on limit and rate checks is carried out within some industrial platforms for the detection of outliers. This contribution will use PE and DR as a reference example, however, one should keep in mind that other model-based or statistical methods may also be used instead. As applications in the process industries rely more and more on large amounts of raw data, which either directly originate from a distributed control system (DCS) or historians, there is an increasing need for automated data pre-processing, e.g., cleaning of data. Soft sensoring, data reconciliation, and parameter estimation are example applications [13] that often require the existence of "clean data". In the case of data reconciliation, the data cleaning has been done simultaneously, see [13]. A new method of de-noising which is based on methods already introduced in [16] is presented in [17]. It is based on the comparison of an information theory criterion which is the "description length" of the data. In [14] and [15] this theoretical approach is applied. The description length is calculated for different subspaces of the basis and the method suggests choosing the noise variance and the subspace for which the description length of the data is minimum. In the work, based on [14], it is shown that the de-noising process can be done simultaneously. The approach presented in [14] and [15] is based on "minimum description length" and that's why it seems to be unsuitable for on-line detection because it needs a relatively large amount of data. Furthermore, our approach is looking for a two step de-noising technique. Simultaneous approaches often suffer from the fact that they are based on iterations which in turn impose heavy CPU load requirements on the applications. As a consequence, these approaches often become unfeasible for real-time applications and therefore have not been widely used in industry. Real-time approaches dealing with outlier detection and/or de-noising and featuring an easy means of implementation have also been developed. Historically, lower and upper limits as well as simple rate change check were used to detect outliers. More recently, in [1], [18] and [19]) the authors have developed a robust version of the Hampel filter where a lower threshold may be set in order to avoid the undesired detection (removal) of noise. The algorithm assumes knowledge of the noise level of the signal, which is often unknown. The aim of this chapter is to develop an approach that addresses outlier detection without de-noising that may be applied in real-time. This is particularly important because some applications require outlier detection without de-noising of the signal. Special attention was given to the fact that the algorithms developed may be tuned automatically during commissioning and may be easily retuned during operation. The strategy consists of using an existing Hampel algorithm (enhanced by [1]) and the newly developed algorithm, based on Wavelets, to determine robustly the noise level of the signal to be processed. The noise level was found to be the parameter of the Hampel algorithm by [1] that cannot be optimally chosen independent of the signal to be processed. The algorithm to estimate the noise is based on Wavelets. The other parameters were optimally fixed using metrics defined by means of simulation studies. Fig. 4.1 shows a block diagram of how a GEDR package (PCK) would interact with gEST, that is, how the GEDR would function on a conceptual level.

4.1.1 Modules

The following list describes the modules (blocks) and the connections in order to define the problem statement in general, the key assumptions to be made, and the implications thereof on the GEDR approaches. Dashed blocks and connections denote additional, desired features to be included on a conceptual level, however, they will not be tested or implemented at this point in time. The scope of this work is the GEDR and Quality Module (QM) where the QM has been given more consideration on a conceptual level.

- **Process and Instrumentation**: The true process values are affected/ changed by sensor (instrumentation) imperfections such as sensor aging or

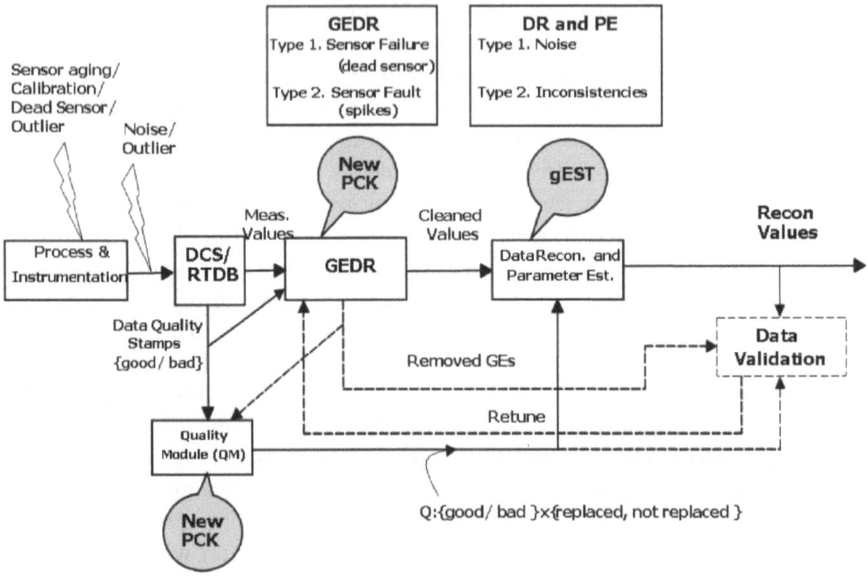

Fig. 4.1. Overview of Modules relevant to GEDR and QM

wrong/poor calibration. Often, these effects are modelled using biases. Usually, biases are assumed to be rather small as opposed to Gross Errors (GEs), which may be caused by dead sensors for example.

- **DCS/RTDB**: The DCS/RTDB logs data and stores them in a data base. The data logged consist of the values from the process and instrumentation block and additional effects: noise, outliers, and possibly interference due to the measurement; possible effects to the anti-aliasing; AD-conversion and quantization noise; and possibly LP filtering. Additional information from sensors such as quality stamps may also be available and used within the GED and the Quality Module (QM). For example, a quality stamp may say good or bad.
- **Quality Module (QM)**: Some of the functionally of the QM resides within the DCS. For example, in "2 out of 3" set-ups, where the same physical property is measured by three different sensors, the measured value with a "bad" stamp is automatically discarded and values from the other two sensors are used. Different scenarios, some of which also include soft sensors, are conceivable. The different scenarios strongly depend on the DCS system used and the configuration chosen by the engineers and operators, respectively. For our purposes, the QM module could provide additional functionality using information from the GEDR. Hence, on input the QM merely combines two streams of information. One stream comes from the DCS/RTDB and the other from the GEDR. Characteristics

assigned to data points are simply merged and relayed with additional information to other blocks: DR, PE, and Data Validation. The set of characteristics generated by the DCS/RTDB may be good/bad whereas the set generated by the GEDR may be replaced/not replaced.

- **GEDR**: Gross error detection and replacement must process the data coming from the DCS/RTDB block. The GEDR cleans the data before data reconciliation (DR) and parameter estimation (PE). The GEs, which were detected and removed, should be analyzed after DR and PE to validate that they actually are GEs and in turn to retune the filter settings. GEDR replaces the GE detected with a value that fits the local behaviour of the variable. For example, the average of the last couple of the valid data points may be used to do that. This module represents the main part of our investigation.
- **Data Recon. And Parameter Est. (DR and PE)**: PE is to estimate the parameters whereas DR reduces the error between predicted and measured variables while ensuring consistent process values using redundant information. For example, if all of the mass fractions were measured they would have to add up to one, which represents redundant information. DR and PE must be done after GEDR. This can be done for both dynamic and steady state data. The DR and PE are done simultaneously within some industrial platforms. The DR is performed by introducing BIAS variables that are considered parameters. They are adjusted in order to make the error (difference between measurement and prediction) disappear. Information provided by the QM regarding the quality and possible alternative measurements should be processed within DR and PE. For example, the quality stamp may be: good/badxreplaced/non replaced whereas possible alternative measurements may be a different sensor or soft sensor. Data points identified as being replaced could be weighted differently in order to weaken their influence upon the PE and DR. Consequently, some provisions regarding the information handling/processing ability of the PE and DR should be made. These may involve, as mentioned, the possibility to change the weights of measured data points in the PE and DR. This is not possible within gEST because the weights for a variable are constant for a given experimental run, that is, no different weights may be assigned for the different data points with respect to time.

4.1.2 Gross Error Types and Examples

Two types of GEs are considered:

- Type 1 GEs, which are caused by, e.g., sensor failures and result in sustained errors in the measurements. Fig. 4.2 shows an example of a type 1 GE. It is assumed that a sensor goes dead generating a noise-free signal with value 0. The sensor failure occurs at sample number 50 and remains for 100 samples. Provisions for detection of type 1 GEs will be made. However, this will require some user interaction such as specifying the number

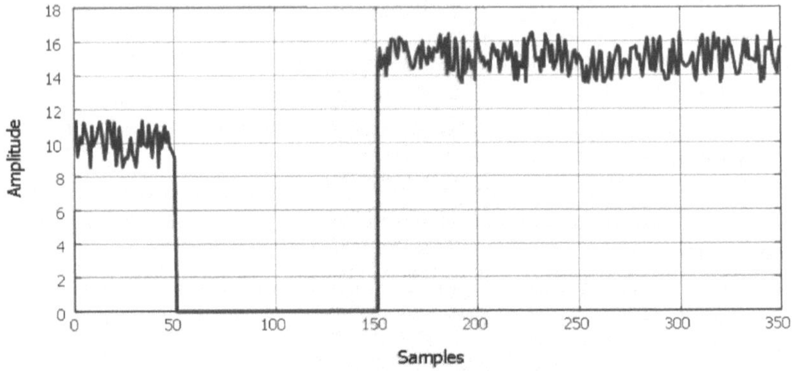

Fig. 4.2. Example of type 1 GE. Computer-generated data with dead sensor defaulting to zero at sample number 50

of constant measurements resulting in a type 1 GE. This option would make sense for variables such as analyzers for which clogging may cause such a behaviour. Henceforth, the specification such as the number of constant measurements as well as a dead band should be done by a process expert.

- Type 2 GEs, which are caused by, e.g. sensor faults, result in short, spike-like errors in the measurements

Fig. 4.3 shows data from a temperature measurement contaminated with outliers and hypothetical limits of the min and max values (horizontal thick lines) of the temperature. The limits can be used to perform an easy check to remove some of the outliers. Obviously, not all of the outliers would be removed using the min/max GE check. Another possible easy check could be implemented with a rate limitation. However, this would involve detailed process knowledge of the time constants and noise levels of the measured variables. The GEDR is to remove and replace all of the spikes shown. After detection and replacement, the cleaned data is shown in the lower part of the figure using a MAD based filter, see [1], combined together with the wavelet based peak noise level estimation algorithm. The chapter is organized as follows. Section 4.2 is devoted to the problem specification. Section 4.3 gives some background elements and presents the state of the art of the noise detection problem. At the end of the section 4.3 a new wavelet based procedure to detect noise variance is presented and validated. Section 4.4 shows an algorithm wavelet based for GED and the MAD algorithm with the noise

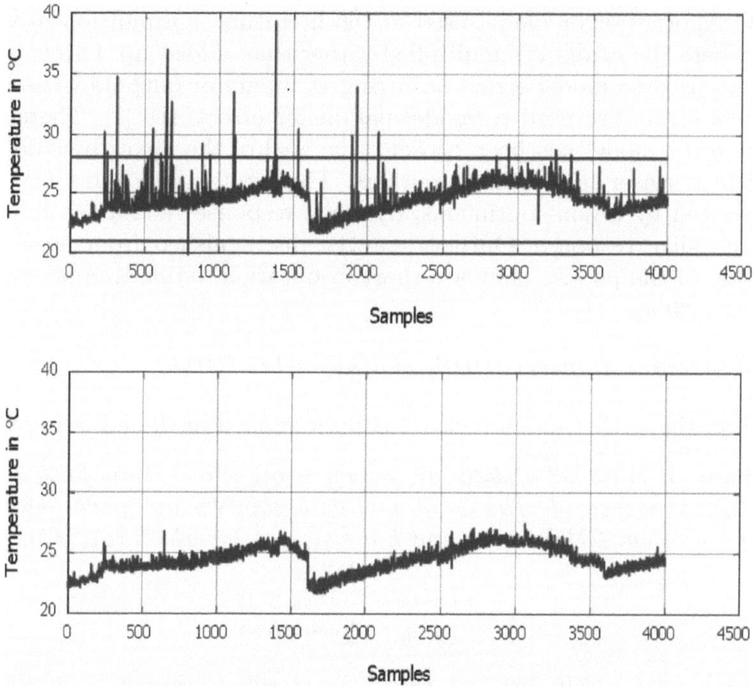

Fig. 4.3. Example of temperature measurements contaminated with outliers (top) and min, max limits. Temperature data cleaned using GEDR (bottom)

parameter estimated by the new wavelet based procedure. In section 4.5 and 4.6 results with computer simulation and real measurements are shown. The conclusions close the chapter.

4.2 Problem Specification

4.2.1 Mathematical Preliminary

Measurement errors in the context of sampled, measured variables are simply the difference between what is measured and what is the true value of the variable, that is,

$$(y)_k = (x)_k + (o)_k + (b)_k + (w)_k \tag{4.1}$$

where: $(y)_k$ is measured, $(x)_k$ the true value, $(e)_k = (o + b + w)_k$ is the error due to the measurement and k denotes the $k - th$ value of the time series. In particular o_k is the outlier, w_k is the noise and b_k represents any other kind of fault such as, for instance, a sustained error in the measurements, (see Fig. 4.2), or systematic errors due to the instrumentation. Equation (4.1) represents an additive error model as an extension of the outlier model presented in [1, 3]. Another error model used in the literature is a multiplicative error model where the errors are multiplied rather than added up. Often, additive errors are used to model errors occurring at inputs or outputs whereas multiplicative errors are used to model parameter errors, see [1]. Therefore, an additive error model has been chosen. The goal of the algorithms developed is to detect sensor fault related outliers. This results in a short GE that is characterized by a non-continuous, dynamic response such as an impulse or an impulse-like response. Outliers may be distinguished from noise by the amplitude of the peaks. They are short in duration, often no more than two points at a time.

Outlier Detection Problem (ODP) and Algorithm (ODA)

Therefore, the outlier index vector has been defined in the following way:

Definition 1 *For a set of data $(y)_k = (x)_k + (o)_k + (b)_k + (w)_k$ for $k = 1, ..., n$ where $(y)_k$ is measured. The number of data points n are given. The outlier detection problem (ODP) is to find $L = (l_1, ...l_n)$ for which $(o)_k \neq 0$, that is:*

$$L = \begin{cases} (l)_k = 1 & (o)_k \neq 0 \\ (l)_k = 0 & elsewhere \end{cases}$$

The index set L itself does not specify what will happen in case an outlier is actually detected. It is supposed to be a mere specification of the detection problem.

The goal of this chapter is not to prove some of the inherent features of a data cleaning filter to solve the ODP. This was done in [1]. However, in order to assess the performance of the filter implemented together with the noise detection and to simplify the further discussion some more definitions are necessary. The implementation of an outlier detection algorithm (ODA) will tag the data points within the time series $(y)_k$ that are believed to be outliers. This may be represented by the following definition of a mapping F. If the data point under consideration satisfies the criteria, as used in the Hampel filter in [1], then the map F returns a value equal to one.

Definition 2 *For a set of data $(y)_k = (x)_k + (o)_k + (b)_k + (w)_k$ for $k = 1, ..., n$ where $(y)_k$ is measured. The number of data points, n, are given. For an implementation of a filter represented by the mapping F, the index vector L^T may be computed as follows:*

$$L^T = \begin{cases} (l^T)_k = 1 & (F(y)_k) = 1 \\ (l)_k = 0 & elsewhere \end{cases}$$

The following part will summarize some of the inherent features.

4.2.2 Noise Level Detection Problem (NLDP) and Algorithm (NLDA)

Definition 3 *For a set of data* $(y)_k = (x)_k + (o)_k + (b)_k + (w)_k$ *for* $k = 1, ..., n$ *where* $(y)_k$ *is measured. The number of data points, n, are given. The noise level detection problem (NLDP) is to find the standard deviation for a normally distributed noise* N_l*. Where* $N_l = \|w\|$ *for* $w = (w_1, w_2, ...w_n)$ *and* $\| \cdot \|$ *is some norm. The noise actually detected will be denoted* N_l^T*.*

The implementation of a noise level detection algorithm (NLDA) calculates a measure of the noise of a time series $(y)_k$. The goal here must be to have an "outlier robust algorithm", that is, the NLDA should not be sensitive to the numbers of outliers present in the time series.

4.2.3 Some Remarks Regarding Wavelet Based Algorithms

Peak Noise Level and Variance

In the context of this investigation, it is referred mostly to the peak noise level estimation. This stems from the fact that the peak noise level estimator reconstructs the peak noise levels of the signal and then calculates their variance. So, when one speaks of a peak noise level estimator, the result will be sometimes close to the standard deviation of the signal and sometimes closer to the actual peak noise level. The goal was to have a robust estimator even for a large range of S/N ratios in the presence of outliers rather than a very accurate estimator for a small range of S/N ratios and no outliers. The proposed technique is wavelet based.

Wavelet Algorithm for GEDR

The developed algorithm estimates the local variance of the Lipschitz constant of the signal over a sliding time horizon. The fault (outlier) is recognized if the local Lipschitz constant lies outside the limit computed. Graphically, a flow pipe for the Lipschitz constant is constructed and if the local Lipschitz constant lies outside the flow pipe the data point is flagged and then replaced.

4.3 Wavelet Based Noise Level Determination

The MAD algorithm uses a wavelet approach to set the variance of the noise in order to initialize the algorithm. In this section this new algorithm, together with the necessary wavelet background, is presented. The developed algorithm estimates the peak noise level. A sliding time horizon is used to estimate a local noise peak. The result for each of the sliding time horizons is stored and used to determine the peak noise level. This is to make the procedure more robust. For an individual time horizon the algorithm selects appropriate wavelet coefficients that represent the noise in the system in the best way.

4.3.1 Background and State of the Art

The wavelet transform, an extension of the Fourier transform, projects the original signal onto wavelet basis functions, thus providing a mapping from the time domain to the time-frequency plane. The wavelet functions are localized in the time and frequency domain and they are obtained from a single prototype wavelet, the mother function $\varphi(t)$, by dilation and translation. Dilation may be viewed as stretching of the mother function with respect to time whereas translation is a mere shifting with respect to time. Thus, by means of dilation and translation operations performed on the mother wavelet $\varphi(t)$ a set of wavelet functions is created with the same shape as the mother wavelets yet of different sizes and locations. In this invention, digital wavelet functions are used to analyze sampled, experimental data. The sets of digital wavelet functions are referred to as wavelet packets. In general, they are characterized by three parameters, two of which may be chosen independently. The parameters depend on the number of the samples considered, that is the data window width, as well as on the Nyquist frequency (half of the sampling frequency). The wavelet packets under consideration are derived from the Haar mother wavelet: Fig. 4.4 shows a set of Haar functions. The Haar functions are iden-

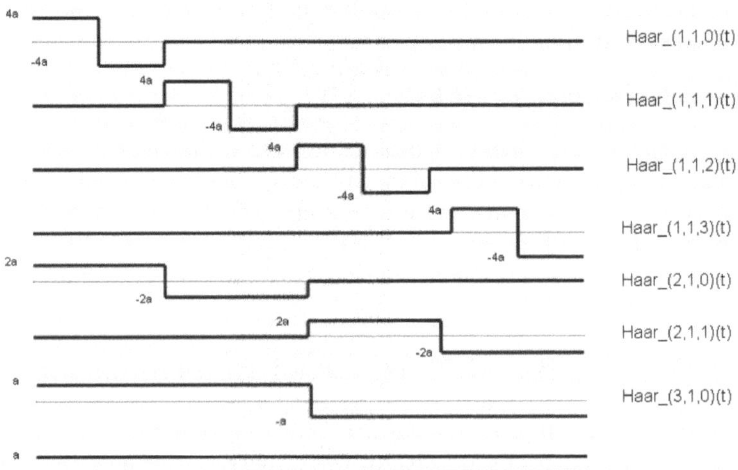

Fig. 4.4. A set of Haar functions

tified using the parameter tupel, (d, j, n). In particular, parameter j is related to the frequency shift, parameter n is related to the time shift, and additional

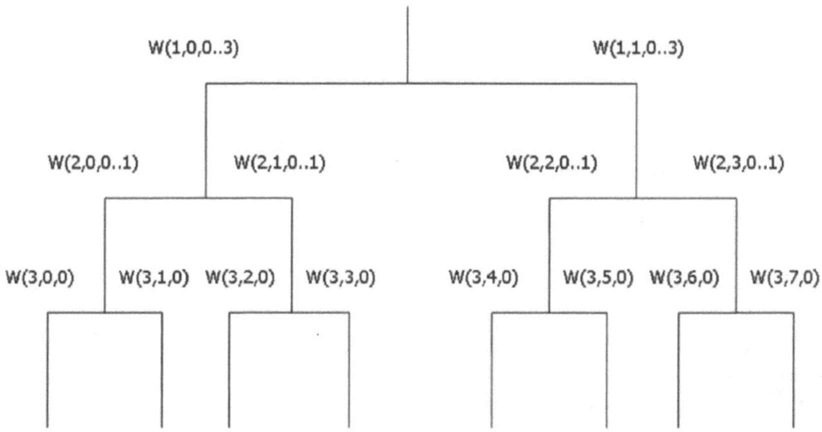

Fig. 4.5. Wavelet coefficients arranged in a tree

parameter d is useful to address the level of the tree which strictly depends on the number of the considered samples. That is, if eight samples were to be analyzed, the level of the tree would be three. In literature, wavelet functions are structured in the form of a tree such that $d = 1$ represents the highest degree of refinement with respect to time. For example, in Fig. 4.4 there are four wavelet functions at $d = 1$ whereas there are only two wavelet functions for $d = 2$. Fig. 4.5 shows the corresponding tree with the wavelet coefficients, $w(d, j, n)$, representing the contribution of each of the wavelets to the signal. The notation $w(1, 0, 0..3)$ denotes the coefficients on the first level on the very left with time shifts 0 through 3.

4.3.2 Noise Level Estimation: State of the Art

Estimation of the noise distribution density is a well-known problem from a theoretical and a practical standpoint. Contributions in this area using wavelets may be found in [2], [3], [4] and [5]. A number of different methods for selecting appropriated threshold values for wavelet de-noising have been proposed. Recent literature [6], with focus on the application domain, addresses the importance of the accurate estimation of noise. In particular, in 3 page 867 a list of threshold criteria is presented. Noise is a phenomenon that affects all frequencies and is present in different forms at different frequencies for different processes. This variability of noise generates difficulty and therefore, heuristic criteria are needed to cope with this. A heuristic used rather often is that the "true" signal tends to dominate the low-frequency band whereas the noise tends to dominate the high-frequency band. The currently used (state of the art) algorithms to estimate noise by using wavelets could be summarized as follows:

- 1 Apply the wavelet transform to a noisy signal and obtain the noise wavelet coefficients.
- 2 Threshold those elements in the wavelet coefficients which are believed to be attributed to noise.
- 3 To reconstruct the noise, apply the inverse wavelet transform to the threshold wavelet coefficients.

In this approach the crucial point is the threshold step, in other words, which wavelet coefficients are believed to be attributed to the noise. In general, wavelet thresholding methods could be classified as soft threshold as in [1], or as hard threshold [8]. Soft thresholds attribute wavelet coefficients, (which are less equal than the soft threshold given), to noise, regardless of the frequencies. Hard thresholds attribute wavelet coefficients, (which are less equal than the hard threshold given), to noise with regard to a minimum frequency given. Both thresholding methods described above may further be sub-divided into two categories: global thresholding and level-dependant thresholding. The two sub-categories relate to the level in the wavelet tree. That is, global thresholding applies only one threshold on the wavelet coefficients of one particular level whereas level-dependant thresholding applies level-dependent thresholds on all levels. The latter technique is suitable when the noise is colored or/and correlated. In [2] in 5.4 a practical example of noise estimation is given. The described algorithm uses the finest scale, that is the first level of the tree ($d = 1$, see Fig. 4.4), of wavelet coefficients to determine the noise. More specifically, the median of the wavelet coefficients at the first level of the tree is used. The use of the median as opposed to the mean makes the estimate more robust because there is a larger number of wavelet coefficients representing noise components of the signal than wavelet coefficients representing "non-noise" components of the signal. The procedure commonly used to detect the noise [2–4] is as follows:

- **Step 1** Choose the best wavelet basis representation of the measured signal by using the algorithm in [7].
- **Step 2** Decompose the signal using the selected wavelet basis.
- **Step 3** Sort the resulting coefficients and choose the threshold level as proposed in [2] or in an adaptive way on the wavelet tree.

4.3.3 The Proposed New Procedure for Peak-Noise Level Detection

The proposed procedure calculates the peak-noise level of a sampled signal. To do this, a method is proposed to represent the noise in the wavelet domain, that is, after applying a wavelet transform. The algorithm looks, at every level of the tree, for the time-frequency cell which describes at this level the incoherent part of the signal as defined in [10] and applies a property demonstrated in [10]. The property is in short the following. If the absolute value of the median over n at the time-frequency cell (d, j) (father wavelet) is not more than the absolute of the median at the time-frequency cell $(d+1, jleft)$ (left child wavelet) plus absolute of the median at the time-frequency cell $(d+1, jright)$ (right child wavelet) , then the absolute value of the father wavelet is not more than the absolute value of the median of the sum of every couple of left and right children belonging to the same branch of the tree at the level $d+n$ (all the deeper), with d and n integer.

Definition 4 *Given the oscillating part of the sequence $y(t)$ as follows*

$$y(t) - \sum_{n=0}^{2^{N_0-d}} s(d,n)\psi(d,n)(t) = \sum_{j=1}^{2^d-1} \sum_{n=0}^{2^{N_0-d}} w(d,j,n)\psi(d,j,n)(t),$$

where $s_{(d,n)} = w_{(d,0,n)}$. Then the time-frequency map of the peak values in wavelet domain is a table of real values specified $\forall d \in \mathsf{N}$ and by the triplet $(d, \hat{\jmath}, n)$ as

$$w_{P_{(d,\hat{\jmath},n)}} = \max_j (w_{(d,j,n)} \|\psi_{(d,j,n)}(t)\|).$$

□

Definition 5 *Given an observed sequence*

$$y(t) = x(t) + e(t).$$

Let $e(t)$ be defined as the incoherent part of the sequence $y(t)$ at every level d of the packet tree as follows:

$$e(t) = \sum_n N_{(d,\hat{\jmath},n)} \psi_{(d,\hat{\jmath},n)}(t),$$

where $\psi_{(d,\hat{\jmath},n)}(t)$ are the wavelet bases and $N_{(d,\hat{\jmath},n)}$ their coefficients. The selected wavelets are characterized by $(d, \hat{\jmath}, n)$ indices such that:

$$\{(d,\hat{\jmath},n)\} = arg\Big(\min_j (median_n\{w_{P_{(d,j,n)}}\}) : \\ \{0 < n \leq 2^{N_0-d}, 0 < j \leq 2^d - 1, \forall d \in \mathsf{N}\}\Big) \quad (4.2)$$

where $median_n$ is the median calculated on the elements which are localized with index n (time localized). The coefficients $w_{P_{(d,j,n)}}$ are the wavelet table coefficients of the sequence $y(t)$ as given in Def. 4. □

It is easy to see how the idea behind Def. 5 is to sort the basis which can illuminate the difference between the coherent and incoherent part of the sequence, where *incoherent* is the part of the signal that either has no information or its information is *contradictory*. In fact, the procedure looks for the subspace characterized either from small components or from opposite components in wavelet domain.

Proposition 1 *Let $y(t)$ be a sequence and $w_{p_{(d,j,n)}}$ its corresponding wavelet sequence on different levels of the tree. At every level d and at every frequency cell j*

$$|median_n(w_{p_{(d,\hat{j},n)}})| \leq |median_n(w_{p_{(d+1,2\hat{j},n)}})| + |median_n(w_{p_{(d+1,2\hat{j}+1,n)}})|.$$

Then

$$|median_n(w_{p_{(d,\hat{j},n)}})| \leq |median_n(w_{p_{(d+m,2\hat{j},n)}})| + |median_n(w_{p_{(d+m,2\hat{j}+1,n)}})|.$$

$\forall\ m \geq 1.$ □

Proof 1 *By observing that, from the orthogonality,*

$$\sum_{(d,j,n)} w_{(d,j,n)}\psi_{(d,j,n)} = \sum_{(d,j,n)} w_{(d+1,2j,n)}\psi_{(d+1,2j,n)} + \sum_{(d,j,n)} w_{(d+1,2j+1,n)}\psi_{(d+1,2j+1,n)}. \quad (4.3)$$

Functions $\psi_{(d,j,n)}$ are organized in packets and they are scaled functions, $\forall$ d, j and $\forall$ $m \geq 1$,

$$\max\left(\psi_{(d,j,n)}\right) \leq max\left(\psi_{(d+m,2j,n)})\right) + max\left(\psi_{(d+m,2j+1,n)}\right)$$

then

$$\max_j\left(w_{(d,j,n)}\psi_{(d,j,n)}\right) \leq \max_j\left(w_{(d+1,2j,n)}\psi_{(d+1,2j,n)}\right) + \max_j\left(w_{(d+1,2j+1,n)}\psi_{(d+1,2j+1,n)}\right). \quad (4.4)$$

From Def. 4, $\forall$ d and $\forall$ $n \geq 1$, it follows:

$$w_{p_{(d,\hat{j},n)}} \leq w_{p_{(d+1,2\hat{j},n)}} + w_{p_{(d+1,2\hat{j}+1,n)}}.$$

Considering that the "median" is a monotonic function:

$$median_n\left(w_{p_{(d,\hat{j},n)}}\right) \leq median_n\left(w_{p_{(d+1,2\hat{j},n)}}\right) + median_n\left(w_{p_{(d+1,2\hat{j}+1,n)}}\right).$$

Thus

$$|median_n(w_{p_{(w,\hat{j},n)}})| \leq |median_n(w_{p_{(d+1,2\hat{j},n)}})| + |median_n(w_{p_{(d+1,2\hat{j}+1,n)}})|.$$

by proceeding $\forall$ $m \geq 1$ the proposition is proven. □

Remark 1 *In other words the proposition says that, if the minimum of the median at level d, according to Def. 4, is not more than the minimum of the median at level $d+1$ (one more in depth) then it is also not more than the minimum of the median at level $d+m$ (all the deeper), with d and m integer.*

The introduced definitions and proposition 1 allow the Building of an efficient procedure in order to estimate the level of the noise like an incoherent part of the signal. One can say that the incoherent part of a signal is the part which doesn't "match" the desired signal. The peak-to-peak noise variance is estimated taking the incoherent part of the signal defined above into consideration.

Procedure. Given a sampled signal of length n.

- **Step 0**: Specify length L as $8, 16, 32$, or 64 (dyadic length) and set the end of shifting window $esw = \mathcal{L}$. Determine the height of the wavelet tree: $h = \log_2(\mathcal{L})$. Take the last data points with respect to the end of the shifting window, "esw", of the sampled signal.
- **Step 1**: Construct the wavelet coefficient tree $W(d, j, n)$ for every $d, j > 1$ and for every n, related to the current shifting data window.
- **Step 2**: Build the wavelet coefficients corresponding to the incoherent part of the signal according to Def. 5. In particular for $d = 1$ (the first level of the wavelet tree, this yields $j = 1$).
- **Step 3**:
  - *(a)*: For all time-frequency intervals such that $j = 1, 2, ..2^d - 1$ with $d > 1$, calculate the absolute of the median for the same time-frequency interval, that is, $Wmed(d,j) = abs(median_n(W(d,j,\forall n)))$.
  - *(b)*: For every "wavelet father" $W(d,j)$ at the node (d,j) with $j > 0$ calculate its left child at the node $(d+1, jLeft)$ and its right child $(d+1, jRight)$. Then calculate the absolute of the medians and denote them as $WmedChildLeft = abs(median_n(Wmed(d, jLeft, \forall n)))$ and $WmedChildRight = abs(median_n(Wmed(d, jRight, \forall n)))$. **While** $(d < h)$ and **For** $j =\leq 2^d - 1$. **If** $Wmed(d,j) \leq WmedChildLeft + WmedChildRight$, then denote $Peak(\hat{d}, \hat{j}) = \inf_{d,j}(Wmed(d,j))$ and limit the tree of the children of this branch **else**
- **Step 4**: Add all of $Peak(\hat{d}, \hat{j})$ of the different levels, that is, $Peak = sum_{(\hat{d},\hat{j})}(Peak(\forall \hat{d}, \forall \hat{j}))$.

4.3.4 Validation of Peak Noise Level Estimation

The two approaches for estimating the variance have been tested.

1. Using the de-noising algorithms in the WaveLab toolbox, see [11].
2. Using the nlevelmed algorithm as an estimate for the variance.

A summary of the test is that approach 1 is not applicable and that approach 2 is a viable approach. The WPDenoise (m-file) from the WaveLab toolbox has been used, see [11], to get a cleaned signal and then reconstruct the noise by taking the difference between the cleaned and the original signal. However, this is not a sound approach. When the variance of the noise is small, the difference between the cleaned and the original signal will be dominated by the "approximation" error. When there is a high variance then much of the noise will also be present in a cleaned signal. Thus the estimated variance will only be correct for a very short rang of variances. This is illustrated in Fig. 4.6, where a sinusoidal signal with the added noise has been constructed. This was repeated for a range of variances, and the results showed that only in a small range of variances the estimator gave correct and reasonable values.

For some signals having small S/N it may be impossible to accurately estimate the variance of the noise. Fig. 4.6 shows an example, a sinusoidal function, where an inaccurate result for small S/N ratios is obtained using a procedure as the one described above.

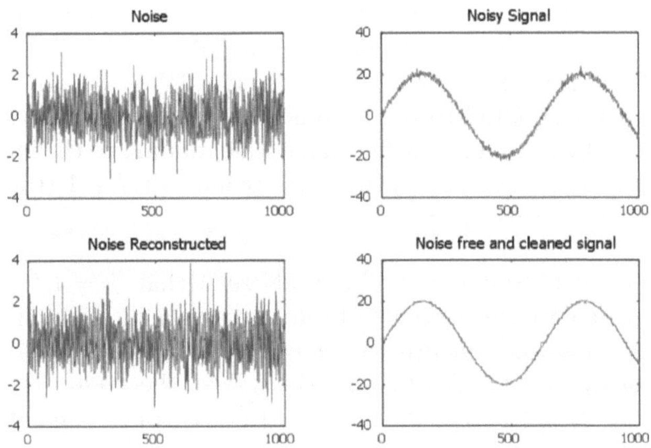

Fig. 4.6. Testing Signal

Using the WPDenoise algorithm from the WaveLab in [11] the results obtained are reported in Fig. 4.7. The same test is done with the here proposed procedure and, as it is possible to see in Fig. 4.7, the results are considerably much better.

Finally, Fig. 4.9 shows a graphical representation of the distribution of the relative error in determining the noise level error of the data sets. The noise level is determined using a wavelet-based algorithm. It almost follows a normal distribution around +10%, that is, the noise level is underestimated by 10% on average with a standard deviation of 13%. The relative error is defined as:

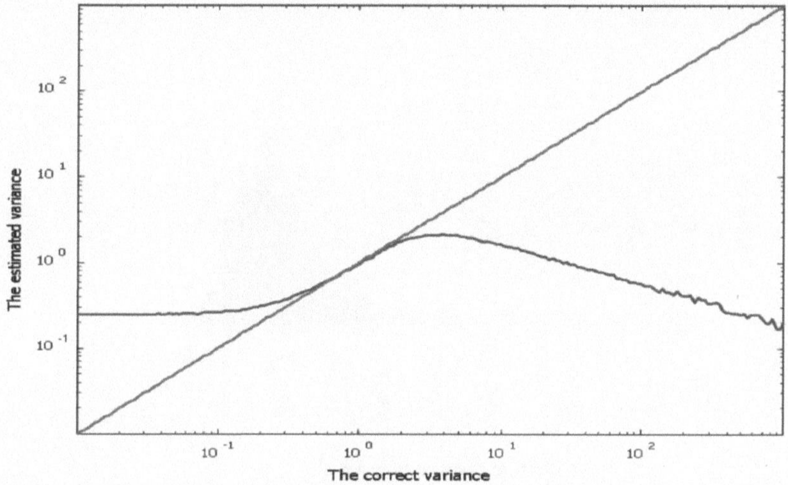

Fig. 4.7. The estimated and true variance using a threshold approach through WPDenoise algorithm from the WaveLab (the worst case)

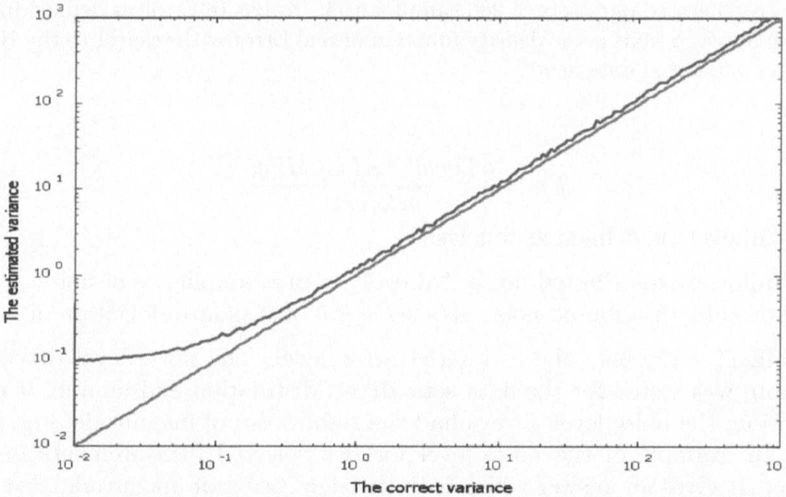

Fig. 4.8. The estimated and true variance using the here proposed procedure

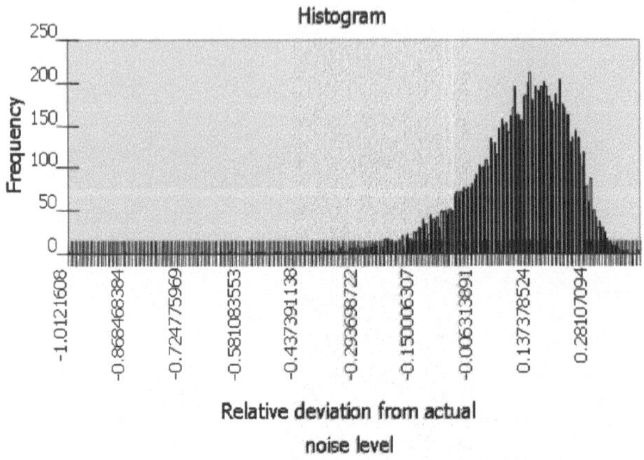

Fig. 4.9. A histogram for Noise Level Detection: the wavelet algorithm for 10000 computer-generated data sets of 200 samples each. X-axis relError as defined in the equation above. Y-axis is the density function of real Error with respect to the 10000 computer generated data sets

$$Er = \frac{nLevel - nLevelEst}{nLevel},$$

where "nLevel" is defined as follows:

- A uniformly distributed noise: "nLevel" = max amplitude of noise.
- A normally distributed noise: $nLevel = 1.5$ of standard deviation of noise.

"nLevelEst" represents the estimated noise level. The noise level detection algorithm was tested for the data sets: dryer, distillation and mining. It does a good job; the noise level always had the right order of magnitude. Fig. 4.10 shows an example of the noise level for two selected measurements in the data set. It gives an answer which is in the right order of magnitude. See also Fig. 4.7 and 4.8 where the estimated noise level is compared with the known variance of a generated noisy signal.

4.4 The Wavelet Algorithm for GEDR

The developed algorithm estimates the local variance of the Lipschitz constant of the signal over a sliding time horizon. The fault (outlier) is recognized if the local Lipschitz constant lies outside the computed boundry. Graphically,

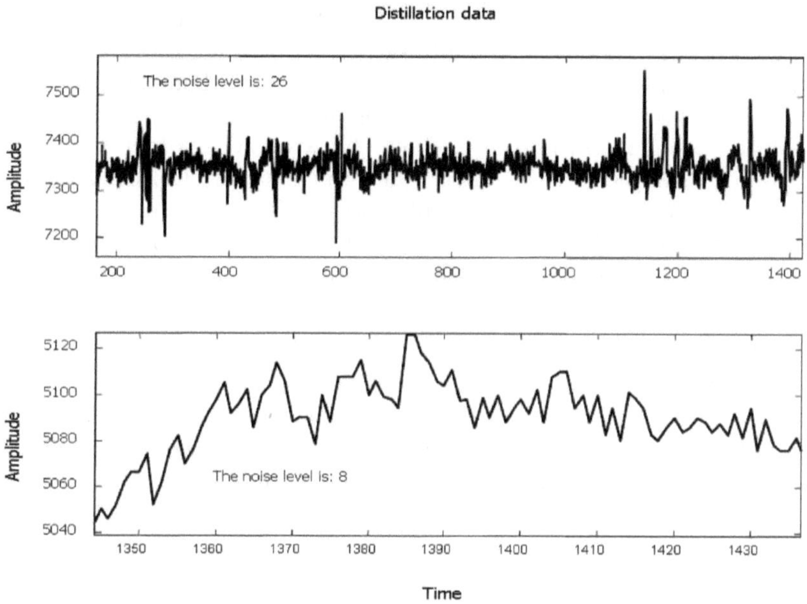

Fig. 4.10. The noise level for two selected measurements form the distillation data set

a flow pipe for the Lipschitz constant is constructed and if the local Lipschitz constant lies outside the flow pipe the data point is flagged and then replaced. One uses a sliding window with length equal to 8 samples in order to compare the results with the procedure proposed in [1]. At each sample time the window is shifted by one sample point. Decompose the signal onto the chosen Haar function. Start building the function for the first seven samples. With c_1, c_2 and c_3 one indicates the confidential constants.

- **Step 1** Calculate the standard deviation σ of the Lipschitz (L) constant for the first consecutive 7 samples and then one calculates the Lipschitz between the 7th and 8th sample.
- **Step 2** If the last Lipschitz constant is less than c_1 (Where c_1 is equal to 2), then it is not an outlier. Add this Lipschitz constant to σ. If the last Lipschitz constant is greater than c_1, then it is an outlier!
- **Step 3** Store the calculated Lipschitz constant and shift the window in order to consider the next sample of the sequence.
- **Step 4** Calculate the Lipschitz constant between the new $7^{th} - 8^{th}$ sample and add it to the stored Lipschitz.
- **Step 5** If the stored L value is less than $c_1\sigma$, then it is a single outlier and add the last Lipschitz constant to σ, see Fig. 4.11. If the stored L value is more than c_1, then check if the last two estimated Lipschitz constants have opposite signs, if no then they are multi-outliers, see Fig. 4.12. If yes,

then check if the stored L value is not less than c_2 (with c_2 equal to 3), if yes, then it is a single inverse multi-outlier, see Fig. 4.13.
If the stored L value is such that $c_1\sigma \leq L \leq c_2\sigma$, then they are multi-outliers as represented in Fig. 4.12 or Fig. 4.14.

- *(a)* In case of multi-outliers, shift the window in order to consider the next sample of the sequence and calculate the Lipschitz constant between the new $7^{th} - 8^{th}$ sample and add it to the stored Lipschitz.
- *(b)* If the stored L value is less than c_1, then it is not an outlier and add the last Lipschitz to σ. Otherwise check a heuristic safety condition on the local dynamic of the signal performed with another confidential constant c_3, if this is not verified, then they are multioutliers.

With regard to the above mentioned heuristic safety condition one can say that it is related to the dominant dynamic of the signal. In other words, one checks if the dominant dynamic of the signal before the suspected jump is close to the dominant dynamic after the jump. Fig 4.15 shows an example where this happens.

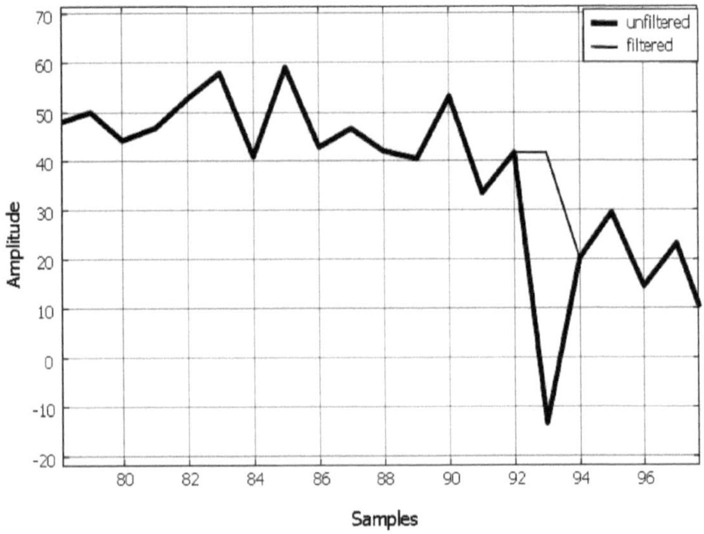

Fig. 4.11. An isolated outlier

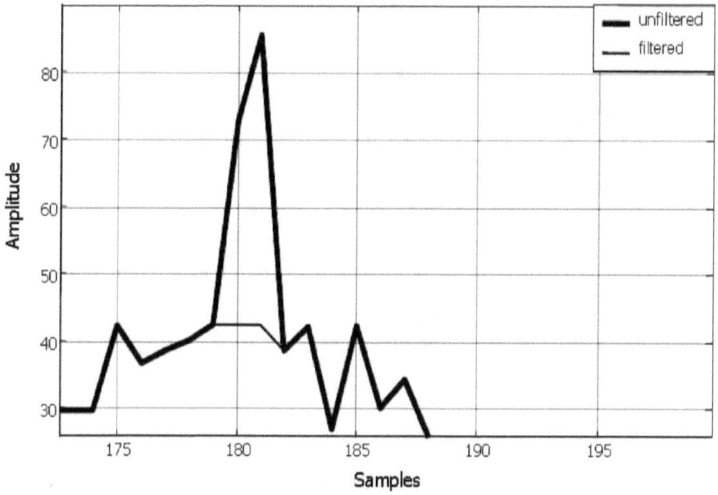

Fig. 4.12. A multiple outlier

4.4.1 Validation and Simulations

In this section one validates the procedure by using artificial data where the position of the outliers is previously known. Fig. 4.16 is related to the classical approach by using a median filter which needs to know the noise level as a priori knowledge. The results are: 94.68% of outliers correctly detected, 3.6% of outliers incorrectly detected in the data. Fig. 4.17 shows performance of the wavelet algorithm where one doesn't need prior knowledge of the noise. The results are: almost 100% of outliers correctly detected, 0% of outliers incorrectly detected in the data. It could happen in the proposed procedure that if the outliers are localized on the first part of the data the percentage of the incorrect outliers detected in the data increases and also the percentage of outliers correctly detected decreases. This is due to the initial very small standard deviation of the local Lipschitz constant during the algorithm initialization. This initialization usually takes $15-20$ samples. The weak points of the procedure are that it is a stochastic approach and it may not be robust in some cases. For instance, when the above mentioned standard deviation is small, the level of the percentage of the correct detection decreases. In Fig. 4.17 it is possible to see incorrectly detected outliers during the initialization of the algorithm, that is, in the first 20 samples.

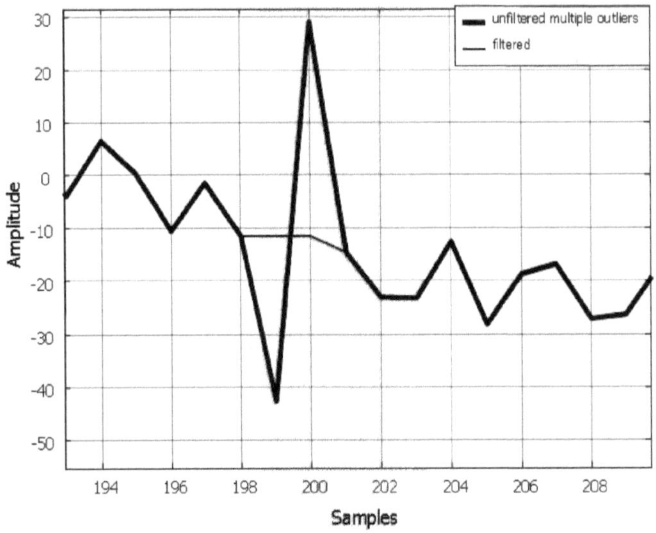

Fig. 4.13. A single inverse outlier

4.4.2 Outlier Detection Algorithm: MAD Algorithm

The algorithm itself is very simple and is described in [1] in detail. Now the original algorithm [1] and the parameters involved are briefly outlined. Basically, the algorithm uses a moving data window to determine whether or not 1) the center data point, (in the case of an off-line application), 2) or the right-most data point, (in the case of an on-line application) locally fits in the dynamics.

Fig. 4.18 shows a graphical representation of the MAD approach with the construction of the MAD flow pipe. For our tuning, the parameters were window width=8, thres=4, and var=4*nlevel. Nlevel represents the peak noise level in the data. This was determined to be the optimal setting such that, the flow pipe is wide enough to accommodate the presence of the noise, however, it is narrow enough to remove a high percentage of outliers. Also, it can be seen that the "radius" of the flow pipe is invariant under outliers and noise, which in turn makes it very robust.

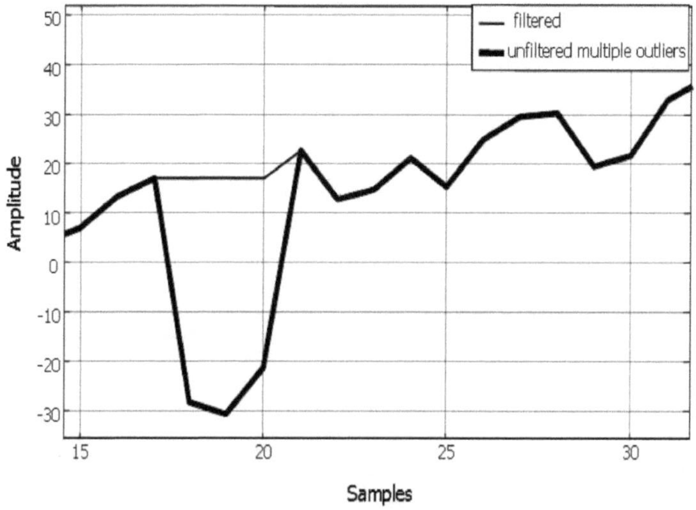

Fig. 4.14. A multiple (3) outlier

4.5 Results

Assessment Criteria The assessment criteria (metrics) are used to do two things:

1. Find optimal parameters for the MAD and the wavelet GEDR algorithms.
2. Assess the performance of the algorithms using the following metrics.
   a) Ratio of the detected and removed outliers referred to as ODR (outliers detected ratio). It can be computed both, for experimental and computer-generated data and therefore it will be the only measurement of sensitivity and robustness for the experimental data because OCDR and OIDR can only be calculated for computer-generated data.
   b) Ratio of outliers correctly detected and removed referred to as OCDR (outliers correctly detected ratio). It can be only computed for computer-generated data, where ♯ of outliers which are present in data is known.
   c) Ratio of outliers incorrectly detected and removed referred to as OIDR (outliers incorrectly detected ratio). It can be only computed for computer-generated data, where ♯ of outliers which are present in data is known.

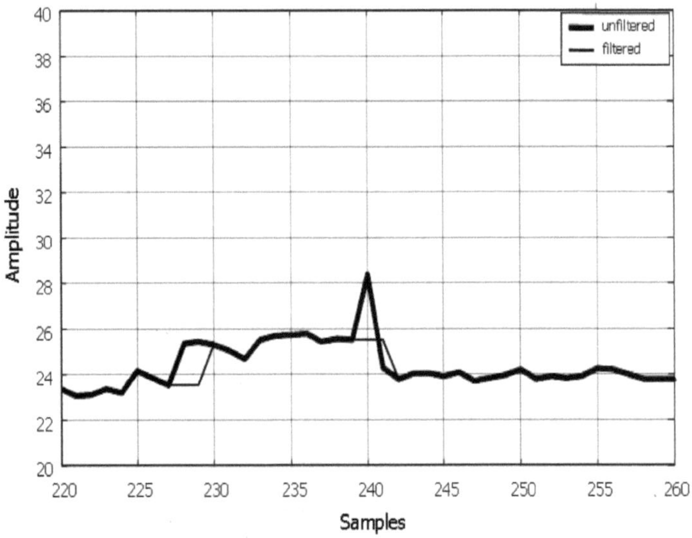

Fig. 4.15. Incorrect detection of multioutliers (on sample number 227 and 228)

d) Introduction of new data using the two-norm over all of the data referred to as SSC (sum of squared changes). It can be computed both for experimental and computer-generated data.
e) Computation Performance (CP). Both algorithms can be computed for MAD and wavelets. Note that the computation is deterministic for both approaches. In the off-line use the computation time is proportional to the number of the data points to be filtered whereas on-line it only depends on the window-width of the filter. This metric was not investigated any further because the algorithms are very fast. For example, for the on-line case using the MAD based GEDR the processing within Matlab takes approximately 1ms whereas the processing time for wavelet based GEDR algorithm is approximately 2 ms.

For the presentation of results, only OIDR and OCDR are used to optimize parametric settings for the MAD approach and the tuning of the wavelet-based approach. On the other hand, the CP is not an issue for the developed univariate signal-based approaches. The same applies to the other metrics, which were not found to be useful/critical in this context.

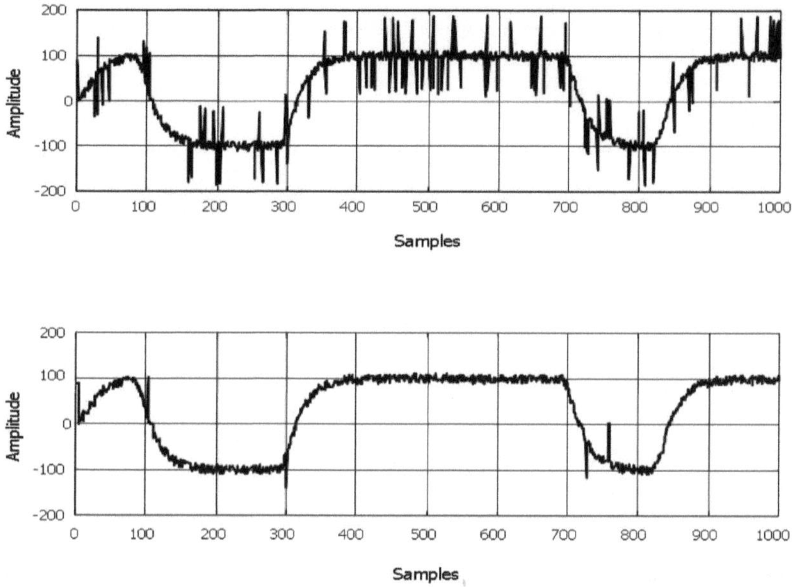

Fig. 4.16. Simulation using median filter (Algower's Algorithm) with a priori knowledge on the noise

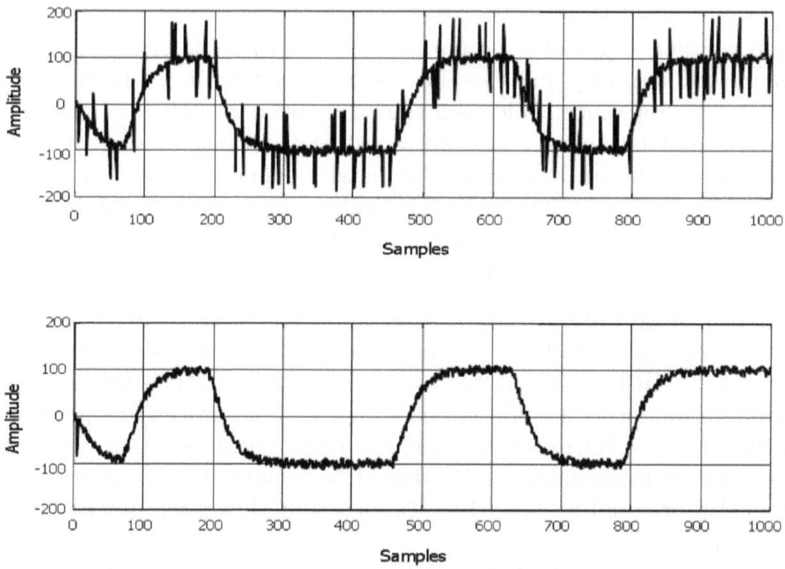

Fig. 4.17. Simulations by using wavelet algorithm without a priori knowledge on the noise

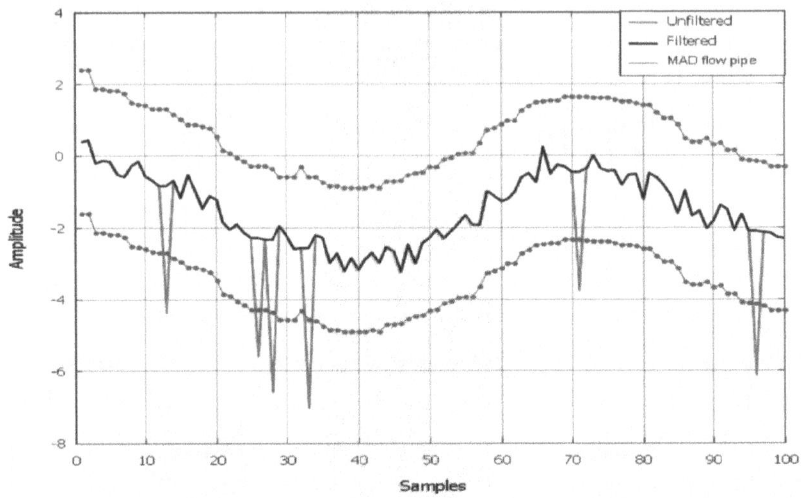

Fig. 4.18. Simulations by using wavelet algorithm without a priori knowledge on the noise

4.5.1 Algorithm Parameterization

For both wavelet approaches no parameters had to be determined. The algorithms are self-adapting and therefore do not need tuning. The goal was the generation of a reference table such that the optimal settings of the filter for the given signal parameters can be found. An optimal setting of the filter parameters was obtained by using the metrics defined above. The OIDR has been minimized while OCDR at a high level has been maintained, that is, sensitivity for robustness has been compromised. The signal parameters available for the computer-generated data are:

- Signal to Noise Ratio, SNR:= (peak to peak signal)/(peak to peak noise) as a dimensionless number. Reasonable SNRs should be between 10 and 100. For a SNR less than 10 the noise with respect to the signal becomes so dominant rendering any kind of meaningful signal processing questionable.
- Outlier to Noise Ratio, ONR:=(peak outlier)/(peak noise) as a dimensionless number. Reasonable ONRs should be greater than 1 (in this application ONRs is used greater than 4) such that they stick out compared to the measurement noise.
- Outlier Probability, OPR: The OPR defines the probability that an outlier occurs at a given time. For all practical purposes, this also represents the ratio of (♯ outliers present)/(♯ data points).
- The fastest time constant ratio, DYNF:=(the fastest time constant)/(tsamp) as a dimensionless number. Reasonable values should be at least 10.

- The slowest time constant ratio, DYNS:=(the slowest time constant)/(tsamp) as a dimensionless number. It should be at least DYNF.
- The number of transients, TRANS:= (the highest frequency of PRBS)/(the sampling frequency). The PRBS is used to generate signals, first order signals with one exponential function, second order signals with two exponential functions, and zero order responses with the PRBS as an input. Reasonable values should be less than 0.5.
- Type: uniformly and normally distributed noise.

Fig. 4.19 shows an example of computer-generated data with uniformly distributed noise. The following figures were obtained by randomly generating

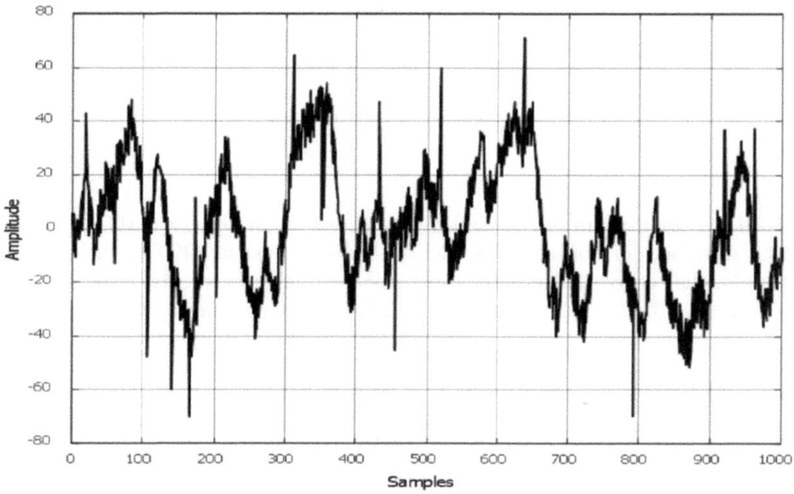

Fig. 4.19. Example of a Computer-Generated Signal: Sampling Time 1, SNR=20, ONR=4,OPR=0.02, DYNF=10, DYNS=20, TRANS=0.1

10000 data sets of 200 samples each where the above mentioned signal parameters were randomly picked out of a pre-specified range as shown in. Furthermore, the MAD filter parameters, which were optimized as presented above. Again, the wavelet algorithm does not have any parameters. Fig. 4.20 shows the accumulated distribution of OCDR using the MAD algorithm. The MAD algorithm shows a good level of sensitivity, that is, in less than 10% (y-axis) of the data sets, the algorithm of the OCDR is less than 85%. Fig. 4.21 shows similar results for the wavelet-based GEDR algorithm. However, for almost 30% of the data sets, the of ODCR is less than 85%. The worst performance is mostly due to the fast dynamics, where 5 was used as a lower limit. As a reminder 10 would be a much better ratio between a sampling rate and fastest time constant. Fig. 4.22 and Fig. 4.23 show a similar representation

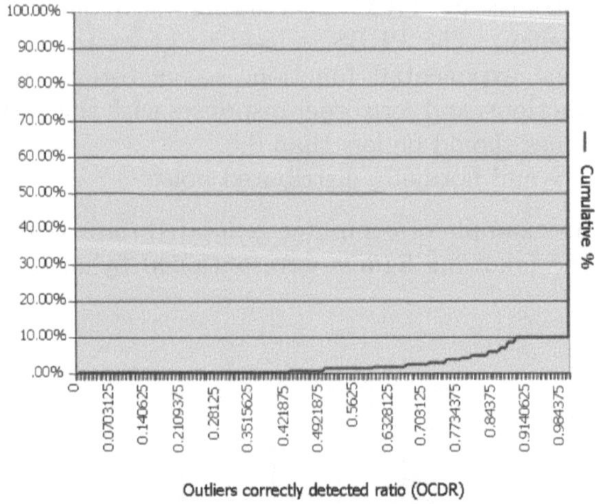

Fig. 4.20. A histogram for OCDR: the MAD algorithm for 10000 computer-generated data sets of 200 samples each. X-Axis shows the OCDR that may range from 0 (no outlier correctly detected) to 1 (all the outliers correctly detected). Y-axis show cumulative % of the data sets for which the "X-property" is true.

for the OIDR. Both algorithms show a very good robustness, that is, very few data points were misclassified. However, the wavelet based GEDR performs considerably worse. In case of the MAD based GEDR algorithm the OIDR is mostly about 99.8% of the data sets analyzed in the case of the wavelet based GEDR algorithm.

4.6 Experimental Data Sources

Data from different process industries were used to evaluate the developed algorithms. The following list represents the people that provided data with a short description of the process, domain, and the variables involved.

1. Separator Train, Oil & Gas : The separator train separates a two-phase flow of gas and liquid. Both phases contain water and oil. The measured quantities are pressures, flow rates, valve openings, and levels, and include process variables, manipulated variables, and disturbances. The variables are strongly coupled and the process shows a strong non-linear behaviour. GEs of type 2 are present in the data.
2. Dryer Section, Pulp & Paper : This process represents a dryer section within a paper mill and uses steam at different pressures for the drying.

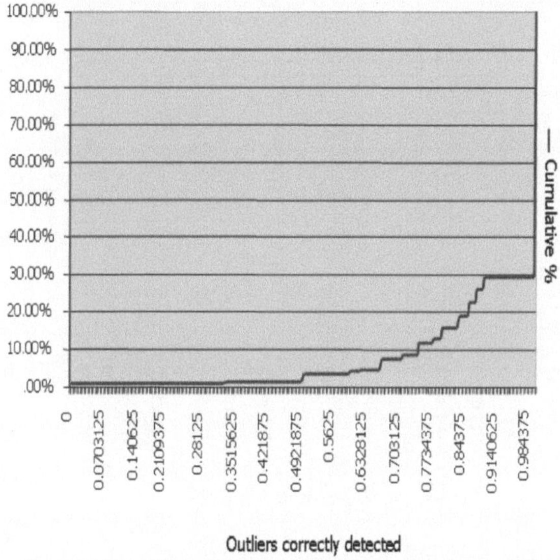

Fig. 4.21. A histogram for the OCDR: the wavelet algorithm for 10000 computer-generated data sets of 200 samples each. X-Axis shows OCDR that may range from 0 (no outlier correctly detected) to 1 (all the outliers detected correctly). Y-axis show cumulative of data sets for which the "X-property" is true.

The measured quantities are pressures, flow rates, moisture, and levels, and include process variables, set point variables, manipulated variables, and disturbances. The variables are coupled and the process shows a non-linear behaviour. GEs of type 2 are present in the data whereas type 1 GEs have not been identified yet.

3. Denox, Power Plant This process reduces the amount of NOX in an effluent stream of a boiler. The measured quantities are flow rates and compositions and include a set point variable, a manipulated variable and two disturbances. The variables are coupled, and the process shows a non-linear behaviour. No GEs are present in the data due to the "2 out of 3" measurement setup.
4. Boiler, Power Plant: This process is a boiler in a power plant. The measured quantities are pressures, temperatures, flow rates, and compositions. The variables are coupled and the process shows a non-linear behaviour. No GEs are present in the data due to the "2 out of 3" measurement set-up.
5. Mining Application: The measured quantities are flow rates.

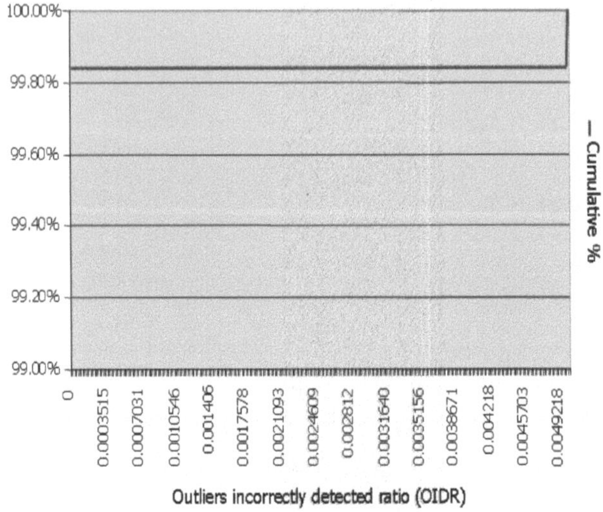

Fig. 4.22. A histogram for the OIDR: the MAD algorithm for 10000 computer-generated data sets of 200 samples each. X-Axis shows OIDR that may range from 0 (no noise misclassified as outliers) to 1 (all noise misclassified as outliers) Y-axis show cumulative % of the data sets for which the "X-property" is true

6. Distillation Column: The process represents a high-purity distillation column with 13 trays and 4 packed beds. The measured quantities are flow rates, temperatures, pressures and levels. Outliers may be present in the data. No non-linear model is available at the time.
7. Discrete Manufacturing Proprietary Information.

4.6.1 Dryer, Distillation and Mining Data with Outliers

Only the dryer data of the examined data sets were clearly contaminated outliers. There were outliers in pressure measurements, which were successfully identified both by the MAD and the wavelet filter. As it is shown in the upper plot in Fig. 4.24, the "ged_mad_filter" (MAD filter) removes the most obvious of the outliers, although there are probably some more of them. However, it does not remove any measurement noise. The "ged_wav_filter" (wavelet filter) removes more of the outliers, but it also removes some points that are clearly measurement noise. Fig. 4.25 shows an example from the distillation where it is uncertain if one is removing outliers or measurement noise. Again it is possible to see that the wavelet filter is more aggressive than the MAD one.

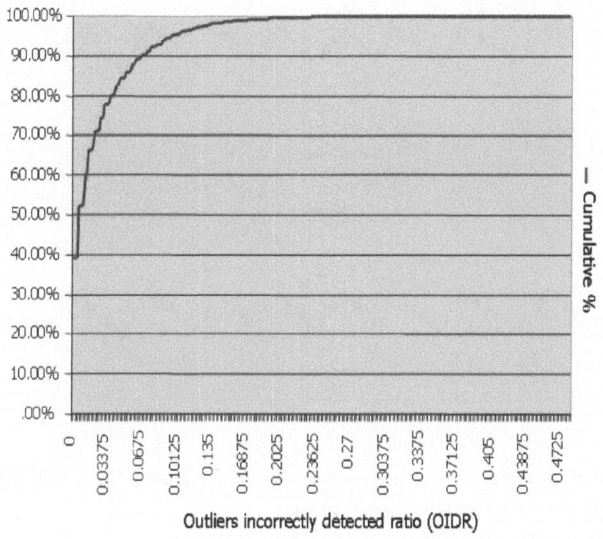

Fig. 4.23. A histogram for the OIDR: the wavelet algorithm for 10000 computer-generated data sets of 200 samples each. X-Axis shows the OIDR that may range from 0 (no noise misclassified as outliers) to 1 (all noise misclassified as outliers) Y-axis shows cumulative % of the data sets for which the "X-property" is true

Noise classified as outliers

As Fig. 4.26 shows there are cases were it is uncertain if it is a noise measurement or if it is an outlier. Fig. 4.26 shows an example where both of the algorithms remove noise in an outlier free data set.

Correlated data and outliers

Since the given algorithms are univariate such effects will not be taken into consideration. In Fig. 4.27 and Fig. 4.28 the original and the filtered data are shown. Figure shows that some outliers are detected, but only in two of the three measurements. However since this peak is present in all three of the variables and that they are strongly correlated, it is unlikely that it is an outlier. In Fig. 4.28 the same results are shown for the wavelet filter, where at least the spikes are removed in all three of the measurements.

Misclassifications

For the gross error detection algorithms to work properly it is required that the sampling period is one tenth of the "dominating" dynamics of the process.

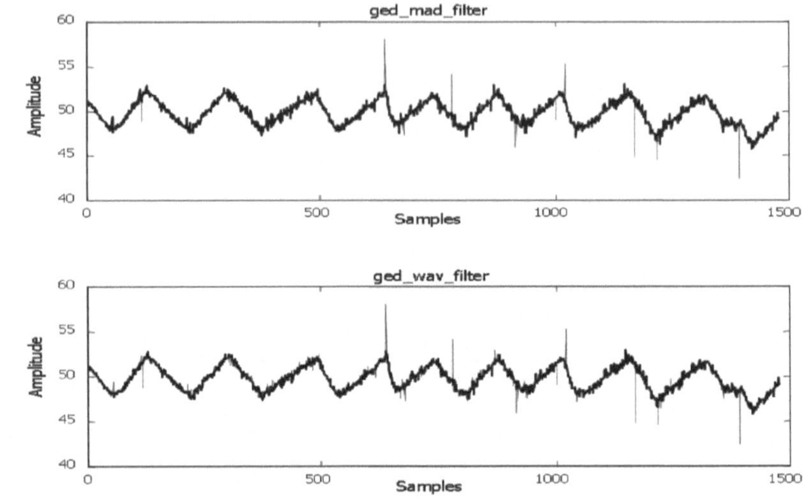

Fig. 4.24. Successful outlier detection in the dryer data set. The fine line is the original data set and the bold is the filtered one

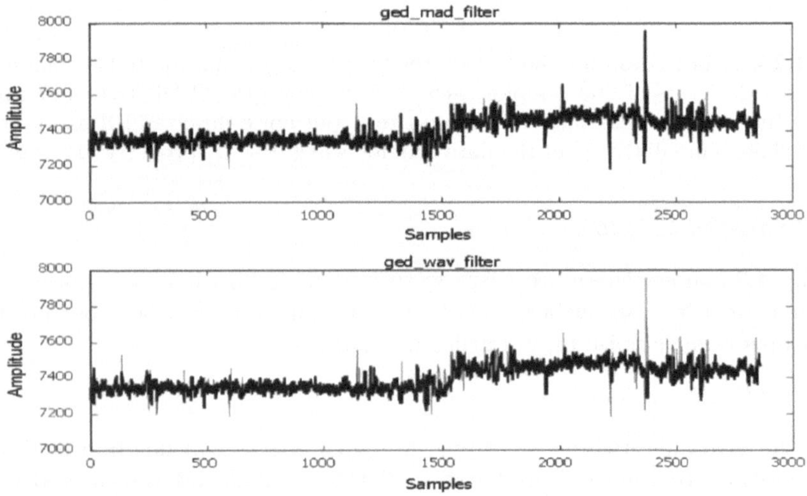

Fig. 4.25. Outlier detection. The fine line is the original data set and the bold is the filtered one

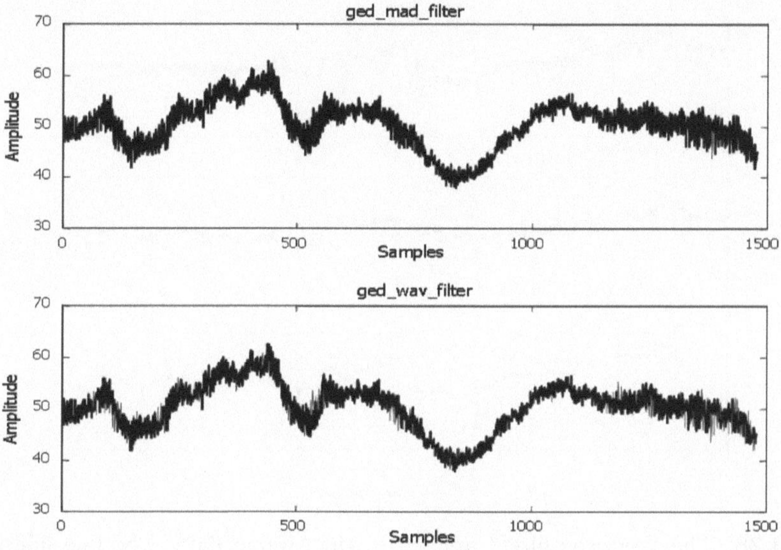

Fig. 4.26. Removal of noise. The fine line is the original data set and the bold is the filtered one

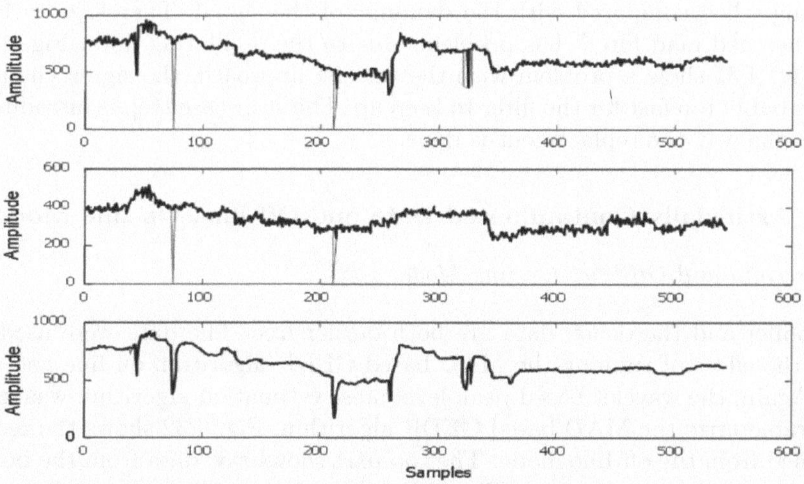

Fig. 4.27. The "ged_mad_filter" applied to the Mining data. The fine line is the original data, and the bold is the filtered one

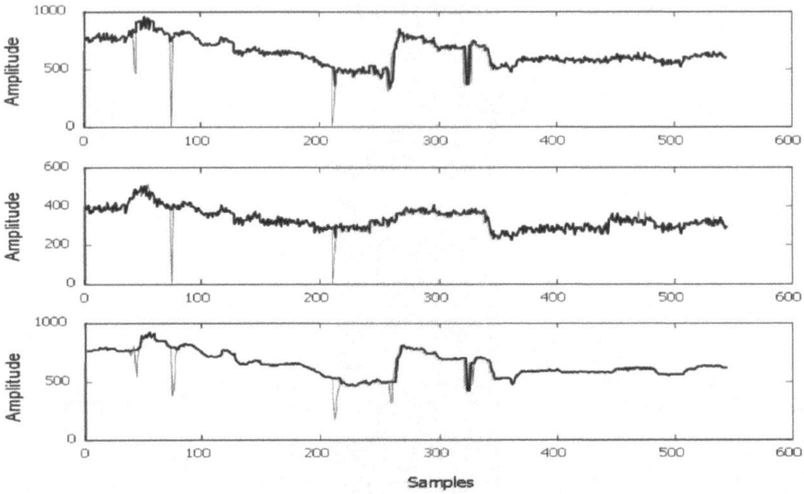

Fig. 4.28. The "ged_wav_filter" applied to the Mining data. The fine line is the original data, and the bold is the filtered one

Below, an example has been shown in which the algorithms have some problems. This may be explained in the following way: the sample time chosen is too long when compared with the dynamic of the signal. Therefore, in Fig. 4.29 the "ged_mad_filter" has problems due to the oscillating data. Fig. 4.30 and Fig. 4.31 show a problem with the wavelet approach, the signal changes are probably too fast for the filter to keep up. The stair case step is introduced due to the way the replacement is done.

4.6.2 Artificially Contaminated Data and Off-line, On-line Mode

Boiler Data and Off-line, On-line Mode

The boiler and the denox data are both outlier free. The data were used to show the effect of running the MAD based GEDR algorithm off-line and on-line. Again, the wavelet based peak level noise estimation algorithm was used to parameterize the MAD based GEDR algorithm. Fig. 4.32 shows the result obtained from the off-line mode. The top part shows raw data from the boiler where no outliers are present. The algorithm does not detect or replace any outliers, that is, the algorithm does the right thing. The same applies to the artificially contaminated data in the bottom part of Fig. 4.32 where all of the outliers are correctly detected and no misclassifications occur. Fig. 4.33 shows the result obtained from on-line mode. The top part shows raw data from the boiler where no outliers are present. The algorithm does detect and replace some outliers, that is, the algorithm misclassifies noise as outliers at around

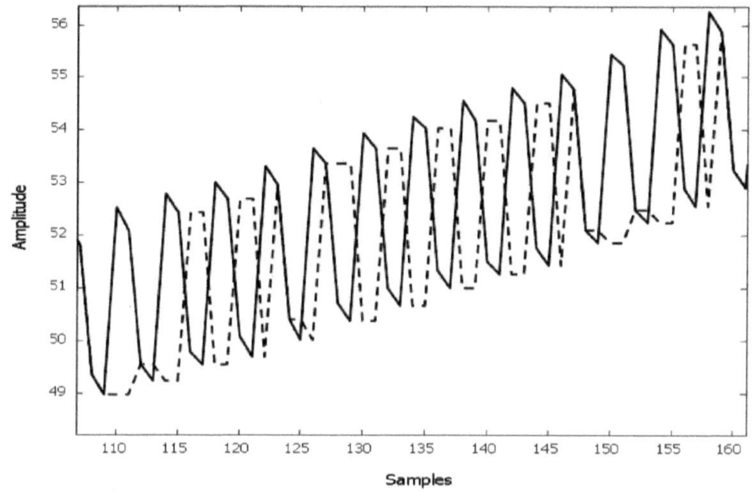

Fig. 4.29. Problems when the dynamics are near the step size for the "ged_mad_filte". The fine line is the unfiltered and the bold is the filtered one

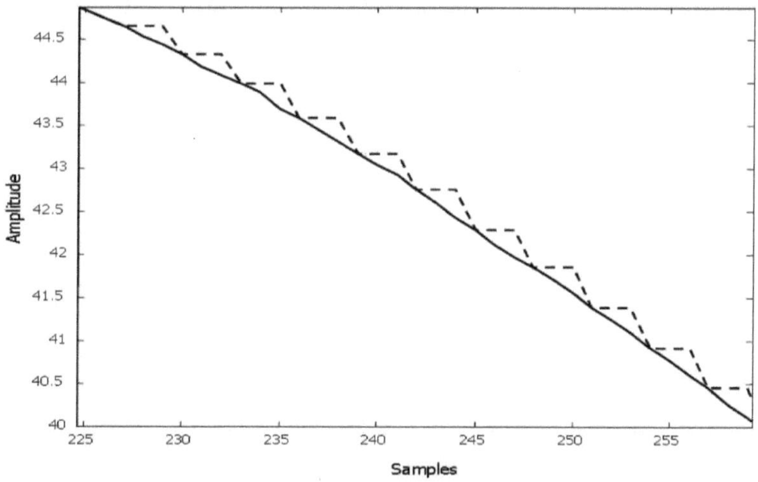

Fig. 4.30. Problems for the wavelet approach

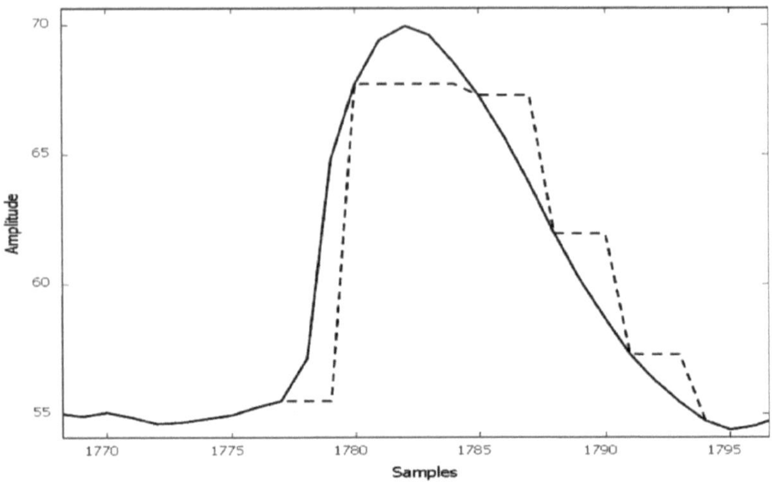

Fig. 4.31. The wavelet filter applied to the distillation case

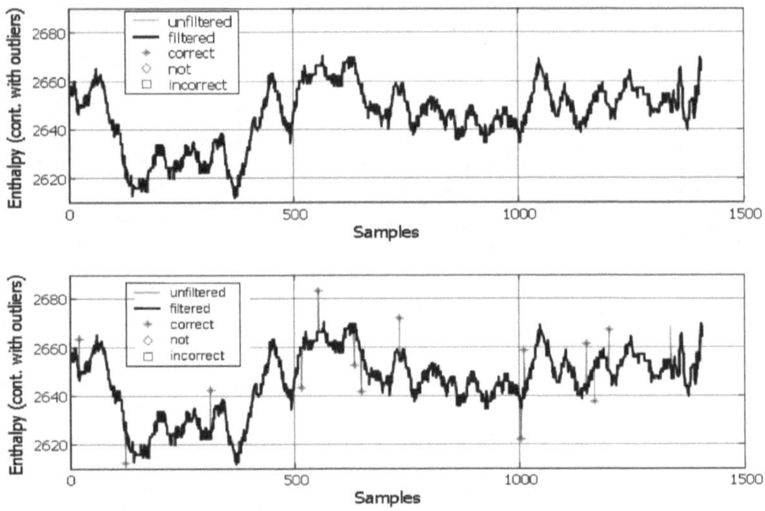

Fig. 4.32. Boiler Data Outlier Detection: The top part, raw data with no outliers due to 2-out-of-3 set-up. The bottom part, artificially contaminated with outliers and filtered using the MAD approach plus noise level detector in the off-line mode

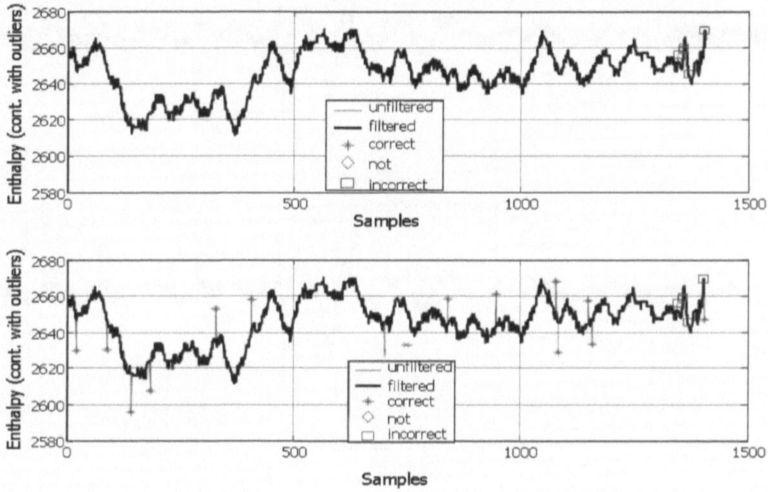

Fig. 4.33. Boiler Data Outlier Detection: The top part, raw data with no outliers due to 2-out-of-3 set-up. The bottom part, artificially contaminated with outliers and filtered using the MAD approach plus noise level detector in on-line mode

sample number 1300. The same applies to the artificially contaminated data in the bottom part of Fig. 4.33 where all of the outliers are correctly detected, however, misclassifications also occur.

Distillation Column

In Fig. 4.34, a data set from the distillation case has been contaminated with the outliers. As it can be seen the MAD algorithm performs much better than the wavelet based one. The MAD algorithm removes all the outliers and does not remove any measurements that are not outliers. On the other hand, the wavelet based approach removes some of the measurement noise and more seriously also some of the signals.

4.7 Summary, Conclusions and Outlook

The conclusions of this investigation are: MAD based GEDR algorithm together with the wavelet based peak noise level estimator:

- show very good performance with regard to robustness and sensitivity in the detected type 2 GEs, that is, outliers.

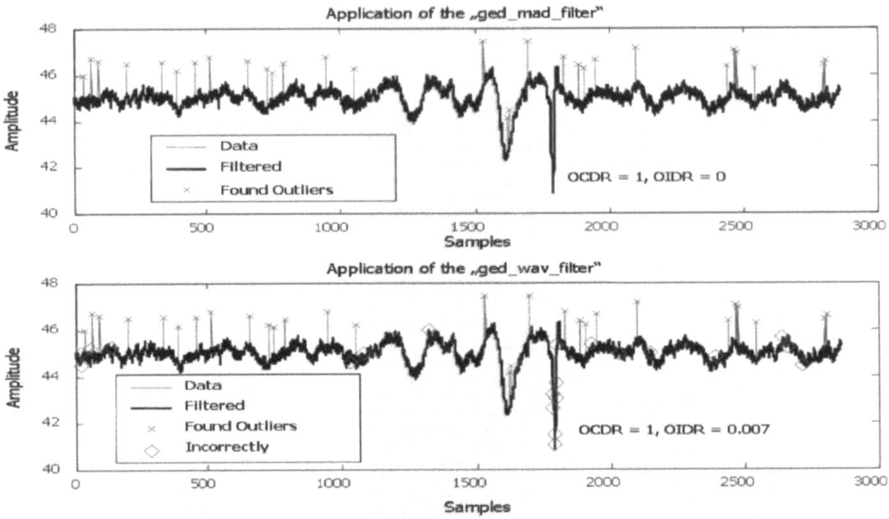

Fig. 4.34. Applications of both algorithms on an artificially contaminated data set

- show very good performance for both artificially contaminated experimental data and real contaminated experimental data.
- may be used both, for off-line and on-line applications, however, the performance disimproves in the case of an on-line application.
- The wavelet based GEDR algorithm performs well and there is still room for improvements, which should be investigated further.

References

1. P.H. Menold, R.K. Pearson, F. Allgwer (1999) Online outlier detection and removal. In: Mediterranean Control Conference, Israel
2. D. L. Donoho, I.M. Johnstone, (1994) Adapting to unknown smoothness via wavelet shrinkage. Journal of the American Statistical Association, 90 (432): 1220–1224
3. D. L. Donoho (1995) Denoising by soft thresholding. IEEE Transaction On Information Theory, 41 (3): 613–627
4. D. L. Donoho, I. M. Johnstone (1994) Ideal spatial adaptation by wavelet shrinkage. Biometrika, 81 (3): 425–455
5. D. L. Donoho, I. M. Johnstone, G. Kerkyacharian, D. Picard (1996) Density estimation by wavelet thresholding. Annals of Statistics, 24 (2): 508-539
6. R. Shao, F. Jia, E.B. Martin, A.J. Morris (1999) Wavelets and non linear principal component analysis for process monitoring. Control Engineering Practice 7: 865–879
7. R. R. Coifman, M. V. Wickerhauser (1992) Entropy based algorithm for best basis selection. IEEE Information Theory 32: 712–718

8. Rolf Isermann (1996) Modellgesttzte berwachung und Fehlerdiagnose Technischer Systeme. Automatisierungstechnische Praxis 5: 9-20
9. P. Mercorelli, M. Rode, P. Terwiesch, (2000) A Wavelet Packet Algorithm for OnLine Detection of Pantograph Vibration, In: Proc. IFAC Symposium on Transportation. Braunschweig, Germany, 13-15 June
10. Mercorelli, A. Frick (2006) "Noise Level Estimation Using Haar Wavelet Packet Trees for Sensor Robust Outlier Detection". In: Lecture Note in Computer Sciences. Springer-Verlag publishing
11. http://www-stat.stanford.edu/ wavelab/
12. I. Daubechies (1995) Ten Lectures on Wavelets. Publisher Society for Industrial and Applied Mathematics, Philadelphia (Pennsylvania)
13. J.S. Albuquerque, L.T. Biegler (1996) Data Reconciliation and Gross-Error Detection for Dynamic Systems. AIChE Journal 42: 2841–2856
14. S. Beheshti, M. A. Dahleh (2002) On denoising and signal representation. In: Proc. of the 10th Mediterranean Control Conference on Control and Automation
15. S. Beheshti, M. A. Dahleh (2003) Noise variance and signal denoising. In: Proc. of IEEE International Conference on Acustic. Speech, and Signal Processing (ICASSP)
16. R. R Coifman, M. V. Wickerhauser (1992) Entropy based algorithm for best basis selection. IEEE Trans. Inform. Theory, 32:712–718
17. R. Shao, F. Jia, E. B. Martin, A. J. Morris (1999) Wavelets and non linear principal component analysis for process monitoring. Control Engineering Practice 7:865–879
18. R.K. Pearson (2001) Exploring process data. J. Process Contr. 11:179–194
19. R.K. Pearson Outliers in Process Modelling and Identification. IEEE Transactions on Control Systems Technology 10: 55–63

5

Immune-inspired Algorithm for Anomaly Detection

Ki-Won Yeom

Information Visualization and Interaction Group
CADCAM Research Center in Korea Institute of Science and Technology
39-1 Hawolgok-dong Seongbuk-gu, Seoul, Korea
pragman@kist.re.kr

The central challenge with computer security is determining the difference between normal and potentially harmful activity. For half a century, developers have protected their systems by coding rules that identify and block specific events. However, the nature of current and future threats in conjunction with ever larger IT systems urgently requires the development of automated and adaptive defensive tools. A promising solution is emerging in the form of Artificial Immune Systems (AIS). Current AIS based Intrusion Detection Systems (IDS) have been successful on test level environments, but the algorithms rely on self-nonself discrimination, as stipulated in classical immunology. Within immunology, new theories are constantly being proposed that challenge current ways of thinking. These include new theories regarding how the immune system responds to pathogenic material. This paper takes relatively new theory: the Danger theory and Dendritic cells, and explores the relevance of those to the application domain of security. In this paper, we introduce an immune based anomaly detection approach from the abstraction of Danger theory. We also present the derivation of bio-inspired anomaly detection from the DC functionality with Danger theory, and depict two examples of how the proposed approach can be applied for computer and network security issues with preliminary results.

5.1 Introduction

The traditional prevention techniques such as user authentication, data encryption, avoiding programming errors and firewalls are used as the first line of defense for computer security. If a password is weak and is compromised, user authentication cannot prevent unauthorized use and firewalls are vulnerable to errors in configuration and ambiguous or undefined security policies as well.

K.-W. Yeom: *Immune-inspired Algorithm for Anomaly Detection*, Studies in Computational Intelligence (SCI) **57**, 129–154 (2007)
www.springerlink.com

They are generally unable to protect against malicious mobile code, insider attacks and unsecured modems. Programming errors cannot be avoided as the complexity of the system is increasing and application software is changing rapidly leaving behind some exploitable weaknesses. Intrusion detection therefore becomes a necessity as an additional wall for protecting systems despite the prevention techniques.

Recently, Intrusion Detection Systems (IDS) have been used in monitoring attempts to break security, which provides important information for timely countermeasures. Intrusion detection is classified into two types: misuse intrusion detection and anomaly intrusion detection. Misuse intrusion detection uses well-defined patterns of the attack that exploit weaknesses in the system and application software to identify the intrusions. These patterns are encoded in advance and used to match against the user behavior to detect intrusion. Anomaly intrusion detection identifies deviations from the normal usage behavior patterns to identify the intrusion.

Over the last few years, Artificial Immune Systems (AIS) have become an increasingly popular computational intelligence paradigm. Inspired by the mammalian immune system, AIS seek to use observed immune components and processes as metaphors to produce systems that encapsulate a number of desirable properties of the natural immune system. These systems are then applied to solve problems in a wide variety of domains [5]. There are a number of motivations for using the immune system as inspiration for data mining; these include recognition, diversity, memory, self-regulation, dynamic protection and learning [4].

Currently, the majority of AIS encompasses two different types of immune inspired algorithms, namely negative selection (T-cell based), and clonal selection with somatic hypermutation (B-cell based) [1, 14, 19, 21].

Within immunology however new theories are constantly being proposed that challenge current ways of thinking. These include new theories regarding how the immune system responds to pathogenic material. Danger theory is one of the emerging theory that is presented by [12] and the application of this theory to AIS was identified and discussed in depth in [16]. This theory encompassed pathogenic detection, where the basis for discrimination was not centered around 'self' or 'non-self', but to the presence or absence of danger signals. It is thought that danger signals are detected and processed through professional antigen presenting cells known as Dendritic cells. Dendritic cells (DCs) are viewed as one of the major control mechanisms of the immune system, influencing and orchestrating T-cell responses, in addition to acting as a vital interface between the innate (initial detection) and adaptive (effector response) immune systems.

DCs are responsible for pathogenic recognition at the initial immune process, sampling the environment and differentiating depending on the concentration of signals or perceived misbehavior in the host tissue cells. Strong parallels can be drawn from this process to the goal of successful anomaly detection. Current anomaly detection systems frequently rely on profiling normal user behavior during a training period. Any subsequent observed behavior that does not match the normal profile (often based on a simple distance metric) classed as anomalous. At this point several alert systems are developed. However, these systems have problems with high levels of false positive errors, as behavior of users on a system changes over a period of time. Anomaly detection systems remain a high research priority as their inherent properties allow for the detection of novel instances, which could not be detected using a signature based approach. AIS featuring negative selection algorithms have been tried and tested for the purpose of anomaly detection [7].

In this paper we review the Danger theory, draw deduction of its significations, and derive an anomaly detection system using Dendritic cells combined with the theory. More specifically, we discuss the background to artificial immune systems including Danger theory immunology and Dendritic cells, and we describe a small literature review concerning the use of Danger theory and Dendritic cells (AIS) in section 5.2. Section 5.3 focuses on Dendritic cells (DCs) with respect to changing morphologies, functions, control of the immune system and outlines of bio-inspired anomaly detection mechanism in terms of the infectious non-self with danger theories. Section 5.4 shows worked examples of how the proposed approach can be used as a signal processor for early detection with complete pseudo-code, and the preliminary results is presented. Section 5.5 includes a brief analysis of the results and details of future work followed by conclusions.

5.2 Background

It is acknowledged that the Danger theory and Dendritic Cells (DCs) are a relatively new area in the realm of artificial immune systems, so to aid the reader's understanding of this paper we would like to discuss some details of both. These details have been simplified and so for a more comprehensive review of this field the reader is directed towards the literature, such as [2,3, 6,10,11,22].

5.2.1 The Danger Theory

The Danger theory attempts to explain the nature and workings of an immune response in a way different to the more traditional and widely held self/nonself viewpoint. This view states that cells of the adaptive immune system are incapable of attacking their host because any cells capable of doing so are deleted during their maturation process. This view, although seemingly elegant and generally easy to understand has come under criticism as it fails to explain a number of observations. Examples of such may be the lack of immune response to injections of inert but foreign proteins, or the failure of the immune system to reject tumors even though nonself proteins are expressed. In [12] author argues a more plausible way to describe the immune response is as a reaction to a stimulus the body considers harmful, not a simple reaction to nonself.

Danger Signals for Immune Revelation

This model allows foreign and immune cells to exist together, a situation impossible in the traditional standpoint. Matzinger hypothesizes that cellsdying unnaturally may release an alarm signal which disperses to cover a small area around that cell, Antigen Presenting Cells[1] (APCs) receiving this signal will become stimulated and in turn stimulate cells of the adaptive immune system. The term 'danger area' was coined by Aickelin and Cayzer in [11] to describe this area, in which the alarm signal may be received by APCs. This simple explanation may provide reasons for the two anomalous observations cited. Foreign proteins in the injection are not harmful and so are ignored, likewise tumor cells are not undergoing necrotic cell death and therefore not releasing alarm signals, hence no immune reaction. The nature of these alarm signals is still under discussion but some possibilities have been empirically revealed. These include elements usually found inside a cell which are encountered outside (pre-packaged signals) or chemicals such as heat shock protein, which are synthesized by cells under stress (inducible signals). As these danger signals only activate APCs, these cells in turn stimulate B and T cells into action according to the following rules:

- Signal one is the binding of an immune cell to an antigenic pattern or an antigen fragment which is presented by an APC.
- Signal two is either a 'help' signal given by a T-helper cell to activate a B-cell, or a co-stimulation signal given by an APC to activate a T-cell.

This co-stimulation signal does not fit well in the traditional self/nonself view and also leads to the question 'if a B-cell requires stimulation from a T-helper cell to become activated, what activates a T-helper cell?' As Matzinger in states "perhaps for this reason co-stimulation was ignored from its creation by Laferty and Cunningham in 1974, until its accidental rediscovery by Jenkins and Schwartz in 1986". The danger model answers this question (often referred to as the primer problem) by stating that T-cells receive their co-stimulation signals from APCs, which in turn have been activated by alarm signals.

Danger vs. Self/Non-Self

There is a criticism of the self/nonself view, which states the thymus is responsible for the negative selection of all autoreactive cells, that as the thymus

[1] An antigen-presenting cell (APC) is a cell that displays foreign antigen complexed with MHC on its surface. T-cells may recognize this complex using their T-cell receptor (TCR).

provides an incomplete description of self, the selection process will impart only a thymus/nonthymus distinction on T-cells. With a few simple laws concerning the described two signal activation mechanisms applied to T-cells we may provide a simple yet plausible explanation of why autoreactive cells are found in the body, yet autoimmune disease is rare:

- A resting T-cell needs two signals to be activated (as described before).
- If a T-cell receives the first signal (a binding of its receptor to an antigen) without the second signal (co-stimulation from an APC) the T-cell is assumed to have matched a host antigen and will die by apoptosis.

Thus Danger theory allows autoreactive cells to roam the body, but if that cell is to bind to a host antigen in the absence of correct antigenic presentation by an APC, the cell will die instead of becoming activated. While a number of functions related to immune response have been considered as the responsibility of the immune system, under the danger model, those are actually thought of the responsibility of the tissues. Firstly, by the second law above, simply by existing and expressing their own proteins, tissue cells induce immune tolerance towards themselves. Secondly, as an immune response is initiated by the tissues, the nature and strength of this response may also be dictated by the tissues. Therefore different types of alarm signal may result in different types of response. It has long been known that in a certain part of the body an immune response of one class may be efficient, but the same class of response in another may severely harm the host. This gives rise to a notion that tissues protect themselves and use the immune system to do so, a proposition which is in stark contrast to the traditional viewpoint whereby it is the immune system's role to protect tissues. There is still much debate in the immunological world as to whether the Danger theory is a plausible explanation for observed immune function, however we are not concerned with the plausibility of the biology. If the Danger theory is a good metaphor on which to base an artificial immune system then it can be exploited. The first steps to this are the identification of useful concepts and the application of these concepts to a suitable problem domain. It is these actions we wish to illustrate throughout the rest of this paper.

5.2.2 Dendritic Cells as Initiator of Primary Immune Response

Dendritic cells (DCs), the pacemakers of the immune response originally identified by Steinman and his colleagues representing [22], are potent antigen presenting cells (APCs) that possess the ability to stimulate naive T cells. They comprise a system of leukocytes widely distributed in all tissues, especially in those that provide an environmental interface. DCs posses a heterogeneous hemopoietic lineage, in that subsets from different tissues have been shown to posses a differential morphology, phenotype and function. The ability to stimulate naive T cell proliferation appears to be shared between these various DC subsets. It has been suggested that the so-called myeloid and lymphoid-derived subsets of DCs perform specific stimulatory or tolerogenic function, respectively. DCs are derived from bone marrow progenitors and circulate in the blood as immature precursors prior to migration into peripheral tissues. Within different tissues, DCs differentiate and become active in the taking up and processing of antigens (Ags), and their subsequent presentation on the cell surface link to major histocompatibility[2] (MHC) molecules. Upon appropriate stimulation, DCs undergo further maturation and migrate to secondary lymphoid tissues where they present Ag to T cells and induce an immune response. DCs are receiving increasing scientific and clinical interest due to their key role in anti-cancer host responses and potential use as biological adjuvants in tumor vaccines, as well as their involvement in the immunobiology of tolerance and autoimmunity [6].

Dendritic Cell Development

One of the most important findings is that DCs are not a single cell type, but a heterogeneous collection of cells that have arisen from distinct, bone marrow-derived, hematopoietic lineages. To date, at least three different pathways have been described. The emerging concepts are that each pathway develops from unique progenitors, that particular cytokine combinations drive developmental events within each pathway and that cells developing within a particular pathway exhibit distinct specialized functions [2].

The pathway of differentiation gives rise to DCs that home to peripheral tissues to take up and process exogenous Antigens (Ags) prior to migrating to the secondary lymphoid tissues to present Ags to naive T cells. Thymic DCs, on the other hand, perform a very different function being involved in the presentation of self-Ag to developing thymocytes and, hence, the subsequent deletion of autoreactive T cells. It would be appropriate for the precursors of thymic DCs to migrate to the thymus in an immature form and undergo development exposed only to self-Ag within the thymus.

---

[2] The major histocompatibility complex (MHC) is a large genomic region or gene family found in most vertebrates. It is the most gene-dense region of the mammalian genome and plays an important role in the immune system, autoimmunity, and reproductive success.
The MHC complex is divided into three subgroups called MHC class I, MHC class II, and MHC class III. The MHC class I encodes heterodimeric peptide binding proteins, as well as antigen processing molecules such as TAP and Tapasin. The MHC class II encodes heterodimeric peptide binding proteins and proteins that modulate peptide loading onto MHC class II proteins in the lysosomal compartment. The MHC class III region encodes for other immune components, such as complement components and some that encode cytokines

Immature and Maturation of DCs

DCs are derived successively from proliferating progenitor cells and non proliferating precursors (especially monocytes). They migrate to and reside as immature DCs at body surfaces and interstitial spaces. Immature DCs have abundant MHC II products within intracellular compartments and respond rapidly to inflammatory cytokines and microbial products to produce mature T cell stimulatory. DCs with abundant surface MHC II proteins eventually lead to apoptotic death. In most tissues, DCs are present in a so-called immature state and are unable to stimulate T cells. Although these DCs lack the requisite accessory signals for T cell activation, they are extremely well equipped to capture Ags in peripheral sites. Once they have acquired and processed the foreign Ags, they migrate to the T cell areas of lymph nodes (LNs) and the spleen, undergo maturation and stimulate an immune response.

The immature DCs have many MIICs but require a maturation stimulus to irreversibly differentiate into active T cell stimulatory, namely mature DCs. [13] has described an *in vitro*[3] system involving monocytes reverse transmigrating across an endothelial monolayer that offers a possible explanation. This type of situation would occur when cells move from tissues to afferent lymph. It is possible that veiled DCs in lymph originate from monocytes in tissue that interact with the lymphoid endothelium to acquire the properties of immature DCs. If the monocytes also phagocytose particles before they reverse transmigrate, then the cells become typical mature DCs; the process occurs within 48 hours. The cells posses several markers that are expressed by mature DCs but are weak or absent in other leukocytes. Dendritic cells that have matured from monocytes in this in vitro system also express very high levels of surface MHC class II and CD86[4] and, in complete contrast to monocytes.

Maturation of DCs is crucial for the initiation of immunity. This process is characterized by reduced Ag-capture capacity and increased surface expression of MHC and co-stimulatory molecules. However, the maturation of DCs is completed only upon interaction with T cells. It is characterized by loss of phagocytic capacity and expression of many other accessory molecules that interact with receptors on T cells to enhance adhesion and signalling (co-stimulation). Expression of one or both of the co-stimulatory molecules on the DCs are essential for the effective activation of T lymphocytes and for

[3] This type of research aims at describing the effects of an experimental variable on a subset of an organism's constituent parts. It tends to focus on organs, tissues, cells, cellular components, proteins, and/or bio-molecules. Overall, it is better suited for deducing the mechanisms of action.

[4] CD86 is a co-stimulatory molecule constitutively expressed by human antigen presenting cells which interacts with CD28 and CTLA-4 expressed by T cells.

IL-2[5] production. These co-stimulatory molecules bind the CD28[6] molecules on T lymphocytes. If this fails to occur at the time of Ag recognition by the TCR (T Cell Receptor), an alternative T lymphocyte function may result, namely induction of anergy [3].

5.3 IDS based on Danger Theory and DCs Properties

In the body the immune system may only become activated within the danger area, likewise we may only need to activate antibodies within some area of interest in an AIS. One of other advantages may be to harness the role of the tissues and assign different responses to different contexts. We have given thought to the implementation of such ideas, although with no such system having been produced to date we have no literature to refer to for guidance. Based on the biology of the Danger theory, we may identify a set of Danger theory characteristics. There are also a number of desirable characteristics exhibited by DCs that we want to incorporate into an algorithm. In order to achieve this, the essential properties, i.e. those that heavily influence immune functions, have to be abstracted from the biological information presented. From this we produce an abstract model of DC interactions and functions, with which we build our algorithm.

5.3.1 Properties of DCs for IDS

DCs are examined from a cellular perspective, encompassing behavior and differentiation of the cells and ignore the interactions on a molecular level and direct interactions with other immune system cells. DCs have a number of different functional properties that we want to incorporate into an algorithm. Bearing this in mind, we can abstract a number of useful core properties, listed below and represented graphically in Figure 5.1 and Figure 5.2.

DCs' first function is to instruct the immune system to act when the body is under attack, policing the tissue for potential sources of damage. DCs can perform a number of different functions, determined by their state of maturation. Modulation between these states is facilitated by the detection of signals within the tissue - namely danger signals, PAMPs (pathogenic associated molecular patterns), apoptotic signals (safe signals) and inflammatory cytokines which are described below. The characteristics of the relevant signals are summarized below as described in [8,17]

[5] Interleukin-2 (IL2) is an interleukin, or hormone of the immune system that is instrumental in the body's natural response to microbial infection and in discriminating between foreign (nonself) and self.

[6] The antigen CD28 is one of the molecules that provide co-stimulatory signals, which are required for T cell activation.

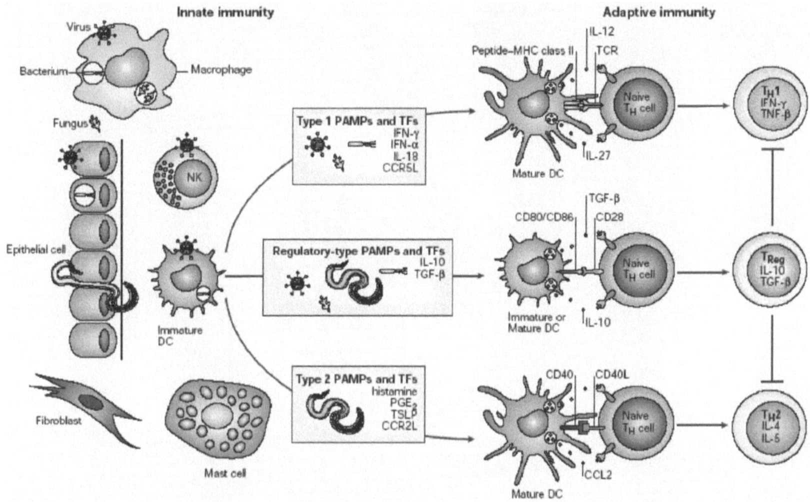

Fig. 5.1. Dendritic-cell polarization is influenced by the type of microorganism that is recognized and the site of activation [18]

- PAMPS are pre-defined bacterial signatures, causing the maturation of immature DCs to mature DCs through expression of 'mature cytokines'.
- Danger signals are released as a result of damage to tissue cells, also increasing mature DC cytokines, and have a lower potency than PAMPs.
- Safe signals are released as a result of regulated cell death and cause an increase in semi-mature DC cytokines, and reduce the output of mature DC cytokines.
- Inflammatory cytokines are derived from general tissue distress and amplify the effects of the other three signals but are not sufficient to cause any effect on immature DCs when used in isolation.

DCs reside in tissue where they are classed as immature DCs (ImmDCs). Whilst in tissue, DCs collect antigen (regardless of the source) and experience danger signals from necrosing cells and safe signals from apoptotic cells. Maturation of DCs occurs in response to the receipt of these signals. On maturation, DCs exhibit the following behavior: collection of antigen ceases; expression of co-stimulatory molecules (necessary for binding to powerful T-lymphocytes) and chemical messengers known as cytokines; migration from the tissue to a lymphatic organ such as a lymph node; and presenting antigen to T-lymphocytes.

As shown in figure 5.2, the context of the tissue (i.e. the type of signals experienced) is reflected in the output chemicals of the DC. If there is a greater concentration of danger signals in the tissue at the time of antigen collection, the DCs will become fully mature DCs with license (LmatDCs), and will express LmatDCs cytokines. Conversely, if the DC is exposed to safe signals, the

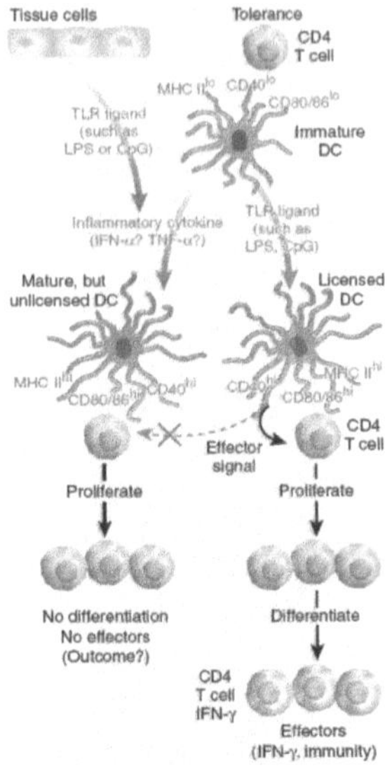

Fig. 5.2. Indirect inflammatory signals cause DC maturation but not license DCs to drive CD4 T-cell effector functions [8]

cell matures differently becoming a unlicensed mature DC, expressing ULmat-DCs cytokines. The LmatDCs cytokines activate T-lymphocytes expressing complimentary receptors to the presented antigen. Any peripheral cells expressing that antigen type are removed through the activated T-lymphocyte. The ULmatDCs cytokines suppress the activity of any matching T-cell, inducing tolerance to the presented antigen. The context of the antigen is assessed based on the resulting cytokine expression of the DC [18].

5.3.2 Abstraction of Anomaly Detection Algorithm

In our model, a combination of infectious non-self model with Danger theory is used to investigate an artificial anomaly detection algorithm. As shown, the orchestration of an adaptive immune response via DCs has many subtleties. Only the essential features of this process are mapped in the first instance as we are interested in building an anomaly detector, not an accurate simulation.

Input data is generated through monitoring environments, behavior of user or system call, and context information by several predefined scan sensors. The collection of input data is also transformed into antigen and signals. Input signals are combined with some information such as data length, name or ID, or process/program service ID etc. This is achieved through using a population of artificial DCs to perform aggregate sampling and data processing. Using multiple DCs means that multiple data items in the form of antigen are sampled multiple times. If a single DC presents incorrect information, it becomes inconsequential provided that the majority of DCs derive the correct context.

The sampling of data is combined with context information received during the antigen collection process. Different combinations of input signals result in two different antigen contexts. Unlicensed mature antigen context implies antigen data was collected under normal conditions, whereas a mature antigen context signifies a potentially anomalous data item.

Our data and method of processing is very different from other AIS, which rely on pattern matching of antigen to drive their systems. In our algorithm, antigen is only used for the labeling and tracking of data, hence we do not have a similarity metric. The representation of the antigen can be a string of either integers or characters. Signals are represented as real-valued numbers, proportional to values derived from the context information of the data-set in use. For example, a danger signal may be an increase in CPU usage of a computer. The value for the CPU load can be normalized within a range and converted into its real-valued signal concentration value. The signal values are combined using a weighted function (Equation 1) with suggested values of the weights derived from empirical data based on immunologists' lab results

The nature of the response is determined by measuring the number of DCs that are fully mature. The closer this value is to 1, the greater the probability that the antigen is anomalous. The value is used to assess the degree of anomaly of a given antigen. By applying thresholds at various levels, analysis can be performed to assess the anomaly detection capabilities of the algorithm.

$$C_{[csm,LmatDCs,ULmatDCs]} = \frac{((W_P * C_P) + (W_S * C_S) + (W_D * C_D))}{W_P + W_S + W_D)} * \frac{(1 + IC)}{2} \tag{5.1}$$

As stated in [15, 17], we are treating DCs as processors of both exogenous and endogenous signal processors. Input signals are categorized either as PAMPs (P), Safe Signals (S), Danger Signals (D) or Inflammatory Cytokines (IC) and represent a concentration of signal. They are transformed to output concentrations of co-stimulatory molecules (csm), ULmatDCs cytokines and LmatDCs cytokines. The signal processing function described in Equation 5.1 is used with the empirically derived weightings presented in Table 5.1.

Additionally, Safe Signals may reduce the action of PAMPS by the same order of magnitude. Inflammatory cytokines are not sufficient to initiate maturation or presentation but can have an amplifying effect on the other signals present. This function is used to combine each of the input signals to derive values for each of the three output concentrations, where C_x is the input concentration and W_x is the weight. These weightings are based on unpublished biological information and represent the ratio of activated DCs in the presence and absence of the various stimuli e.g. approximately double the number of DCs mature on contact with PAMPs as opposed to Danger Signals.

Inflammatory cytokines are not sufficient to initiate maturation or presentation but can have an amplifying effect on the other signals present. This function is used to combine each of the input signals to derive values for each of the three output concentrations, where C_x is the input concentration and W_x is the weight.

Each DC transforms each value of using the following equations with suggested values for weightings given in equation 5.2 and 5.3. Both the equations and weights are derived from observing experiments performed on natural DCs.

Unconditional Packet Transforming (UPT)

$$LFP = \frac{packetsactuallyforwarded}{packetstobeforwarded} = \frac{\#^L_{(m;M)} - \#^L_{([m];M)}}{\#^L_{(M;m)} - \#^L_{(M;[m])}} \tag{5.2}$$

over a sufficiently long time period L.

Selective Packet Transforming (SPT)

$$SPT = \frac{packetsfromsourcesactuallybeingforwarded}{packetsfromsourcestobeforwarded} = \frac{\#^L_{([s];m;M)}}{\#^L_{([s];M;m)} - \#^L_{([s];M;[m])}} \tag{5.3}$$

over a sufficiently long time period L.

We now describe some notations of statistics:

- $\#_{(\&;m)}$ - the number of incoming packets on the monitored node m.
- $\#_{(m;\&)}$ - the number of outgoing packets from the monitored node m.
- $\#_{([m];\&)}$ - the number of outgoing packets of which the monitored node m is the source.
- $\#_{(\&;[m])}$ - the number of incoming packets of which the monitored node m is the destination.
- $\#_{([s];M;m)}$ - the number of packets that are originated from s and transmitted from M to m.

- $\#_{([s];M;[m])}$ - the number of packets that are originated from s and transmitted from M to m, of which m is the final destination.

Table 5.1. Weights for the signal processing function

Weight factors	csm	ULmatDCs	LmatDCs
PAMPs(P)	2	0	2
Danger Signals	1	0	1
Safe Signals	2	3	-3

In nature, DCs sample multiple antigens within the same section of tissue. To mirror this, we create a population of DCs to collectively form a pool from which a number of DCs are selected for the sampling process, in a similar manner to [17]. An aggregate sampling method should reduce the amount of false positives generated, providing an element of robustness. For such a system to work, a DC can only collect a finite amount of antigen. Hence, an antigen collection threshold must be incorporated so a DC stops collecting antigen and migrates from the sampling pool to a virtual lymph node. In order to achieve this we will use a fuzzy threshold, derived in proportion to the concentration of co-stimulatory molecules expressed.

In order to add a stochastic element, this threshold is within a range of values, so the exact number of antigens sampled per DC varies in line with the biological system. On migration to the virtual lymph node, the antigens contained within an individual DC are presented with the DC's maturation status. If the concentration of mature cytokines is greater than the ULmatDC cytokines, the antigen is presented in a 'mature' context. It is possible to count how many times an antigen had been presented in either context to determine if the antigen is classified as anomalous. In order to crystallize these concepts, a worked example and details of a basic implementation are given in the next section.

5.4 DCs based Implementation of Practical Applications

To illustrate the use of this immune inspired approach to the artificial immune systems, we introduce two cases such as Denial-of-Message (DoM) attack and port scanning. The DoM nowadays is frequently attacked by malicious user using included bomb script, vicious contents, or embedded virus and port scan attack to search whether there is any vulnerable port is used as the first step of intrusion and manifold attacks affecting crucial effects on the computer system. In following subsection we describe how our research is able to adopt to those cases more detail.

5.4.1 A Detection of DoM Attack

We propose a more secure sampling system which an algorithm by which a broadcasting base station detects the failure of nodes to receive its broadcasts. We presume the attackers goal is to deny the broadcast to as many nodes as possible while remaining undetected by the base station. Our goal, then, is to limit the attacker's capacity to achieve this, so that the attacker's increased disruption of broadcasts results in an increased probability that he is detected, irrespective of the strategy he pursues. During this section we use a number of small sections of pseudocode to illustrate our example. In this pseudocode we use AB to refer to an initially empty set of artificial antibodies.

Signals and Antigen

The system we propose works over two distinct stages. The first will be an initialization and training stage with the second as a running stage. During this second stage the system will detect a disruptive adversary and in the meantime, reduce the number of acknowledgments sent to the base station, by having a subset of recipients acknowledge each broadcast, where this subset is computed deterministically but in a way that is hidden from the attacker. During the first stage, a summary of which is given in pseudocode 1, the system must generate an initial collection of antibodies, and so for a given amount of time the system may observe user actions when confronted with a new broadcast message. If the cryptographic mechanisms we employ cannot be broken by the attacker, then the attacker can ascertain whether an uncompressed node should acknowledge a broadcast only by observing the node produce the acknowledgment. This, of course, is too late for the attacker to disrupt the node from receiving this broadcast. Moreover, if the attacker disrupts the acknowledgment, then this provides evidence of his presence to the base station.

```
PROCEDURE InitializationOfTrain()
  While (size of AB ¡ a threshold)
    If (user expresses DISINTEREST in an message item from net-
work packets)
      create new antibody (ab) from broadcast message
      add newly generated antibody (ab) to AB
      For (ab ∈ AB)
        clone and mutate ab to maximize affinity with ab
        add best n clones to AB
```

Pseudocode 1. Initialization and Training

When the repertoire of immune cells has reached a given size the system may run on new data as described in pseudocode 2. The AIS will convert all incoming message into a format such that affinity between it and antibodies can be evaluated. Conceptually therefore the broadcast message is an antigen. The system makes a distinction between the terms antigen and message, as follows. Antigen is the name used to refer to a processed message which contains just a generalized representation of the original destination and the class assigned to the message. The system starts initializing an antigen count to 0. When this count reaches a certain number (K) of antigens, we use the latest K antigens to perform clonal selection and mutation with the set of antibodies as described in the Update_Population procedure.

Note that it is important that this procedure is performed only after we have a reasonable number of duly processed antigens, to avoid antibodies adapted to just one antigen and preserving generality. One of the main advantages of the immune inspired approach is this built in ability to adapt. The use of the clonal selection principle here has the effect of allowing our antibody set to change over time. The clonal selection procedure will lead to an increase in the size of set AB over time and so to counter this the final line of this procedure will remove the w most unhelpful antibodies in the set AB.

Detection algorithm

The danger signal should signal something is wrong but there is still time to recover, and should come from something the antigens have little or no control over. There is also no reason why the danger signal must indicate danger in the real world; it could be a positive signal as long as it signals that something of significance to the system is taking place in a particular area. In this instance we considered several possibilities for such a danger signal, such as abnormal frequency and/or size of messages. Although the system might work with a combination of danger signals, each of them requiring a somewhat different response, in this paper we focus on a single danger signal based on the idea that network packet has too many messages. Hence, this part of the pseudocode works as follows. First, the system computes the degree of danger. If the system has no messages waiting to be scan then we may not have a danger signal at all. If however there are many messages a danger signal may be raised. Every time a new message arrives the danger area should be reevaluated based on the current state of the network. Although in the natural immune system the danger area is spatial, in our system we are not so constrained.

The nature of this danger area must be decided upon based on what we want it to signal or how we want to react to it. Hence in this example we propose a temporal danger zone, the size of which will vary according to a measured value of the danger signal. Thus the messages in the packets which may have been let through previously may become candidates for removal on receipt of a danger signal. This temporal danger area will therefore stretch into the past and make the system even more adaptive: the larger the degree of danger, the larger the size of the danger area. Once the danger area has been computed, the system has to decide whether the message is abnormal or not.

This decision consists of predicting the class of the network, based solely on the affinity between each message and the antibodies. Hence, the system computes the affinity between each message in the danger area and each antibody.

```
PROCEDURE Continuous_adaptation()
  antigen_count ← 0
  LOOP
    receive incoming message from network packets
    ag ← preprocess message into antigen
    antigen_count ← antigen_count + 1
    IF(antigen_count = K)
      AG ← last K antigens
      Update_Population(AG)
      antigen_count ← 0
    compute degree of danger()
    WHILE(danger is high)
      compute temporal danger zone
      AG ← all message in the danger zone
    FOREACH(ag ∈ AG)
      FOREACH(ab ∈ AB)
        compute affinity (ab,ag)
      high_aff ← highest affinity value
      IF(high_aff > a threshold)
        remove the network packet

PROCEDURE Update_Population(AG)
  FOREACH(ab ∈ AB)
    FOREACH(ag ∈ AG)
      compute affinity(ab,ag)
      U ← (ag | ag's class is uninteresting)
      I ← (ag | ag's class is interesting)
      quality_ab ← ∑_{i∈U} aff(ab, ag_i) − ∑_{j∈I} aff(ab, ag_j)
    clone and mutate ab in proportion to quality_ab
    remove from AB the w antibodies with the lowest value of
quality_ab
```

Pseudocode 2. More secure mail adaptation

For each message in the danger area, if the affinity between that message and the most similar antibody is greater than a threshold, then the message is considered uninteresting (means that it is normal network packet or message). Otherwise the message is considered interesting (means it is malicious message) and remove the packet from the network. This is an analogy to the natural situation as a recognition of an antigen by an antibody is signal one. There is a correspondence in the natural system between the release of a danger signal and the activation of APCs which in return supply the co-stimulatory signal to activate T-cells. Just the presence of danger in this context may therefore take the place of signal two.

For clarification and to illustrate that by our definition this system is truly danger inspired we may draw the parallels described in Table 5.2.

Table 5.2. Parameters for secure e-mail

Description	Values
Tissue	Ethernet packet receiver
Danger Signal 1	High affinity between antibody and antigen(message)
Danger Signal 2	Receipt of danger signal
Source of danger	High number of undefined messages
Danger area	Temporal
Immune response	Move message to temporary storage
Localized response	Only messages in the danger area are considered to be remove from temporary storage

5.4.2 Experiments and Results

To achieve the desired realistic behavior, we have performed all of our simulations with radio noise accumulation. We use the default values for the remaining network device parameters. We find that the teachability behavior of our simulated broadcasts is similar to the behavior observed in experiments conducted on actual sensor nodes. For each data point, we use at least 20 simulation runs.

At the beginning of each simulation run, we allow sufficient time for the table exchanges necessary in this distance-vector protocol to stabilize. During each individual experiment, the topology remains constant, minimizing the need to send routing updates.

For each experiment, we first randomly generate a network topology. Then, we ran simulations with no attackers to derive the parameter, as defined in Equation 5.1. Note that parameters obtained in this way will account for the natural loss rate, including any message loss due to collision or contention from the acknowledgment routing protocol overhead. After obtaining parameters, we ran many simulation rounds with varying percentages of randomly compromised nodes. Each compromised node sends back an acknowledgement if sampled, but compromised nodes do not forward any broadcast messages.

The general analysis of the optimal attacker behavior demonstrates the tradeoffs an attacker makes when trying to achieve his goals. In particular, finding the optimal set of nodes the attacker should compromise is intimately tied to a particular network topology. Our analysis on the proposed approach is based on the number of deprived nodes, rather than the number of attacking nodes. This serves to decouple our detection results from the attacker model and broadcast protocol we used. For our experiments, the network density was rather low. Few nodes had more than five neighbors within radio communication range. This case benefits the attacker, since a commonly considered goal of the attacker is to create a vertex cut set of compromised nodes, partitioning the network.

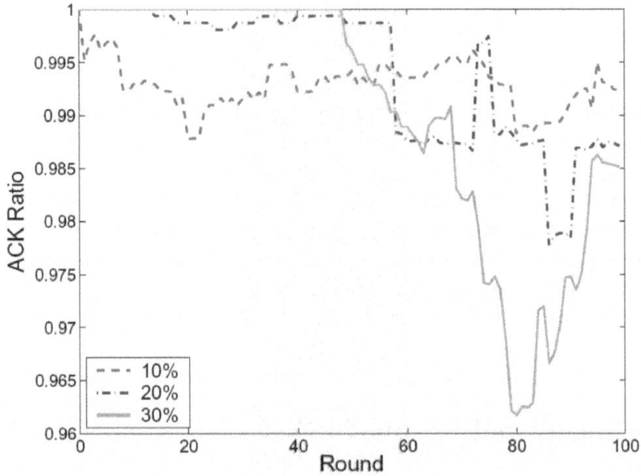

Fig. 5.3. The ratio of received to expected acknowledgements computed at the base station

Results and Analysis

In this section, we study the impact of the DoM attack on broadcast protocols. We are interested in the case where the attacker has compromised sufficiently many nodes to have a significant impact on how many legitimate nodes receive the broadcast message. Since flooding is the most general and widely used broadcast algorithm, we consider flooding in this section. As a sufficient number of nodes are compromised, their impact becomes evident. As the number of victim nodes in mid-sized networks (hundreds of nodes) increases, the attackers impact on the network becomes detectable by SIS. First, we present a naive example to convey the intuition behind detection. Figure 5.3 illustrates the ratio of received to expected acknowledgments as computed at the base station.

After the 49th broadcast, some percentage of nodes are randomly chosen to be malicious. The impact on the ratio of received to expected acknowledgements is evident, as is the unique level of natural loss for each topology. Thus, an effective threshold that can be used to detect an attack is dependent on the natural loss characteristics of the particular topology, and configuring a scheme to detect attackers in sensor network broadcast is best done post-deployment.

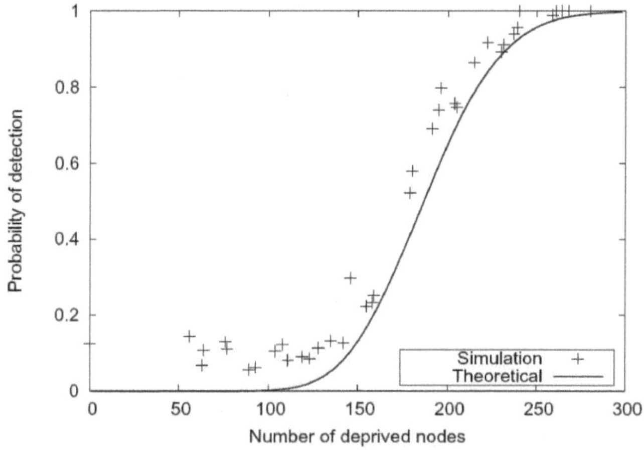

Fig. 5.4. The Probability of detection of the DoM attack based on DCs

Figure 5.4 illustrates the detection capabilities of our approach in simulation, and compares those results with the theoretically expected performance, based on the parameters (defined in Equation 5.1) observed in 0-attacker broadcast rounds. We performed the experiment on a simulated network containing 600 nodes. We used the same randomly generated topology for multiple simulation runs, while varying which nodes were malicious. Figure 5.5 depicts the false positive rate we observed. The x-axis represents, the theoretical false positive rate. is used to pick the threshold of detection. The y-axis shows the simulated false positive rate (the number of false positives over 2000 rounds of broadcast to 599 nodes).

5.4.3 A Detection of Port Scan Attack

Port scanning is a technique for discovering hosts' weaknesses by sending port probes. Although sometimes used by system administrators for network exploration, port scanning generally refers to scans carried out by malicious users seeking out network vulnerabilities. The negative effects of port scans are numerous and range from wasting resources, to congesting the network, to enabling future, more serious, attacks.

While a port scan is not an intrusion per se, it is a 'hacker tool' used frequently during the information gathering stage of an intrusion. This can reveal the topology of a network, open ports and machine operating systems.

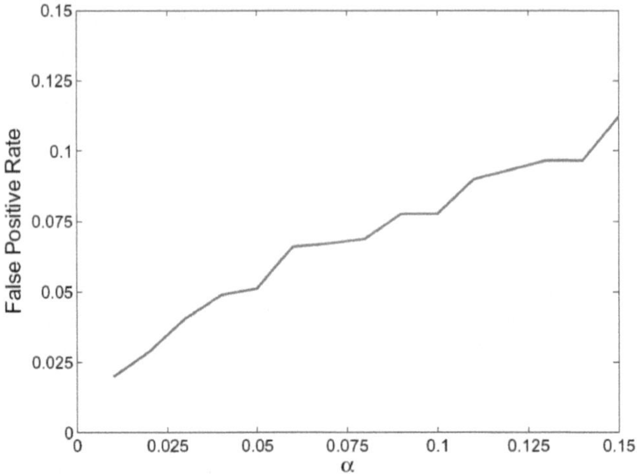

Fig. 5.5. Simulated false positive rate

The behavior of outgoing port scans provide a small scale model of an automated attack. The algorithm is applied to the detection of an outgoing port scan across a range of IP addresses, based on the ICMP 'ping' protocol.

There is a plethora of tools that aim to determine a system's weaknesses and determine the best method for an attack. The best known and documented tool is nmap. Nmap uses a variety of active probing techniques and changes the packet probe options to determine a host's operating system. Nmap offers its users the ability to randomize destination IPs and change the order of and timing between packets. This functionality can obscure the port scanning activity and thus fool intrusion detection systems. Other port scanners include queso, checkos, and SS. However, these tools do not provide all the capabilities of nmap and thus are not as popular. In this second case study we use the nmap to simulate the port scan attacks.

Signals and Antigen

Data from the monitored system are collected for the duration of a session. These values are transformed into signal values and written to a log file. Each signal value is a normalized real-number, based on a pre-defined maximum value. For this experiment the signals used are PAMPs, danger and safe signals. Inflammatory cytokines (Si4) do not feature as they are not relevant for this particular problem. PAMPs are represented as the number of 'unreachable destination' errors. When the port scan process scans multiple IP addresses indiscriminately, the number of these errors increases, and therefore is a positive sign of suspicious activity. Danger signals are represented as the number of outbound network packets per second. An increase in network traffic could imply anomalous behavior. The safe signals in this experiment are the inverse rate of change of network packets per second. This is based on the assumption that if the rate of sending network packets is highly variable, the machine is behaving suspiciously. None of these signals are enough on their own to indicate an anomaly. In these experiments the signals are used to detect the port scan (Table 5.3).

Table 5.3. Parameters for detecting port scanning

Description	Values
Tissue	Monitored data-set
Danger Signal 1	PAMP, unreachable destination
Danger Signal 2	The number of outbound network packets
Safe Signal	The inverse rate of changes of network packets
Source of danger	High network traffic
Antigen	Process ID
Immune response	Detection of port scan events
Localized response	Protect the scanned port

Detection Algorithm

In this case we apply the similar procedure as described in 5.4.1. Signals, though interesting, are inconsequential without antigen. To a DC, antigen is an element which is carried and presented to a T-cell, without regard for the structure of the antigen. Antigen is the data to be classified, and works well in the form of an identifier, be it an anomalous process ID or the ID of a data item. At this stage, minimal antigen processing is performed and the antigen presented is an identical copy of the antigen collected. Detection is performed through the correlation of antigen with signals. Following pseudocode shows the procedure of detection of port scan.

5.4.4 Experiments and Results

For the purposes of our analysis, we define a port scan as all anomalous messages sent from a single source during the trace period. To test the local warning system, we use a worm model developed in the packet level network simulator - Georgia Tech Network Simulator [20]. Because this is a real packet level simulator (which includes network router congestion, TCP latency, and other stochastic events) every run takes time. A small network was simulated with a realistic packet level model instead of using a partial analytical model.

```
PROCEDURE InitializationOfStatus()
  While (size of AB < a threshold)
    If (detector expresses NOSYMTOMS in network)
      create new antibody (ab) from log file
      add newly generated antibody (ab) to AB
      For (ab ∈ AB)
        clone and mutate ab to maximize affinity with ab
        add best n clones to AB

PROCEDURE Continuous_adaptation()
  antigen_count ← 0
  LOOP
    receive incoming network packets
    ag ← preprocess packets into antigen
    antigen_count ← antigen_count + 1
    IF(antigen_count = K)
      AG ← last K antigens
      Update_Population(AG)
      antigen_count ← 0
    compute degree of danger()
    WHILE(danger is high)
      compute temporal danger zone
      AG ← all packets in the danger zone
    FOREACH(ag ∈ AG)
      FOREACH(ab ∈ AB)
        compute affinity (ab,ag)
      high_aff ← highest affinity value
      IF(high_aff > a threshold)
        block the scanned port
```

Pseudocode 3. Initialization and Detection

Our motivation for using packet level network simulation is to validate our local early warning system and the results of our analytical models in an Internet-like setting. We used a hybrid network topology with clustering backbone and hierarchical sub-networks. The address space was populated with uniformly random hosts. Then, the vulnerable hosts were selected from the population in a uniform random way. Since the simulated network is small compared to the whole Internet, we set the time required to detect a worm as negligible. Thus, whenever a host in a monitored space became infected, it was detected and the number of infected hosts in the entire network was recorded.

We use three different monitored network sizes to see how fast we could detect random scan worms. We ran our experiment 15 times for each monitored network size and computed the average and variance. The result is shown in Table 5.4 (early warning time based on local victim information).

Table 5.4. Network simulator environments

Monitor Space	2^5	2^6	2^7
Avg Scan Rate	3.44	3.51	3.79
Avg Detection Time (%)	0.675	0.872	1.526
Variance	0.373	0.706	2.522

We can clearly see that using a smaller network, the detection time will be longer and the variance will be larger. Further, the detection time of the network simulation experiments matched the output of our analytical model when given identical input parameters.

Results and Analysis

With the information provided by the proposed detection algorithm, we can automatically take immediate and accurate responses that block victim port so as to effectively stop the propagation. The policy decision to block local ports can be accomplished entirely within the local network in realtime. This contrasts with global response strategies which require complex and time consuming coordination between CDC-like authority and Internet routers (Fig. 5.6).

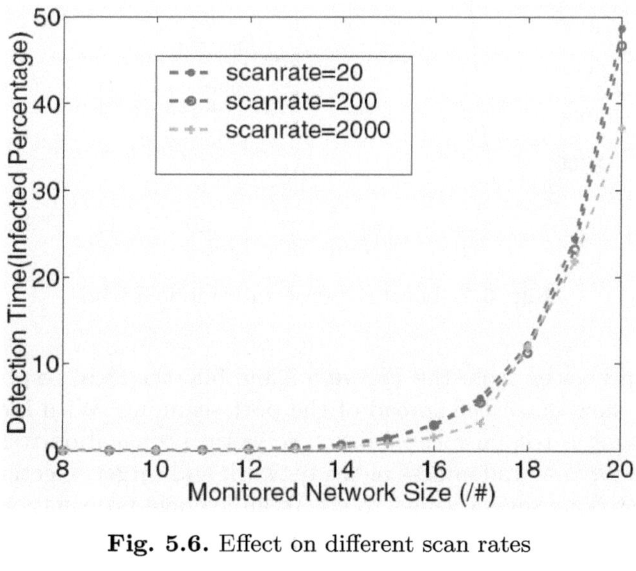

Fig. 5.6. Effect on different scan rates

Our local response is kind of Williamson's idea which limits the rate on the host by limiting the number of new connections [9]. Local response can be more effective, since local administrators know details about the victim machines and take more accurate action to block (not rate limit) the outgoing connections of victims (not all hosts) at that port (not all ports). This local response policy can be deployed on every host as [9]. We call this approach 'host level local response with random'. But it is probably more effective to only deploy on the edge router of LAN (network level local response) so that every LAN only need one position to deploy such a quarantine policy. We can also imagine a local response with preference because network level local response can only prevent victims from infecting outside network hosts but cannot prevent worm propagation within local network, which is worse with local preference scans.

For the network level response, the size of the local networks was set to be /27 in size. When a host is infected, all worm packets going out of the same /27 network were blocked, and vulnerable hosts inside could still be infected. The test was done for both uniform random and local preference scanning (scan policy: 0.25 in same /27, 0.25 in same /24, 0.25 in same /21, 0.25 in same /18). The result is shown in Figure 5.7 and 5.8.

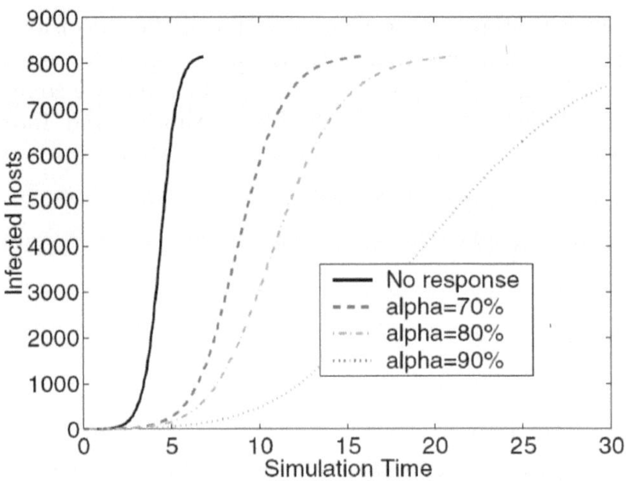

Fig. 5.7. Local response with random scan

As it can be seen from the Figure 5.7 and 5.8, the local response method was able to slow down the spread of the port scanning. With large numbers of hosts having a response capability, the worm propagation was slower. To compensate for the randomness of the network and target selection variation, each run was done several times to create an average propagation.

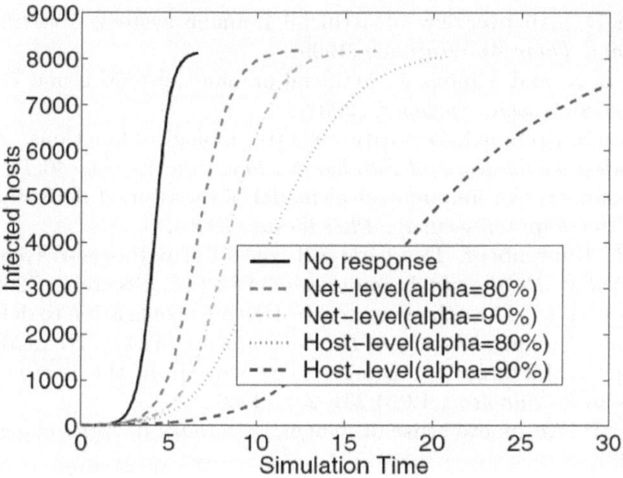

Fig. 5.8. Local response with preference scan

5.5 Conclusion

In this paper we have presented a detailed description of dendritic cells and the antigen presentation process from which an algorithm was abstracted. We have also discussed the relatively new Danger theory and given examples as to how Danger theory principles may be used in the field of AIS. We have also presented a worked example and prototype implementation based on this abstraction. The preliminary results are encouraging as both data orders produced low rates of false positive errors.

Many aspects of this algorithm remain unexplored such as the sensitivity of the parameters and scalability in terms of number of cells and number of input signals. For instance, we did not include any inflammatory cytokines in our examples due to data constraints. It would be interesting to explore their proposed amplifying effects on the other signals and on the behavior across a population of DCs. Additionally, an improved baseline for comparison must be developed in order to make serious claims regarding the effectiveness of the algorithm as an anomaly detector.

References

1. Forrest S. Hofmeyr S. Somayaji A. Computer immunology. *Communications of the ACM*, 40(10), (1997).
2. Shortman K. Caux C. Dendritic cell development: multiple pathways to nature's adjuvants. *Stem Cells*, 15:409–419, (1997).
3. Shortman K. Caux C. Dendritic cells and t lymphocytes: developmental and functional interactions. *Ciba Found Symp*, 204:130–141, (1997).

4. Dasgupta D. An overview of artificial immune systems. *Artificial Immune Systems and Their Applications*, (1999).
5. deCastro L. N. and Timmis J. Artificial immune systems a new computational intelligence approach. *Springer*, (2002).
6. Satthaporn S. EreminO. Dendritic cells (i) : biological functions. *Journal of the Royal College of Surgeons of Edinburgh : Scientific Review*, 46(1), (2001).
7. Steven Hofmeyr. An immunological model of distributed detection and its application to computer security. *PhD thesis*, (1999).
8. Martien L. Kapsenberg. Dendritic-cell control of pathogen-driven t-cell polarization. *Nature Reviews in Immunology*, 3:984–993, December (2003).
9. Williamson M. M. Throttling viruses. restricting propagation to defeat malicious mobile code. *Technical Report of HP Laboratories Bristol,*, HPL-2002-172, 2002.
10. Anderson C. Matzinger P. Danger: The view from the bottom of the cliff. *Seminars in Immunology*, 12(3):231–238, 2000.
11. Matzinger P. An innate sense of danger. *Seminars in Immunology*, 10(5):399–415, (1998).
12. Matzinger P. The danger model: A renewed sense of self. *Science*, 296:301–305, (2002).
13. G. J. et al. Randolph. Differentiation of monocytes into dendritic cells in a model of transendothelial trafficking see comments. *Science*, 282:480–483, (1998).
14. Percus Jerome K. Ora E. Alan S. Predicting the size of the t-cell receptor and antibody combining region from consideration of efficient self non-self discrimination. *Proceedings of the National Academy of Sciences of the United States of America*, 90:1691–1695, (1993).
15. Greensmith J. Aickelin U. Jamie T. Detecting danger. applying a novel immunological concept to intrusion detection systems. *6th International Conference in Design and Manufacture*, (2004).
16. Aickelin U. and Cayzer S. The danger theory and its application to artificial immune systems. *In proceedings of The First International Conference on Artificial Immune Systems (ICARIS 2002)*, pages 141–148, (2002).
17. Jamie T. Aickelin U. Towards a conceptual framework for innate immunity. *Proceedings of the 4th International Conference on Artificial Immune Systems*, (2004).
18. William R. Heath Jose A. Villadangos. No driving without a license. *Nature Immunology*, 6(2):125–126, (2005).
19. Jansen W. Intrusion detection with mobile agents. *Computer Communications*, 25(15), (2002).
20. Riley G. F. Sharif M. I. Lee W. Simulating internet worms. *Proceedings of IEEE/ACM MASCOTS 2004*, (2004).
21. Ki-Won Yeom and Ji-Hyung Park. An immune system inspired approach of collaborative intrusion detection system using mobile agents in wireless adhoc network. *Lecture Notes in Artificial Intelligence*, 3802:204–211, (2005).
22. Steinman R. Cohn Z. Identification of a novel cell type in peripheral lymphoid organs of mice. *The Journal of Experimental Medicine*, 137:1142–1162, (1973).

6

How to Efficiently Process Uncertainty within a Cyberinfrastructure without Sacrificing Privacy and Confidentiality

Luc Longpré and Vladik Kreinovich

Department of Computer Science, University of Texas at El Paso, 500 W. University, El Paso, TX 79968, USA, longpre@utep.edu, vladik@utep.edu

In this Chapter, we propose a simple solution to the problem of estimating uncertainty of the results of applying a black-box algorithm – without sacrificing privacy and confidentiality of the algorithm.

6.1 Cyberinfrastructure and Web Services

6.1.1 Practical Problem: Need to Combine Geographically Separate Computational Resources

In different knowledge domains in science and engineering, there is a large amount of data stored in different locations, and there are many software tools for processing this data, also implemented at different locations. Users may be interested in different information about this domain.

Sometimes, the information required by the user is already stored in *one of* the *databases.* For example, if we want to know the geological structure of a certain region in Texas, we can get his information from the geological map stored in Austin. In this case, all we need to do to get an appropriate response to the query is to get this data from the corresponding database.

In other cases, different pieces of the information requested by the user are *stored at different locations.* For example, if we are interested in the geological structure of the Rio Grande Region, then we need to combine data from the geological maps of Texas, New Mexico, and the Mexican state of Chihuahua. In such situations, a correct response to the user's query requires that we access these pieces of information from different databases located at different geographic notations.

In many other situations, the appropriate answer to the user's request requires that we not only collect the relevant data $x_1, \ldots, x_n$, but that we

L. Longpré and V. Kreinovich: *How to Efficiently Process Uncertainty within a Cyberinfrastructure without Sacrificing Privacy and Confidentiality*, Studies in Computational Intelligence (SCI) **57**, 155–173 (2007)
www.springerlink.com

also use some *data processing* algorithms $f(x_1, \ldots, x_n)$ to process this data. For example, if we are interested in the large-scale geological structure of a geographical region, we may also use the gravity measurements from the gravity databases. For that, we need special algorithms to transform the values of gravity at different locations into a map that describes how the density changes with location. The corresponding data processing programs often require a lot of computational resources; as a result, many such programs reside on computers located at supercomputer centers, i.e., on computers which are physically separated from the places where the data is stored.

The need to combine computational resources (data and programs) located at different geographic locations seriously complicates research.

6.1.2 Centralization of Computational Resources – Traditional Approach to Combining Computational Resources; Its Advantages and Limitations

Traditionally, a widely used way to make these computational resources more accessible was to move all these resources to a *central location*. For example, in the geosciences, the US Geological Survey (USGS) was trying to become a central depository of all relevant geophysical data. However, this centralization requires a large amount of efforts: data are presented in different formats, the existing programs use specific formats, etc. To make the central data depository efficient, it is necessary:

- to reformat all the data,
- to rewrite all the data processing programs – so that they become fully compatible with the selected formats and with each other,
- etc.

The amount of work that is needed for this reformatting and rewriting is so large that none of these central depositories really succeeded in becoming an easy-to-use centralized database.

6.1.3 Cyberinfrastructure – A More Efficient Approach to Combining Computational Resources

Cyberinfrastructure technique is a new approach that provides the users with the efficient way to submit requests without worrying about the geographic locations of different computational resources – and at the same time avoid centralization with its excessive workloads. The main idea behind this approach is that *we keep all (or at least most) the computational resources*

- *at their current locations,*
- *in their current formats.*

To expedite the use of these resources:

- we supplement the local computational resources with the "metadata", i.e., with the information about the formats, algorithms, etc.,
- we "wrap up" the programs and databases with auxiliary programs that provide data compatibility into *web services*,

and, in general, we provide a cyberinfrastructure that uses the metadata to automatically combine different computational resources.

For example, if a user is interested in using the gravity data to uncover the geological structure of the Rio Grande region, then the system should automatically:

- get the gravity data from the UTEP and USGS gravity databases,
- convert them to a single format (if necessary),
- forward this data to the program located at San Diego Supercomputer Center, and
- move the results back to the user.

This example is exactly what we are designing under the NSF-sponsored Cyberinfrastructure for the Geosciences (GEON) project; see, e.g., [1–3, 8, 9, 27, 37, 45, 47, 48, 58, 62, 63]. This is similar to what other cyberinfrastructure projects are trying to achieve.

6.1.4 What Is Cyberinfrastructure: The Official NSF Definition

According to the final report of the National Science Foundation (NSF) Blue Ribbon Advisory Panel on Cyberinfrastructure, "a new age has dawned in scientific and engineering research, pushed by continuing progress in computing, information, and communication technology, and pulled by the expanding complexity, scope, and scale of today's challenges. The capacity of this technology has crossed thresholds that now make possible a comprehensive 'cyberinfrastructure' on which to build new types of scientific and engineering knowledge environments and organizations and to pursue research in new ways and with increased efficacy.

Such environments and organizations, enabled by cyberinfrastructure, are increasingly required to address national and global priorities, such as understanding global climate change, protecting our natural environment, applying genomics-proteomics to human health, maintaining national security, mastering the world of nanotechnology, and predicting and protecting against natural and human disasters, as well as to address some of our most fundamental intellectual questions such as the formation of the universe and the fundamental character of matter."

6.1.5 Web Services: What They Do – A Brief Summary

In different knowledge domains, there is a large amount of data stored in different locations; algorithms for processing this data are also implemented at different locations. Web services – and, more generally, cyberinfrastructure – provide the users with an efficient way to submit requests without worrying about the geographic locations of different computational resources (databases and programs) – and avoid centralization with its excessive workloads [21]. Web services enable the user to receive the desired data $x_1, \ldots, x_n$ and the results $y = f(x_1, \ldots, x_n)$ of processing this data.

6.2 Processing Uncertainty Within a Cyberinfrastructure

6.2.1 Formulation of the problem

The data x_i usually come from measurements or from experts. Measurements are never 100% accurate; as a result, the measured values $\widetilde{x}_i$ are, in general, somewhat different from the actual (unknown) values x_i of the corresponding quantities. Experts can also only provide us with approximate values of the desired quantities.

As a result of this measurement or expert uncertainty, the result $\widetilde{y} = f(\widetilde{x}_1, \ldots, \widetilde{x}_n)$ of data processing is, in general, different from the actual value $y = f(x_1, \ldots, x_n)$ of the desired quantity. It is desirable to gauge this difference. To do that, we must have some information about the errors of direct measurements.

Bounds on the measurement errors. What do we know about the errors $\Delta x_i \stackrel{\text{def}}{=} \widetilde{x}_i - x_i$ of direct measurements? First, the manufacturer of the measuring instrument must supply us with an upper bound Δ_i on the measurement error. If no such upper bound is supplied, this means that no accuracy is guaranteed, and the corresponding "measuring instrument" is practically useless. In this case, once we performed a measurement and got a measurement result $\widetilde{x}_i$, we know that the actual (unknown) value x_i of the measured quantity belongs to the interval $\mathbf{x}_i = [\underline{x}_i, \overline{x}_i]$, where $\underline{x}_i = \widetilde{x}_i - \Delta_i$ and $\overline{x}_i = \widetilde{x}_i + \Delta_i$.

Case of probabilistic uncertainty. In many practical situations, we not only know the interval $[-\Delta_i, \Delta_i]$ of possible values of the measurement error; we also know the probability of different values Δx_i within this interval. This knowledge underlies the traditional engineering approach to estimating the error of indirect measurement, in which we assume that we know the probability distributions for measurement errors Δx_i.

These probabilities are often described by a normal distribution, so in standard engineering textbook on measurement, it is usually assumed that the distribution of Δx_i is normal, with 0 average and known standard deviation σ_i; see, e.g. [20, 46].

In general, we can determine the desired probabilities of different values of Δx_i by comparing the results of measuring with this instrument with the results of measuring the same quantity by a standard (much more accurate) measuring instrument. Since the standard measuring instrument is much more accurate than the one use, the difference between these two measurement results is practically equal to the measurement error; thus, the empirical distribution of this difference is close to the desired probability distribution for measurement error.

Case of interval uncertainty. There are two cases, however, when in practice, we do not determine the probabilities:

- First is the case of cutting-edge measurements, e.g., measurements in fundamental science. When a Hubble telescope detects the light from a distant galaxy, there is no "standard" (much more accurate) telescope floating nearby that we can use to calibrate the Hubble: the Hubble telescope is the best we have.
- The second case is the case of measurements on the shop floor. In this case, in principle, every sensor can be thoroughly calibrated, but sensor calibration is so costly – usually costing ten times more than the sensor itself – that manufacturers rarely do it.

In both cases, we have no information about the probabilities of Δx_i; the only information we have is the upper bound on the measurement error.

In this case, after we performed a measurement and got a measurement result $\widetilde{x}_i$, the only information that we have about the actual value x_i of the measured quantity is that it belongs to the interval $\mathbf{x}_i = [\widetilde{x}_i - \Delta_i, \widetilde{x}_i + \Delta_i]$. In such situations, the only information that we have about the (unknown) actual value of $y = f(x_1, \ldots, x_n)$ is that y belongs to the range $\mathbf{y} = [\underline{y}, \overline{y}]$ of the function f over the box $\mathbf{x}_1 \times \ldots \times \mathbf{x}_n$:

$$\mathbf{y} = [\underline{y}, \overline{y}] = \{f(x_1, \ldots, x_n) \,|\, x_1 \in \mathbf{x}_1, \ldots, x_n \in \mathbf{x}_n\}.$$

The process of computing this interval range based on the input intervals $\mathbf{x}_i$ is called *interval computations*; see, e.g., [24, 26].

Case of fuzzy uncertainty. Often, knowledge comes in terms of uncertain expert estimates. In the fuzzy case, to describe this uncertainty, for each value of estimation error Δx_i, we describe the degree $\mu_i(\Delta x_i)$ to which this value is possible.

For each degree of certainty α, we can determine the set of values of Δx_i that are possible with at least this degree of certainty – the *α-cut* $\{x \,|\, \mu(x) \geq \alpha\}$ of the original fuzzy set. In most cases, this α-cut is an interval.

Vice versa, if we know α-cuts for every α, then, for each object x, we can determine the degree of possibility that x belongs to the original fuzzy set [7, 30, 38, 40, 41]. A fuzzy set can be thus viewed as a nested family of its α-cuts.

So, if instead of a (crisp) interval $\mathbf{x}_i$ of possible values of the measured quantity, we have a fuzzy set $\mu_i(x)$ of possible values, then we can view this information as a family of nested intervals $\mathbf{x}_i(\alpha)$ – α-cuts of the given fuzzy sets.

6.2.2 Description of uncertainty: general formulas

In this chapter, we will only consider a typical situation in which the direct measurements and/or expert estimates are accurate enough, so that the resulting approximation errors Δx_i are small, and terms which are quadratic (or of higher order) in Δx_i can be safely neglected. In such situations, the dependence of the desired value $y = f(x_1, \ldots, x_n) = f(\widetilde{x}_1 - \Delta x_1, \ldots, \widetilde{x}_n - \Delta x_n)$ on Δx_i can be safely assumed to be linear.

Comment. There are practical situations when the accuracy of the direct measurements is not high enough, and hence, quadratic terms cannot be safely neglected (see, e.g., [26] and references therein). In this case, the problem of error estimation for indirect measurements becomes computationally difficult (NP-hard) even when the function $f(x_1, \ldots, x_n)$ is quadratic [34,60]. However, in most real-life situations, the possibility to ignore quadratic terms is a reasonable assumption, because, e.g., for an error of 1% its square is a negligible 0.01%.

When approximation errors are small, we can simplify the expression for $\Delta y = \widetilde{y} - y = f(\widetilde{x}_1, \ldots, \widetilde{x}_n) - f(x_1, \ldots, x_n)$ if we expand the function f in Taylor series around the point $(\widetilde{x}_1, \ldots, \widetilde{x}_n)$ and restrict ourselves only to linear terms in this expansion. As a result, we get the expression

$$\Delta y = c_1 \cdot \Delta x_1 + \ldots + c_n \cdot \Delta x_n, \tag{6.1}$$

where by c_i, we denoted the value of the partial derivative $\partial f / \partial x_i$ at the point $(\widetilde{x}_1, \ldots, \widetilde{x}_n)$:

$$c_i = \frac{\partial f}{\partial x_i}_{|(\widetilde{x}_1, \ldots, \widetilde{x}_n)}. \tag{6.2}$$

Case of probabilistic uncertainty. In the statistical setting, the desired measurement error Δy is a linear combination of independent Gaussian variables Δx_i. Therefore, Δy is also normally distributed, with 0 average and the standard deviation

$$\sigma = \sqrt{c_1^2 \cdot \sigma_1^2 + \ldots + c_n^2 \cdot \sigma_n^2}. \tag{6.3}$$

Comment. A similar formula holds if we *do not* assume that Δx_i are normally distributed: it is sufficient to assume that they are independent variables with 0 average and known standard deviations σ_i.

Case of interval uncertainty. In the interval setting, we do not know the probability of different errors Δx_i; instead, we only know that $|\Delta x_i| \le \Delta_i$. In this case, the sum (6.1) attains its largest possible value if each term $c_i \cdot \Delta x_i$ in this sum attains the largest possible value:

- If $c_i \ge 0$, then this term is a monotonically non-decreasing function of Δx_i, so it attains its largest value at the largest possible value $\Delta x_i = \Delta_i$; the corresponding largest value of this term is $c_i \cdot \Delta_i$.

- If $c_i < 0$, then this term is a decreasing function of Δx_i, so it attains its largest value at the smallest possible value $\Delta x_i = -\Delta_i$; the corresponding largest value of this term is $-c_i \cdot \Delta_i = |c_i| \cdot \Delta_i$.

In both cases, the largest possible value of this term is $|c_i| \cdot \Delta_i$, so, the largest possible value of the sum Δy is

$$\Delta = |c_1| \cdot \Delta_1 + \ldots + |c_n| \cdot \Delta_n. \tag{6.4}$$

Similarly, the smallest possible value of Δy is $-\Delta$.

Hence, the interval of possible values of Δy is $[-\Delta, \Delta]$, with Δ defined by the formula (6.4).

Case of fuzzy uncertainty. We have already mentioned that if instead of a (crisp) interval $\mathbf{x}_i$ of possible values of the measured quantity, we have a fuzzy set $\mu_i(x)$ of possible values, then we can view this information as a family of nested intervals $\mathbf{x}_i(\alpha)$ – α-cuts of the given fuzzy sets.

Our objective is then to compute the fuzzy number corresponding to this the desired value $y = f(x_1, \ldots, x_n)$. In this case, for each level α, to compute the α-cut of this fuzzy number, we can apply interval computations to the α-cuts $\mathbf{x}_i(\alpha)$ of the corresponding fuzzy sets. The resulting nested intervals form the desired fuzzy set for y.

So, e.g., if we want to describe 10 different levels of uncertainty, then we must solve 10 interval computation problems – i.e., apply the formula (6.4) 10 times.

In many practical situations, there is no need to perform 10 computations. For example, it is often reasonable to assume that all the membership functions $\mu_i(x_i)$ have the same shape and only differ by a scaling parameter, i.e., all have the form $\mu_i(\Delta x_i) = \mu_0(\Delta x_i/\Delta_i)$ for some fixed function $\mu_0(x)$ (e.g., triangular or Gaussian). In this case, as it is well known [14–16,25], the membership function for Δy has a similar form $\mu_0(\Delta y/\Delta)$, where Δ is determined by the formula (6.4).

Comment. These formulas correspond to the case when in the *extension principle* that describe how the uncertainty in Δx_i transforms into the uncertainty in Δy, we interpret "and" as min, i.e., if we consider

$$\mu(\Delta y) = \max_{\Delta x_1, \ldots, \Delta x_n} \min(\mu_1(\Delta x_1), \ldots, \mu_n(\Delta x_n)), \tag{6.5}$$

where maximum is taken over all the values $\Delta x_1, \ldots, \Delta x_n$ for which the expression (6.1) for Δy leads to the given value of Δy. In general, we can use a different t-norm to combine the values $\mu_i(\Delta x_i)$. For example, we may use the product and describe the resulting membership function as

$$\mu(\Delta y) = \max_{\Delta x_1, \ldots, \Delta x_n} \mu_1(\Delta x_1) \cdot \ldots \cdot \mu_n(\Delta x_n). \tag{6.6}$$

In this case, if we assume that all the membership functions $\mu_i(\Delta x_i)$ are Gaussian, i.e., have the form $\mu_i(\Delta x_i) = \mu_0(\Delta x_i/\sigma_i)$, where $\mu_0(z) = \exp(-z^2)$, then the resulting membership function for Δy is also Gaussian $\mu(\Delta y) = \mu_0(\Delta y/\sigma)$, where σ is determined by the formula (6.3); see, e.g., [14–16,25].

6.2.3 Error Estimation for the Results of Data Processing: A Precise Computational Formulation of the Problem

As a result of the above analysis, we get the following explicit formulation of the problem: given a function $f(x_1, \ldots, x_n)$, n numbers $\widetilde{x}_1, \ldots, \widetilde{x}_n$, and n positive numbers $\sigma_1, \ldots, \sigma_n$ (or $\Delta_1, \ldots, \Delta_n$), compute the corresponding expression (6.3) or (6.4).

Let us describe how this problem is solved now.

6.2.4 How This Problem Is Solved Now

Textbook case: the function f is given by its analytical expression. If the function f is given by its analytical expression, then we can simply explicitly differentiate it, and get an explicit expression for (6.3) and (6.4). This is the case which is typically analyzed in textbooks on measurement theory; see, e.g., [20, 46].

A more complex case: automatic differentiation. In many practical cases, we do not have an explicit analytical expression, we only have an *algorithm* for computing the function $f(x_1, \ldots, x_n)$, an algorithm which is too complicated to be expressed as an analytical expression.

When this algorithm is presented in one of the standard programming languages such as Fortran or C, we can let the compute perform an explicit differentiation; for that, we can use one of the existing automatic differentiation tools (see, e.g., [6, 22]). These tools analyze the code of the program for computing $f(x_1, \ldots, x_n)$ and, as they perform their analysis, they produce the "differentiation code", i.e., a program that computes the partial derivatives c_i.

Once we know an algorithm that computes f in time T, automatic differentiation (AD) enables us to compute all partial derivatives in time $\leq 3T$, hence we can compute σ or Δ in time $O(T + n)$.

6.3 Need for Privacy Makes the Problem More Complex

Privacy situation: description. In cyberinfrastructure, the owners of the program f may not want to disclose its code; instead, they may only allow to use f as a black box.

Real world example. For example, to find places where oil can be found, it is important to know the structure of the Earth crust at different locations. One of the main techniques for determining this structure is the use of seismic data: we use the measured travel-times of the natural (or artificial) seismic signals as they go from their source to their on-surface destination, and then we solve the corresponding *inverse problem* to find the Earth structure; see, e.g., [2–5, 11, 13, 23, 44, 64].

Because of the importance of the seismic inverse problem, oil companies are heavily investing in developing algorithms for solving such problems. This is a very competitive area of research, so the algorithms are kept confidential. Since most major algorithms have similar efficiency and accuracy, a company is willing to allow other users – researchers or competitors from other companies – to actually use their code. However, companies are very reluctant to let the users have actual access to their code. The reason for this reluctance is that different companies have achieved the similar level of efficiency and accuracy by using different ideas and programming improvements. As a result, a company that unilaterally discloses its code would thus let competitors use its ideas (and thus, improve their code) without getting any benefits in return.

Thus, the companies are allowing to use their code, but only as a black box.

What was known before. The problem of preserving the code's privacy and confidentiality is a particular case of a general problem of privacy preservation in data processing; see, e.g., [10, 12, 17–19, 28, 29, 35, 39, 42, 43, 49–57, 61]. At present, privacy-preserving algorithms are mainly concerned with privacy of *data*; privacy of *code* is a problem for which few results are known.

Related difficulty. If we do not know the code of f, then we cannot apply AD to compute all n partial derivatives $c_i = \dfrac{\partial f}{\partial x_i}$.

A straightforward method of solving this problem: numerical differentiation. The most straightforward algorithm for solving this problem is to compute the derivatives c_i one-by-one, and then use the corresponding formula (6.3) or (6.4) to compute the desired σ. To compute the i-th partial derivative, we change the i-th input x_i to $\widetilde{x}_i + h_i$ for some h_i, and leave other inputs unchanged, i.e., we take $\delta_i = h_i$ for this i and $\delta_j = 0$ for all $j \neq i$. Then, we estimate c_i as

$$c_i = \frac{f\left(\widetilde{x}_1, \ldots, \widetilde{x}_{i-1}, \widetilde{x}_i + h_i, \widetilde{x}_{i+1}, \ldots, \widetilde{x}_n\right) - \widetilde{y}}{h_i}. \tag{6.7}$$

This algorithm is called *numerical differentiation.*

We want the change h_i to be small (so that quadratic terms can be neglected); we already know that changes of the order σ_i are small. So, it is natural to take $h_i = \sigma_i$ (or, correspondingly, $h_i = \Delta_i$). In other words, to compute c_i, we use the following values: $\delta_1 = \ldots = \delta_{i-1} = 0$, $\delta_i = \sigma_i$ (or $\delta_i = \Delta_i$), $\delta_{i+1} = \ldots = \delta_n = 0$.

Problem: sometimes, numerical differentiation takes too long. Very often, the program f requires a reasonable time to compute (e.g., in the geological applications, computing f may involve solving an inverse problem). In this case, applying the function f is the most time-consuming part of this algorithm. So, the total time that it takes us to compute σ or Δ is (approximately) equal to the running time T for the program f multiplied by the number of times N_f that we call the program f.

For numerical differentiation, $N_f = n$ (we call f n times to compute n partial derivatives). Hence, if the program f takes a long time to compute, and n is huge, then the resulting time $T \cdot n$ (which is $\gg T + n$) may be too long. For example, if we are determining some parameters of an oil well from the geophysical measurements, we may get n in the thousands, and T in minutes. In this case, $T \cdot n$ may take several weeks. This may be OK for a single measurement, but too long if we want more on-line results.

6.4 Solution for Statistical Setting: Monte-Carlo Simulations

Monte-Carlo simulations: main idea. In the statistical setting, we can use straightforward (Monte-Carlo) simulation, and drastically save the computation time. In this approach, we use a computer-based random number generator to simulate the normally distributed error. A standard normal random number generator usually produces a normal distribution with 0 average and standard deviation 1. So, to simulate a distribution with a standard deviation σ_i, we multiply the result α_i of the standard random number generator by σ_i. In other words, we take $\delta_i = \sigma_i \cdot \alpha_i$.

As a result of N Monte-Carlo simulations, we get N values $c^{(1)} = \mathbf{c} \cdot \boldsymbol{\delta}^{(1)}, \ldots, c^{(N)} = \mathbf{c} \cdot \boldsymbol{\delta}^{(N)}$ which are normally distributed with the desired standard deviation σ. So, we can determine σ by using the standard statistical estimate

$$\sigma = \sqrt{\frac{1}{N-1} \cdot \sum_{k=1}^{N} \left(c^{(k)}\right)^2}. \tag{6.8}$$

Computation time required for Monte-Carlo simulation. The relative error of the above statistical estimate depends only on N (as $\approx 1/\sqrt{N}$), and not on the number of variables n. Therefore, the number N_f of calls to f that is needed to achieve a given accuracy does not depend on the number of variables at all.

The error of the above algorithm is asymptotically normally distributed, with a standard deviation $\sigma_e \sim \sigma/\sqrt{2N}$. Thus, if we use a "two sigma" bound, we conclude that with probability 95%, this algorithm leads to an estimate for σ which differs from the actual value of σ by $\le 2\sigma_e = 2\sigma/\sqrt{2N}$.

This is an error with which we estimate the error of indirect measurement; we do not need too much accuracy in this estimation, because, e.g., in real life, we say that an error is ±10% or ±20%, but *not* that the error is, say, ±11.8%. Therefore, in estimating the error of indirect measurements, it is sufficient to estimate the characteristics of this error with a relative accuracy of, say, 20%.

For the above "two sigma" estimate, this means that we need to select the smallest N for which $2\sigma_e = 2\sigma/\sqrt{2N} \le 0.2 \cdot \sigma$, i.e., to select $N_f = N = 50$.

In many practical situations, it is sufficient to have a standard deviation of 20% (i.e., to have a "two sigma" guarantee of 40%). In this case, we need only $N = 13$ calls to f.

On the other hand, if we want to guarantee 20% accuracy in 99.9% cases, which correspond to "three sigma", we must use N for which $3\sigma_e = 3 \cdot \sigma/\sqrt{2N} \leq 0.2 \cdot \sigma$, i.e., we must select $N_f = N = 113$, etc.

For $n \approx 10^3$, all these values of N_f are much smaller than $N_f = n$ required for numerical differentiation.

So, if we have to choose between the (deterministic) numerical differentiation and the randomized Monte-Carlo algorithm, we must select:

- a deterministic algorithm when the number of variables n satisfies the inequality $n \leq N_0$ (where $N_0 \approx 50$), and
- a randomized method if $n \geq N_0$.

Additional advantage: parallelization. In Monte-Carlo algorithm, we need 50 calls to f. If each call requires a minute, the resulting time takes about an hour, which may be too long for on-line results. Fortunately, different calls to the function f are independent on each other, so we can run all the simulations in parallel.

The more processors we have, the less time the resulting computation will take. If we have as many processors as the required number of calls, then the time needed to estimate the error of indirect measurement becomes equal to the time of a single call, i.e., to the time necessary to compute the result $\widetilde{y}$ of this indirect measurement. Thus, if we have enough processors working in parallel, we can compute the result of the indirect measurement *and* estimate its error during the same time that it normally takes just to compute the result.

In particular, if the result $\widetilde{y}$ of indirect measurement can be computed in real time, we can estimate the error of this result in real time as well.

6.5 Solution for Interval and Fuzzy Setting: New Method Based on Cauchy Distribution

Can we use a similar idea in the interval setting? Since Monte-Carlo simulation speeds up computations, it is desirable to use a similar technique in interval setting as well.

There is a problem here. In the interval setting, we do not know the exact distribution, we may have different probability distributions – as long as they are located within the corresponding intervals. If we only use one of these distributions for simulations, there is no guarantee that the results will be valid for other distributions as well.

In principle, we could repeat simulations for several different distributions, but this repetition would drastically increase the simulation time and thus, eliminate the advantages of simulation as opposed to numerical differentiation.

Yes, we can. Luckily, there is a mathematical trick that enables us to use Monte-Carlo simulation in interval setting as well. This trick is based on using *Cauchy distribution* – i.e., probability distributions with the probability density

$$\rho(z) = \frac{\Delta}{\pi \cdot (z^2 + \Delta^2)}; \tag{6.9}$$

the value Δ is called the *scale parameter* of this distribution, or simply a *parameter*, for short.

Cauchy distribution has the following property that we will use: if $z_1, \ldots, z_n$ are independent random variables, and each of z_i is distributed according to the Cauchy law with parameter Δ_i, then their linear combination $z = c_1 \cdot z_1 + \ldots + c_n \cdot z_n$ is also distributed according to a Cauchy law, with a scale parameter $\Delta = |c_1| \cdot \Delta_1 + \ldots + |c_n| \cdot \Delta_n$.

Therefore, if we take random variables δ_i which are Cauchy distributed with parameters Δ_i, then the value

$$c = f(\widetilde{x}_1 + \delta_1, \ldots, \widetilde{x}_n + \delta_n) - f(\widetilde{x}_1, \ldots, \widetilde{x}_n) = c_1 \cdot \delta_1 + \ldots + c_n \cdot \delta_n \tag{6.10}$$

is Cauchy distributed with the desired parameter (6.4). So, repeating this experiment N times, we get N values $c^{(1)}, \ldots, c^{(N)}$ which are Cauchy distributed with the unknown parameter, and from them we can estimate Δ.

The bigger N, the better estimates we get.

There are two questions to be solved:

- how to simulate the Cauchy distribution;
- how to estimate the parameter Δ of this distribution from a finite sample.

Simulation can be based on the functional transformation of uniformly distributed sample values:

$$\delta_i = \Delta_i \cdot \tan(\pi \cdot (r_i - 0.5)), \tag{6.11}$$

where r_i is uniformly distributed on the interval $[0, 1]$.

In order to estimate σ, we can apply the Maximum Likelihood Method $\rho(d^1) \cdot \rho(d^2) \cdot \ldots \cdot \rho(d^n) \to \max$, where $\rho(z)$ is a Cauchy distribution density with the unknown Δ. When we substitute the above-given formula for $\rho(z)$ and equate the derivative of the product with respect to Δ to 0 (since it is a maximum), we get an equation

$$\frac{1}{1 + \left(\frac{c^{(1)}}{\Delta}\right)^2} + \ldots + \frac{1}{1 + \left(\frac{c^{(N)}}{\Delta}\right)^2} = \frac{N}{2}. \tag{6.12}$$

The left-hand side of (6.12) is an increasing function that is equal to $0 (< N/2)$ for $\Delta = 0$ and $> N/2$ for $\Delta = \max\left|c^{(k)}\right|$; therefore the solution to the equation (6.12) can be found by applying a bisection method to the interval $\left[0, \max\left|c^{(k)}\right|\right]$.

It is important to mention that we assumed that the function f is reasonably linear within the box

$$[\widetilde{x}_1 - \Delta_1, \widetilde{x}_1 + \Delta_1] \times \ldots \times [\widetilde{x}_n - \Delta_n, \widetilde{x}_n + \Delta_n]. \tag{6.13}$$

However, the simulated values δ_i may be outside the box. When we get such values, we do not use the function f for them, we use a normalized function that is equal to f within the box, and that is extended linearly for all other values (we will see, in the description of an algorithm, how this is done).

As a result, we arrive at the following algorithm (described, for a somewhat different problem, in [32, 33, 36, 59]):

Algorithm.

- Apply f to the results of direct measurements: $\widetilde{y} := f(\widetilde{x}_1, \ldots, \widetilde{x}_n)$;
- For $k = 1, 2, \ldots, N$, repeat the following:
  - use the standard random number generator to compute n numbers $r_i^{(k)}$, $i = 1, 2, \ldots, n$, that are uniformly distributed on the interval $[0, 1]$;
  - compute Cauchy distributed values $c_i^{(k)} := \tan(\pi \cdot (r_i^{(k)} - 0.5))$;
  - compute the largest value of $|c_i^{(k)}|$ so that we will be able to normalize the simulated measurement errors and apply f to the values that are within the box of possible values: $K := \max_i |c_i^{(k)}|$;
  - compute the simulated measurement errors $\delta_i^{(k)} := \Delta_i \cdot c_i^{(k)}/K$;
  - compute the simulated measurement results $x_i^{(k)} := \widetilde{x}_i + \delta_i^{(k)}$;
  - apply the program f to the simulated measurement results and compute the simulated error of the indirect measurement:

$$c^{(k)} := K \cdot \left(f\left(x_1^{(k)}, \ldots, x_n^{(k)}\right) - \widetilde{y} \right);$$

- Compute Δ by applying the bisection method to solve the equation (6.12).

When is this randomized algorithm better than deterministic numerical differentiation? To determine the parameter Δ, we use the maximum likelihood method. It is known that the error of this method is asymptotically normally distributed, with 0 average and standard deviation $1/\sqrt{N \cdot I}$, where I is Fisher's information:

$$I = \int_{-\infty}^{\infty} \frac{1}{\rho} \cdot \left(\frac{\partial \rho}{\partial \Delta}\right)^2 \,\mathrm{d}z.$$

For Cauchy probability density $\rho(z)$, we have $I = 1/(2\Delta^2)$, so the error of the above randomized algorithm is asymptotically normally distributed, with a standard deviation $\sigma_e \sim \Delta \cdot \sqrt{2/N}$. Thus, if we use a "two sigma" bound, we conclude that with probability 95%, this algorithm leads to an estimate for Δ which differs from the actual value of Δ by $\leq 2\sigma_e = 2\Delta \cdot \sqrt{2/N}$. So, if we want to achieve a 20% accuracy in the error estimation, we must use the smallest N for which $2\sigma_e = 2\Delta \cdot \sqrt{2/N} \le 0.2 \cdot \Delta$, i.e., to select $N_f = N = 200$.

When it is sufficient to have a standard deviation of 20% (i.e., to have a "two sigma" guarantee of 40%), we need only $N = 50$ calls to f. For $n \approx 10^3$, both values N_f are much smaller than $N_f = n$ required for numerical differentiation.

So, if we have to choose between the (deterministic) numerical differentiation and the randomized Monte-Carlo algorithm, we must select:

- a deterministic algorithm when the number of variables n satisfies the inequality $n \le N_0$ (where $N_0 \approx 200$), and
- a randomized algorithm if $n \ge N_0$.

Comment. If we use fewer than N_0 simulations, then we still get an approximate value of the range, but with worse accuracy – and the accuracy can be easily computed by using the above formulas.

This algorithm is naturally parallelizable. Similarly to the Monte-Carlo algorithm for statistical setting, we can run all N simulations in parallel and thus, speed up the computations.

Conclusion. When we know the code for f, then we can use AD and compute Δ and σ in time $O(T + n)$. If the owner of the program f only allows to use it as a black box, then we cannot use AD any more. In principle, we can compute each of n derivatives $\partial f/\partial x_i$ by numerical differentiation, but this would require computation time $T \cdot n \gg T + n$.

For probabilistic uncertainty, one can use Monte-Carlo simulations and compute σ in time $O(T) \ll T \cdot n$. We have shown that for interval uncertainty, we can also compute Δ in time $O(T)$ by using an artificial Monte-Carlo simulations in which each Δx_i is Cauchy distributed with parameter Δ_i – then simulated Δy is Cauchy distributed with the desired parameter Δ.

Remark: the problem of non-linearity. In the above text, we assumed that the intervals $\mathbf{x}_i$ are narrow. In this case, terms quadratic in Δx_i are negligible, and so, we can safely assume that the desired function $f(x_1, \ldots, x_n)$ is linear on the box

$$\mathbf{x}_1 \times \ldots \times \mathbf{x}_n.$$

In practice, some intervals $\mathbf{x}_i$ may be wide, so even when restricted to the box, the function $f(x_1, \ldots, x_n)$ is non-linear. What can we do in this case?

Usually, experts (e.g., designers of the corresponding technical system) know for which variables x_i, the dependence is non-linear. For each of these variables, we can *bisect* the corresponding interval $[\underline{x}_i, \overline{x}_i]$ into two smaller subintervals – for which the dependence is approximately linear. Then, we estimate the range of the function f separately on each of the resulting subboxes, and take the union of these two ranges as the range over the entire box.

If one bisection is not enough and the dependence of f on x_i is non-linear over one or several subboxes, we can bisect these boxes again, etc.

This bisection idea has been successfully used in interval computations; see, e.g., [24,26].

6.6 Summary

In different knowledge domains, there is a large amount of data stored in different locations; algorithms for processing this data are also implemented at different locations. Web services – and, more generally, cyberinfrastructure – provide the users with an efficient way to submit requests without worrying about the geographic locations of different computational resources (databases and programs) – and avoid centralization with its excessive workloads [21]. Web services enable the user to receive the desired data $x_1, \ldots, x_n$ and the results $y = f(x_1, \ldots, x_n)$ of processing this data.

The data x_i usually come from measurements or from experts. Measurements are never 100% accurate; as a result, the measured values $\widetilde{x}_i$ are, in general, somewhat different from the actual (unknown) values x_i of the corresponding quantities. Experts can also only provide us with approximate values of the desired quantities.

As a result of this measurement or expert uncertainty, the result $\widetilde{y} = f(\widetilde{x}_1, \ldots, \widetilde{x}_n)$ of data processing is, in general, different from the actual value $y = f(x_1, \ldots, x_n)$ of the desired quantity. It is desirable to gauge this difference.

Traditional methods for estimating the resulting uncertainty in y are based on the assumption that we know the code of the function f. In cyberinfrastructure, owners of the program f may not want to disclose its code; instead, they may only allow to use f as a black box. In this case, traditional techniques are not applicable.

There exist techniques for processing uncertainty under such a "black-box" situation, but these techniques require much longer computation time. In this chapter, we describe new Monte-Carlo-type techniques that process uncertainty in such privacy-protecting black-box situations and that require the same amount of computation time as the traditional non-privacy-protecting techniques.

Acknowledgments

This work was supported in part by NASA under cooperative agreement NCC5-209, NSF grants EAR-0225670 and DMS-0532645, Star Award from the University of Texas System, and Texas Department of Transportation grant No. 0-5453. The authors are thankful to the anonymous referees for valuable suggestions.

References

1. Aguiar MS, Dimuro GP, Costa ACR, Silva RKS, Costa FA, Kreinovich V (2004) The multi-layered interval categorizer tesselation-based model, In: Iochpe C, Câmara G (eds), IFIP WG2.6 Proceedings of the 6th Brazilian Symposium on Geoinformatics Geoinfo'2004, Campos do Jordão, Brazil, November 22–24, 2004, pp. 437–454. ISBN 3901882200
2. Aldouri R, Keller GR, Gates A, Rasillo J, Salayandia L, Kreinovich V, Seeley J, Taylor P, Holloway S (2004) GEON: Geophysical data add the 3rd dimension in geospatial studies. In: Proceedings of the ESRI International User Conference 2004, San Diego, California, August 9–13, 2004, Paper 1898
3. Averill MG, Miller KC, Keller GR, Kreinovich V, Araiza R, Starks SA (2005) Using expert knowledge in solving the seismic inverse problem, In: Proceedings of the 24nd International Conference of the North American Fuzzy Information Processing Society NAFIPS'2005, Ann Arbor, Michigan, June 22–25, 2005, pp. 310–314
4. Averill MG, Miller KC, Keller GR, Kreinovich V, Araiza R, Starks SA (2007) Using Expert Knowledge in Solving the Seismic Inverse Problem. International Journal of Approximate Reasoning, to appear
5. Bardossy G, Fodor J (2004) Evaluation of uncertainties and risks in geology. Springer Verlag, Berlin
6. Berz M, Bischof C, Corliss G, Griewank A (1996), Computational differentiation: techniques, applications, and tools. SIAM, Philadelphia
7. Bojadziev G, Bojadziev M (1995) Fuzzy sets, fuzzy logic, applications. World Scientific, Singapore
8. Ceberio M, Ferson S, Kreinovich V, Chopra S, Xiang G, Murguia A, Santillan J (2006) How to take into account dependence between the inputs: from interval computations to constraint-related set computations, with potential applications to nuclear safety, bio- and geosciences. In: Proceedings of the Second International Workshop on Reliable Engineering Computing, Savannah, Georgia, February 22–24, 2006, pp. 127–154
9. Ceberio M, Kreinovich V, Chopra S, Ludäscher B (2005) Taylor model-type techniques for handling uncertainty in expert systems, with potential applications to geoinformatics. In: Proceedings of the 17th World Congress of the International Association for Mathematics and Computers in Simulation IMACS'2005, Paris, France, July 11–15, 2005
10. Dalenius T (1986) Finding a needle in a haystack – or identifying anonymous census record. Journal of Official Statistics 2(3):329–336
11. Demicco R, Klir G, eds (2003) Fuzzy logic in geology. Academic Press
12. Denning DERD (1982) Cryptography and data security. Addison-Wesley, Reading, Massachusetts
13. Doser DI, Crain KD, Baker MR, Kreinovich V, Gerstenberger MC (1998) Estimating uncertainties for geophysical tomography. Reliable Computing 4(3):241–268
14. Dubois D, Prade H (1978) Operations on fuzzy numbers. International Journal of Systems Science 9:613–626
15. Dubois D, Prade H (1979) Fuzzy real algebra: some results. Fuzzy Sets and Systems 2:327–348
16. Dubois D, Prade H (1980) Fuzzy sets and systems: theory and applications. Academic Press, New York, London

17. Duncan G, Lambert D (1987) The risk of disclosure for microdata. In: *Proc. of the Bureau of the Census Third Annual Research Conference*, Bureau of the Census, Washington, DC, 263–274
18. Duncan G, Mukherjee S (1991) Microdata disclosure limitation in statistical databases: query size and random sample query control. In: Prof. 1991 IEEE Symposium on Research in Security and Privacy, Oakland, CA, May 20–22, 1991
19. Fellegi I (1972) On the question of statistical confidentiality. Journal of the American Statistical Association 7–18
20. Fuller WA (1987) Measurement error models. J. Wiley & Sons, New York
21. Gates A, Kreinovich V, Longpré L, Pinheiro da Silva P, Keller GR (2006) Towards secure cyberinfrastructure for sharing border information. In: Proceedings of the Lineae Terrarum: International Border Conference, El Paso, Las Cruces, and Cd. Juárez, March 27–30, 2006
22. Griewank A (2000) Evaluating derivatives: principles and techniques of algorithmic differentiation. SIAM, Philadelphia
23. Hole JA (1992) Nonlinear high-resolution three-dimensional seismic travel time tomography. J. Geophysical Research 97(B5):6553–6562.
24. Jaulin L, Kieffer M, Didrit O, Walter E (2001) Applied interval analysis. Springer Verlag, London
25. Kauffman A, Gupta MM (1985) Introduction to fuzzy arithmetic: theory and applications. Van Nostrand, New York
26. Kearfott RB, Kreinovich V, eds (1996) Applications of interval computations. Kluwer, Dordrecht
27. Keller GR, Hildenbrand TG, Kucks R, Webring M, Briesacher A, Rujawitz K, Hittleman AM, Roman DJ, Winester D, Aldouri R, Seeley J, Rasillo J, Torres T, Hinze WJ, Gates A, Kreinovich V, Salayandia L (2006) A community effort to construct a gravity database for the United States and an associated Web portal. In: Sinha AK (ed), Geoinformatics: Data to Knowledge, Geological Society of America Publ., Boulder, Colorado, pp. 21–34
28. Kim J (1986) A method for limiting disclosure of microdata based on random noise and transformation. In: Proceedings of the Section on Survey Research Methods of the American Statistical Association 370–374
29. Kirkendall N et al. (1994) Report on Statistical Disclosure Limitations Methodology. Office of Management and Budget, Washington, DC, Statistical Policy Working Paper No. 22
30. Klir G, Yuan B (1995) Fuzzy sets and fuzzy logic. Prentice Hall, New Jersey
31. Kreinovich V, Beck J, Ferregut C, Sanchez A, Keller GR, Averill M, Starks SA (2007) Monte-Carlo-type techniques for processing interval uncertainty, and their potential engineering applications. Reliable Computing 13(1):25–69.
32. Kreinovich V, Bernat A, Villa E, Mariscal Y (1991) Parallel computers estimate errors caused by imprecise data. Interval Computations (2):21–46
33. Kreinovich V, Ferson S (2004) A new cauchy-based black-box technique for uncertainty in risk analysis. Reliability Engineering and Systems Safety 85 (1–3):267–279
34. Kreinovich V, Lakeyev A, Rohn J, Kahl P (1998) Computational complexity and feasibility of data processing and interval computations. Kluwer, Dordrecht
35. Kreinovich V, Longpré L, Starks SA, Xiang G, Beck J, Kandathi K, Nayak A, Ferson S, Hajagos J (2007) Interval versions of statistical techniques, with

applications to environmental analysis, bioinformatics, and privacy in statistical databases. Journal of Computational and Applied Mathematics 199(2):418–423
36. Kreinovich V, Pavlovich MI (1985) Error estimate of the result of indirect measurements by using a calculational experiment. Measurement Techniques 28(3):201–205
37. Longpré L, Kreinovich V, Freudenthal E, Ceberio M, Modave F, Baijal N, Chen W, Chirayath V, Xiang G, Vargas JI (2005) Privacy: protecting, processing, and measuring loss. In: Abstracts of the 2005 South Central Information Security Symposium SCISS'05, Austin, Texas, April 30, 2005, p. 2
38. Moore RE, Lodwick WA (2003) Interval analysis and fuzzy set theory. Fuzzy Sets and Systems 135(1):5–9
39. Morgenstern M (1987) Security and inference in multilevel database and knowledge base systems. In: Proc. of the ACM SIGMOD Conference 357–373
40. Nguyen HT, Kreinovich V (1996) Nested intervals and sets: concepts, relations to fuzzy sets, and applications, In: Kearfott RB, Kreinovich V (eds) Applications of interval computations. Kluwer, Dordrecht, pp. 245–290.
41. Nguyen HT, Walker EA (2005) First course in fuzzy logic. CRC Press, Boca Raton, Florida
42. Office of Technology Assessment (1993) Protecting privacy in computerized medical information. US Government Printing Office, Washington, DC
43. Palley M, Siminoff J (1986) Regression methodology based disclosure of a statistical database. In: Proceedings of the Section on Survey Research Methods of the American Statistical Association 382–387
44. Parker RL (1994) Geophysical inverse theory. Princeton University Press, Princeton, New Jersey
45. Platon E, Tupelly K, Kreinovich V, Starks SA, Villaverde K (2005) Exact bounds for interval and fuzzy functions under monotonicity constraints, with potential applications to biostratigraphy. In: Proceedings of the 2005 IEEE International Conference on Fuzzy Systems FUZZ-IEEE'2005, Reno, Nevada, May 22–25, 2005, pp. 891–896
46. Rabinovich S (2005) Measurement errors and uncertainties: theory and practice. American Institute of Physics, New York
47. Schiek CG, Araiza R, Hurtado JM, Velasco AA, Kreinovich V, Sinyansky V (2006) Images with uncertainty: efficient algorithms for shift, rotation, scaling, and registration, and their applications to geosciences. In: Nachtegael M, Van der Weken D, Kerre EE (eds), Soft computing in image processing: recent advances, Springer Verlag
48. Sinha AK (2006) Geoinformatics: data to knowledge. Geological Society of America Publ., Boulder, Colorado
49. Su T, Ozsoyoglu G (1991) Controlling FD and MVD inference in multilevel relational database systems. IEEE Transactions on Knowledge and Data Engineering 3:474–485
50. Sweeney L (1996) Replacing personally-identifying information in medical records, the scrub system. Journal of the American Medical Informatics Association 333–337
51. Sweeney L (1997) Weaving technology and policy together to maintain confidentiality. Journal of Law, Medicine and Ethics 25:98–110
52. Sweeney L (1997) Guaranteeing anonymity when sharing medical data, the datafly system. Journal of the American Medical Informatics Association 51–55

53. Sweeney L (1997) Computational disclosure control for medical microdata. In: Proceedings of the Record Linkage Workshop, Bureau of the Census, Washington, DC
54. Sweeney L (1998) Commentary: researchers need not rely on consent or not. New England Journal of Medicine 338(15)
55. Sweeney L (1998) Towards the optimal suppression of details when disclosing medical data, the use of sub-combination analysis. In: Proceedings of MEDINFO'98, International Medical Informatics Association, Seoul, Korea, North-Holland, p. 1157
56. Sweeney L (1998) Three computational systems for disclosing medical data in the year 1999. In: Proceedings of MEDINFO'98, International Medical Informatics Association, Seoul, Korea, North-Holland pp. 1124–1129
57. Sweeney L (1998) Datafly: a system for providing anonymity in medical data. In: Lin TY, Qian S (eds.) Database security XI: status and Prospects. Elsevier, Amsterdam
58. Torres R, Keller GR, Kreinovich V, Longpré L, Starks SA (2004) Eliminating duplicates under interval and fuzzy uncertainty: an asymptotically optimal algorithm and its geospatial applications. Reliable Computing 10(5):401–422
59. Trejo R, Kreinovich V (2001) Error estimations for indirect measurements: randomized vs. deterministic algorithms for 'black-box' programs. In: Rajasekaran S, Pardalos P, Reif J, and Rolim J (eds), Handbook on randomized computing. Kluwer, Dordrecht, 673–729
60. Vavasis SA (1991) Nonlinear optimization: complexity issues, Oxford University Press, New York
61. Willenborg L, De Waal T (1996) Statistical disclosure control in practice. Springer Verlag, New York
62. Wen Q, Gates AQ, Beck J, Kreinovich V, Keller JR (2001) Towards automatic detection of erroneous measurement results in a gravity database. In: Proceedings of the 2001 IEEE Systems, Man, and Cybernetics Conference, Tucson, Arizona, October 7–10, 2001, pp. 2170–2175
63. Xie H, Hicks N, Keller GR, Huang H, Kreinovich V (2003) An IDL/ENVI implementation of the FFT based algorithm for automatic image registration. Computers and Geosciences 29(8):1045–1055
64. Zelt CA, Barton PJ (1998) Three-dimensional seismic refraction tomography: A comparison of two methods applied to data from the Faeroe Basin. J. Geophysical Research 103(B4):7187–7210.

7

Fingerprint Recognition Using a Hierarchical Approach

Chengfeng Wang, Yuan Luo, Marina L. Gavrilova and Jon Rokne

University of Calgary, Calgary, Alberta, Canada
cwang,yluo,marina,rokne@cpsc.ucalgary.ca

This chapter introduces a topology-based approach to fingerprint recognition utilizing both global and local fingerprint features. It also describes a new hierarchical approach to fingerprint matching which can be used to accelerate the speed of fingerprint identification in large databases. The proposed matching scheme consists of two stages. In a coarse filtering stage, two fingerprint images are compared using the Singular Points method. Since the method does not resort to image enhancement and minutiae extraction, coarse matching is very efficient and can dramatically reduce the number of candidate fingerprints for minutia-level matching. During fine matching stage, Delaunay triangulation is used to speed up the minutia-level verification. This chapter also proposes to apply Radial Basis Functions to model fingerprint's elastic deformation, which greatly increases the system's tolerance to distorted images. Experimental results confirm that the proposed hierarchical matching algorithm achieves very good performance with respect to both speed and accuracy.

7.1 Introduction

Biometric technologies are automated methods of verifying or recognizing the identity of a person based on a psychological or behavioral characteristic. The recent developments in the areas of electronic banking, e-commerce, and security, has necessitated the development of reliable automatic personal authentication. Examples of physiological and behavioral characteristics commonly used for automatic identification include fingerprints, iris, retina, hand, face, handwriting, keystroke and voice. Fingerprint identification is one of the most reliable methods of personal authentication due to the number of properties, such as robustness, distinctiveness, availability and accessibility, that fingerprints posses. Compared to other biometrics, fingerprint recognition is a

C. Wang et al.: *Fingerprint Recognition Using a Hierarchical Approach*, Studies in Computational Intelligence (SCI) **57**, 175–199 (2007)
www.springerlink.com

well-researched area, which experiences significant growth in the recent years. Existence of large fingerprint databases, for which it became very computationally expensive to repeat the matching procedures for each fingerprint image, drives the research further in this area. In this chapter, a scheme is presented which achieves much better matching performance for larger fingerprint databases in the presence of fingerprints with elastic distortions.

A fingerprint is formed from an impression of the pattern of ridges on a finger. A ridge is defined as a single curved segment, and a valley is the region between two adjacent ridges. In Fig.7.1 some of the ridges enter from the bottom-left of the image, loop around a common center point, and exit on the left. There are a lot of features in fingerprints, some of which are introduced below. Local ridge discontinuities, known as minutiae, have little effect on the

(a) (b)

Fig. 7.1. (a) A fingerprint; (b) A magnified view of fingerprint ridge and valleys [2]

global ridge-valley pattern. However, it is the existence and location of these minutiae that embody much of a fingerprint's individuality. There are two kinds of minutiae: ridge endings and bifurcations. Ridge endings are places where ridges terminate and bifurcations are locations where a single ridge separates into two ridges (see Fig.7.2). Singular points have both global and local properties. There are two types of fingerprint singular points: cores and deltas. Locally, a core is the turning point of an inner-most ridge and a delta is a place where two ridges running side-by-side diverge. Core and delta points are illustrated by example (see Fig.7.2).

Automatic fingerprint identification systems typically include feature extraction and matching components. Here, we provide a brief overview about general methods of feature extraction, that will be more formally introduced in the later sections. To match two fingerprint images, one typically matched the feature pattern extracted from the images. Although many new features were proposed overtime for fingerprint identification, minutiae is still the most important one due to its good performance and distinctiveness. However, in practice, a fingerprint image may not always be well defined due to the ele-

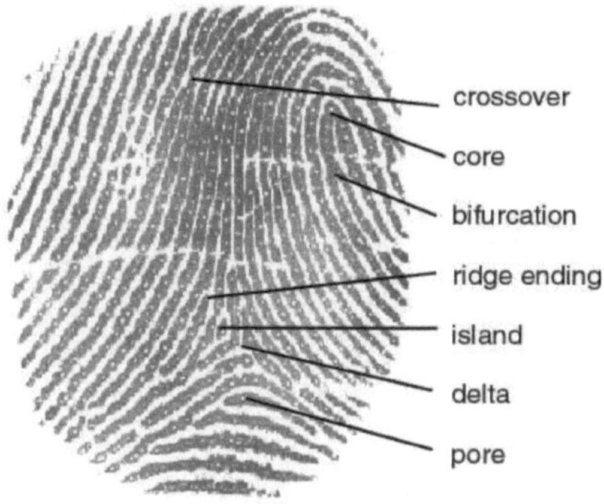

Fig. 7.2. Fingerprint features [2]

ments of noise that corrupt the clarity of the ridge structures. This corruption may occur due to variations in skin and impression conditions such as scars, humidity, dirt, and non-uniform contact with the fingerprint capturing device. Thus, image enhancement techniques are often employed to reduce the noise and to recover the topological structure of ridges and valleys from the noisy images. Fig.7.3 shows a complete flowchart for fingerprint image preprocessing and feature extraction systems. The description of the detailed steps for preprocessing and extrcation together with our contributions to the methodology will be presented in the Section 2 and Section 3 of this Chapter.

Majority of the fingerprint identification methods that appeared in the literature in recent years mainly differ in the features extracted from the fingerprint image. These features determine the performance of fingerprint matching that we coarsely classify into three families:

1. Correlation-based matching. Two fingerprint images are superimposed and the correlation between corresponding pixels is computed for different alignments [1];
2. Minutiae-based matching. This is the most popular and widely used technique. Minutiae are extracted from two fingerprints and stored as sets of points in the two dimensional plane [3, 15, 16];
3. Ridge-feature-based matching. These features can be the orientation map, the ridge frequency and so on [4].

Using the same amount of information, minutiae-based matching is considered to be the best approach, with the highest matching capability. Usually, algorithms for this family of matching stores minutiae's characteristics, including their coordinates, directions, shape and so on. In the method presented in this chapter, we employ minutiae features for local matching, and in addition propose to use some global features such as singular points and orientation

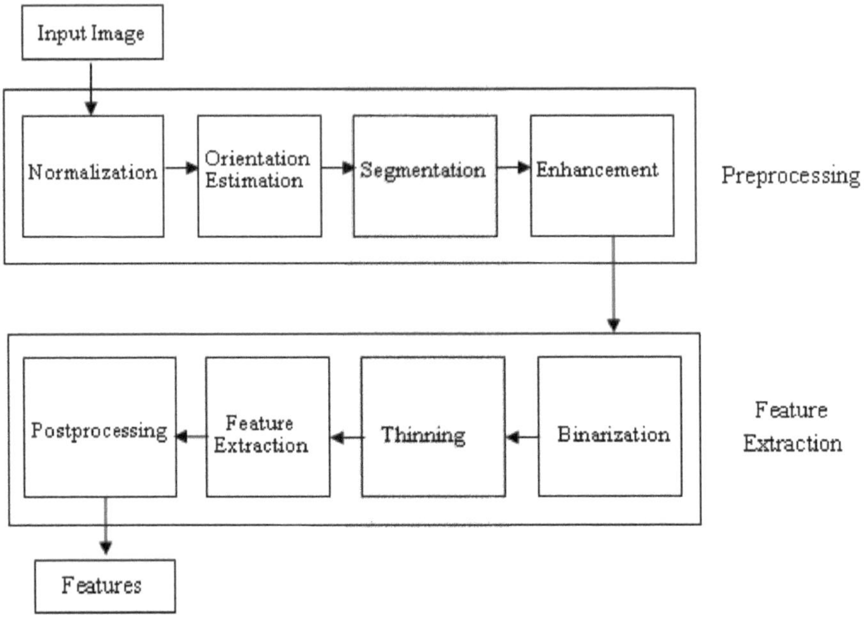

Fig. 7.3. The flowchart of our fingerprint feature extraction process

fields to improve the performance of current minutiae matching methods. The approach is radically different from other matching methods [3, 4, 15, 16], in that we apply the singular points and minutiae algorithms on two matching levels. This allows us to address the following challenges: (1) Missing and spurious features that can occur in poor quality images can be dealt with efficiently by our method; (2) Computationally expensive general minutiae-based matching methods are replaced by the fast and efficient two-level scheme which at the same time preserves matching accuracy; (3) The main difficulty for minutiae-base matching is the nonlinear deformation of a fingerprint which is explicitly modeled by our method in order to achieve a perfect alignment of the point sets.

Our Automatic Fingerprint Identification System can operate either in verification mode to comfirm the user who he says he is (one-to-one comparison) or identification mode to establish the identity of the user (one-to-many comparisons). The overall approach is described below. First, the system establishes whether the two fingerprints could be from the same finger, using the Singular Points (SPs) information extracted. We call this procedure the *Coarse Matching Stage*. As a global feature of fingerprints, SPs can be extracted much easier and quicker than minutiae points, which is why the SPs structure is used at the first filtering stage. By coarsely matching a fingerprint to the images stored in the database, approximately 30% of candidate fingerprints are excluded before the minutia-level verification stage, significantly improving the system performance. If a fingerpint and a canditate fingerprint from the database pass the first filtering stage, a topology-matching algorithm is applied at the *Fine Matching Stage*, where the two fingerprints are matched based on their local minutiae features. The method is based on building the Delaunay triangulation for the set of minutiae points, the fundamental computational geometry data structure that provides efficient representation for nearest-neighbour information. Compared with the traditional minutia matching methods, the proposed topology matching method differs in three main aspects. First, we use the Delaunay triangulation edges rather than minutiae or triangles build on minutiae points as a matching index to improve accuracy. Second, we apply a deformation model, which helps to deal with elastic finger deformations that present significant challenges in fingerprint verification and thus improve reliability. Third, we apply a maximum bipartite matching scheme to achieve a better matching performance and improve the accuracy.

In the following sections, we describe the developed hierarchical matching algorithm in detail. Section 2 presents the main steps of *Coarse Matching Stage*, including Singular Points extraction and matching method based on the structure of SPs. The novel topology-matching algorithm is described in Section 3 as part of the *Fine Matching Stage*. Experimental results are provided in Section 4. Finally, Section 5 contains the summary and conclusions.

7.2 Coarse Fingerprint Matching

The purpose of coarse fingerprint matching is to reduce the number of matches for fingerprint identification at the minutia-level. The traditional way to achieve this is fingerprint classification [5,7]. Basically, fingerprints are divided into several principal classes, such as: Arch, Tented Arch, Left Loop, Right Loop and Whorl. Then, the identification procedures are applied to one particular class, which obviously reduces the number of candidate fingerprints. Unfortunately, the problem with this classification technique is that the number of principal classes is small and the fingerprints are not uniformly distributed (31.7%, 33.8%, 27.9%, 3.8%, 2.9% for R, L, W, A and T) [8]. Moreover, there are many fingerprints whose exclusive membership can not be reliably determined even by human experts.

Based on the recent results on singular points [9, 10], we developed a different approach for coarse fingerprints matching . The main steps of coarse matching algorithm include: foreground segmentation, singular points extraction and matching. Compared with the traditional classification method, the proposed coarse matching algorithm is more precise and efficient, as will be demonstrated in the experimental section.

7.2.1 Fingerprint Foreground Segmentation

An important step in an automatic fingerprint recognition system before SPs extraction is the segmentation of fingerprint images (see Fig.7.3). A captured fingerprint image usually consists of two components, called the *foreground* and the *background*. The foreground is the component that originates from the contact of a fingertip with the sensor. The noisy area at the borders of the image is called the background. Accurate segmentation is especially important for the reliable extraction of features. Most feature extraction algorithms would not perform well when applied to the noisy background [2]. Therefore, the main goal of the segmentation is to discard the background, and thus reduce the number of falsely detected features.

Typical segmentation methods require computing the orientation fileds and coherence of each pixel, which is an expensive procedure [4, 15]. To better serve the purpose of fast coarse matching, we introduce a segmentation method based on an improved Canny Edge detector [20] and morphological operators. The main steps are as follows:

1. Normalization: Normalize the original image I to image G with a desired mean M_0 and variance VAR_0, which are optimized in different fingerprint databases by experimental data. The actual mean and variance of the original image is M and VAR. The normalization can be described by:

$$G(i,j) = \begin{cases} M_0 + \sqrt{\frac{VAR_0 \times (I(i,j)-M)^2}{VAR}}, & if\ I(i,j) > M \\ M_0 - \sqrt{\frac{VAR_0 \times (I(i,j)-M)^2}{VAR}}, & otherwise \end{cases} \tag{7.1}$$

   After normalization, we use Gaussian smoothing to reduce the influence of noise.
2. Noise Reduction: Compute the gradient altitude

$$M(i,j) = \sqrt{\partial_x^2(i,j) + \partial_y^2(i,j)} \tag{7.2}$$

   and the gradient orientation at each pixel,

$$O(i,j) = tan^{-1}\frac{\partial_y(i,j)}{\partial_x(i,j)} \tag{7.3}$$

   where the gradients $\partial_x(i,j)$ and $\partial_y(i,j)$ are the x and y components of the gradient.
3. Non-maxima suppression: we divide the gradient orientation $O(i,j)$ into four directions (labeled as Fig.7.4), and each direction has its own orientation template for comparison. For example, if $O(i,j)$ belongs to district 2, we compare its gradient altitude $M(i,j)$ with its two neighbors in district 2. If $M(i,j)$ is not the local maximum, assign $M(i,j)$ to zero; otherwise *I(i,j)* is a potential edge pixel.

2	3	4
1	$I(i,j)$	1
4	3	2

Fig. 7.4. Orientation district template

4. Edge hysteresis: we obtain the histogram of those potential edge pixels in non-maximum suppressed gradient matrix $M(i,j)$. Due to its double-apex shape, Otsu thresholding approach [6] is utilized to determine the high threshold $h1$ and the low threshold $h2$ is set to be $0.5 \times h1$. Once the two thresholds are determined, we follow the original hysteresis process of Canny algorithm [20] to extract the initial foreground boundary of the fingerprint image.
5. Boundary refinement: morphological operators are applied as the post-processing steps to refine the initial foreground boundary. First, hole blocks that are incorrectly assigned to the foreground are removed by means of an *opening* operation. Next, noise blocks that are incorrectly assigned to the background are removed by a *closing* operation.

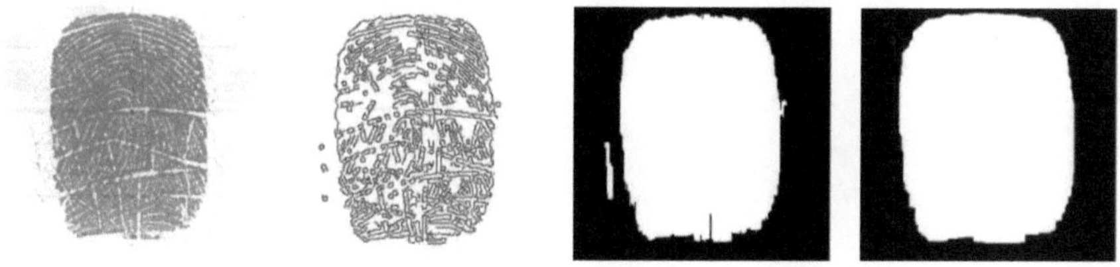

(a) (b) (c) (d)

Fig. 7.5. Results of segmentation algorithm on a fingerprint (300×300). (a) Input fingerprint image; (b) Initial result of Canny Edge segmentation; (c) Foreground (white area) before morphology; (d) Foreground after morphology.

We record the edge pixel position as the foreground of fingerprint images. Fig.7.5 shows the results of our segmentation algorithm. Visual inspection demonstrates that this method provides satisfying results.

7.2.2 Singular Points Extraction

Once the foreground is segmented from the original fingerprint image, Singular Points are only extracted from the blocks that belong completely to the foreground areas. There are two types of Singular Points, core and delta, which denote the discontinuities in the orientation field in fingerprint images [11].

We locate every 3×3 block area containing SPs by calculating Poincare Index (PI) in block orientation field. PI represents the accumulation of orientation changes in a counter-clockwise close contour around an interested area (pixel or block). The value of PI will be π if there exists a core in this area, $-\pi$ if there is a delta, and 0 if no SPs exists [19]. In each block area, the type of the SPs is determined and if it is a core, the orientation of it will be calculated. Then, SPs are extracted according to local intensity gradient. The specific algorithm is presented below.

Computing Orientation Fields

We use the standard method presented in [5] to compute the orientation fields of fingerprints. We divide the input fingerprint image into 15×15 blocks.

$$\theta(i,j) = \frac{1}{2}\tan^{-1}\left(\frac{\sum_{u=i-\lfloor W/2 \rfloor}^{i+\lfloor W/2 \rfloor}\sum_{v=j-\lfloor W/2 \rfloor}^{j+\lfloor W/2 \rfloor} 2\partial_x(u,v)\partial_y(u,v)}{\sum_{u=i-\lfloor W/2 \rfloor}^{i+\lfloor W/2 \rfloor}\sum_{v=j-\lfloor W/2 \rfloor}^{j+\lfloor W/2 \rfloor}(\partial_x^2(u,v) - \partial_y^2(u,v))}\right) \tag{7.4}$$

where ∂_x, ∂_y are the same x and y component of gradient that we get during segmentation. Also, the Gaussian smoothing operation is performed at the block level (the filter size is set to 21×21 according to experimental data). (Fig.7.6).

Fig. 7.6. Fingerprint image (the left image) and block orientation field (the right image)

PI Calculating and SPs Position Locating

Next, the PI is calculated for those blocks in the foreground of the image. In each block, the PI is defined and computed as follows:

$$PI(i,j) = \frac{1}{2\pi}\sum_{k=0}^{N-1}\Delta(k) \tag{7.5}$$

$$\Delta(k) = \begin{cases} \delta(k) & \text{if } |\delta(k)| < \frac{\pi}{2} \\ \pi + \delta(k) & \text{if } \delta(k) < -\frac{\pi}{2} \\ \pi - \delta(k) & \text{otherwise} \end{cases}$$

$$\delta(k) = \theta(X(k'), Y(k')) - \theta(X(k), Y(k))$$
$$k' = (k+1) \bmod N$$

where $\theta(i,j)$ is the orientation of the block (i,j), and $X(k)$, $Y(k)$ are the coordinates of the blocks that are in the counter-clockwise of N blocks. In our experiment, N stands for its 8-neighbor blocks. Since different types of SPs (core and delta) have different orientation fields (Fig.7.8), we can determine whether SPs is existed in the block as well as the SP's type according to the block PI : if the PI is 1/2 (Fig.7.7(a)), this block is a core block; if the PI is -1/2 (Fig.7.7(b)), this block is a delta block; if this block contains two core as a whorl, the PI is 1 (Fig.7.7(c)); if no SPs existed, the PI would be 0 (Fig.7.7(d)).

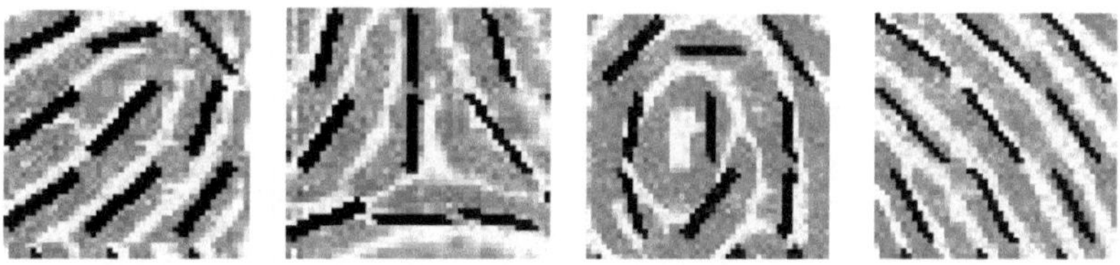

(a) Core area (b) Delta area (c) Whorl area (d) Normal orientation field

Fig. 7.7. Block orientation field around different types of SPs.

If there exists a SPs in the block (i,j), we will locate it to a pixel-level precision within the block. Taking the similar approach proposed in [11], we treat the pixel whose directional gradients equal to zero as the exact location of SPs. However, in our system, the time-consuming pixel-level detection is only performed in a block which contains SPs, not on the whole fingerprint image as [11] did. And we do not need to take additional steps to determine the type of SPs in pixel-level, because SPs type is already determined at the block level. All these improvements lead to significant computational efficiency.

Singular Point Orientation Estimation

If the type of SPs is core, we propose to use a Block Orientation Template to coarsely estimate the orientation of the core. The detailed method is described below:

1. Select the most coherent block from the 8 neighbors of the considered block containing a core. This is achieved by computing the difference between the orientation of each neighbor block and the orientation in

the corresponding position of a Block Orientation Template (shown in Fig.7.8(b)). The block with the least difference is marked (shown in the Fig.7.9(a), 7.9(b) as gray block).

2. Calculate the orientation of the core according to orientation in both the marked block and adjacent blocks to the marked one. If the marked block is in the 4-neighborhood of the block containing a core, we use the mean value of the orientations of six blocks as the orientation of the core (see Fig.7.9(b); otherwise the mean value of the orientations of four blocks is used (see Fig.7.9(a)).
3. The orientation of the core is normalized to the range $(-\pi/2, \pi/2)$ as the last step.

We take into account both the orientation of the marked block and of its adjacent blocks. Since the orientations of ridge lines are more consistent than those in the block containing the core, a robust computation of the orientation of the core can be attained in this way.

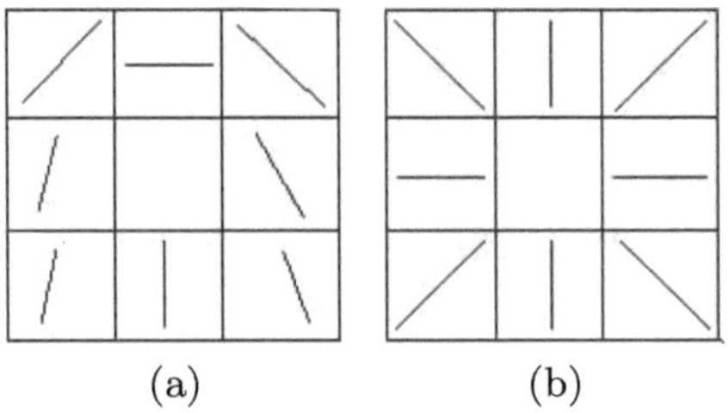

Fig. 7.8. (a) Typical block orientations in a 3×3 blocks with the center one containing a core, (b) Block orientation template

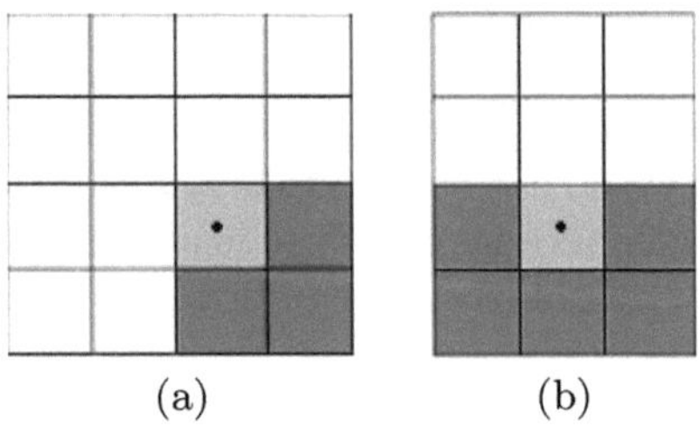

Fig. 7.9. Blocks involved when computing the orientation of core. The block containing a core is marked with a black dot, the marked block is filled with gray, and those adjacent blocks considered are filled with black. (a) the case if the marked block is in the 8-neighborhood but not in the 4-neighborhood of the block containing a core (b) the case if the marked block is a 4-neighbor of the block containing a core

7.2.3 Singular Points Matching

Based on the number, the position and the orientation of SPs, fingerprints are matched at SPs level. If these two fingerprints even fail to be matched at SPs level, we do not need to extract their minutia points for detail matching, which skips the most time-consuming step for fingerprint identification. Fig.7.10 shows the four basic structures of SPs, our similarity measurement is applied based on the distance between two SPs, the angle difference between the orientation of the core and the line connecting two SPs. More specific coarse matching rules are presented in Table 7.1:

SPs No.	2 Cores	1 Core & 1 Delta	1 Core
2 Cores	*SM	*DM	*DM
1 Cores & 1 Delta	*DM	*SM	*DM
1 Core	*DM	*DM	PASS

Table 7.1. Coarse Matching Rules

- *SM (Similarity Measure): if the differences of distance between two SPs and the angle between the orientation of core with connecting line are less than two specific thresholds, make a positive decision that these two fingerprints match.
- *DM (Distance Measure): if the minimum distance from the core to the background in one fingerprint image is greater than the distance between two cores in another fingerprint image, make a negative decision that these two fingerprints do not match.

The basic principle of our coarse matching is that we only reject those fingerprints that definitely are not of the same type. This lets all fingerprints with similarities pass. For example, in Fig.7.10, structure (a) and (c) might possibly have to be matched, because there might be a delta in (a), which is out of the fingerprint foreground due to the enrolment position. Here we should compare the minimum distance D_{min} from that core to the background with the distance D_{sp} between two SPs: if D_{min} is greater than D_{sp}, which means that the second core cannot exist in that structure, we reject these two fingerprints. If more than two cores are detected in a fingerprint, it is usually due to the influence of noise. All these fingerprint images including those without any SPs are passed to the minutiae-level verification.

7.3 Topology-based Fine Matching

The motivation for seeking a topological description of fingerprints is provided by the elasticity of the human skin since successive rolled impressions from the

(a) (b) (c) (d)

Fig. 7.10. Four structures of Singular Points (Core marked by circle, delta marked by cross)

same finger will invariably suffer a degree of relative distortion (translation, rotation and stretching). A topology based system would be less affected by the negative effects of plastic distortion.

Our developed hierarchical fingerprint matcher utilizes two distinct sets of fingerprint information: singular points and minutiae points. In above section, we proposed a fingerprint coarse matching scheme to avoid matching all pairs of fingerprints in a large database. Many potential candidate fingerprints can be excluded during the *Coarse Matching Stage*. However, for the fingerprints that are passed, the SPs are not distinctive enough to identify which one is from the same finger. In the fingerprint fine matching stage, we proposed a novel topology-based method for identifying such fingerprints.

In order to determine if two fingerprint images are from the same finger, we have to compare them based on their minutiae. We align the input fingerprint with the template fingerprint represented by its minutia pattern. But even if an input image comes from the same finger as the template image in the database, there are transformations such as translation, rotation and scaling that might make the matching difficult. To match two fingerprint images, we need to estimate these transformation parameters to align input minutiae with the template minutiae (see Fig.7.11).

Under a rigid transformation $(\Delta\theta, \Delta x, \Delta y)$, a point from the input image can be transformed to the corresponding point of the template image after rotating $\Delta\theta$ and translating $(\Delta x, \Delta y)$. The assumption is that the scaling factor between input and template images is identical since both images are captured with the same device.

The flow diagram for the fingerprint fine matching algorithm is shown in Fig.7.12. We first use edges of the Delaunay triangle as a matching index. Second, we use a deformation model, which helps to deal with the elastic finger deformations that sometimes are very damaging for correct fingerprint verification. Third, to achieve a better performance, we consider singular points and minutiae sets for matching and apply a new matching scheme. The key benefit for using topology-based approach for fingerprint matching is an exceptionally good performance of the technique even in the presence of bad quality input images. We will discuss this in the following sections.

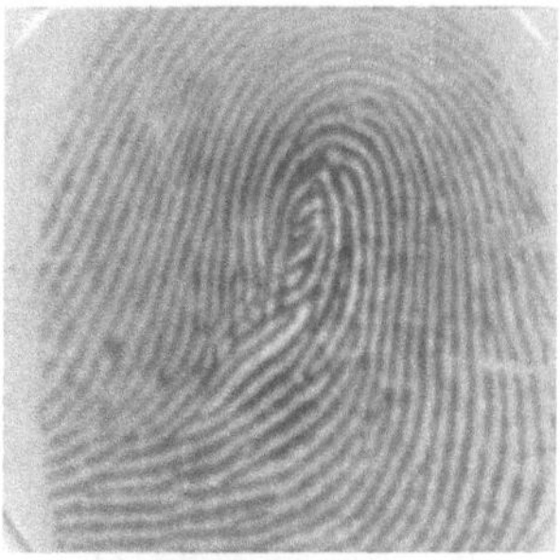

(a) Input image

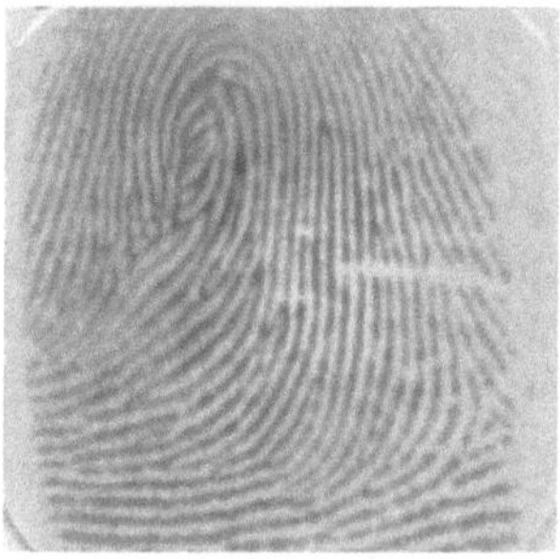

(b) Template image

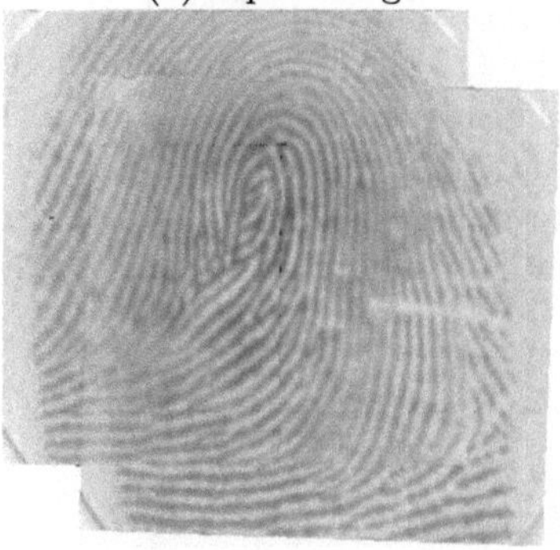

$(\Delta\theta, \Delta x, \Delta y) = (2^o, -41pixels, -63pixels)$

(c) Registration result

Fig. 7.11. Registration of fingerprint image

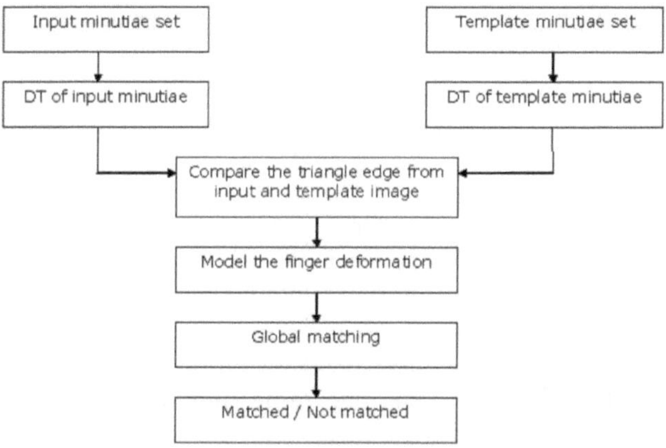

Fig. 7.12. Fingerprint fine matching flowchart

7.3.1 Delaunay Triangulation of Minutiae Set

The Voronoi Diagram and Delaunay Triangulation (DT) are closely related data structures [18], that can be used to describe the topology of the fingerprint, which is considered to be the most stable information for fingerprint matching purposes (Fig. 7.13).

Voronoi region associated with a feature is a set of points closer to that feature than to any other feature. Given a set S of points $p_1, p_2,...,p_n$, the Voronoi diagram decomposes the 2D space into regions around each point p_i such that all points in the region around p_i are closer to it than they are to any other point from S. Let $V(S)$ be a Voronoi diagram of a planar point set S. Consider the straight-line dual $D(S)$ of $V(S)$, i.e., the graph embedded in the plane obtained by adding a straight-line segment between each pair of points in S whose Voronoi regions share an edge. The dual of an edge in $V(S)$ is an edge in $D(S)$. $D(S)$ is a triangulation of the original point set, and is called the Delaunay triangulation after Delaunay who proved this result in 1934.

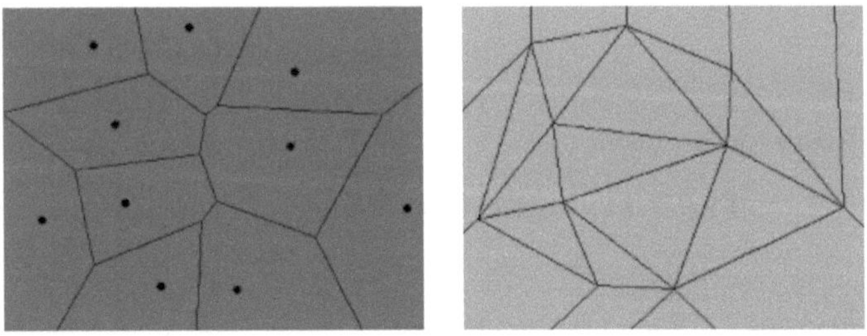

(a) (b)

Fig. 7.13. The Voronoi Diagram and Delaunay Triangulation of a set of points

The purpose of fingerprint local matching is to determine the transformation parameters $(\Delta\theta, \Delta x, \Delta y)$ between an input fingerprint and a template fingerprint. Features such as minutiae, singular points, ridges and orientation field can be used to determine the transformation.

We attempt to acquire the best registration of two fingerprint images, by first assuming that they are from the same finger. To register two images, we propose to use the Delaunay minutiae triangle as an important topological structure of the fingerprint images. A novel method basinged on the Delaunay triangulation is applied to match minutiae pairs.

We proposed to use DT for the alignment of the minutiae set for several important reasons: (1) The DT is uniquely identified by the set of minutiae points. (2) Inserting a new minutiae point or dropping a minutiae point in a set of points in most cases affects the triangulations locally, which means the algorithm can tolerate some error in minutiae extraction. (3) The number of triangles in DT is only $O(n)$. (4) As a very useful topology structure, DT can tolerate some fingerprint deformation. Fig.7.14 shows a Delaunay minutiae triangulation of two corresponding images and it also shows the reason for using triangle edges to compare rather than the whole triangle. It is more likely to find a matched triangle edge pair than a whole triangle pair in corresponding images.

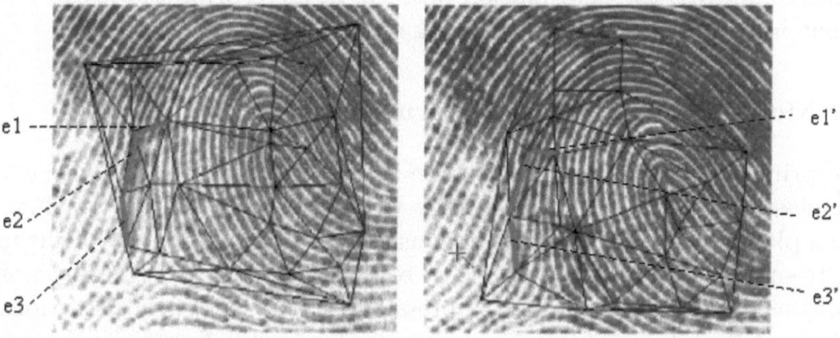

Fig. 7.14. Delaunay minutiae triangle (three triangle edges are matched, where $e1$, $e2$, and $e3$ match $e1'$, $e2'$, and $e3'$ respectively, but not a single whole triangle is matched)

Let $Q = ((x_1^Q, y_1^Q, \theta_1^Q)...(x_n^Q, y_n^Q, \theta_n^Q))$ denote the set of n minutiae points in the input image ((x, y):minutiae location; θ: orientation field of minutiae) and $P = ((x_1^P, y_1^P, \theta_1^P)...(x_m^P, y_m^P, \theta_m^P))$ denote the set of m minutiae points in template image. When two edges match successfully, their lengths and their angles related to the orientation field should be identical with some tolerance. The transformation $(\Delta\theta, \Delta x, \Delta y)$ is obtained by comparing the minutiae pairs of these two edges:

$$\begin{aligned} \Delta\theta &= \theta_i^Q - \theta_j^P \\ \Delta x &= x_i^Q - x_j^P \cos\Delta\theta + y_j^P \sin\Delta\theta \\ \Delta y &= y_i^Q - x_j^P \sin\Delta\theta - y_j^P \cos\Delta\theta \end{aligned} \tag{7.6}$$

(In the two edges, minutia $m_i^Q = (x_i^Q, y_i^Q, \theta_i^Q)$ match minutia $m_j^P = (x_j^P, y_j^P, \theta_j^P)$.)

If one edge in the input image matches another edge in template image, we obtain one transformation satisfying the above criteria. Yet, we might have more than one matched triangle edge pairs. For example, there are three matched pairs in Fig.7.14.

For a certain range of translation and rotation dispersion, we put each transformation $(\Delta\theta_i, \Delta x_i, \Delta y_i)$ as a point in the transformation space, detect the peak where there are maximum number of transformations in the range, and record those transformations that are neighbors of the peak in transformation space.

Through comparing the minutiae Delaunay triangle edges, we can record a group of transformations $(\Delta\theta_i, \Delta x_i, \Delta y_i)$. These transformations should be identical, if the input image and the template image are from the same finger and there is no non-linear deformation. Yet, non-linear deformation always exists and can affect the fingerprint matching accuracy. We propose a new method to address the deformation problem in fingerprint and it can improve the matching accuracy in the following section.

7.3.2 Modeling Fingerprint Deformation

Fingerprint deformation problems arise because of the flexibility of the finger. Some elastic distortion will necessarily result from the skin of the finger not being a planar surface. Other distortions arise from the pliability of the underlying tissue. Consequently, pressing or rolling it against a flat surface induces distortions which vary from one impression to another. Such distortions lead to relative translations of features when comparing one fingerprint image with another.

A good fingerprint identification system will always compensate for these deformations. We develop a simple framework aimed at approximately quantifying and modeling the local, regional and global deformation of the fingerprint. We propose to use a Radial Basis Function (RBF), which represent a practical solution to the problem of modeling of a deformable behavior. The application of RBFs has been explored in medical image matching and image morphing [14, 17]. To the best of our knowledge, it has not been applied for fingerprint verification. In the followings, we propose a RBF based method and apply it to fingerprint matching.

When registering images from rigid regions, a spatial transformation is normally used in the form of a rigid or affine transformation. These transformations account for global differences, but ignore non-rigid deformations. When matching non-rigid images, it is important to use a spatial model which adequately represents the spatial differences that exist between them. If the selected transformation does not truly model the spatial differences, an overall accurate match will not be obtained, even when accurate control point correspondences are found by matching the DT edges in the local matching.

When two images are registered, we select a couple of corresponding minutiae point pairs and obtain a group of transformations from these corresponding minutiae point pairs. We call these points *control points*. For our fingerprint matching algorithm, the deformation problem can be described as: knowing the consistent transformations of some points which we call *control points* in the minutiae set of input image, how to interpolate the transformation of other minutiae which are not control points? We do not consider all the transformations obtained by the local matching. We pick those consistent transformations which form large clusters. This allows us to consider both global deformation and local deformation.

In the following, we briefly describe the proposed RBF based method in two dimensions. Given the coordinates of a set of corresponding points (control points) in two images: $\{(x_i, y_i), (u_i, v_i) : i = 1, ..., n\}$, determine function $f(x, y)$ with components $f_x(x, y)$ and $f_y(x, y)$ such that:

$$\begin{aligned} u_i &= f_x(x_i, y_i) \\ v_i &= f_y(x_i, y_i) \quad i = 1, ..., n \end{aligned} \tag{7.7}$$

A spatial transformation is considered rigid if the distance between any two points is preserved. Rigid transformation can be decomposed into a translation and a rotation. In 2D, a simple affine transformation can represent rigid transformation as:

$$f_k(\mathbf{x}) = a_{1k} + a_{2k}x + a_{3k}y \quad k = 1, 2 \tag{7.8}$$

where $\mathbf{x} = (x, y)$. There are only three unknown coefficients in a rigid transformation. The RBF is used for a non-rigid transformation appropriate for fingerprint matching. In the two dimensional case, this transformation is determined by $n + 3$ coefficients in each dimension:

$$f_k(\mathbf{x}) = a_{1k} + a_{2k}x + +a_{3k}y + \sum_{i=1}^{n} A_{ik}g(r_i) \quad k = 1, 2 \tag{7.9}$$

The first three terms is an affine transformation. The last term is the sum of a weighted elastic or nonlinear basis function $g(r_i)$, which is related to the distance between $\mathbf{x}$ and the ith control point. The coefficients of the function $f_k(\mathbf{x})$ are determined by requiring that $f_k(\mathbf{x})$ satisfy the interpolation conditions:

$$\begin{aligned} f_1(\mathbf{x_i}) &= u_i \qquad \text{and} \\ f_2(\mathbf{x_i}) &= v_i \quad \text{for} \quad i = 1, 2, ..., n \end{aligned} \tag{7.10}$$

where n is the number of control points. Giving n linear equations together with the additional compatibility constrains:

$$\sum_{i=1}^{n} A_{ik} = \sum_{i=1}^{n} A_{ik}x_i = \sum_{i=1}^{n} A_{ik}y_i = 0 \quad k = 1, 2 \tag{7.11}$$

These conditions guarantee that the RBF is affinely reducible, i.e. the RBF is purely affine whenever possible. There are $2(n + 3)$ unknown coefficients and $2(n + 3)$ constraints in Eq.7.9 and Eq.7.10. The $2(n + 3)$ unknown coefficients of the basis function and the polynomial are solved in [14, 17].

The application of the deformation model can help us to match two corresponding images with a higher accuracy, as shown in Fig.7.15. Fig.7.15(c) shows the alignment of two images Fig.7.15(a) and (b) when we consider the transformation of input image to be a rigid transformation. We tried every transformation in the rigid transformation space and found the maximum number of matching minutiae pairs between input image and template image to be 6 (labeled by dash circle in Fig.7.15(c)). Circles denote minutiae of input image after transformation, while squares denote minutiae of template image. Knowing the transformation of five minutiae (*control points*) in the input image, we apply the RBF to model the non-rigid deformation, which is visualized by the deformed grid in Fig.7.15(d). The number of matching minutiae pairs is 10 which greatly increases the matching scores of these two corresponding images. Experiments presented in Fig.7.15 show that we can detect more matching minutiae pairs if we regard there is a non-rigid transformation between input and template images, and apply the RBF to model the deformation. A considerably better matching score can be obtained after we apply this technique.

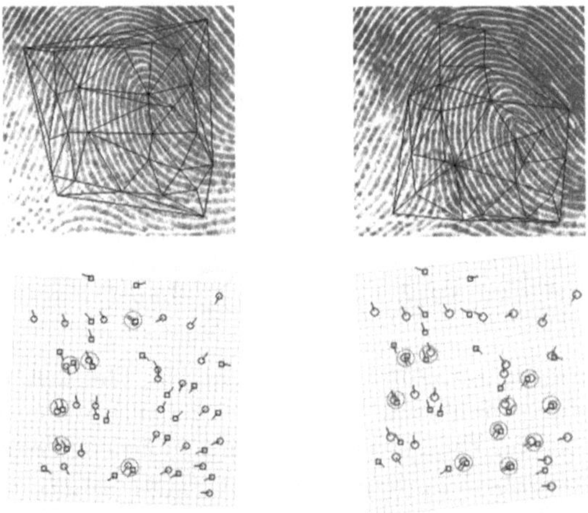

(a) Input image (b) Template image
(c) Rigid transformation (d) Non-rigid transformation

Fig. 7.15. Comparison of rigid transformation and non-rigid transformation in fingerprint matching

7.3.3 Maximum Bipartite Matching

In the above two sections, we used DT to register two fingerprint images, and applied RBF to model the non-linear deformation of fingerprints. Fig.7.11 shows the registration result ignoring the non-linear deformation, and Fig.7.15 (d) show the registration result considering the non-linear deformation. Now, we need to compare the fingerprints similarity after registration.

In this section, we discuss the fingerprint global matching. The purpose of global matching is to find the number of matching minutiae of two fingerprint images after registration. If the number of matched minutiae pairs is greater than a threshold, we can conclude that these two images are corresponding images from the same finger. Based on theory about maximum flow problem, we propose a novel maximum bipartite matching scheme for this purpose. Maximum flow is a traditional problem to compute the greatest rate at which substance can be transferred from the source to the sink without violating any capacity constrains [13]. To the best of our knowledge, this is the first application of maximum flow to fingerprint minutiae matching.

We need to match minutiae pairs under a known transformation. If one minutia from the input image and one minutia from template image fall into the same tolerance box after transformation, they are defined as matched. However, to obtain an optimal pairing, which maximize the number of matched point pairs, is not as easy as it looks.

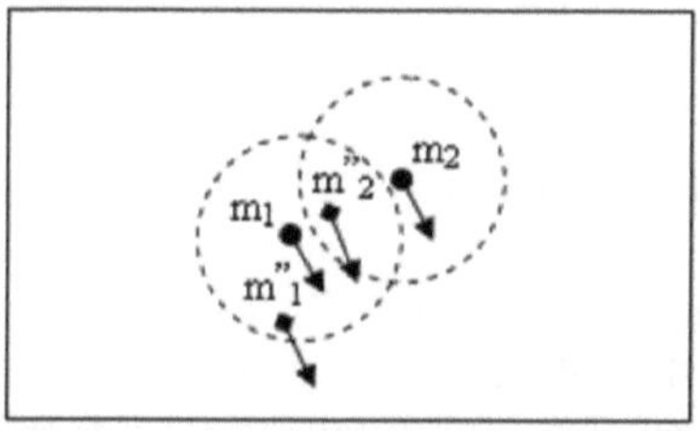

Fig. 7.16. Strategy of matching minutiae pairs. In this example, if m_1 was matched with m_2'' (the closest minutiae), m_2 would remain unmatched; however, pairing m_1 with m_1'', allows m_2 to be matched with m_2'', thus maximizing the matching pairings.

The most important rule of matching feature points is to guarantee that one minutia from the input image can match to at most one minutia from the template image. To comply with this constraint, one can mark the minutiae that have already been matched to avoid matching them twice or more. However, it is hard to find the optimal pairing of the feature points. For example, in Fig.7.16. the best pairing is the configuration that can maximize the final number of matched minutia pairs. A more sophisticated method should be used to obtain this optimum pairing.

We propose to utilize a bipartite weight graph $B = (Q, P)$ (see Fig.7.17) to match the minutiae pairs, where Q denotes the set of n minutiae in the input image and P denote the set of m minutiae in template image. For minutiae q in Q and minutiae p in P, if their locations and angles are identical with some tolerance, an edge is built between them. The capacity for these edges is set 1. For edges from source s to q and p to sink t, we set the capacity to 1. In this setting, the problem of finding the maximum number of matched point pairs between the input and template images is turned into finding the maximum flow in this bipartite graph. We can use Ford-Fulkerson algorithm [13] to find the maximum flow in such bipartite graph.

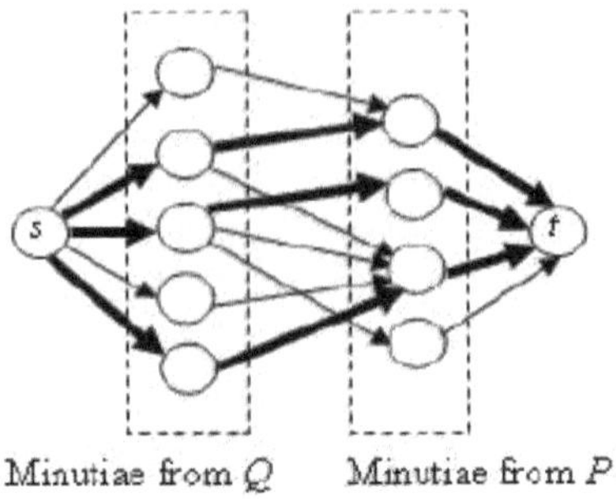

Fig. 7.17. Point pattern matching problem. The edge from s to Q has capacity 1. The edge from P to t has capacity 1. Minutiae from Q and P are connected if these two minutiae are paired, and capacity of that edge is 1. The occurrence arrow corresponds to those in a maximum flow or maximum matching pairs of the input and template images.

7.4 Experimental Results

The system was implemented using C++ programming environment and tested on a Pentium 4 2.8GHZ CPU, 512 RAM computer. Experiments have been done on three different fingerprint databases. Database1 is from FVC2000 for SPs detection. Database2 includes 448 fingerprint images of size 300×300 captured by a Veridicom COMS sensor. Database3 is from Biometric System Laboratory at University of Bologna. Database3 consists of 21×8 fingerprint images, where there are 21 fingers and each fingers contains 8 images captured in different time.

In order to compare our SPs detection results with other algorithms, tests are first performed in the second database of FVC2000. Fig.7.18 shows some samples of SPs detection. Comparison is made among the performances of algorithm in [11] and the developed approach, see Table 7.2. Obviously, our SPs detection methods are much better than the methods in [11] both in efficiency and accuracy

Alg.	No. of False SPs	Ratio of False SPs	Ratio of Missed SPs	Proc Time
[11]	2.67	0.73	0.55	1.66 s
Ours	0.31	0.11	0.17	0.45 s

Table 7.2. Performance comparison between [11] and the proposed algorithms on database1

Coarse matching is tested on database2. Our test strategy is: True Matching algorithm is applied to fingerprint images from a same finger to measure the FRR; False Matching algorithm is applied to the fingerprints from different fingers to measure the FAR. We conduct the True Matching 1016 times,

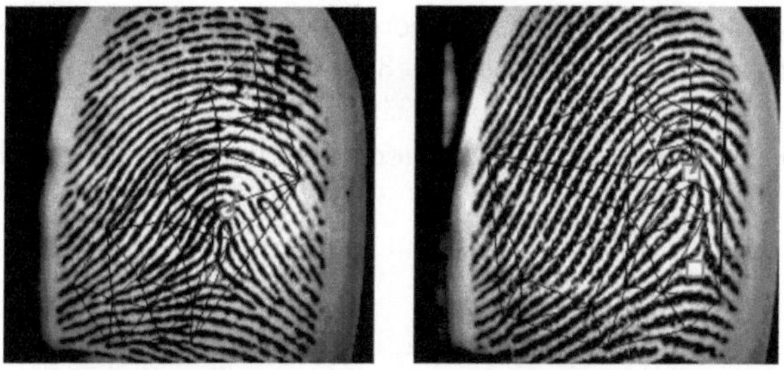

Fig. 7.18. Results of SPs detection (the orientation of the core is indicated by a line) and Delaunay minutiae triangulation

with 36 regarded as different fingerprints; False Matching 2016 times, 629 times to successfully reject fingerprints from different fingers, that is to say, to search a fingerprint in a database, 31.2% fingerprints can be excluded in the coarse matching level.

When identifying a fingerprint in a large database, we can save time by excluding part of them at the coarse matching level. In the *Fine Matching Stage*, we match the input image with all other fingerprints in the database on minutiae-level except the excluded part. Table 7.3 shows the computational

Delaunay Triangulation	Deformation	Global Matching	Total
$O(nlogn)$	$O(n^2)$	$O(n^2)$	$O(n^2)$

Table 7.3. Computation complexity of our *Fine Matching Stage*

complexity of fingerprint matching on minutiae level. There are three steps in *Fine Matching Stage*: local matching, deformation model and global matching. Let n refer to the number of minutiae. The computational complexity for Delaunay triangulation construction by the sweep-line method is $O(nlogn)$. In the worst case (there are n control points), the calculation of RBF function and registration takes $O(n^2)$ time. Finally, it takes $O(n^2)$ time to match point pairs in global matching. We sum the computation complexity of them. Thus, the computation complexity of our algorithm is $O(n^2)$. It is $O(n^3)$ for most minutiae or minutiae and ridge based methods [15, 16]. Experimental results also prove that our algorithm performs faster than standard methods. Average time of one matching is about 27ms (feature extraction time is not included). Average time cost is 98ms for standard methods [15, 16] using the same computer and the same minutiae sets.

When evaluating a biometric identification system, there are two important criteria. The FAR (False Accepted Rate) is a statistical measurement of the number of impostors likely to be accepted by a biometric system. The FRR (False Refused Rate) is a statistical and empirical measurement of the likelihood of genuine users being rejected by a system. We tested our hierarchical matching algorithm on database3 from University of Bologna. The FRR and FAR are 6.53% and 0.16% respectively for our hierarchical approach. We also test the other minutiae based fingerprint identification algorithms [15,16].

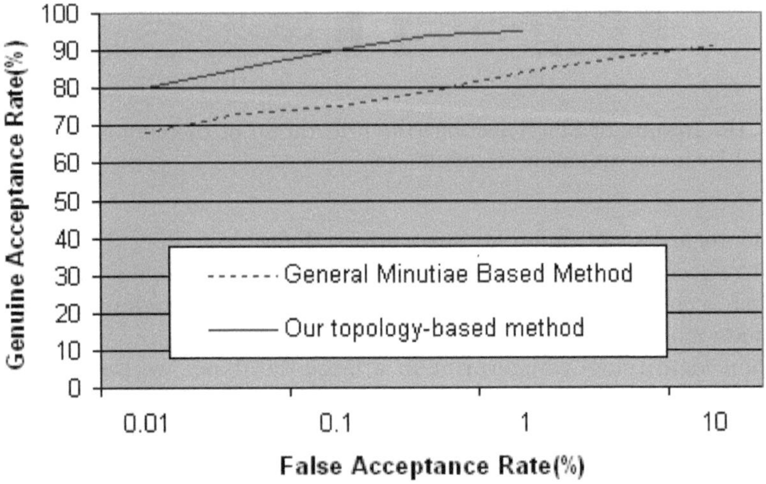

Fig. 7.19. ROC curves of general method vs. our method

Receivers operating characteristic (ROC) curves can provide the relationship between FAR and FRR. For convenience of comparison, we use genuine acceptance rate, which is an often used criterion for testing of matching algorithms. It is equal to $1-FRR$. The comparison between our algorithm and the general minutiae based matching method are shown in Fig.7.19. Dotted curve and solid curve represent general minutiae-based method and our topology-based method respectively. To explain the figure, we can take some points as an example. When FAR is 0.1% for both methods, the genuine acceptance rate of our method is 88%, and that of general minutiae-based method is 74%. The performance of our matching algorithm is much better.

The following tests are directed to study our methods tolerance to distortion. One of the main difficulties in matching two fingerprint samples of the same finger is to deal with the non linear distortions (see Fig.7.20), often produced by an incorrect finger placement over the sensing element. This would make a global rigid comparison not feasible. To illustrate that our algorithm is robust against the non-linear distortion, we conduct further experiments on the distorted images. Thus, we apply a mathematical model to model the real distortion [12].

We assumed that the original images in test database were images without non-linear distortion. In the plastic distortion model, there are three distinct

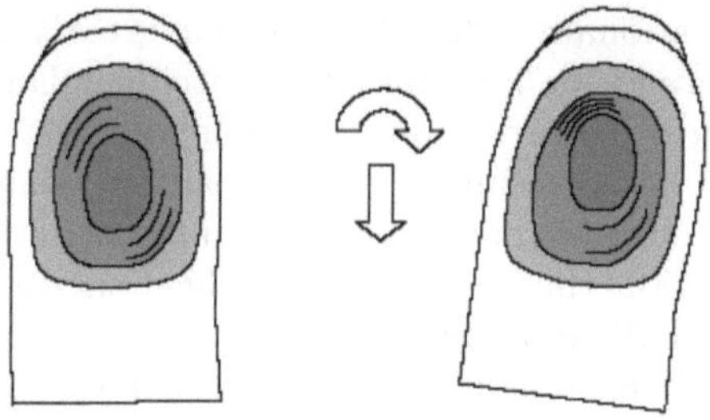

Fig. 7.20. View of a finger before and after the application of traction and torsion forces [12]

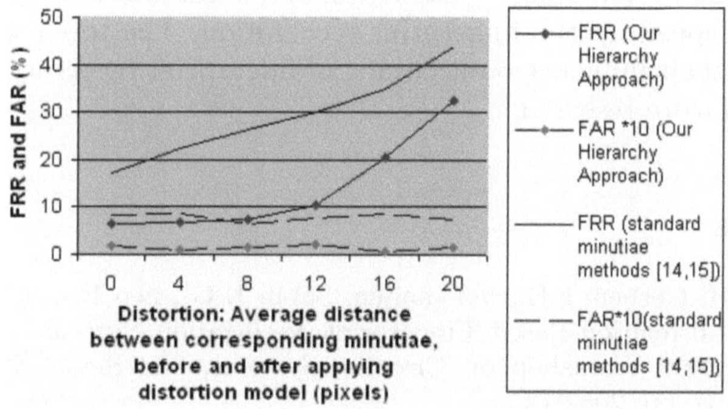

Fig. 7.21. Performance of topology methods vs. standard methods [15, 16] under distortion

regions in the fingerprint (see the three gray areas in Fig.7.20). A close-contact region does not allow any skin slippage. For minutiae points in other two regions, their locations will be changed. The average distance in Fig.7.21 represents the average distance change of minutiae points in pixels after we distort the images. When there is no distortion, the location of minutiae points does not change and the average distance is 0. Fig.7.21 shows that when the distortion is not very significant, i.e., average distance is less than 8 pixels, the accuracy of our matching algorithm remains almost the same. However, the FRR of the standard minutiae methods [15, 16] will increase in a much faster manner than ours. In other words, our algorithm is more robust against the non-linear distortion than the standard methods.

7.5 Conclusions

This chapter introduced a new hierarchical approach for fingerprint identification by using both the global and local features of fingerprints. During *Coarse Matching Stage*, SPs information is extracted as the global feature of that fingerprint. Without image enhancement and minutiae extraction, coarse matching is computationally efficient and can greatly reduce the number of candidate fingerprints for minutia-level matching. During *Fine Matching Stage*, we use the local features: minutiae. They are determined using a Delaunay triangulation based technique which leads to triangle matching a fast process. This overcomes the non-linear deformation problems. To overcome the relative deformation that is present in the fingerprint image pairs, we also propose a novel RBF model that is able to deal with elastic distortions. The proposed maximum bipartite matching schema can optimize the global matching. Our experiments show that fingerprint images can be well matched using the hierarchical matching method even on fingerprint database with some nonlinear distortion. The investigation of other features and their application in coarse matching to exclude more candidate fingerprints in large database will speed up the fingerprint recognition. The topological features, methods and their influence on accuracy of fingerprint recognition can be the direction for future research.

References

1. Asker M.B, Gerben T.B., Verwaaijen, Sabih H.G., Leo P.J. and Berend J.Z. 2000, A Correlation-Based Fingerprint Verification System, Proceedings of ProRISC 2000 Workshop on Circuits, Systems, and Signal Processing, the Netherlands, pp. 205-213.
2. Maltoni, D. Jain A.K., and Maio D. 2003, Handbook of Fingerprint Recognition, Springer-Verlag.
3. Bebis G., Deaconu T., and Georiopoulous M. 1999, Fingerprint identification using Delaunay triangulation, ICIIS99, Maryland, pp. 452-459.
4. Jain A.K., Salil P., Hong L., and Pankanti S. 2000, Filterbank-Based Fingerprint Matching, IEEE Transactions On Image Processing, vol. 9, pp. 846-859.
5. Kamijo M. 1993, Classifying Fingerprint Images using Neural Network: Deriving the Classification State, IEEE International Conference on Neural Network, vol. 3, pp. 1932-1937.
6. Otsu N. 1979, A Threshold Selection Method from Gray-Level Histograms, IEEE Transactions on Systems, Man, and Cybernetics, vol. 9, pp. 62-66.
7. Rao K. and Balck K. 1980, Type Classification of Fingerprints: A Syntactic Approach, TPAMI, vol. 2, pp. 223-231.
8. Bhanu B. and Tan X. 2003, Fingerprint Indexing Based on Novel Features of Minutiae Triplets, TPAMI, vol. 25, pp. 616-622.
9. Wang C.F. and Gavrilova M.L. 2004, A Multi-Resolution Approach to Singular Point Detection in Fingerprint Images, Proceedings of the International Conference on Artificial Intelligence, vol. 1, Las Vegas, Nevada, USA, pp. 506-514.
10. Wang F., Zou X., Luo Y., and Hu J. 2004, A Hierarchy Approach for Singular Point Detection in Fingerprint Images, Proceedings of the first International Conference on Biometric Authentication (ICBA04), vol. 1, HongKong, pp. 359-365.
11. Ramo P., Tico M., Onnia V., and Saarinen J. 2001, Optimized singular point detection algorithm for fingerprint images, Proceedings. International Conference on Image Processing, vol. 3, pp. 242-245.

12. Cappelli R., Maio D. and Maltoni D. 2001, Modelling Plastic Distortion in Fingerprint Images, ICAPR, LNCS 2013, pp. 369-376.
13. Cormen T., Leiserson C., Rivest R., and Stein C. 2002, Introduction to Algorithm, The MIT Press.
14. Fornefett M., Rohr K. and Stiehl H.S. 2001, Radial basis functions with compact support for elastic registration of medical images, Image and Vision Computing, vol. 19, pp. 87-96.
15. Jain A.K., Hong L., and Bolle R. 1997, On-line fingerprint verification, TPAMI, vol. 19, pp. 302-314.
16. Ratha N.K., Karu K., Chen S., and Jain A.K. 1996, A Real-Time Matching System for Large Fingerprint Databases, TPAMI, vol. 18, pp. 799-813.
17. Wang C.F. and Hu Z.Y. 2003, Imange Based Rendering under Varying Illumination, the Journal of High Technology Letters, vol. 9, pp. 6-11.
18. Okabe A., Boots B., and Sugihara K. 1992, Concepts and Applications of Voronoi Diagrams, Wiley Publishing, Chichester, England.
19. Karu K. and Jain A.K. 1996, Fingerprint Classification, Pattern Recognition, vol. 18, pp. 389-404.
20. Canny J. 1986, A Computational Approach to Edge Detection, TPAMI, vol. 8, pp. 679-698.

8

Smart Card Security

Kostas Markantonakis[1], Keith Mayes[1], Michael Tunstall[1], and Damien Sauveron[2] Fred Piper[3]

[1] Smart Card Centre, Information Security Group, Royal Holloway, University of London, Egham, Surrey TW20 0EX, UK
(k.markantonakis, keith.mayes, m.j.tunstall)@rhul.ac.uk

[2] XLIM — UMR CNRS 6172, University of Limoges, 123 avenue Albert Thomas — 87060 LIMOGES Cedex, FRANCE
damien.sauveron@unilim.fr

[3] Information Security Group, Royal Holloway, University of London, Egham, Surrey TW20 0EX, UK
f.piper@rhul.ac.uk

In recent years smart cards have become one of the most common secure computing devices. Their uses include such diverse applications such as: providing secure wireless communication framework, banking and identification. Direct threats to smart card security can be invasive (attacks that alter the chip inside the card), analysis of a side channel, induced faults, as well as more traditional forms of attack. This chapter will describe the various attacks that can be applied to smart cards, and the subsequent countermeasures required in software to achieve a secure solution. Invasive attacks are considered beyond the scope of this chapter and can be mitigated with software countermeasures. A case study on the various generations of the European mobile telephone networks is given as an example of how the deployment of countermeasures has changed due to the various attacks described in this chapter.

8.1 Introduction

Smart cards are used to ensure secure execution and secure storage. A secure smart card has to guarantee that secret information cannot be retrieved or modified by an unauthorised entity. This information could be secret keys used in cryptographic algorithms, an e-purse balance, etc. Smart cards are generally not vulnerable to classical software attacks implemented against PCs, but their unique form as led to different methods of attack. These attacks are briefly described below; a more thorough description including a description of the necessary countermeasures is given in subsequent sections.

K. Markantonakis et al.: *Smart Card Security*, Studies in Computational Intelligence (SCI) **57**, 201–233 (2007)
www.springerlink.com

Side Channel Attacks: Sensitive systems that are based on smart cards use well-known cryptosystems. Generally, these cryptosystems have been subject to rigorous mathematical analysis in the effort to uncover weaknesses in the system. The cryptosystems used in smart cards are therefore not usually vulnerable to these types of attacks. Since smart cards are small, physical objects that can actually be carried around in ones pocket, adversaries turned to different mechanisms of attack. Side channel analysis is a class of attacks that seek to deduce information in a less direct manner. This is achieved by extracting secret information held inside devices, such as smart cards, via the monitoring of information that leaks naturally during its operation.

The first attack of this type was a timing attack against RSA and some other public-key cryptographic primitives [42]. This was followed by attacks using the power consumption [43] and electromagnetic emanations [28] as a side channel to extract information.

Fault Attacks: Since the publication of an fault attack against the RSA in 1997 [12] fault analysis has come to the foreground as a possible means of attacking devices such as smart cards. The vast majority of papers written on this subject involved very specific faults (usually a bit-flip in a specific variable) that are extremely difficult to produce. However, there have been several attacks published that have enough freedom in the type of fault required that they can be realised with current fault injection methods.

Fault attacks have been implemented against chips that could be used in smart cards [4]. This has meant that countermeasures against this type of attack need to be implemented in embedded devices to protect against this class of attack.

Multi-Application Security: Modern smart cards usually incorporate a java virtual machine that is capable of interpreting multiple applets present in the smart card's Non-Volatile Memory (NVM). There are therefore two security considerations that need to be taken into account when implementing a java virtual machine: secure bytecode interpretation and secure resource partitioning. Bytecode needs to be implemented such that the security of the smart card cannot be compromised by a malicious applet writer, and resources managed such that one applet cannot access the secrets of another applet. This is a relatively new problem in smart card operating system design, as previously it was not possible to load arbitrary applications into smart cards.

A secure smart card needs to include all of the above security considerations. There are also other security aspects that need to be taken into account to have a totally secure solution. There are hardware features that are used to prevent reverse engineering of a chip, its operating system, and any secret keys held within. A discussion of this topic is beyond the scope of this chapter but needs to be taken into account when choosing a chip for use as a smart card.

This can be mitigated by including more stringent software countermeasures to achieve a secure solution when inadequate hardware security is present.

The specific case of side channel and fault attacks applied to smart cards is covered in Section 8.2. This is followed by a discussion of multi-application platform security in Section 8.3. In Section 8.4 a case study is given detailing the changes in security in the European mobile telecommunications standards in response to various smart card based attacks. This is followed by a summary in Section 8.5.

8.2 Smart Card Specific Attacks

Sensitive systems that are based on smart cards use protocols and algorithms that have usually been subjected to rigorous analysis by the cryptographic community. Attackers have therefore sought other means to circumvent the security of protocols and algorithms used in smart card based systems. As smart cards are small, portable devices they can easily be put in a situation where their behaviour can be observed and analysed. The simplest form of analysis of this type is intercepting all the communication between a smart card and its reader. Tools are readily available on the Internet [62] that allow the commands to and from a smart card to be logged. This problem is relatively easy to solve with secure sessions, as proposed in [31], but highlights the ease of man-in-the-middle type attacks against smart cards.

More complex attacks can be realised by monitoring information that leaks naturally during a smart cards processing of information, referred to as side channel attacks. Smart cards can also be attacked by inducing a fault during their normal processing to change the chips behaviour. Comparing a faulty result with a standard response can then be used to make deductions about secrets held by a smart card. These attacks are referred to as fault attacks. The following sections describe these two classes of attack in more detail.

8.2.1 Side Channel Attacks

The first example of a side channel attack was proposed in [42]. This involved observing the differences in the amount of time required to calculate a RSA signature for different messages to derive the secret key. This attack was conducted against a PC implementation but a similar analysis could be applied to smart card implementations. It would be expected to be more efficient against a smart card as more precise timings can be achieved with an oscilloscope or proprietary readers. An example of equipment capable of acquiring this sort of information is shown in Figure 8.1. The I/O trace can be seen on the oscilloscope as the yellow trace that will allow the exact amount of clock cycles taken by a given command can be determined (the smart card being analysed is in the reader in the top left corner of the image). For this reason, the amount of time taken for a command is often constant or, if variable, not related to any secret information.

The most common form of side channel attack is the analysis of the power consumption, originally proposed in [43]. This is measure by placing a resistor in series with a smart card, between the card and earth. The potential difference across this resistor is measured with an oscilloscope. This gives a measurement of current in the circuit between the chip and the resistor. The equipment shown in Figure 8.1 is capable of taking these sort of measurements. The current at each point in time during a command is shown as the blue trace on the oscilloscope. An enlargement of this trace is shown on the computer screen, where the minimum, average and maximum of the points represented by one pixel can be seen. This is referred to as the power consumption in the literature, and the attacks based on these measurements are referred to as power analysis attacks. There are two main types of power attack; these are simple power analysis (SPA) and differential power analysis (DPA).

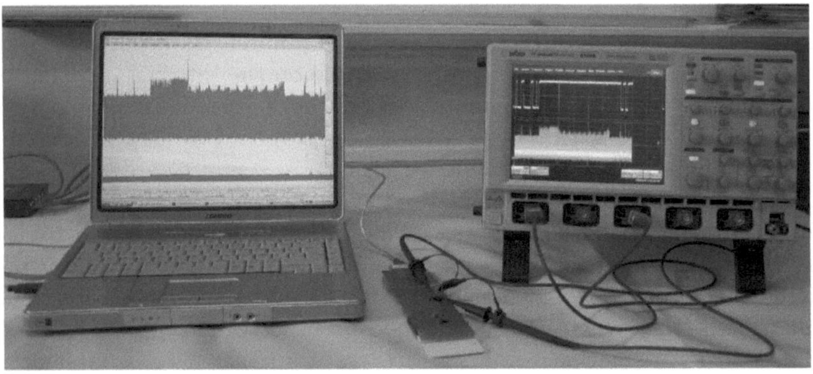

Fig. 8.1. Data Acquisition Tools.

Simple Power Analysis.

This is the analysis of one, or the comparison of a few, power consumption traces. An attacker will look for repetitive patterns and distinctive events to reverse engineer the algorithm being used and to try and derive any secrets being manipulated. Figure 8.2 shows a power consumption trace taken during the execution of an AES implementation, using equipment similar to that shown in Figure 8.1, where the 10 rounds of the DES algorithm can be observed. A pattern that repeats 9 times can be seen, with a much smaller tenth pattern, that correspond to the rounds of an AES.

Much smaller differences can be seen in the power consumption. By zooming in on specific areas the difference of 1 or more clock cycles can be used

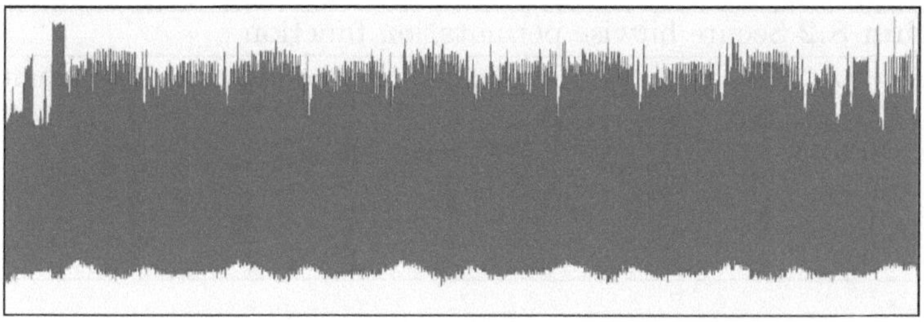

Fig. 8.2. Power consumption during the execution of an AES implementation.

to identify secret information. For example, A compiler can implement the bitwise permutations used in DES in an insecure manner. An example of an algorithm that could be produced is given in Algorithm 8.1, where each bit in buffer X (a buffer of n bits, where $(x_0, x_1, \ldots, x_{n-1})_2$ is the binary representation) is tested to determine whether a bit in buffer Y should be set.

Algorithm 8.1 Unsecure bitwise permutation function

Input: $X = (x_0, x_1, \ldots, x_{n-1})_2$, $P[\cdot]$ containing the indexes of the permutation
Output: $Y = (y_0, y_1, \ldots, y_{n-1})_2$
1. $Y := 0$;
2. for $i = 0$ to $n - 1$
3.   if $(x_{P[i]} = 1)$ then $y_i := 1$;
4. return Y;

end.

This is not a secure implementation due to the conditional test present in step 3. The setting of one bit in buffer Y will take a certain amount of time when the tested bit in buffer X is equal to 1. If a power consumption trace for this function is compared to a trace taken during the manipulation of a known key, for example all zeros, the point where the two traces differ will reveal the first manipulated bit that is not equal to the known key. In the case where the known key is all zeros, the power consumptions diverge where the manipulated bit of the unknown key is equal to one. All previous bits were therefore equal to 0. The power consumption trace can then be shifted to take into account the time difference due to this bit and the process repeated for the rest of the key.

This problem can be avoided by implementing the bitwise permutation as shown in Algorithm 8.2, where each bit is taken separately for buffer X and assigned to buffer Y. This is usually more time consuming than Algorithm 8.1 as each bit needs to taken from the relevant machine word.

It is not always possible to change an algorithm in this manner. For example, the modular exponentiation used in RSA is exceedingly difficult to secure against this type of attack, as branches in the algorithm cannot be avoided.

Algorithm 8.2 Secure bitwise permutation function

Input: $X = (x_0, x_1, \ldots, x_{n-1})_2$, $P[\cdot]$ containing the indexes of the permutation
Output: $Y = (y_0, y_1, \ldots, y_{n-1})_2$
1. for $i = 0$ to $n - 1$
2. $\quad y_i := x_{P[i]}$;
3. return Y;
end.

An example of a power consumption trace that shows the square-and-multiply algorithm is given in [40].

Differential Power Analysis.

In this type of attack the smart card is forced to run the same command numerous times with different inputs. For each input the corresponding power consumption is captured and an analysis is performed on the data gathered to make inferences on the secret information held inside the card. These techniques require many more acquisitions than with simple power analysis and a certain amount of treatment *a posteriori.*

Differential power attacks [43] are based on the relationship between the current consumption of the hamming weight of the data manipulated at a particular point in time. The variation in current consumption is extremely small and acquisitions cannot be interpreted individually as the information is very small when compared to the usual amount of noise in a given acquisition. These differences can be seen in Figure 8.3, and can be amplified by using DPA.

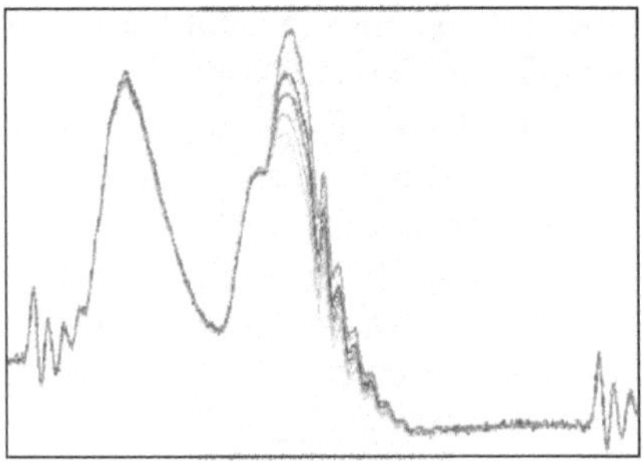

Fig. 8.3. Superimposed acquisitions of one clock cycle showing the data dependence of the power consumption.

If two average curves are produced (one where the bit is equal to one and the other where it is not), the effect of the modified bit on the current consumption becomes visible in the difference between the two averages, an example of this is shown in Figure 8.4. A peak in this waveform will occur at the moment the bit used to divide the acquisitions into two sets was manipulated. This occurs because the rest of the data can be assumed to be random and will therefore provide the same average. Figure 8.4 shows a sample DPA trace showing five different points in time where the bit used to divide the acquisitions is manipulated.

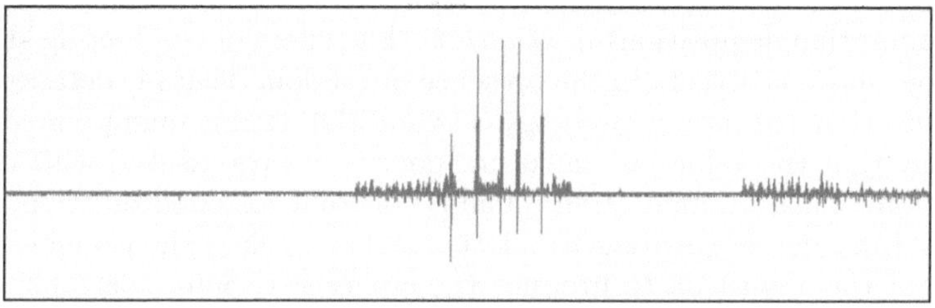

Fig. 8.4. A DPA trace.

This can be used to break a secret key algorithm if an attacker forms a hypothesis about the exit of a given function e.g. an S-box. This technique can be used to verify the hypothesis as peaks, as shown in Figure 8.4, will be present. This would appear to be analogous to an exhaustive search but the output of 1 S-box in DES depends on 6 bits of the secret key. An attacker therefore only needs to test hypotheses on these 6 bits to determine what they are. The process could then be repeated for each S-box to derive the entire key.

A method of improving this method to reduce the existence of false positives is given in [14].

Countermeasures.

There are several different countermeasures for protecting algorithms against attacks based on the power consumption, although some of the countermeasures merely complicate the attack process. The combination of all the countermeasures below makes attacking a secure algorithm exceedingly difficult, as each countermeasure needs to be taken into account.

Constant Execution: As described above, the time taken by an algorithm should remain constant, so that no deductions on secret information can be made. This extends to individual processes being executed by a smart card. If a process takes different lengths of time depending on some secret information and the difference in time is made up by a dummy function, there is a good chance that this will be visible in the power consumption as detailed in Section 8.2.1. It is therefore important that an algorithm is written so that the same code is executed for all the possible values that the secret information being manipulated can take.

Random Delays can be inserted at different points in the algorithm being executed i.e. a dummy function that takes a random amount of time to execute can be called. The algorithm can no longer be said to comply with the constant execution criteria given above, but any variation is completely independent to any secret information. This does not provide a countermeasure, but creates an extra step for an attacker. In order to conduct any power analysis an attacker needs to synchronise the power consumption acquisitions *a posteriori*. The effect of conducting statistical power analysis attacks in the presence of random delays is detailed in [18].

Randomisation (or data whitening) is where the data is manipulated in such a way that the value present in memory is always masked with the same random. This randomisation remains constant for one execution, but will vary from one acquisition to another. This mask is then removed at the end of the algorithm to produce the ciphertext. Some ideas for building countermeasures were proposed in [15], and an example of this sort of implementation can be found in [3].

The size of the random is generally limited as S-boxes need to be randomised before the execution of the random so that the input and output values of the S-box leak no information. This is done using an algorithm such as Algorithm 8.3. As shown the random used for masking the input data can be no larger than n, and the random used for the output value can be no larger that x.

Algorithm 8.3 Randomising S-box Values

Input: $S = (s_0, s_1, s_2, \ldots, s_n)_x$ containing the S-box, $\mathbf{R}$ a random $\in [0, n]$, and r a random $\in [0, x)$

Output: $RS = (rs_0, rs_1, rs_2, \ldots, rs_n)_x$ containing the randomised S-box

1. for $i = 0$ to n
2. $\quad rs_i := s_{(i \oplus \mathbf{R})} \oplus r$;
3. return RS;

end.

In the case of AES both $\mathbf{R}$ and r will be one byte, which means that the random mask during the calculation is likely to be one byte. There are theoretical attacks against this protection method [52] but an actual implementation has yet to be published.

This is not possible in the case of RSA, where the calculation methods do not facilitate the method described above. A method for randomising the calculation of RSA is given in [40], where the signature generation (i.e. $s = m^d \pmod{n}$ where $n = p \times q$) can be replaced by:

$$s = \left((m + r_1 n)^{d + r_2 \phi(n)} \pmod{r_3 n} \right) \pmod{n} \tag{8.1}$$

where $\phi(\cdot)$ is Euler's totient function and, r_1, r_2 and r_3 are small random values. This does not provide a totally secure algorithm as the modular exponentiation itself also has to be secured against SPA attacks. A discussion of these algorithms is given in [17].

Randomised Execution is the manipulation of data in a random order so that an attacker does not know what is being manipulated. In the case of Algorithm 8.1 an attacker would not know which bit is being manipulated at any given point in time. Given the $n!$ possible combinations, an attack by simple power analysis would be extremely difficult.
This also inhibits any statistical analysis of the power consumption, as this relies on the same unknown variable being treated at the same point in time. As an attacker cannot know the order in which the data has been treated, this provides an extremely efficient countermeasure when combined with randomisation. An example of this technique applied to DES is described in [51].

8.2.2 Fault Attacks

The first example of a theoretical fault attack was presented in 1997 [12] as a method of injecting faults in to RSA signature generation when using the Chinese Remainder Theorem. This was followed by several other attacks on public key algorithms [5,41] and an equivalent for private key algorithms was proposed in [9], usually based on the effect of a 1 bit error during the computation of the algorithm under study. As no implementations were forthcoming interest in this type of attack waned. One of the first publications describing an implementation of this type of attack was presented in 2002 [4], and described an implementation of the attack against the RSA signature scheme and some countermeasures. This revitalised interest in this type of attack, which is now an important aspect of smart card security.

Modelling the Effect of a Fault.

There are several known mechanisms for injecting faults into microcontrollers. These include variations in the power supply to create a glitch or spike [11], white light [63], laser light [6] and eddy currents [60]. The models for the faults that can be created by these effects can be summarised as follows:

Data randomisation: the adversary could change an arbitrary amount of data to a random value. However, the adversary does not control the random value and the new value of the data is unknown to the adversary.

Resetting Data: the adversary could force the data to the blank state, i.e., reset a given byte, or bytes, of data back to `0x00` or `0xFF`, depending on the logical representation.

Modifying opcodes: the adversary could change the instructions executed by the chip's CPU. This will often have the same effect as the previous two types of attack. Additional effects could include removal of functions or the breaking of loops. The previous two models are algorithm dependent, whereas the changing of opcodes is implementation dependent.

These three types of attack cover everything that an attacker could hope to do to an algorithm. In most cases it is not usually possible for an attacker to create all of these possible faults. Nevertheless, it is important that algorithms are able to tolerate all types of fault, as the fault injection methods that may be realisable on a given platform are unpredictable. While an attacker might only ever have a subset of the above attacks available, if that attack is not taken into account it may have catastrophic consequences the security of a protocol or algorithm.

All of these models assume that a fault injected into a chip will change a machine word rather than a single bit. This model was based on studies of faults caused by the effects of the radiation present in the upper atmosphere [75], which is important in the design of vehicles designed to travel in the upper atmosphere or space. The effects of radiation are random, whereas an injected fault is intentional and will target a given point in time.

There have not been any publications that detail practical implementations of attacks that exploit the change of 1 bit of information in a chip. It is possible to use a lasers simulate the effect of radiation in a circuit [38], but smart cards will generally use scrambled/randomised layouts so the possibilities are limited. More success has been achieved by targeting larger areas to provoke a large fault.

Injecting Faults in Algorithms.

In this section two of the most widely known attacks will be described. The first fault attack published in [12] and implemented in [4], which is detailed below:

If we assume that the calculation of an RSA signature i.e. $s = m^d \pmod{n}$, where $n = p \times q$, the two primes upon which the security of RSA is based. If this is calculated using the Chinese remainder theorem the following values are calculated,

$$\begin{aligned} s_p &= m^{(d \pmod{p-1})} \pmod{p} \\ s_q &= m^{(d \pmod{q-1})} \pmod{q} \end{aligned} \tag{8.2}$$

which can be combined to form the RSA signature s using the formula $s = a \times s_p + b \times s_q \pmod{N}$, where:

$$\begin{cases} a \equiv 1 \pmod{p} \\ a \equiv 0 \pmod{q} \end{cases} \text{and} \begin{cases} b \equiv 0 \pmod{p} \\ b \equiv 1 \pmod{q} \end{cases}$$

This can be calculated using the following formula:

$$s = s_q + \left((s_p - s_q) \times q^{-1} \pmod{p}\right) \times q \tag{8.3}$$

If this is calculated correctly, and then recalculated but with a fault injected during the computation of s_p or s_q, information can be derived on one of the primes used to create n.

If s_q is changed to s'_q, then the signature generated will be of the form $s' = a \times s_p + b \times s'_q \pmod{n}$. The difference between s and s' gives:

$$\begin{aligned}\Delta &\equiv s - s' \\ &\equiv (a \times s_p + b \times s_q) - (a \times s_p + b \times s'_q) \\ &\equiv b(s_q - s'_q) \pmod{n}\end{aligned} \tag{8.4}$$

As $b \equiv 0 \pmod{p}$ and $b \equiv 1 \pmod{q}$ it follows that $\Delta \equiv 0 \pmod{p}$ (but $\Delta \not\equiv 0 \pmod{q}$) meaning that Δ is a multiple of p (but not of q). Hence, a GCD calculation gives the secret factors of n, i.e. $p = \gcd(\Delta \pmod{n}, n)$ and $q = n/p$.

This attack could potentially be achieved with any of the fault models given above. This is because any fault in one of the modular exponentiations given in Equations 8.2 will be enough to conduct the attack. This includes modifying the variable entering the equation before the start of the computation of the modular exponentiation.

Another attack that was proposed shortly after the RSA attack was published targeting DES [9]. This assumed a 1-bit fault injected into the last few of rounds of a DES implementation. This was generalised to allow for a larger fault and became a frequently cited attack within the smart card industry. The most popular form of this attack is described in [6], and is repeated here.

DES can be considered as a transformation of two 32 bit variables (L_0, R_0), i.e. the message, though sixteen iterations of the function as shown in Figure 8.5 to produce the ciphertext (L_{16}, R_{16}). The Expansion and P permutations are bitwise permutations, and are generally not considered when studying DES. This means that the round function can be simplified to:

$$\begin{aligned}R_n &= S(R_{n-1} \oplus K_n) \oplus L_{n-1} \\ L_n &= R_{n-1}\end{aligned} \tag{8.5}$$

where $S(\cdot)$ is the S-box function. The existence of the initial and final permutation is also ignored as they do not contribute to the security of the algorithm.

The last round, as described above, can therefore be expressed in the following manner:

$$\begin{aligned}R_{16} &= S(R_{15} \oplus K_{16}) \oplus L_{15} \\ &= S(L_{16} \oplus K_{16}) \oplus L_{15}\end{aligned} \tag{8.6}$$

If a fault occurs during the execution of the fifteenth round, i.e. R_{15} is randomised by a fault to become R'_{15}, then:

$$\begin{aligned}R'_{16} &= S(R'_{15} \oplus K_{16}) \oplus L_{15} \\ &= S(L'_{16} \oplus K_{16}) \oplus L_{15}\end{aligned} \tag{8.7}$$

If we XOR R_{16} and R'_{16} we get:

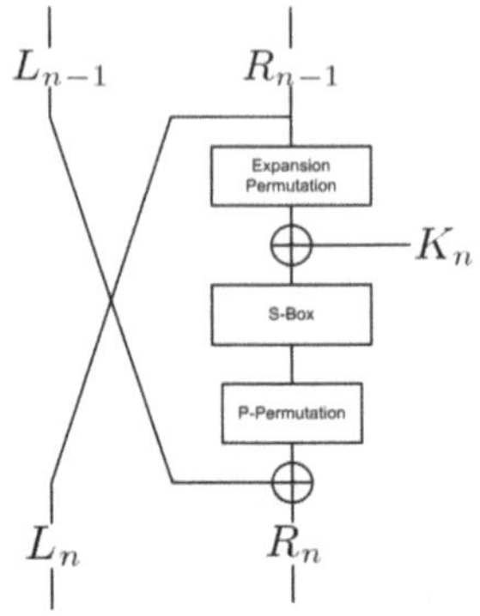

Fig. 8.5. The DES round function for round n.

$$\begin{aligned} R_{16} \oplus R'_{16} &= S(R_{15} \oplus K_{16}) \oplus L_{15} \oplus S(R'_{15} \oplus K_{16}) \oplus L_{15} \\ &= S(R_{15} \oplus K_{16}) \oplus S(R'_{15} \oplus K_{16}) \\ &= S(L_{16} \oplus K_{16}) \oplus S(L'_{16} \oplus K_{16}) \end{aligned} \tag{8.8}$$

This provides an equation where only the last subkey, K_{16}, is unknown. All of the other variables are visible in the ciphertext. This equation holds for each S-box in the last round, which means it is possible to search for key hypotheses in sets of six bits. All 64 possible key values corresponding to the XOR just before each S-box are exhausted to generate a list of possible key values for these key bits. After this, all the possible combinations of the hypotheses can be searched though with the extra 8 key bits that are not included in the key to find the entire key.

If R'_{15} becomes random then the expected number of hypotheses that are generated can be predicted. For a given input and output difference there are certain number of values that could create the pair of differences, as described in [8]. The expected number of hypotheses for the last subkey will be around 2^{24}, giving an overall expected keyspace of 2^{32}.

If the attack is repeated to acquire two faulty ciphertexts the intersection of the two keyspaces can be taken. This will greatly reduce the keyspace that will need to be searched through to derive the key. It would be expected that two faulty ciphertexts with the properties described above would give around 2^6 hypotheses for the last subkey, leading to an exhaustive search of around 2^{14} for the entire key.

An implementation of this is described in [30] where two faulty ciphertexts are acquired, leading to a small exhaustive search to find the DES key used. Again, it would be expected that any of the fault models given would enable this attack to be conducted. As any fault during the calculation of R_{15} will give a ciphertexts with the correct properties.

This attack is also extended in [30] to allow for any faults from the eleventh round onwards, the details of which are beyond the scope of this article.

Another attack [32] attempts an attack of a similar nature but using faults at the beginning of the DES algorithm.

This gives two different attacks against commonly used algorithms. There are a plethora of other attacks that target other algorithms, but they cannot be listed here. These two attacks were chosen to represent the fault attacks that have been proposed and implemented.

Countermeasures.

There are various types of countermeasure that can be implemented to defend against fault attacks. These are usually based on existing techniques used for integrity purposes. It would normally be expected that anomaly sensors would be implemented in a smart card that would detect an attempted fault attack. However, this cannot be relied upon as a new fault injection technique may be able to circumvent these.

A detailed list of the possible fault resistant hardware implementations is given in [6]. Some of the software countermeasures that can be used are listed below.

Checksums can be used to verify the integrity of data at all times. This is especially important when data is being moved from one memory area to another. For example, a fault injected in an RSA variable can compromise the secret key, as described above.

Variable redundancy is the reproduction of a variable in memory and functions are performed on each variable independently. The result will be known to be correct if both answers are identical.

Execution redundancy is the repetition of the same function, part of the function or it's inverse. In the case of the DES algorithm it would be prudent to repeat the first 3 rounds to protect against the attack described in [32], and the last 5 rounds to protect against the attack described in [9, 30]. This greatly increases the execution time but is less costly than repeating the entire algorithm.

In the case of RSA the simplest solution to protect the signature generation function is to verify the result with the public key. This is efficient as signature verification is extremely fast when compared to signature generation. In some standards it is not always possible to have access to the public key, countermeasures have therefore been proposed that verify certain conditions after signature generation has taken place. An example of this type of countermeasure, and an account of previous methods, is given in [16].

Execution Randomisation: If all the functions are conducted in a random order it is not be possible to determine exactly where a fault needs to be injected. The presents a similar problem to that described in Section 8.2.1, as an attacker is unsure of what function is being attacked. However, this merely slows an attacker as an attack can be repeated until successful. An example of this type of countermeasure on the context of fault attacks is given in [55].

Ratification counters and baits: A countermeasure described in [6] involves including small functions in sensitive code that perform a calculation and then verify the result. When an incorrect result is detected a fault in known to have been injected. The reaction to such an event would be to decrement a counter. When this counter reaches 0 the smart card would then cease to function. When combined with random delays, as described in Section 8.2.1, this can be a very effective countermeasure.

In order to achieve a secure implementation the above countermeasures would need to be combined with those presented in Section 8.2.1. This implies a significant overhead when implementing cryptographic algorithms for smart cards but is necessary to defend against modern attack techniques.

8.3 Smart Card Platform Security

The aim of this section is to provide an overview of the issues, apart from the underlying hardware functionality, affecting the overall concept of smart card security. Therefore, it will explore the functionality offered by multi-application smart card platforms and it will also highlight the main security concerns behind them.

8.3.1 The Evolution of Smart Card Platforms

The evolution of smart card operating systems (SCOS) and platforms is very closely coupled with the various improvements of the underlying smart card hardware. A typical smart card of the mid 90's had approximately 1-3 Kbytes of ROM, less than 128 bytes of RAM and 3-6 Kbytes of Electrically Erasable Programmable Read Only Memory (EEPROM). Within this very restrictive processing environment the role of the smart card operating system was naturally limited.

In most cases, the role of these early smart card operating systems (i.e. first generation) was to mainly offer the basic functionality defined within the various smart card standards [35–37] and, more specifically, to present a protected file system. This was because as smart card microprocessors were mainly perceived as secure storage devices. The main drawback of these operating systems was that the smart card applications developers had very little flexibility in terms of developing a smart card application, as the functionality offered was mostly implemented in the Read Only Memory (ROM) of the card. Although, some of these operating systems [29, 56] claimed that they offered multi-application smart card functionality, the reality was different.

The next generation (i.e. second generation) of smart card operating systems, e.g. Multos [47] emerged between 1995 and 1999 with the introduction of more powerful microprocessors consisting of 6-8 Kbytes of ROM, 128 bytes of RAM and 6-12 Kbytes of EEPROM. The smart card microprocessors were capable of hosting what it often referred to as the "monolithic" [34, 48] smart card operating systems. The main characteristic of the "monolithic" operating system was that its functionality was enhanced compared to the previous generation due to a greater underlying set of commands. Smart card applications and operating systems were still very closely coupled together within the ROM memory of the card and, although they were more advanced, portability was a major concern.

The third generation of smart card operating systems came into existence from approximately 1999 onwards. The market requirements were more mature in terms of application interoperability, hardware independence, and post issuance capabilities. Among the main characteristics of these operating systems was that applications and operating system functionality were written to the EEPROM and ROM respectively, see Figure 8.6. Moreover, the presence of a Hardware Abstraction Layer (HAL) ensured that application portability should no longer be an issue.

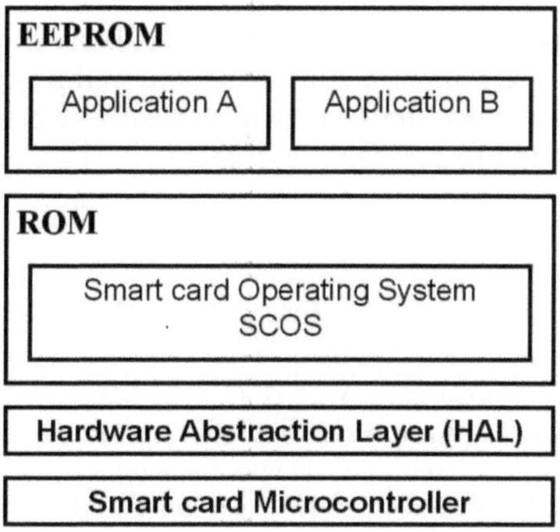

Fig. 8.6. The typical architecture of a third generation SCOS.

8.3.2 The Different Multi-application smart card Platforms

Among the main characteristics of secure Multi-application smart card platforms was application interoperability and inter application isolation. Smart card issuers would like to be able to develop smart card applications that would be interoperable irrespectively of the underlying smart card hardware. Similarly, they were looking for an underlying platform that would be able securely isolate one application from another by performing efficient memory management.

The two principal Multi-application smart card standards are Java Card [21, 67–69] and MULTOS [47]. However, there are two other technologies namely the Smartcard.NET [39, 64] of Hive-Minded and Multi-application BasicCard ZC6.5 [74] of ZeitControl. The main Multi-application smart card platforms are presented below, excluding the Java Card that will be discussed in the next section.

WfSC.

Around 1998, Microsoft launched a new business proposition under the name of Windows for Smart Cards (WfSC). The whole concept was defined in [45], where the underlying platform was capable of performing dynamic application management and the applications were developed in Visual Basic (VB) or C. The philosophy of the proposed architecture [61] was that smart card programming should become part of mainstream programming rather than to force programmers to acquire new competences and new models of programming. In fact, under this framework a WfSC project was simply another type of Visual BASIC or Visual C++ project. A couple of years afterwards, Microsoft dismantled the WfSC team as it became evident that it was no longer directly supporting it. The most notable observation about WfSC was that the overall security functionality was reviewed and guaranteed by Microsoft, which, along with the fact that it was a "closed" system, created barriers in terms of the technology adoption.

Multos.

The Multos [47] smart card operating system was specifically designed for smart card microprocessors. Security was taken into serious consideration right from its conceptualisation. An industrial consortium, named MAOSCO, was created in order to promote MULTOS as a multi-application operating system for smart cards, to manage the specifications MULTOS, and to provide the licences and certifications of MULTOS services. Multos is an operating system that also offers dynamic application management and secure application isolation right at the operating system kernel. Some of the key elements of the MULTOS platform are:

- a highly protected architecture;
- the possibility of having several applications on the same card;
- independence of the applications compared to the underlying platform;
- compatibility with industrial standards such ISO 7816 and EMV [12].

Applications can be developed in a language optimised for smart cards: MEL (Multos Executable Language) based on the Pascal P-codes and the virtual machine described in [72]. From a security point of view, Multos achieved one of the highest certified security levels (i.e. EL6 of ITSEC [44] which corresponds to EAL7 of Common Criteria [20] security evaluations) ever achieved by a commercial product. Although this provided some security reassurance it also increased the overall cost of the platform.

The ZeitControl BasicCard.

The concept of BasicCard came into existence around 1996 with the first card being released in 2004, i.e. MultiApplication BasicCard ZC6.5 [74]. The BasicCard was programmed in ZeitControl BASIC and it also contained a virtual machine. The main difference to the other card was that it was the only one capable of supporting floating numbers [61]. Among the main advantage of the Basic card was that it was a low cost card and it offered a relatively good starting point for smart card programmers

8.3.3 Java Card

Java was among the very first programming languages that took security into account and, at the same time, enjoyed wider acceptance. The general Java language can be described as an "object-oriented (with single inheritance), statically typed, multithreaded, dynamically linked and has automatic garbage collection [where] application developers can use a number of Application Programmers Interfaces (API) when developing new applications" [27]. It is generally accepted that the Java programming language provides a secure model that prevents programs from gaining unauthorised access to sensitive information. At the same time, since Java is based on a runtime byte code interpreter, the portability issue is successfully addressed.

Although Java Card may be perceived as another flavour of traditional Java, the reality is different. The first Java Card API (version 1.0) [65] was released in October 1996 as the initial attempt to bring the benefits of the Java language into the smart card world. Since then, with major enhancements taking place around 1999 with the release of the Java Card API version 2.0 [66], further versions were released incorporating major improvements. Java Card preserves the object-oriented scope of the Java programming language, but it is essentially a subset of normal Java and does not support such things as: Dynamic class loading, security manager, object cloning, threads, and large primitive data types.

The internal architecture of the Java Card specification is illustrated on the right hand side of Figure 8.7. At the bottom layer of the smart card architecture is the hardware (i.e. the actual smart card microprocessor). In the next layer, we encounter the smart card operating system (SCOS). On top of the SCOS, we have the Java Card Virtual Machine (VM). Both the Java Card VM and the SCOS are written in the native language of the microprocessor. This abstraction hides the manufacturer's proprietary technology with a common language interface.

The steps for creating a Java application (the so called applet), download it and execute it on the smart card are the following: First of all, the application programmer must take into account the Java Card APIs and develop an application by using a standard Java development environment. Since there is rarely a byte code verifier in the card, the newly created Java code classes should be verified externally. This implies that the Java Card VM may rely on a digital signature before accepting code from an external source. An external converter is also utilised in order to reduce the size of the actual application. As soon as the Java classes are verified, the application code is ready for loading onto the card via the external application loader. The whole procedure is summarised on the left hand side of Figure 8.7.

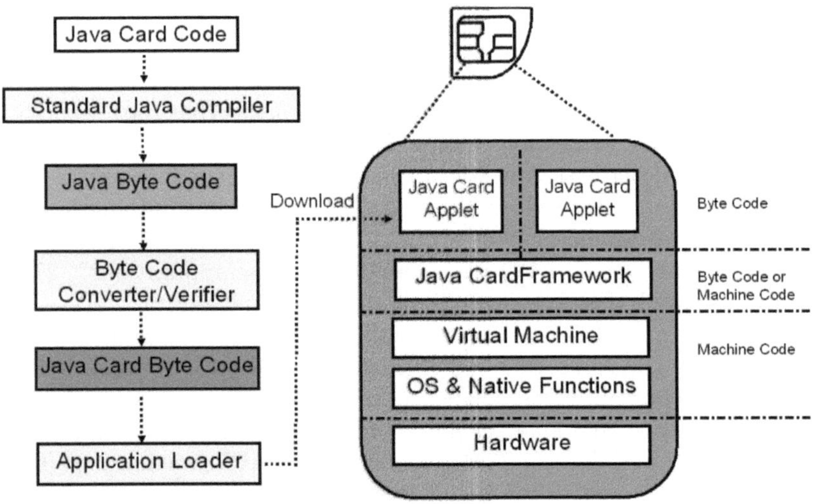

Fig. 8.7. The Java Card application development cycle and the Java Card architecture.

The Java Card platform security feature are inherited from the Java language and several enhancements are also provided through the Java Card framework and run-time environment. The fact that certain Java functionality was removed from the Java Card framework in order to ensure that the technology is ported to smart cards introduced some potential security problems.

The discussion around the security issues introduced is very closely coupled with the security functionality offered by the underlying platform. Therefore, in the next few sections the reader will be presented with the differences and additional functionality introduced by the Java Card platform in relation to the most notable Java Card attacks and vulnerabilities.

8.3.4 Java Card Security

Although there are many attacks directed against the Java Card platform, it is not possible to provide their details here, as they have been explained extensively in the academic literature. However, our discussion on the Java Card risks will also include particular references to the differences with traditional Java that remove certain security problems.

The lack of dynamic class loading within Java Card makes the overall static code verification process simpler. However, it is not possible to guarantee the type safety of classes that are about to be loaded. Furthermore, the fact that only a single Java Card application can be active and executing at any point in time (single threading) further simplifies the overall Java Card security model. Java Card has been extensively criticised in that the programming functionality is seriously restricted within the Java Card API, which can be particularly restraining when new powerful methods need to be implemented, e.g. cryptographic operations. However, the existence of a mechanism that will bypass the Java language syntax and type safety features could allow a programmer to execute arbitrary code, which would have been a major concern for smart card vendors and issuers as it could create serious access violations. Among the most notable properties related to the Java Card security model are the following [49, 73]:

- Garbage collection
- CAP verification
- Applet loading
- The firewall mechanism
- Transaction atomicity
- Native methods
- Exceptions

The lack of a garbage collection mechanism, which is an optional feature, can be a major concern. The idea in normal Java is that unused space is collected and reused. This process is achieved though the existence of a complex mechanism. The so-called "garbage collector" is a relatively large in terms of code size and it was therefore not possible to be defined as a mandatory object within the Java Card specifications. This opens the opportunity to create a malicious Java Card applet that leaks memory resulting in a denial-of-service attack.

The CAP file is a representation of the original Java binary file, modified due to the limited resources available to a given smart card. It is stated in [73] that since the CAP file is restricted in size, it has also lost some of the inherent Java language security features. It is considered to be possible to modify the CAP file and change it in a way that will be able to threaten the security of the underlying platform. Therefore, it is very important to be able to verify that the CAP file adheres to the Java Card security rules. Although Java Card developers initially believed that it was impossible to include a Java

Card verifier on a Java card, due to the various resource constraints. Over the last few years on-card Java verifiers have been developed [19, 46, 59]. Some of these solutions "allow to bypass the signature step of the application without jeopardising the card security". However, the existence of a defensive VM that will dynamically execute the byte codes may be preferable to a stand alone verifier will statically check during the load time [61].

Application downloads, especially at post issuance, lies within the core concept of a multi-application smart card. If applications can be added deleted or modified at post issuance, then the issuers are presented with a relatively powerful and flexible platform that will meet their future requirements. However, the existence of an application downloading mechanism within the Java Card API is not defined. The Global Platform [31] specification comes into play, in order to handle the security sensitive operation of post issuance application downloading. The details of the Global Platform specification are widely known and beyond the scope of this analysis. The Global Platform specification is an important element of Java Card security as it takes care of further security risks. For example, it strengthens the applet firewall (to be explained below) by the existence of a card manager. The Global Platform card manager is responsible, among others, for ensuring that only trusted code (bearing the appropriate credentials) can be downloaded to the card. Therefore, the existence of trusted applications (that have been previously verified) facilitates the existence of an elaborate application sandbox. This in turn requires the existence of thorough application testing and verification process.

The Java Card firewall mechanism is responsible for isolating applications from each other. It ensures that no application is allowed to access another application or its data in an authorised way. The firewall is considered of paramount importance in a multi-application smart card environment, as there is always the risk of inter-application attacks. By default the Java Card [73] applications are only allowed to access data objects that they own. The mapping between what is owned by an application is done during the application downloading and installation process. It is obvious that this process requires the existence of memory protection mechanisms that will perform the necessary checks. The role of the firewall is also to control authorised data sharing. In terms of firewall security there are plenty of attacks that have been described in the literature. "Two attacks against the firewall mechanism (AID) impersonation and illegal reference casting [54] that provides access to all interface methods of a class, along with other attacks originating from problems in the specifications [73]" [22]. A further type of attacks is based on "type confusion" and it is further described in [7].

Transaction atomicity is a further important concept that aims to improve the overall stability of the Java Card platform. Smart cards are often operating in environments in which a user can remove a smart card from its reader at any point in time. In all cases, smart cards, and particularly Java cards, should be in a position to recover when the normal operation is disrupted, e.g.

when power is lost. Transaction atomicity guarantees that any updates to a single persistent object or class will be atomic (i.e. either not performed at all or fully performed). Automatic transaction integrity describes how the virtual machine behaves when power is lost during the update of a field in an object. On the other hand, block transaction integrity describes the virtual machine behaviour when the application programmer identified a specific part of his code that needs to be executed in one piece. For example consider a code example presented in [70] (*CommitBuffer.begin*(), ATC++, ATC++, *CommitBuffer.commit*()). The ATC is a simple Application Transaction Counter. If such a block is defined (*CommitBuffer.begin*()) and not ended with (*CommitBuffer.commit*()) an automatic rollback is performed at the end of the application execution or when card is powered back on.

Native methods are not allowed by the Java Card specifications. However, Java Card application developers would love to be able to access specific calls that will allow them to perform low level tasks, e.g. by obtaining direct access to the any cryptographic coprocessors. As mentioned before, implementing some of the Java Card API functionality requires using native methods. The existence of native methods will completely violate the language based encapsulation model [50] and this one of the main reasons that smart card vendors decided not to offer such functionality. The existence of native methods would also compromise portability.

Exceptions are thrown by the Java Card VM (when internal runtime problems are encountered) or they can be thrown programmatically (Checked exceptions, Unchecked exceptions). The Java Card VM catches the exceptions that are not caught by the application. The Java Card application programmers can define their own exceptions by declaring subclasses of the class Exception. The wide utilisation of exceptions is another interesting topic within Java card. It is mentioned in [49] that unhandled exceptions may cause denial of service attacks. The significance of exception handling was recognised in the Java Card API version 2.1 when the "Exception" class was redesigned.

Java Card CAP file reverse engineering is a potentially serious type attack. It requires obtaining access to a CAP file and running it through a de-compiler, to find out as much information as possible about the application. The next step requires importing certain changes and minor modifications, in order to change the behaviour of the application. However, although minor changes in the CAP file create further changes in the complete file, it is possible to perform pattern changes [23] that will reveal certain information of the application execution.

8.4 GSM and 3G Security

Smart cards play an important role in mobile telecommunications where their security functionality combined with tamper-resistance, underpins the control and viability of the industry. However this was not always the case and the use of smart cards in 2G & 3G systems (such as GSM and UMTS) is largely due to problems that occurred in early systems. The lessons learned form the evolution of the security in mobile phones have played a large part in defining the security issues discussed in Sections 8.2 and 8.3.

8.4.1 1G - TACS

Total Access Communication System (TACS) - was an analog cellular phone systems working at 900MHz and initially deployed within various European countries including UK, Italy, Austria and Spain. TACS was derived from the US cellular system known as Advanced Mobile Phone Service (AMPS). The first commercial TACS system was deployed in the UK and was put into service by 1985.

Early mobile phone systems such as TACS started life using analogue technology and handsets that did not contain smart cards. There was however a security solution embedded in the handsets and network, designed to ensure that only valid users were allowed access to communications services and that they were billed appropriately. This solution was limited to some extent by the available technology but it is probably fair to say that the technical emphasis of the day was primarily to get the radio system to work rather than on its security.

When an analogue phone was switched on, or made a call request, it transmitted (in clear) two important pieces of information; the SNB and the ESN. The SNB was the Subscriber Number (telephone number) and the ESN was the handset Electronic Serial Number. When the network received this information it checked that the transmitted ESN matched the stored value within the network for that particular SNB. In theory the ESN was fixed per phone and the SNB could only be changed by the operator, however many phones were easily tampered with, meaning that both fields could be reprogrammed to create clones.

Typically, an attacker would purchase a radio scanner to monitor the frequencies used by the analogue networks and set up near to an area where analogue phones were likely to be used e.g. motorway services or airport. ESN-SNB pairs could then be captured and this information programmed into another handset. The network would then permit the use of the handset and the charges for the calls made on the "cloned" phone would appear on the legitimate customer's bill. The end result was a major headache for the mobile operators who not only lost revenue because of cloning but also from customers making false claims of cloning to dispute their bills. By 1995 the problems in the UK had become so severe that Members of Parliament were briefed on the issues [57].

8.4.2 2G - GSM

The industry reaction was to ensure that the next generation of digital mobile communication system (GSM) being standardised by ETSI [24] would be designed to avoid clear transmissions of security keys and not be reliant on the discredited tamper-resistant capabilities of handsets. The answer was the introduction of smart cards in the form of Subscriber Identity Modules (SIMs) that are described in detail by GSM 11.11 [25]. From an access and security viewpoint the SIM had to provide integrity, authentication and help ensure confidentiality.

The identity that is important to the GSM network is the International Mobile Subscriber Identity (IMSI) stored in the SIM. It is not the users telephone number but a unique number associated with a customer account. It cannot be altered but is otherwise not very well concealed, as when a user first arrives on a network the IMSI is transmitted in clear over the radio interface. Thereafter a Temporary Mobile Subscriber Identity (TMSI) is allocated and used but this just adds a little extra inconvenience for the attacker. The IMSI can also be read by putting a SIM card in a smart card reader (providing it is not PIN-locked).

The stronger authentication security comes from an algorithm and a 128 bit secret key (Ki) that is stored in the SIM and mimicked in a network system called the Authentication Centre (AuC). Logically the AuC is part of the Home Location Register (HLR) that holds a mapping of all the IMSIs and Kis. The AuC generates the necessary challenges and expected responses to ensure that the SIM card is genuine. The algorithm is sometimes referred to as A3/8 and is in fact two algorithms, one to produce the authentication result (A3) and the other (A8) to produce a radio cipher key. A basic diagram is shown in Figure 8.8.

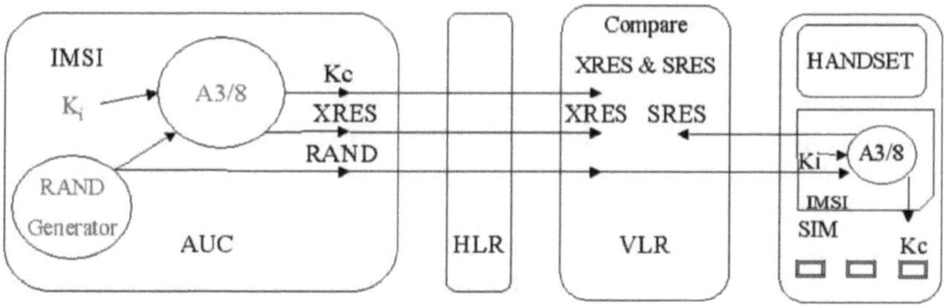

Fig. 8.8. GSM authentication.

When the user wishes to access the GSM network it is not sufficient just to present the IMSI, but also to prove that a valid SIM card is present. This works as detailed in Algorithm 8.4.

This introduction of the tamper-resistant SIM solution was a huge leap forward and overcome the problems so prevalent in the analogue phone system i.e. it prevented simple cloning and offered some confidentiality on the radio interface. History has shown that the SIM has done a pretty good job at maintaining mobile communications over a period of many years, although

Algorithm 8.4 Accessing the GSM network

1. The AuC Generates a 128 bit random number (RAND).
2. The AuC runs the algorithm which uses RAND and secret key (Ki) as inputs.
3. The expected result is produced (XRES) plus a cipher key (Kc) for encrypting radio transmissions.
4. RAND, XRES and Kc. The triplet is handed to a Visitor Location Register (VLR) that is in the same location area as the user.
5. RAND is passed to the SIM via the radio network and the handset.
6. The SIM runs its copy of the algorithm and generates a result (SRES) and a cipher key (Kc).
7. SRES is sent back to the VLR.
8. If SRES equals XRES the user is authenticated.
9. Kc is then used for ciphering subsequent communications.

end.

cannot be claimed to be perfect and its weaknesses have been well publicised. Like all smart cards it can be subject to the security attacks described in earlier sections, but the smart card industry has been vigilant in this respect and modern cards implement robust good countermeasures and are difficult to tamper with.

However, this does not help if a poorly designed security algorithm is used that will reveal its secrets to a logical or brute force attack. There is one particular algorithm called COMP128-1 that is invariably given as an example of, not only a weak authentication algorithm, but also to illustrate the perils of "security by obscurity". Whilst the 128 bit Ki should have held off brute force attacks, the leakage of the design documentation was swiftly followed by researchers at Berkley [13] extracting a secret Ki in about 130K-160K algorithm executions. The researchers had spotted a flaw that made the algorithm vulnerable to collisions i.e. different RAND challenges giving the same output. The algorithm used multiple rounds of compression based on the Ki and RAND values, but referring to Figure 8.9 one can see that the output bytes of the second round are only dependent on 4 input bytes. The final gift to the attackers was that a collision occurring at the second round propagates to the final result. This is described in more detail within the literature [13].

The immediate response was to introduce an authentication retry count into the SIM card so that it would cease to function before the required number of executions had been performed. However, techniques were developed to attack the subsequent rounds of the compression that brought the number of required trials down to 20'000. Side channel leakage was also investigated at IBM [58] bringing the number of executions down to the range of 16 to 1000 to extract the key using a particular form of SPA.

The inescapable conclusion is that there is no reasonable excuse for a GSM network to still issue COMP128-1 SIMs. A common misunderstanding is that it is "the" GSM algorithm whereas it was really only an example made

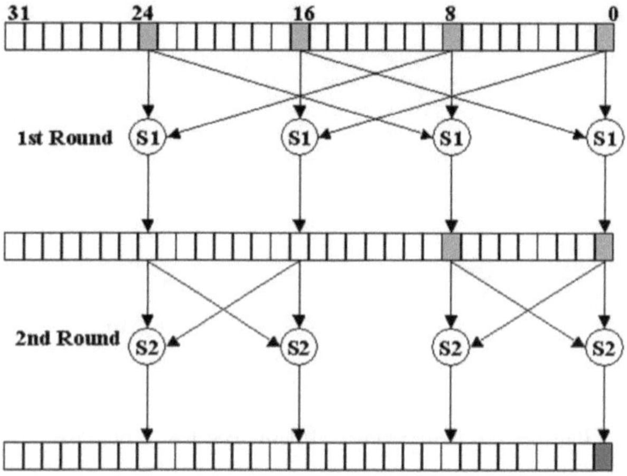

Fig. 8.9. Compression Rounds in Comp128-1.

available to members of the GSM association. Standardisation only defined the method of authentication and network operators are free to specify or develop their own algorithms. The standards also permit the use of AuCs that support multiple authentication algorithms which means that networks can introduce new algorithms whilst gradually phasing out legacy SIMs

Whilst the problems inherent to COMP128-1 were perhaps avoidable there were other security gaps due to the overall GSM security architecture. This is not to say that GSM was badly designed but just that it was a solution to the problems of the previous generation of phones, notably to stop a cloning mechanism, prevent simple radio eavesdropping and make it harder to discover the subscriber's identity. Because of this, GSM is still vulnerable to a type of attack called man-in-the-middle, which becomes possible when a false base station is used as illustrated in Figure 8.10. The attack is described is described in GSM and UMTS [33], and is possible because there is no mutual authentication and the user has no way of proving that the network or base station is legitimate.

Another vulnerability relates to the ciphering that is actually carried out in the handset using an algorithm usually called A5/1 (others are available) using the cipher key (Kc) generated by the A8 algorithm in the SIM. There are academic papers [10] describing attacks on the A5/1 algorithm but this seems not to have resulted in a practical criminal exploitation. This is partly due to the difficulty in collecting sequences of ciphertext and corresponding plaintext but more likely because there are easier ways to access the transmitted information.

Aside from the false base station method, there are other opportunities because ciphering is only applied between the user and the base station, everywhere else in the network the information is unprotected. This means that

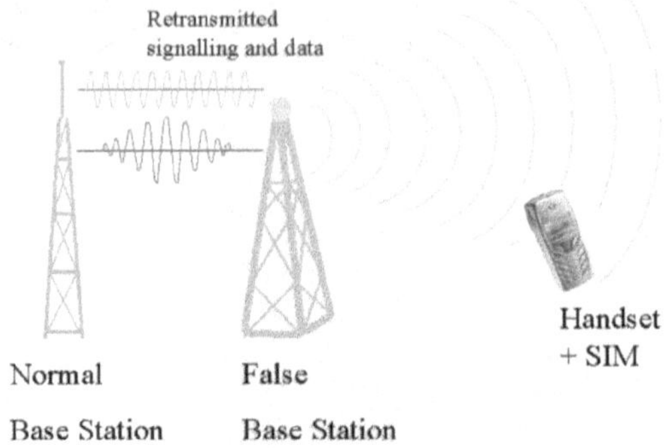

Fig. 8.10. False Base Station Attack Scenario.

a hacker needs physical access to the network but there are miles of cabling to attack as well as microwave links used between some base stations and core network equipment.

8.4.3 3G - UMTS

In designing the successor to GSM the goal was to further exploit the capabilities of the SIM (referred to as a UMTS SIM or USIM in this context) to remove security weaknesses in the GSM architecture and its algorithms, especially in the light of massive advances in computing power. This work was originally driven by ETSI but was eventually handed over to the 3GPP organisation [1]. The most important change was the introduction of mutual authentication between the user and the network based on a strong algorithm. Whereas the description of COMP128-1 was intended to be a secret, the ETSI SAGE team [26] defined a default 3G algorithm (Milenage) based on AES that has been open to expert public scrutiny. The architecture is also improved compared to GSM. The cipher key is longer and there are new integrity (IK) and anonymity (AK) keys plus a sequence number (SQN) to provide replay attack protection. The complete solution at the network end is shown in Figure 8.11 with further detail being found in [2].

The basic authentication process from the user/USIM perspective is as follows: The AuC generates an Authentication Vector (AV), this containing a secret value (MAC) that is generated with knowledge of the subscribers key (K) and the sequence number (SQN). This process is shown in Figure 8.11. To check the MAC the USIM has to follow the sequence shown in Figure 8.12. The USIM generates AK from its own K and the RAND sent from the AuC using function f5. AK is then used to recover the SQN, if this is valid the USIM will then generate the expected value of MAC called XMAC. If XMAC and MAC are identical the network is authenticated. The USIM can then

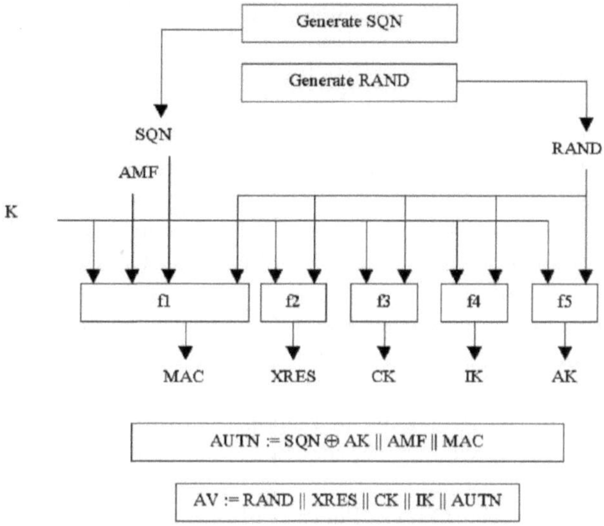

Fig. 8.11. UMTS Network Authentication Vector Generation - source [2].

calculate RES, CK and IK that are used to authenticate the USIM to the network.

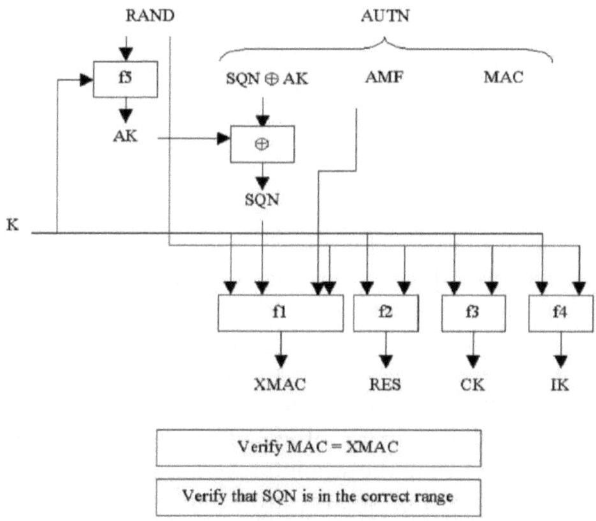

Fig. 8.12. USIM Authentication Vector Handling - source [2].

This solves many of the anticipated problems, but whilst ciphering no longer terminates at the base station and protects some signalling information, it is not end-to-end. The end-point is in fact further back in the network than GSM at a node called the Radio Network Controller (RNC) and this safeguards a lot more of the back-haul, the core interconnect remains unencrypted. Another problem is that users may dynamically switch between GSM and UMTS networks based on coverage, performance and service issues. The hand-over between the networks is regarded by some as a security weak point and researchers have sought to use combinations of GSM attacks to try and undermine UMTS [53].

The above text has introduced and compared only the most fundamental security functions of GSM and UMTS that are underpinned by the tamper-resistant functionality of SIM and USIM smart cards, and shown how the security has improved over time in response to fraud. However the (U)SIM provides a range of auxiliary security functions including PIN management, file access control, remote file and application loading and management either locally or Over The Air (OTA). There is also increasing interest in providing APIs (such as JSR177 [71]) in handheld devices so that the security functionality may be accessed and exploited to help protect handset applications and services. In fact with the introduction of Java card and Global Platform functionality the USIM in common with other UICC based smart cards, has become a powerful multi-application Platform. Whilst this brings new opportunities it also beings new challenges and possibilities for attack.

8.5 Summary

Smart card security can be analysed like many other systems as a stack of services each relying on the security of the lower levels. For example, the application developers should be able to rely on the security of the underlying platform. The platform developer relies on the security of the cryptographic algorithms and the hardware being used.

This chapter has provides an overview of the potential problems on smart card security and how these can be avoided. A case study on the European mobile telephone standards is used as a case study to show how security requirements have evolved over time due to the system being exploited.

Smart cards are often considered the magic solution to a problem; often this just removes a weak link in a protocol. The rest of the process also needs to be analysed to ensure that the whole process is secure.

Research in the domain of smart card security is a cyclic process. New attacks are developed against algorithm implementations, standards, APIs, etc. and countermeasures are proposed. The modifications are then reviewed for potential vulnerabilities and further countermeasures proposed if required. The aim of this process is to remain sufficiently ahead of what can be achieved by an individual attacker that smart cards remain secure throughout the period they are active.

References

1. 3GPP organisation. `http://www.3gpp.org/`
2. 3GPP organisation. 3GPP TS 33.102 3G Security; Security Architecture (Release 99) V3.13.0 (2002-12), 2002
3. M.-L. Akkar and C. Giraud. An implementation of DES and AES secure against some attacks. In Ç. K. Koç, D. Naccache, and C. Paar, editors, *Cryptogaphic Hardware and Embedded Systems — CHES 2001*, volume 2162 of *Lecture Notes in Computer Science*, pp. 309–318. Springer-Verlag, 2001
4. C. Aumüller, P. Bier, P. Hofreiter, W. Fischer, and J.-P. Seifert. Fault attacks on RSA with CRT: Concrete results and practical countermeasures. In B. S. Kaliski Jr., Ç. K. Koç, and C. Paar, editors, *Cryptogaphic Hardware and Embedded Systems — CHES 2000*, volume 2523 of *Lecture Notes in Computer Science*, pages 260–275. Springer-Verlag, 2002
5. F. Bao, R. H. Deng, Y. Han, A. Jeng, A. D. Narasimhalu and T. Ngair. Breaking Public Key Cryptosystems on Tamper Resistant Devices in the Presence of Transient Faults, the Proceedings of the *5th Workshop on Secure Protocols*, volume 1361 of *Lecture Notes in Computer Science*, Springer-Verlag, pp. 115–124, 1997
6. H. Bar-El, H. Choukri, D. Naccache, M. Tunstall, and C. Whelan. The sorcerers apprentice guide to fault attacks. *Proceedings of the IEEE: Special Issue on Cryptography and Security*, 94(2):370–382, IEEE, 2006
7. G. Betarte, E. Gimenez, B. Chetali, and C. Loiseaux. FORMAVIE: Formal Modeling and Verification of Java Card 2.1.1 Security Architecture, In Proceedings of E-Smart 2002, pp. 215–229, 2002
8. E. Biham and A. Shamir. Differential cryptanalysis of DES-like cryptosystems. In A. Menezes and S. Vanstone, editors, *Advances in Cryptology — CRYPTO '90*, volume 537 of *Lecture Notes in Computer Science*, pp. 2–21. Springer-Verlag, 1991
9. E. Biham and A. Shamir. Differential fault analysis of secret key cryptosystems. In B. S. Kaliski Jr., editor, *Advances in Cryptology — CRYPTO '97*, volume 1294 of *Lecture Notes in Computer Science*, pp. 513–525. Springer-Verlag, 1997
10. A. Biryukov, A. Shamir, and D. Wagner. Real time cryptanalysis of A5/1 on a PC, In B. Schneier, editor, *Fast Software Ecryption — FSE 2000*, volume 1978 of *Lecture Notes in Computer Science*, pp. 1–18, Springer-Verlag, 2000
11. J. Blömer and J.-P. Seifert. Fault based cryptanalysis of the advanced encryption standard (AES). In R. N. Wright, editor, *Financial Cryptography*, volume 2742 of *Lecture Notes in Computer Science*, pp. 162–181. Springer-Verlag, 2003
12. D. Boneh, R. A. DeMillo, and R. J. Lipton. On the importantce of checking computations. In W. Fumy, editor, *Advances in Cryptology — EUROCRYPT '97*, volume 1233 of *Lecture Notes in Computer Science*, pages 37–51. Springer-Verlag, 1997
13. M. Briceno, I. Goldberg, and D. Wagner. GSM Cloning. 20 April 1998. `http://www.isaac.cs.berkeley.edu/isaac/gsm.html`

14. E. Brier, C. Clavier and F. Olivier. Correlation power analysis with a leakage model. In M. Joye and J.-J. Quisquater, editors, *Cryptographic Hardware and Embedded Systems — CHES 2004*, volume 3156 of *Lecture Notes in Computer Science*, pp. 16–29. Springer-Verlag, 2004
15. S. Chari, C. S. Jutla, J. R. Rao, and P. Rohatgi. Towards approaches to counteract power-analysis attacks. In M. Wiener, editor, *Advances in Cryptology — CRYPTO '99*, volume 1666 of *Lecture Notes in Computer Science*, pp. 398–412, Springer-Verlag, 1999
16. M. Ciet and M. Joye. Practical fault countermeasures for chinese remaindering based RSA. In L. Breveglieri and I. Koren, editors, *Workshop on Fault Diagnosis and Tolerance in Cryptography 2005 — FDTC 2005*, pp. 124–131, 2005
17. B. Chevallier-Mames, M. Ciet, and M. Joye. Low-cost solutions for preventing simple side-channel analysis: side-channel atomicity. *IEEE Transactions on Computers*, 53(6):760–768, IEEE, 2004
18. C. Clavier, J.-S. Coron, and N. Dabbous. Differential power analysis in the presence of hardware countermeasures. In Ç. K. Koç and C. Paar, editors, *Cryptographic Hardware and Embedded Systems — CHES 2000*, volume 1965 of *Lecture Notes in Computer Science*, pp. 252–263. Springer-Verlag, 2000
19. R. Cohen. The defensive Java virtual machine specification, Technical Report, Computational Logic Inc., 1997
20. Common Criteria. `www.commoncriteria.org`
21. Z. Chen. Java Card Technology for Smart Cards : Architecture and Programmer's Guide, Addison-Wesley, 2000
22. S. Chaumette and D. Sauveron. Some Security Problems Raised by Open Multi-application Smart cards, 10th Nordic Workshop on Secure IT-systems — NordSec 2005, 2005
23. S. Chaumette and D. Sauveron, An efficient and simple way to test the security of Java cards, In Proceedings of the 3rd International Workshop on Security in Information Systems — WOSIS 2005, pp. 331–341. INSTICC Press, 2005
24. European Technical Standards Institute, `http://www.etsi.org/`
25. European Technical Standards Institute. GSM 11:11 - Digital cellular telecommunications system (phase 2+); Specification of the Subscriber Identity Module - Mobil Equipment (SIM-ME) interface, Version 8.3.0, 1999
26. European Technical Standards Institute, Security Algorithms Group of Experts (SAGE). `http://portal.etsi.org/sage/sage_tor.asp`
27. Europay International. MAOS Paltforms Status Technical Report, `www.europay.com`
28. K. Gandolfi, C. Mourtel, and F. Olivier. Electromagnetic analysis: concrete results. In Ç. K. Koç, D. Naccache and C. Paar, editors, *Cryptographic Hardware and Embedded Systems — CHES 2001*, volume 2162 of *Lecture Notes in Computer Science*, pp. 251–261. Springer-Verlag, 2001
29. Gemplus. MPCOS Multi Application Payment Chip, Reference Manual Ver 4.0, 1994
30. C. Giraud and H. Thiebeauld. A survey on fault attacks. In Y. Deswarte and A. A. El Kalam, editors, *Smart Card Research and Advanced Applications VI — 18th IFIP World Computer Congress*, pp. 159–176. Kluwer Academic, 2004
31. Global Platfom. Global Platform Card Specification, Version 2.1, 2001, `http://www.globalplatform.org`

32. L. Hemme. A differential fault attack against early rounds of (triple-)DES. In M. Joye and J.-J. Quisquater, editors, *Cryptographic Hardware and Embedded Systems — CHES 2004*, volume 3156 of *Lecture Notes in Computer Science*, pp. 254–267. Springer-Verlag, 2004
33. F. Hilebrand. GSM & UMTS, Wiley 2002
34. Intercede Group plc. OpenPlatform, `http://www.intercede.com/Technology-OpenPlatform.htm`
35. International Standard Organisation. ISO/IEC 7816, Information technology — Identification cards — Integrated circuit(s) cards with contacts — Part 4: Interindustry commands for interchange, 1995
36. International Standard Organisation. ISO/IEC 7816, Information technology — Identification cards — Integrated circuit(s) cards with contacts — Part 5: Numbering system and registration procedure for application identifiers, 1994
37. International Standard Organisation. ISO/IEC 7816, Information technology — Identification cards — Integrated circuit(s) cards with contacts — Part 6: Inter-industry data elements, 1996.
38. D.H Habing. The use of lasers to simulate radiation-induced transients in semiconductor devices and circuits, In *IEEE Transactions On Nuclear Science*, volume 39, pp. 1647–1653, IEEE, 1992
39. Hive-Minded. Smartcard.NET, `www.hiveminded.com`
40. M. Joye and F. Olivier. Side-channel attacks. In H. van Tilborg, editor, *Encyclopedia of Cryptography and Security*, pp. 571–576. Kluwer Academic Publishers, 2005
41. M. Joye, J.-J. Quisquater, F. Bao, and R.H. Deng. RSA-type signatures in the presence of transient faults, In M. Darnell, editor, *Cryptography and Coding*, volume 1355 of *Lecture Notes in Computer Science*, pp. 155–160, Springer-Verlag, 1997
42. P. Kocher. Timing attacks on implementations of diffe-hellman, RSA, DSS, and other systems. In N. Koblitz, editor, *Advances in Cryptology — CRYPTO '96*, volume 1109 of *Lecture Notes in Computer Science*, pp. 104–113. Springer-Verlag, 1996
43. P. Kocher, J. Jaffe, and B. Jun. Differential power analysis. In M. J. Wiener, editor, *Advances in Cryptology — CRYPTO '99*, volume 1666 of *Lecture Notes in Computer Science*, pp. 388–397. Springer-Verlag, 1999
44. ITSEC. `http://www.ssi.gouv.fr/site_documents/ITSEC/ITSEC-fr.pdf`
45. T. M. Jurgensen and S. B. Guthery. Smart cards : the developer's toolkit, Prentice Hall, 2002
46. X. Leroy. Bytecode verification for java smart card. Software Practice & Experience, volume 32, pp. 319–340, 2002
47. MAOSCO Ltd. The MULTOS^{TM}Specification, `http://www.multos.com/`
48. C. Markantonakis. The case for a secure multi-application smart card operating system. In E. Okamoto, G. I. Davida, and M. Mambo, editors, *Information Security Workshop 97 — ISW '97)*, volume 1396 of *Lecture Notes in Computer Science*, pp. 188–197. Springer-Verlag, 1997
49. G. McGraw and E. W. Felten. Securing java, J. Wiley & Sons, 1999
50. G. McGraw, K. Ayer, and E. W. Felten. Jave Security meets smart cards, security enhancements in java card 2.1.1 will help multi-application smart cards take off in U.S. markets, Information Security Magazin, `http://www.infosecurity.com/articles/march01/cover.shtml`, 2001

51. T. S. Messerges. *Power Analysis Attacks and Countermeasures for Cryptographic Algorithms.* PhD thesis, University of Illinois, Chicago, 2000
52. T. S. Messerges. Using second-order power analysis to attack DPA resistant software. In Ç. K. Koç and C. Paar, editors, *Cryptogaphic Hardware and Embedded Systems — CHES 2000*, volume 1965 of *Lecture Notes in Computer Science*, pp. 71–77. Springer-Verlag, 2000
53. U. Meyer and S. Wetzel, On the Impact of GSM Encryption & Man-in-the-Middle Attacks on the Security of Interoperating GSM/YMTS Networks. In Proceedings of IEEE International Symposium on Personal, Indoor and Mobile Radio Communications — PIMRC 2004, volume 4, pp. 2876–2883, IEEE, 2004.
54. M. Montgomery, K. Krishna. Secure object sharing in Java card, In proceedings of the USENIX Workshop on Smart Card Tehnology — Smartcard '99, USENIX, 1999
55. D. Naccache, P. Q. Nguyễn, M. Tunstall, and C. Whelan. Experimenting with faults, lattices and the DSA. In S. Vaudenay, editor, *Public Key Cryptography — PKC 2005*, volume 3386 of *Lecture Notes in Computer Science*, pp. 16–28. Springer-Verlag, 2005
56. General Information Systems Ltd. OSCAR, Specification of a smart card filling system incorporating data security and message authentication, `http://www.gis.co.uk/oscman1.htm`
57. Parliamentary Office of Science and Technology. Mobile Telephone Crime. In POST Briefing Note 64, 1995
58. J. R. Rao, P. Rohatgi, H. Scherzer, and S. Tinguely. Partitioning attacks: or how to rapidly clone some GSM cards. In Proceedings of IEEE Symposium on Security and Privacy, pp. 31–41, IEEE, 2002
59. E. Rose and K. H. Rose. Lightweight bytecode verification. In Formal Underpinnings of Java — OOPSLA '98, ACM, 1998
60. D. Samyde, S. P. Skorobogatov, R. J. Anderson, and J.-J. Quisquater. On a new way to read data from memory. In *Proceedings of the First International IEEE Security in Storage Workshop*, pp. 65–69, IEEE, 2002
61. D. Sauveron. *Étude et réalisation d'un environnemet d'expérimentation et de modélisation pour la technologie java card^{TM}. application à la sécurité.* PhD thesis, University of Bordeaux, Bordeaux, 2004
62. Season 2 Interface. `http://www.maxking.co.uk/`
63. S. P. Skorobogatov and R. J. Anderson. Optical fault induction attacks. In B. S. Kaliski Jr. and Ç. K. Koç and C. Paar, editors, *Cryptogaphic Hardware and Embedded Systems — CHES 2002*, volume 2523 of *Lecture Notes in Computer Science*, pp. 2–12. Springer-Verlag, 2002
64. SmartCard Trends. .NET brings web services to smart cards, April/May Issue, 2004
65. Sun Microsystems. Java Card API Ver 1.0, `www.javasoft.com/javacard/`
66. Sun Microsystems. Java Card API Ver 2.0, `www.javasoft.com/javacard/`
67. Sun Microsystems. Java Card 2.2.1 Application Programming Interface, 2003
68. Sun Microsystems. Java Card 2.2.1 Runtime Environment (JCRE) Specificqtion, 2003
69. Sun Microsystems. Java Card 2.2.1 Virtual Machine Specification, 2003
70. Sun Microsystems. Java Card API 2.2.1 Reference Implementation, 2002, `http://www.javasoft.com/products/javacard/`
71. Sun Microsystems. JSR 177 Expert Group. Security and Trust Services API (SATSA) for J2ME V1.0, 2004

72. D. A. Watt and D. F. Brown. Programming Language Processors in java: compilers and interpreters, Prentice Hall, 2000
73. M. Witteman, Java Card Security, Information Security Bulletin 8, pp. 291–298, 2003
74. ZeitControl. BasicCard. `http://www.basiccard.com/`
75. J. Ziegler. Effect of Cosmic Rays on Computer Memories, *Science*, volume 206, pp. 776–788, 1979

9

Governance of Information Security: New Paradigm of Security Management

Sangkyun Kim

SOMANSA, Seoul, Korea saviourkim@somansa.com; saviour@yonsei.ac.kr

Governance refers to the process whereby elements in society wield power and authority, and influence and enact policies and decisions concerning public life, and economic and social development [13]. There are three kinds of governance concept which should be considered in corporate environments: enterprise governance, IT governance, and security governance. The success factors of the governance are summarized: Adequate participation by business management; Clearly defined governance processes; Clarify stakeholders' roles; Measure the effectiveness of governance; Facilitate the evolution of governance; Clearly articulated goals; Resolution of cultural issues.

The approaches of security management, which manage an organization's security policy by monitoring and controlling security services and mechanisms, distribute security information, and report security events, are related with the purpose of security governance. However, studies on enterprise governance or IT governance, and security management lack in the provision of detailed framework and functionalities when considering the success factors of the governance described above. For example, BS7799, which is one of the most famous standards of security management in the world, provides general guidance on the wide variety of information security. Nevertheless, it takes the broad-brush approach. Accordingly, BS7799 does not provide definitive or specific materials on any topic of the security management and certainly could be useful as a high-level overview of information security topics that could help senior management to understand the basic issues involved in each of the topic areas.

This chapter provides a structured approach of security governance to corporate executives. Previous studies on the governance and security management are summarized to explain the components and requirements of a governance framework for corporate security. Finally, a governance framework for corporate security, which consists of four domains and two relationship categories, is provided. The domains have several objects respectively. The objects

S. Kim: *Governance of Information Security: New Paradigm of Security Management*, Studies in Computational Intelligence (SCI) **57**, 235–254 (2007)
www.springerlink.com

consist of components that should be resolved or provided to govern the issues of corporate security. The domains include a community (shareholder and management; media and customer; employee and supplier; government), security (control; enterprise strategy), performance (resource; competitive value), and information (owner; value; risk). The relationship among the objects of the security governance framework has two categories of harmonization and flywheel. The harmonization category governs the relationship among a community, performance, and security domain. The harmonization category deals with the problems of social, organizational, and human factors of corporate security. The flywheel category governs the relationship between a performance domain and security domain. The flywheel category deals with the virtuous cycle of corporate security. With this framework, corporate executives could create greater productivity gains, cost efficiencies, and a safer business community internally, for their customers and others interconnected throughout the critical infrastructure.

9.1 Introduction

Data security became computer security, computer security became IT security, IT security became information security and information security became business security because of the better understanding of the business impact and associated risk of not properly protecting a company's electronic resources [55]. However, it's difficult to resolve the problems of business security with the traditional approaches of corporate security management because traditional approach of corporate security management dose not focus on the governance of business impacts but concentrates on the value of information assets and risk analysis [32, 33, 36, 51, 60].

The need for governance exists anytime a group of people come together to accomplish an end. Most agree that the central component of governance is decision-making. It is the process through which this group of people make decisions that direct their collective efforts [01-20]. Recent studies of Conner and Coviello, IT Governance Institute, Moulton and Coles, Solms and Solms, Solms, Solms, Swindle and Conner show that the problems of business security should be resolved with the concept of security governance [8, 24, 41, 55–58].

This chapter provides an integrated framework of security governance. In the following sections, the literature on governance and security management are reviewed to explain fundamental perspectives of governance concept and business security. Second, the success factors of governance of corporate security are described according to the background and implications of governance concept. Third, the components of a governance framework for corporate security are explained along with the common approaches of governance frameworks and the limitations of security management. Finally, a governance framework for corporate security is provided with the domain, object, methods and practices of security governance.

9.2 Rise of the Governance

9.2.1 Definitions of the Governance

Governance refers to the process whereby elements in society wield power and authority, and influence and enact policies and decisions concerning public life, and economic and social development [13]. Yahoo's dictionary defines governance as "the act, process, or power of governing." There are three kinds of governance: enterprise governance, IT governance, and security governance. Related research and definition are summarized as described below.

Category: Enterprise Governance

Research institution or researcher: IT Governance Institute

Definition: Set of responsibilities and practices exercised by the board and executive management with the goal of providing strategic direction, ensuring that objectives are achieved, ascertaining that risks are managed appropriately and verifying that the enterprise's resources are used responsibly [23].

Research institution or researcher: OECD

Definition: It involves a set of relationships between a company's management, its board, its shareholders, and other stakeholders. It also provides the structure through which the objectives of the company are set, and the means of attaining those objectives and monitoring performance are determined. It should provide proper incentives for the board and management to pursue objectives that are in the interests of the company and its shareholders and should facilitate effective monitoring [45].

Category: IT Governance

Research institution or researcher: Allen

Definition: The actions required to align IT with enterprise objectives and ensure that IT investment decisions and performance measures demonstrate the value of IT toward meeting these [1].

Research institution or researcher: Appel

Definition: The principles, processes, people, and performance metrics enacted to enable freedom of thinking and action without compromising the overall objectives of the organization [3].

Research institution or researcher: IT Governance Institute

Definition: The leadership, organizational structures, and processes that ensure that the enterprise's IT sustains and extends the enterprise's strategies and objectives [23].

Research institution or researcher: Neela and Mahoney

Definition: IS(Information System) governance is the mechanism for assigning decision rights and creating an accountability framework that drives desirable behavior in the use of IT [42].

Category: Security Governance

Research institution or researcher: Moulton and Coles

Definition: It is the establishment and maintenance of the control environment to manage the risks relating to the confidentiality, integrity and availability of information and its supporting processes and systems [41].

9.2.2 Implications of the Governance

The benefits gained from introducing effective governance strategy to an IT department are considerable. The holistic view of IT governance integrates principles, people, processes, and performance into a seamless operational whole [3]. IT governance specifies the decision-making authority and accountability to encourage desirable behaviors in the use of IT. IT governance provides a framework in which the decisions made about IT issues are aligned with the overall business strategy and culture of the enterprise [9].

Solms stated that 'Information security is a direct corporate governance responsibility and lies squarely on the shoulders of the Board of the company [56].' Conner and Coviello stated that 'Corporate governance consists of the set of policies and internal controls by which organizations, irrespective of size or form, are directed and managed [8]. Information security governance is a subset of organizations' overall (corporate) governance program.' According to Solms, in the process of establishing environments for information security governance, companies are realizing that is it preferable to follow some type of internationally recognized reference framework for establishing an information security governance environment, rather than doing it ad hoc [57]. By using the information security governance framework, CEOs and boards of directors will create a safer business community internally and for their customers and others interconnected throughout the critical infrastructure. In aggregate, such measures serve as an executive call to action that will also help better protect our nation's security. Information security governance will provide organizations far greater benefits than just legal or regulatory compliance. Robust security serves as a catalyst to even greater productivity gains and cost efficiencies for businesses, customers, citizens and governments during times of crisis and normal operations [58].

IT Governance Institute summarized the necessities of security governance: Risks and threats are real and could have significant impact on the enterprise; Effective information security requires coordinated and integrated action from the top down; IT investments can be very substantial and easily misdirected; Cultural and organizational factors are equally important; Rules and priorities need to be established and enforced; Trust needs to be demonstrated toward trading partners while exchanging electronic transactions; Trust in reliability in system security needs to be demonstrated to all stakeholders; Security incidents are likely to be exposed to the public; Reputational damage can be considerable [24].

9.2.3 Success Factors of the Governance

Dallas proposed six steps which should be took by CIOs to improve IT governance effectiveness. Dallas's model includes: 1. Match the governance initiative with the decision-making style of the enterprise; 2. Align decision-making authority with the domain; 3. Integrate the governance mechanisms - that is, the structures and processes - and evolve them; 4. Clarify stakeholders' roles; 5. Measure the effectiveness of IT governance; 6. Facilitate the evolution of IT governance [10].

Avoid the common causes of IT governance failure by collaborating with business leaders to create processes designed to achieve governance goals tailored to your enterprise's culture and management style. This failure to implement usually results from one or more of the following three factors: inadequate participation by business management; a lack of clearly articulated goals; a lack of clearly defined governance processes [15].

An organization's culture is the common values and beliefs through which it views its people, customers, opportunities and decisions. Organizational culture is a powerful influence on the success or failure of governance mechanisms. Cultures that allow bad governance usually rely more on power than process [42].

To sum up, the success factors of the governance are summarized as described below:

No.1 Adequate participation by business management (align decision-making authority with the domain)

No.2 Clearly defined governance processes (integrate the governance mechanisms and evolve them)

No.3 Clarify stakeholders' roles

No.4 Measure the effectiveness of governance

No.5 Facilitate the evolution of governance

No.6 Clearly articulated goals

No.7 Resolution of cultural issues

9.3 Why the Security Management Fails

9.3.1 What the Security Management Can Do

Bitpipe defines security management as "the management of an organization's security policy by monitoring and controlling security services and mechanisms, distributing security information, and reporting security events." Table 9.1 summarizes the functions and descriptions of the security management.

Among the studies on the security management, BS7799 is the most widely recognized security standard in the world which includes the whole realm of security management. BS7799 is the most widely recognized security standard in the world. Although it was originally published in the mid-nineties, it was revised in May 1999 which really put it on to the world stage. Ultimately, it evolved into BS EN ISO17799 in December 2000. BS7799(ISO17799) is comprehensive in its coverage of security issues containing a significant number of control requirements. Compliance with BS7799 is consequently far from trivial task, even for the most security consciousness of organizations. It consists of ten major sections, each covering a different topic or area.

Business Continuity Planning: The objectives of this section are summarized: To counteract interruptions to business activities and to critical business processes from the effects of major failures or disasters.

Access Control: The objectives of this section are summarized: 1) To control access to information, 2) To prevent unauthorized access to information systems, 3) To ensure the protection of networked services 4) To prevent unauthorized computer access, 5) To detect unauthorized activities, and 6) To ensure information security when using mobile computing and tele-networking.

System Development and Maintenance: The objectives of this section are summarized: 1) To ensure security is built into operational systems, 2) To prevent loss, modification or misuse of user data in application systems, 3) To protect the confidentiality, authenticity and integrity of information, 4) To ensure IT projects and support activities are conducted in a secure manner, and 5) To maintain the security of application system software and data.

Physical and Environmental Security: The objectives of this section are summarized: 1) To prevent unauthorized access, damage and interference to business premises and information, 2) To prevent loss, damage or compromise of assets and interruption to business activities, and 3) To prevent compromise or theft of information and information processing facilities.

Table 9.1. What the security management can do

Function	Related research	Description
Auditing of information security management system	BS7799; ISO17799 [22]	10 domains of evaluation and auditing factors of information security management system
Evaluation of information security management system	Kim and Leem [28]; NIST [44]; Tudor [59]	Evaluation factors of information security management system
Life cycle of security activities	Henze [18]; ISO13335 [21]; Kim et al. [34]; NIST [43]	Process model of enterprise security management
Security strategy planning	Kim and Leem [29]; Kim and Leem [32]	Methodology for strategy planning of enterprise security systems
Domain and capability of security engineering	SEI [54]	Maturity model of information security management system
Risk analysis	Kim and Leem [31]; Kim and Leem [33]; Rex, Charles and Houston [49]; Ron [50]	Analysis and assessment of risks including asset value, vulnerability, and threat
Evaluation of security products	Gilbert [01-16]; Kim and Leem [30]; Lynch and Stenmark [37]; Polk and Bassham [46]	Evaluation and selection process of security controls
Classification, decision and comparison factors of security controls	TCSEC; ITSEC; CC; Beall and Hodges [5]; Firth, et al. [11]; Fites et al. [12]; Hutt [19]; Kavanaugh [26]; Kim and Leem [30]; Kim [35]; Krutz and Vines [36]; Schweitzer [52]; Vallabhaneni [60]	Evaluation and selection factors of security controls
Concept of economic justification	Blakley [6]; Geer [14]; Malik [38]; Scott [53]	Framework of economic justification of security investment
Empirical study on economic justification	Bates [4]; Power [48]	Process of economic justification of security investment
Cost factors of security systems	Harris [17]; Roper [51]; Witty [62]	Cost factors of economic justification of security investment

Compliance: The objectives of this section are summarized: 1) To avoid breaches of any criminal or civil law, statutory, regulatory or contractual obligations and of any security requirements, 2) To ensure compliance of systems with organizational security policies and standards, and 3) To maximize the effectiveness of and to minimize interference to/from the system audit process.

Personnel Security: The objectives of this section are summarized: 1) To reduce risks of human error, theft, fraud or misuse of facilities, 2) To ensure that users are aware of information security threats and concerns, and are equipped to support the corporate security policy in the course of their normal work, and 3) To minimize the damage from security incidents and malfunctions and learn from such incidents.

Security Organization: The objectives of this section are summarized: 1) To manage information security within the Company, 2) To maintain the security of organizational information processing facilities and information assets accessed by third parties, and 3) To maintain the security of information when the responsibility for information processing has been out-sourced to another organization.

Computer and Network Management: The objectives of this section are summarized: 1) To ensure the correct and secure operation of information processing facilities, 2) To minimize the risk of systems failures, 3) To protect the integrity of software and information, 4) To maintain the integrity and availability of information processing and communication, 5) To ensure the safeguarding of information in networks and the protection of the supporting infrastructure, 6) To prevent damage to assets and interruptions to business activities, and 7) To prevent loss, modification or misuse of information exchanged between organizations.

Asset Classification and Control: The objectives of this section are summarized: To maintain appropriate protection of corporate assets and to ensure that information assets receive an appropriate level of protection.

Security Policy: The objectives of this section are summarized: To provide management direction and support for information security.

BS7799 provides general guidance on the wide variety of topics listed above. It takes the broad-brush approach. So BS7799 does not provide definitive or specific materials on any topic of the security management and certainly could be useful as a high-level overview of information security topics that could help senior management to understand the basic issues involved in each of the topic areas.

9.3.2 What the Security Management Cannot Do

By 'elevating' Information Security to Business Security, it will get the extra focus and attention it needs to ensure the prolonged existence of the company, and to integrate all the present efforts as far as such protection is concerned [55]. However, the functions of the security management have some limitations to solve the problems of business security as described in the sequel.

Success factors of governance no.1

Description: Adequate participation by business management (align decision-making authority with the domain)

Related research based on the security management: Kim and Leem [29]; Kim and Leem [30]; Kim and Leem [32]

Limitations (Additional requirements of the security governance): Focuses on the provision of decision-making factors; Concentrates on the technology-oriented factors.

Success factors of governance no.2

Description: Clearly defined governance processes (integrate the governance mechanisms and evolve them)

Related research based on the security management: ISO13335; Henze [18]; Kim et al. [34]; NIST [43]

Limitations (Additional requirements of the security governance): Focuses on the provision of lifecycle of security controls; Lacks in the integration of the governance mechanism and various participants of enterprise security.

Success factors of governance no.3

Description: Clarify stakeholders' roles

Related research based on the security management: Not available

Limitations (Additional requirements of the security governance): Does not provide the relationship and responsibility of stakeholders.

Success factors of governance no.4

Description: Measure the effectiveness of governance

Related research based on the security management: Bates [4]; Blakley [6]; Geer [14]; Harris [17]; Malik [38]; Power [48]; Roper [51]; Scott [53]; Witty [62]

Limitations (Additional requirements of the security governance): Lacks in the provision of strategic benefits and overall performance.

Success factors of governance no.5

Description: Facilitate the evolution of governance

Related research based on the security management: SEI [54]

Limitations (Additional requirements of the security governance): Only provides the maturity model of security engineering (doest not provide a virtuous cycle of enterprise security).

Success factors of governance no.6

Description: Clearly articulated goals

Related research based on the security management: Kim and Leem [29]; Kim and Leem [31]; Kim and Leem [32]; Kim and Leem [33]; Rex, Charles and Houston [49]; Ron [50]

Limitations (Additional requirements of the security governance): Lacks in the alignment of security strategy with business strategy.

Success factors of governance no.7

Description: Resolution of cultural issues

Related research based on the security management: Not available

Limitations (Additional requirements of the security governance): Does not consider the relationship among the participants of enterprise security.

9.4 Governance of Corporate Security

9.4.1 General Frameworks for the Governance

There are various frameworks on the governance. Table 9.2 summarizes the existing studies on a governance framework including the enterprise governance, IT governance, and security governance.

There are also three frameworks which show a fundamental concept of security governance as described in table 9.2. However, these frameworks on the enterprise governance, IT governance, and security governance do not sufficiently meet the requirements of the success factors of security governance. Table 9.3 summarizes the limitations of general frameworks which lack in the provision of an essential architecture, domain, and presentation for the governance of corporate security.

9.4.2 Integrated Framework for the Governance of Corporate Security

Table 9.4 summarizes the key factors of a governance framework for corporate security considering the success factors of the governance, what the security management cannot do, and the limitations of general frameworks.

Table 9.2. General frameworks for the governance

Category	Research	Component
Enterprise governance	Connell et al. [7]	Conformance processes; Performance processes; Accountability and Assurance; Value creation and Resource utilization
IT governance	Weill and Woodham [61]	Component: Business objectives (desirable behavior), IT governance style (IT governance mechanisms), Business performance goals (metrics), IT domains (IT principles, IT infrastructure, IT architecture, IT investment); Relationship: harmonize what, harmonize how
	CobiT [40]	IT processes; IT information criteria; IT resources
	Mercury Interactive Corporation	Demand management; Portfolio management; Program management; Software demand; Operational demand; Deployment; Change management
Security governance	Moulton and Coles [41]	Security responsibilities and practices; Strategies/objectives for security; Risk assessment and management; Resource management for security; Compliance with legislation, regulations, security policies and rules; Investor relations and communications activity in relation to security
	IT Governance Institute [24, 25]	Strategic Alignment; Value Delivery; Risk Management; Performance Measurement
	Posthumus and Solms [47]	Component: Business issues, Legal/regulatory, IT infrastructure, Standards/best practices, Management (vision, strategy, mission, policy); Relation: direct, control, information risk directive

Table 9.3. Limitations of general frameworks

Previous framework	Limitations
Connell et al. [7]	Does not provide the detailed process and community model of enterprise security
Weill and Woodham [61]	Security issues are treated as a partial element of IT management
CobiT [40]	Fundamentally, its a IT auditing methodology; Lacks in the alignment with security strategy with enterprise strategy
Mercury Interactive Corporation	Lacks in the consideration of business security; Focuses on the lifecycle management of enterprise information systems
Moulton and Coles [41]	Performance management and detailed description of security controls are required
IT Governance Institute [24]	Management of the relationships among the participants of enterprise security is required; Factors of investment and performance should be described in further detail
Posthumus and Solms [47]	Relationship among the participants of enterprise security, detailed descriptions of governance components, investment and performance management are required

Table 9.4. Key factors of the governance framework for corporate security

View	Key factors
Architecture	Domain which adjusts the partitions of every object of enterprise security; Clear relationship among domains
Domain	Every participants of enterprise security should be considered; Characteristics of business information; Performance management should provide the cost and benefit factors; Security controls and strategies should be subdivided in further detail
Presentation	Illustration which shows a bird-eye view of the security governance framework should be provided; Structured table which describes every object and component of domain, and their characteristics should be provided

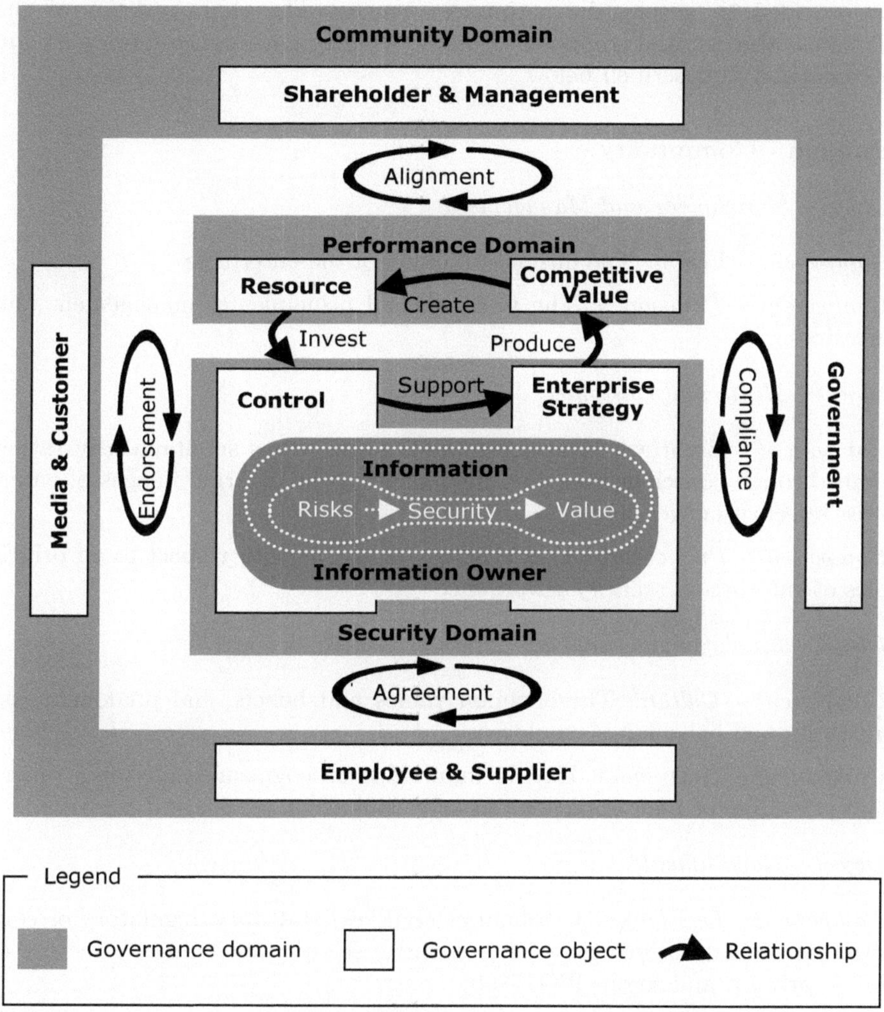

Fig. 9.1. Security governance framework

Considering the key factors of table 9.4, a framework of the security governance is provided as illustrated in figure 9.1.

The governance framework for corporate security consists of four domains and two relationship categories. The domains include: Community, Security, Performance, and Information. With the domains, every participants of enterprise could be considered in governing information security. The domains are correlated with a fundamental consideration of performance management including the cost and benefit factors. The domains have several objects respectively. The objects consist of some components that should be resolved or provided to govern the corporate security. The characteristics of the domains, objects, and components of the security governance framework are summarized as described below.

Domain - Community

Object - Shareholder and Management

Component - Vision: The ultimate mission of the enterprise

Component - Philosophy: The philosophical principles of management and investors

Object - Media and Customer

Component - Emotion: Mental response, which arises spontaneously rather than through conscious effort, of customers on the external images of enterprise's security governance

Component - Public opinion: Public consensus, as with respect to an principles of enterprise's security governance

Object - Employee and Supplier

Component - Culture: The common values and beliefs, and predominating attitudes and behavior of employees

Component - Awareness: Having knowledge or cognizance on the strategy and principles of their enterprise security [60]

Object - Government

Component - Legislation: Criminal or civil law, statutory, regulatory or contractual obligations related with enterprise security including protection of asset, privacy, and so on (ISO17799)

Component - Standard: Set of agreements produced by many organizations; BS7799/ISO17799, ISO13335, ISO7816, ISO9798, ISO10736, ISO11577, ISO14765, ISO9594, ISO10164, ISO10181, ISO8372, ISO9796, ISO10118, and ISO13888 are widely recognized standards of information security

Domain - Security

Object - Control

Component - Policy: General statement produced by senior management to dictate what type of role security plays within the organization [17]

Component - Standard: Specification of the use of specific technologies in a uniform way [36]

Component - Procedures: The detailed steps that are followed to perform a specific task; The detailed actions that personnel must follow [36]

Component - Guideline: Discretionary statement or other indication of security policy or procedure by which to determine a course of action

Component - Responsibilities: A duty, obligation, or burden that each member of the enterprise should keep to achieve the goal and objectives of enterprise's security strategy

Component - System: Technical controls including an access control, encryption, backup, auditing, and intrusion detection

Component - Facility: HVAC and physical access control

Object - Enterprise Strategy

Component - Business strategy: It's the basis for the organization's existence and includes a mission statement which defines the business goals and objectives that typically include references to revenue growth, increased production and performance, competitive advantage, increased quality and service levels, and decreased costs and time-to-market [59]

Component - Information strategy: It's the part of business strategy concerned with deploying the enterprise's information systems resources, including people, hardware, and software [2]

Component - Security strategy: It defines the mission, goals, and objectives of enterprise security management

Domain - Performance

Object - Resource

Component - Facility: Building, geographic location, environmental control and physical safety [36]

Component - Organization: A group of persons organized for security governance; A structure through which individuals cooperate systematically to govern the enterprise security

Component - System: Network, server, and applications

Component - Data: Business data, information, knowledge, and intelligence

Component - Time: Priority of tactical or operational planning

Component - Motivation: Provision of an incentive to improve the agreement and awareness of security principles

Object - Competitive Value [27]

Component - Operational benefits: Enhanced efficiency of organization's operations including cost saving, added profitability, enhanced decision-making and enhanced business function

Component - Strategic benefits: Enhanced competitive advantages including the reduced threat of rivalry, enhanced supplier relationship, and enhanced customer relationship

Domain - Information

Object - Owner

Component - Personnel: Member of the enterprise who creates, processes, memorizes, and communicates the business information

Component - Facility: Media; Environments for media or personnel and system

Component - System: Assembly of network, server, and applications which is used to create, process, store, and share the business information

Object - Value

Component - Official value: Market value of information asset including the cases of exclusive possession, liability, book value, and expert appraisal

Component - Utility: Potential value of information asset for the operational or strategic utilities

Object - Risk [39]

Component - Confidentiality: Protection from unauthorized disclosure

Component - Integrity: Protection from unauthorized, unanticipated, or unintentional modification

Component - Availability: Available on a timely basis to meet mission requirements or to avoid substantial losses.

The relationship among the domains of the security governance framework has two categories of harmonization and flywheel. The characteristics of each relationship are summarized as described below.

Category of Relationship - Harmonization

Relationship - Alignment: The ultimate goal of business security is not a safeguard for the enterprise's information but an improvement of competitive value from the effective support for the enterprise's strategies [35]

Relationship - Agreement: In spite of the importance of support and conformation of employees and suppliers, they seldom have an opportunity to design and implement the enterprise security strategy; Mutual agreement between the management body of enterprise security and employees and suppliers should be established [17]

Relationship - Endorsement: Principles of security governance should not go against the stream of popular opinion [35]

Relationship - Compliance: To avoid breaches of any criminal or civil law, statutory, regulatory or contractual obligations, it should be checked continuously (BS7799; ISO17799)

Category of Relationship - Flywheel

Relationship - Invest: Abundant resources delivered from the enhanced competitiveness are invested to plan, implement, and operate the security controls

Relationship - Support: Security controls achieve the goal of security strategy which is a subset of the business and information strategy [29,32]

Relationship - Produce: Achievement of the goal of security strategy produces an enhanced competitiveness including operational and strategic benefits [27]

Relationship - Create: Enhanced competitiveness resulted from the security governance delivers more abundant resources for security governance.

The harmonization category governs the relationship among the performance, security, and community domain. The harmonization category deals with the problems of social, organizational, and human factors of corporate security. The flywheel category governs the relationship between the performance domain and security domain. The flywheel category deals with the virtuous cycle of corporate security.

9.5 Summary

This chapter provides a governance framework of corporate security. Following studies are summarized: 1) Definition of the enterprise governance, IT governance, and security governance. 2) Literatures on the security management. 3) General frameworks for the governance. Finally, a framework for the security governance, which consists of four domains including eleven objects and two relative categories including eight relationships, is provided.

Managerial implications of the security governance framework are summarized as follow: 1) It integrates strategies, security controls, performance, and communities into a seamless operational whole. 2) It specifies the decision-making authority and accountability to encourage desirable behaviors. 3) Robust governance of enterprise security creates greater productivity gains and cost efficiencies for businesses.

References

1. Allen J (2005) An introduction to governing for enterprise security. Software Engineering Institute, Carnegie Mellon University in Pittsburgh
2. Alter S (1999) Information systems: a management perspective. Addison-Wesley, New York
3. Appel W (2005) Redefining IT governance readiness. META Group
4. Bates RJ (1991) Disaster recovery planning. McGraw-Hill, New York
5. Beall S, Hodges R (2002) Protection and security: software comparison columns. Gartner Inc., Stamford
6. Blakley B (2001) Returns on security investment: an imprecise but necessary calculation. Secure Business Quarterly 1
7. Connell B, Rochet P, Chow E, Savino L, Payne P (2004) Enterprise governance: getting the balance right. International Federation of Accountants
8. Conner FW, Coviello AW (2004) Information security governance: a call to action. National Cyber Security Summit Task Force
9. Dallas S, Bell M (2004) The need for IT governance: now more than ever. Gartner Inc., Stamford
10. Dallas S (2002) Six IT governance rules to boost IT and user credibility. Gartner Inc., Stamford
11. Firth R, Fraser B, Konda S, Simmel D (1998) An approach for selecting and specifying tools for information survivability. Software Engineering Institute, Carnegie Mellon University in Pittsburgh
12. Fites PE, Kratz MPJ, Brebner AF (1989) Controls and security of computer information systems. Computer Science Press, Rockville
13. GDRC (2005) The global development research center (http://www.gdrc.org)
14. Geer DE (2001) Making choices to show ROI. Secure Business Quarterly 1
15. Gerrard M (2003) Creating an effective IT governance process. Gartner Inc., Stamford
16. Gilbert IE (1989) Guide for selecting automated risk analysis tools (SP 500-174). NIST, Gaithersburg
17. Harris S (2003) CISSP all-in-one exam guide 2nd edition. McGraw-Hill, New York
18. Henze D (2000) IT baseline protection manual. Federal Agency for Security in Information Technology, Germany
19. Hutt AE (1988) Management's roles in computer security. In: Hutt AE (eds) Computer security handbook. Macmillan, New York
20. Institute on Governance (2005) What is governance?: getting to a definition (http://www.iog.ca)
21. ISO13335-1: information technology - guidelines for the management of IT security - part 1: concepts and models for IT security. International Organization for Standardization, Geneva
22. ISO17799: information technology - security techniques - code of practice for information security management. International Organization for Standardization, Geneva
23. IT Governance Institute (2001) Board briefing on IT governance. IT Governance Institute, Rolling Meadows
24. IT Governance Institute (2004) Information security governance. IT Governance Institute, Rolling Meadows

25. IT Governance Institute (2001) Information security governance: guidance for boards of directors and executive management. IT Governance Institute, Rolling Meadows
26. Kavanaugh K (2001) Security services: focusing on user needs. Gartner Inc., Stamford
27. Kim S, Lee HJ (2005) Cost-benefit analysis of security investments: a methodology and case study. Lecture Notes in Computer Science 3482: 1239-1248
28. Kim S, Leem CS (2004) An evaluation methodology of enterprise security management systems. Fifth International Conference on Operations and Quantitative Management, Seoul
29. Kim S, Leem CS (2004) An information engineering methodology for the security strategy planning. Lecture Notes in Computer Science 3482: 597-607
30. Kim S, Leem CS (2004) Decision supporting method with the analytic hierarchy process model for the systematic selection of COTS-based security control. Lecture Series on Computer Science and on Computational Sciences 1: 896-899
31. Kim S, Leem CS (2004) Implementation of the security system for instant messengers. Lecture Notes in Computer Science 3314: 739-744
32. Kim S, Leem CS (2004) Information strategy planning methodology for the security of information systems. ICCIE 2004, Cheju
33. Kim S, Leem CS (2005) Security of the internet-based instant messenger: risks and safeguards. Internet Research: Electronic Networking Applications and Policy 15: 88-98
34. Kim S, Choi SS, Leem CS (1999) An integrated framework for secure e-business models and their implementation. INFORMS'99, Seoul
35. Kim S (2002) Security consultant training handbook. HIT, Seoul
36. Krutz RL, Vines RD (2001) The CISSP prep guide: mastering the ten domains of computer security. John Wiley and Sons, New York
37. Lynch G., Stenmark I (1996) A methodology for rating security vendors. Gartner Inc., Stamford
38. Malik W (2001) A security funding strategy. Gartner Inc., Stamford
39. Marianne S (1998) Guide for developing security plans for information technology systems. NIST, Gaithersburg
40. Mingay S, Bittinger S (2002) Combine CobiT and ITIL for powerful IT governance. Gartner Inc., Stamford
41. Moulton R, Coles RS (2003) Applying information security governance. Computers and Security 22: 580-584
42. Neela AM, Mahoney J (2003) Work with, not against, your culture to refine IT governance. Gartner Inc., Stamford
43. NIST (1995) An introduction to computer security: the NIST handbook. NIST, Gaithersburg
44. NIST (2001) Security self-assessment guide for information technology systems. NIST, Gaithersburg
45. OECD (1999) OECD principles of corporate governance. Organization for Economic Cooperation and Development
46. Polk WT, Bassham LE (1992) A guide to the selection of anti-virus tools and techniques (SP 800-5). NIST, Gaithersburg
47. Posthumus S, Solms RV (2004) A framework for the governance of information security. Computers and Security 23: 638-646
48. Power R (2002) CSI/FBI computer crime and security survey. Federal Bureau of Investigation, Washington

49. Rex RK, Charles SA, Houston CH (1991) Risk analysis for information technology. Journal of Management Information Systems 8
50. Ron W (1988) EDP auditing: conceptual foundations and practice. McGraw-Hill, New York
51. Roper CA (1999) Risk management for security professionals. Butterworth-Heinemann, Boston
52. Schweitzer JA (1983) Protecting information in the electronic workplace: a guide for managers. Reston Publishing Company, Reston
53. Scott D (1998) Security investment justification and success factors. Gartner Inc., Stamford
54. SEI (1999) A systems engineering capability maturity model version 2. Software Engineering Institute, Carnegie Mellon University in Pittsburgh
55. Solms BV, Solms RV (2005) From information security to business security?. Computers and Security 24: 271-273
56. Solms BV (2001) Corporate governance and information security. Computers and Security 20: 215-218
57. Solms BV (2005) Information security governance: CobiT or ISO 17799 or both?. Computers and Security 24: 99-104
58. Swindle O, Conner B (2004) The link between information security and corporate governance. Computerworld
59. Tudor JK (2000) Information security architecture: an integrated approach to security in the organization. Auerbach, New York
60. Vallabhaneni R (2000) CISSP examination textbooks. SRV Professional Publications, Los Angeles
61. Weill P, Woodham R (2002) Don't just lead, govern: implementing effective IT governance. Center for Information Systems Research, Sloan School of Management, Massachusetts Institute of Technology in Cambridge
62. Witty RJ, Girard J, Graff JW, Hallawell A, Hildreth B, MacDonald N, Malik WJ, Pescatore J, Reynolds M, Russell K, Wheatman V, Dubiel JP, Weintraub A (2001) The price of information security. Gartner Inc., Stamford

Author Index

MIX
Papier aus verantwortungsvollen Quellen
Paper from responsible sources
FSC® C105338

If you have any concerns about our products,
you can contact us on
ProductSafety@springernature.com

In case Publisher is established outside the EU,
the EU authorized representative is:
Springer Nature Customer Service Center GmbH
Europaplatz 3, 69115 Heidelberg, Germany

Printed by Libri Plureos GmbH
in Hamburg, Germany

Experimental Fluid Mechanics

Series Editors

Wolfgang Merzkirch
Donald Rockwell
Cameron Tropea

For further volumes:
http://www.springer.com/series/3837

Zhengji Zhang

LDA Application Methods

Laser Doppler Anemometry for Fluid Dynamics

Springer

Zhengji Zhang
Oberhasli Hydroelectric Power Company
3862 Innertkirchen
Switzerland
zha@kwo.ch

ISBN 978-3-642-26458-0 ISBN 978-3-642-13514-9 (eBook)
DOI 10.1007/978-3-642-13514-9
Springer Heidelberg Dordrecht London New York

Softcover reprint of the hardcover 1st edition 2010

Cover design: WMXDesign GmbH, Heidelberg

Printed on acid-free paper

Springer is part of Springer Science+Business Media (www.springer.com)

Preface

Experimental flow measurements, as the indispensable measure to investigate and improve engineering flows and flow processes, have been greatly advanced, as the laser methods have found applications in this area. Contrary to the traditional methods of using mechanical probes, the laser method obviously provides the most effective and accurate tools for non-intrusive flow measurements. Nowadays, the laser method for flow measurements has become very fashionable, mainly because of a lot of fashionable applications of laser techniques everywhere. The most widely applied laser methods for flow measurements are doubtlessly the laser Doppler anemometry (LDA), also known as the laser Doppler velocimetry (LDV), and particle image velocimetry (PIV). While the PIV method is suitable to quantitatively image the flow distribution, the LDA method is mostly applied to accurately diagnose and quantify all types of flows. The subject of treatment in the current book is the LDA method.

Since the first successful test of LDA principles, especially during the last 20 years, the LDA technique has been developed to be a high-level standard method for flow measurements. It has been acknowledged as being the most successful and widely applied measurement technique in both the scientific and engineering flow investigations. The advanced laser and computer technologies have greatly contributed to the development of advanced LDA technology. As a very effective and reliable measurement technique, the LDA method demonstrates its established significance not only in the field of mechanical engineering but also extensively in the fields of chemical and biological engineering, as well as in many other fields.

With regard to the uninterrupted developments of LDA technology up until now, it is worth noting that the majority of developments are mainly restricted to LDA principles under application of optics on one hand, and to improved soft- and hardware on the other hand. These developments lead to establishment of standard LDA instruments that have been commercial products and are easily obtainable for applications. Corresponding professional publications mostly concern the basic principle of the LDA method and the related developments named above. Only few investigations and developments have been conducted with regard to the integration of LDA optical facilities into flow mechanics. As has been perceived for a long time, it clearly lacks a supportive reference for LDA users in the practical applications.

The integration of LDA optical facilities into flow mechanics is designated as the LDA application methods. It stands for the methods to improve the optical conditions and to enhance the measurement accuracies. It also provides the guidelines for simplifying the measurements and correcting measurement errors as well as for clarifying the application limits and extending the application areas of LDA techniques. Based on corresponding developments in the last 15 years, the author of this book tries to summarize all important methods related to the aspects listed above and to make a useful reference for LDA users. As a practical reference, the book also contains all other basic knowledge of LDA technology. It is therefore suitable for all LDA users in universities, research institutes and industries. It also supports the further developments of both the hard- and software of LDA instrumentations.

The author highly esteems and thanks his lovely wife Nan for her great spiritual support and the great patient she has shown for many years. He also thanks Sulzer Markets & Technology Ltd for supporting the research works of applying LDA for flow measurements during the years 1990–2003.

Innertkirchen, Switzerland Dr.-Ing. Zhengji Zhang
March 2010

Contents

Symbols

Symbol	Unit	Description
a	m	Half length of the measurement volume
a	m/s^2	Acceleration (velocity gradient in the time domain)
a	m/s	Velocity gradient in the phase angle domain
a	1/s	Velocity gradient in the nozzle and diffuser flows
A	m^2	Area
A_f	m/s	Amplitude of the oscillation carrying flow
c	m/s	Light speed
c_D	–	Drag coefficient in viscous flow
d_{mv}	m	Measurement volume thickness
d	m	Window thickness
d_p	m	Particle diameter
E	–	Wave amplitude
f	m	Lens focal length
f_w	–	weighting factor
F	m^4/s^4	Flatness (velocity)
F	N	Forces
I	–	Light intensity
$\dot{J}$	N/m^2	Momentum flux
k	m^2/s^2	Turbulent kinetic energy
k	1/m	Angular wavenumber of light
k_m	1/m	Modulation wavenumber
k_{mv}	–	Measurement volume shift ratio
k_{vel}	–	Velocity correction factor concerning the fringe spacing
$\vec{l}$	–	Directional unit vector
m_p	kg	Particle mass
n	–	Refractive index
N	–	Sample size
N	–	Fringe number in measurement volume
N_S	–	Stokes number
p	Pa	Pressure
p	1/(m/s)	Probability density function of velocity
$\dot{Q}$	m^3/s	Volumetric flow rate
r	–	Radial coordinate
$\vec{r}$	–	Unit vector
R	m	Circular pipe radius
R	m	Curvature radius of the wave front

Symbol	Unit	Description
R	–	Transformation matrix
R	–	Correlation coefficient
Re	–	Reynolds number
rms	m/s	Root mean square (velocity)
s	m	Distance
s_{tt}	s^2	Statistical quantities
s_{ut}	m	Statistical quantities
s_{uu}	m^2/s^2	Statistical quantities
s_{uv}	m^2/s^2	Statistical quantities
$s_{\mathrm{u}\varphi}$	m/s	Statistical quantities
$s_{\varphi\varphi}$	–	Statistical quantities
S	m^3/s^3	Skewness (velocity)
t	s	Time
tbd	s	Time between data
T	s	Period
Tu	–	Turbulence intensity
u	m/s	Velocity component
$\bar{u}$	m/s	Mean velocity
$\hat{u}$	m/s	Regression velocity
u'	m/s	Fluctuation velocity
$\bar{u}_{\mathrm{app}}$	m/s	Apparent mean velocity
$\bar{u}_{\mathrm{bias}}$	m/s	Biased mean velocity
$\bar{u}_{\mathrm{E}}$	m/s	Mean velocity in energy flux
$\bar{u}_{\mathrm{J}}$	m/s	Mean velocity in momentum flux
u_{p}	m/s	Particle velocity
u_{sh}	m/s	Shift speed of fringes in measurement volume
v	m/s	Velocity component
v_{sh}	m/s	Velocity shift arising from LDA alignment error
w	m/s	Velocity component
w	m	Half thickness of a Gaussian beam
w_0	m	Half thickness of the beam at the beam waist
x	–	Coordinate
y	–	Coordinate
z	–	Coordinate
z_{R}	m	Rayleigh length
α	deg	Half intersection angle between laser beams
α	deg	Angle between coordinates
α	–	Energy flux correction factor
β	deg	Angle between coordinates
β	–	Momentum flux correction factor
γ	deg	Angle between coordinates
γ	–	Fringe distortion number
δ	deg	Bias angle in LDA alignment
ε	deg	Incident and refraction angle
θ	deg	Divergence angle of the Gaussian beam
λ	m	Wavelength
μ	Pas	Dynamical viscosity
ν	Hz	Light wave frequency

Symbol	Unit	Description
ν	m^2/s	Kinematic viscosity
ν_D	Hz	Doppler frequency
ν_{sh}	Hz	Shift frequency
ξ	–	Coordinate
ρ	kg/m^3	Density
σ	m/s	Standard deviation of velocity
σ_{mn}	m^2/s^2	Reynolds stress components
σ_{11}	m^2/s^2	Normal stress in the main flow direction
σ_{22}	m^2/s^2	Normal stress perpendicular to the main flow
τ	deg	Bias angle (error parameter in DMM)
τ	s	Relaxation time
τ_{max}	m^2/s^2	Maximum shear stress
τ_{mn}	m^2/s^2	Shear stress component
φ	deg	Directional angle
$\overline{\varphi}$	deg	Mean flow angle
φ_{LDA}	deg	Off-axis angle of LDA head
φ_m	deg	Directional angle of the principal normal stress
ψ	deg	Bias angle in LDA alignment
ω	1/s	Angular frequency
$\overline{\omega}$	1/s	Mean angular frequency
ω_m	1/s	Modulation angular frequency
Δx	m	Fringe spacing
$\Delta x_{m,s}$	m	Astigmatic difference

Subscript

11	Normal stress in the main flow direction
22	Normal stress perpendicular to the main flow direction
a	Air
app	Apparent velocity and turbulence intensity
b	Biased velocity and turbulence intensity
D	Doppler frequency; drag force
E	Energy flux
f	Flow
g	Glass
J	Momentum flux
m	Meridian focal point; Modulation wavenumber
max	Maximum shear stress
mn	Reynolds stress component
m,s	Distance between meridian and sagittal focal points (astigmatic difference)
mv	Measurement volume
p	Particle
s	Sagittal focal point
sh	Shift frequency, velocity shift
tt	Variance of time series
ut	Covariance between velocity and time
uu	Variance of velocity component u
uv	Covariance between velocity components u and v
uφ	Covariance between velocity component u and phase angle φ
vel	Velocity correction factor
w	Water; Weighting factor

Chapter 1
Introduction

1.1 Flows and Flow Measurements

In industrial applications as well as in scientific research, fluid flows are often utilized to serve diverse functions. The associated physical processes such as those in thermal and fluid engineering, as well as in chemical and biological process controls, constantly require accurate quantifications and optimizations, especially as concerns flow dynamics. The complex flows encountered in diverse industrial applications usually comprise various varieties of turbulent flows, three-dimensional and non-stationary flows, flows with separation and relative eddies, multiphase flows and so forth. To some extent it even deals with non-Newtonian fluid flows. Depending on the application areas and process specifications, most flows are further specified by flow rate, Reynolds number, velocity distribution, turbulence intensity and other relevant flow dynamical parameters. For the flows in heat exchangers, for instance, both the Reynolds number and the related flow state are crucial for the thermal efficiency of the apparatus. In treating flows in aerodynamics the most relevant flow dynamical parameters are directly related to the turbulent boundary layers. Obviously each engineering flow has individual specifications with corresponding flow dynamical parameters. Amongst all of these flows, the flow turbulence acts as the most important and complex phenomenon.

Because of the complexities of most industrial and natural flows, theoretical flow analysis that relies on fluid mechanics mostly appears to be inefficient and unable to quantify the respective flows. This is the case even if simplifications are used. Although nowadays the method of computational fluid dynamics (CFD) has been found to be of wide application in evaluating complex flows and improving the flow processes, its general reliability and applicability still need to be enhanced, especially through experimental validations. Moreover, the CFD method is unable to provide the online analysis of the flow process of interest. For these reasons, experimental flow measurements have often been taken into account, aiming to investigate the flows and thus to optimise the related flow processes.

Zh. Zhang, *LDA Application Methods*, Experimental Fluid Mechanics,
DOI 10.1007/978-3-642-13514-9_1, © Springer-Verlag Berlin Heidelberg 2010

1.2 Traditional Methods of Flow Measurements

Traditional flow measurements in the field of flow dynamics basically include the measurements of velocity and pressure distributions in the flow. The most familiar methods for local flow measurements are using the Pitot tubes to measure total pressure and the Prandtl tubes to gauge dynamical pressure. Because such pressure probes all require insertion into the flow, the flow is disturbed. Both the Pitot and Prandtl tubes are usually restricted to stable or quasi stable flow measurements. They are generally not used for measuring flow turbulence or any high frequency flow fluctuations, mainly because of the delay of pressure signals in the pressure probes. This can happen due to the compressibility of the gaseous flow or other causes. Also, the associated velocity fluctuations cannot be calculated from the measured pressure fluctuations in a straightforward way. In addition, it should be kept in mind that the pressure probes are not applicable to the boundary layers nor to flows with streamline curvatures.

A much more appropriate method for turbulent flow measurements are the hot wire anemometers (HWA). This method makes use of the relationship between the heat transfer on the thin hot wire surface and the flow velocity. The hot wire is only a few micrometers in diameter and is usually made of platinum or tungsten. Such a thin hot wire ensures the rapid response of the hot wire temperature and hence of the electrical signals to the fluctuations in the local flow velocity. Thus the method can be well applied to measurements of most turbulent flows, including those in the turbulent boundary layers. The downside of the method is that the hot wire has to be calibrated prior to each application. The thin hot wire also demands that the flow does not contain any hard particles, which could damage the wire. Because of the considerable flow resistance exerted on the hot wire the method is not applicable to water flows of high velocity. In practice, the hot wire anemometer has found wide application in air flow measurement.

There are many other methods that are used in the practical flow measurements. Most of them require that the flow has to be arranged so that it is mechanically accessible for probe insertion. Another well-known method without flow disturbance is using ultrasonic waves. It has found wide application in pipe flows, where neither mechanical nor optical access to the flow is possible.

1.3 Laser Methods and Laser Doppler Anemometry (LDA)

Modern measurement techniques for flow investigations are doubtless marked by use of laser techniques. The laser method obviously provides the far and wide perspective of measuring flows both more accurately and informatively. On one side, traditional measurement methods (Sect. 1.2) will be replaced where high measurement accuracies are demanded and special flow parameters should be quantified. On the other side, the significant progress of making the laser method to be standard and hence to be easily handled has largely extended the application area of flow measurements. The well-known laser methods being extended in the practical applications are Particle Image Velocimetry (PIV) for flow field measurements and the Laser Doppler Anemometry (LDA), also known as Laser Doppler Velocimetry (LDV), for measurements of local flow velocities. In principle, both methods complement each other.

The Particle Image Velocimetry (PIV) is a method of using a laser light sheet to illuminate particles that are seeded and suspended in the flow. Based on measurements of particle displacements in the image of the visualized flow, the flow distribution in the illuminated flow area can be quantitatively evaluated. This can be conducted by means of efficient evaluation software and high-speed computers, which have all reached a very high standard. The PIV method helps to identify the flow pattern which could comprise flow separation and relative eddies. In comparison with conventional methods of flow visualization for instance using smokes, the PIV method additionally provides quantitative flow information. Also to be mentioned is that this advanced property of PIV measurements has often been insufficiently put to use. As known, the flow measurement merely provides data for further analyses and hence behaves as the prerequisite of flow investigation. In many application examples, quantitative velocities measured by the PIV method are barely exploited to show the flow pattern. For the most part, no further intensive and extensive analyses have been or could be completed. Hence, the flow pattern that is outlined through PIV measurements does not provide any more useful flow information than that provided through the simple flow visualization, for instance by using smoke or color. This comparison implies that the flow investigation is not simply restricted to the stage of flow measurements. More about this aspect will be explained in Sect. 1.4.

The Laser Doppler Anemometry (LDA), of which the functionality and application methods constitute the subject of this book, is probably the most effective and widest applied non-intrusive method in experimental investigations of flows and flow dynamics. It represents an optical, state of the art method commonly with high measurement accuracy. Since the first application of the LDA method by Yeh and Cummins (1964), the method has been continuously developed and extended, so that it nowadays becomes a standard instrument for flow measurements in both industrial applications and model flow investigations. The fundamental development in LDA technology includes the progressive hard- and software developments, which are achieved mainly based on the progressive developments of laser and computer technologies. In general, the LDA technology can be categorized into two areas:

LDA fundamentals
LDA applications

In the aspect of LDA applications, significant achievements have been made since the nineties of the last century, as the LDA method had then found its wide application. In accordance with more and more LDA applications at that time and for the purpose of exchanging the application experiences, many local and global associations have been grounded like EALA (European Association for Laser Anemometry, no longer active) and GALA (German Association for Laser Anemometry).

1.3.1 Developments of LDA Fundamentals and Instrumentations

Since the first LDA measurement was successfully tested by Yeh and Cummins (1964), the method is undergoing continuous development. These developments are mainly restricted to the enhancement of opto-electronic performance of LDA system and the hard- and software improvements. Such developments have enabled the LDA nowadays to be a mature and important measurement instrument which is also commercially available. Within the framework of this book, it is not the purpose to make a historic review of LDA developments. For informative LDA developments, the readers are referred to the earlier work of Durst et al. (1981) as well as to the recent work of Albrecht et al. (2003).

Developments on LDA fundamentals also include investigations of diverse optical and flow-specific aspects that are tightly related to LDA applications and could considerably influence the measurement accuracies. Corresponding influencing factors are known, for instance, as the effect of fringe distortion in LDA measurement volumes, the velocity and angular bias effects, the effect of the time and spatial velocity gradients and so on. Because of their importance in LDA measurements, correspondingly extended investigations have been carried out at an earlier time. They are mentioned here, as follows:

(A) *Velocity bias effect*: The velocity bias arises from the effect that the velocity sampling rate in LDA measurements is not time-equidistant but depends on the velocity itself. In concrete terms, high velocities are proportionally more frequently sampled than low velocities. This effect generally exists both in non-stationary flow and turbulent flow with velocity fluctuations. Because of its dependence on flow velocity, the velocity bias effect is indeed a flow phenomenon. It was firstly recognized by McLaughlin and Tiederman (1973) and later had been broadly investigated by a great number of researchers (for instance Buchhave 1975, Erdmann and Tropea 1981). Corresponding investigations are mainly restricted to the correction of the related measurement error. Based on numerical calculations the velocity bias errors were extensively characterized by Nobach (1998) for three-dimensional flow turbulence. Fully analytical specifications of velocity bias and their dependence on the turbulence intensity have been completely accomplished by Zhang (2000, 2002), for three-dimensional flow turbulence and the turbulence intensity covering the range from zero to infinity. It should be here remarked that the velocity bias does not always represent the measurement error. This viewpoint is completely described in Chap. 17 of this book.

(B) *Fringe distortion*: As an optical phenomenon, the fringe distortion in the LDA measurement volume is mainly caused by improper optical layout or by irregular i.e. asymmetrical laser beam refractions on the medium interface. The fringe distortion in the measurement volume due to improper optical layout was carefully dimensioned by Hanson (1973, 1975). By use of a magnified image the distorted fringe pattern in the measurement volume was visualized for instance by Miles and Witze (1994, 1996). The fringe distortion in the LDA measurement volume as a result of laser beam refractions was confirmed to be the effect of astigmatism by Zhang (1995) and Zhang and Eisele (1995a, b). It represents a particularly crucial phenomenon when the laser beam refraction takes place on the inclined plane surface or the surface of a circular pipe (Zhang 2004a, 2004b). The outcome of all types of fringe distortion in the LDA measurement volume is the deterioration of the measurement accuracy. Measurement errors, for instance, are mainly interpreted in the overestimation of all relevant turbulence quantities. For this reason the associated effect is also called the broadening effect in turbulence measurements (Hanson 1973). Quantitative evaluations of the fringe distortion effect on the flow measurement accuracy were performed by Zhang and Eisele (1997, 1998c) for the case of linear fringe distortion in the measurement volume. For other types of fringe distortion in the measurement volume, the respective effect cannot yet be well-estimated.

(C) *Spatial velocity gradient effect*: For flows with spatial velocity gradient, such as the flows in the turbulent boundary layers, LDA measurements suffer from measurement errors in both mean and fluctuation velocities. The reason for these errors is the non-uniform velocity distribution within the length of the LDA measurement volume. Because the standard LDA optics is unable to resolve the velocity distribution within the measurement volume, both the mean velocity and especially turbulence quantities suffer from measurement errors. Corresponding investigations have been carried out by Durst et al. (1996, 1998). As a matter of fact, the existence of the spatial velocity gradient within the LDA measurement volume also leads to ambiguity in representing the measurement result because of the effect of the involved velocity bias.

(D) *Non-stationary flow measurements*: In the enforced non-stationary turbulent flows, flow fluctuations comprise both the enforced velocity variation because of the flow periodicity, for instance, and the stochastic velocity fluctuations because of the flow turbulence. To evaluate such flows based on LDA measurements, the appropriate data processing has to be worked out. Usually it refers to the method of resolving the stochastic from composite fluctuations. Corresponding investigations on the evaluation method have been carried out by Zhang et al. (1996, 1997) as well as by Jakoby et al. (1996).

1.3.2 Developments of LDA Application Methods

As a result of fundamental developments and the developments of hard- and software, the LDA method has been established to be a very efficient optical technique for flow measurements, especially for investigations of complex turbulent flows. Correspondingly the LDA instrument has become a mature commercial product and found the widest applications. Because of this, the LDA technique has been commonly considered as an available tool, although complicated, for direct application in the flow measurements.

In reality and as experienced from applications, the knowledge of LDA fundamentals and the instrument functionalities does not ensure any fully-correct measurement of flows that are practically of countless varieties with respect to the flow itself, the flow arrangement and the related optical specifications. This viewpoint is objective and true, as it has been well-known for instance that the particle size, the velocity bias and the fringe distortions could largely influence the measurement accuracies. Although there have been countless LDA applications in almost all possible flows, only few attempts have been made to improve the optical conditions and enhance the measurement accuracies, to simplify the measurements and correct measurement errors as well as to clarify the application limits and extend the application areas of LDA techniques. Aside from the concerned velocity bias and fringe distortion, practical applications of LDA method in effect suffer from much more undesirable, partly unknown optical phenomena. The most significant disturbing factor in LDA applications is related with refractions of laser beams for internal flow measurements. In facing this situation and to suppress the occurrence of any optical aberration, the refractive index matching method has been occasionally applied. This method, however, acts only as a passive method and is actually not always applicable. The problem arising from the laser beam refractions will be enlarged, if the beam refractions take place on a curved surface like that of a circular pipe. Obviously the LDA technique with respect to its applications and application optimizations still needs to be developed.

(A) *Optical aberration and astigmatism*: In LDA applications, the optical aberrations generally exist in each measurement of internal flows, where the laser beams must transmit through at least one optical window and hence undergo refractions. The most remarkable optical aberration was confirmed to be astigmatism which in worst cases would lead to total interruption of measurements (Zhang 1995, Zhang and Eisele 1995a, b). The phenomenon and the associated disturbances on both the signal quality and the signal rate become crucial, if the LDA optical axis is aligned off-axis i.e. not coincident with the normal of the optical window. Some LDA users might have experienced that at the mentioned optical configuration either no signal or very bad signals will be received. The reason for signal disappearance is the non-intersection of laser beams after refractions on the air-glass and glass-flow interfaces. The reason for bad signals is mostly the bad intersection between laser beams and the deterioration of receiving aperture of the receiving optics. Another issue of astigmatism is the fringe distortion in the measurement volume that could lead to measurement errors (Zhang and Eisele 1996b). Detailed descriptions of influences of astigmatism on LDA measurements and the guideline for correct measurements with correction method can be found in the mentioned references. To minimize the effect of astigmatism the method of configuring the receiving optics is described in Zhang and Eisele (1996a, 1998b).

(B) *Three-component flow measurements in circular pipes*: Another most common case of LDA measurements is referred to flow measurements in circular pipes. In this case, the optical aberration associated with the laser beam refractions is much more complex and serious than that at a plane surface. In reality, measurements could not run without any aid, either of matching the refractive index of the flow, or by the exact calculation of the laser beam intersection in the flow. Because the refractive index matching method is not available each time, direct measurements of pipe flows must often be carried out. Accordingly great difficulties in obtaining signals of high quality in such measurements would have been encountered by lots of LDA users. Some users tried a great deal to track the laser beams in both the pipe wall and the flow (Boadway and Karahan 1981). As a matter of fact, the optical aberrations in the receiving optics and the corresponding most serious influence on both the signal strength and quality have often been overlooked. As shown by Zhang (2004a, b) the optical performance of direct flow measurement can be greatly enhanced by making the outside surface of the circular pipe to be plane (Fig. 1.1). This contributes not only to the reduction of optical aberrations in both the transmitting and receiving optics, but also to the simplification of calculating the laser beam intersection in the flow for measurements of all three velocity components.

(C) *Dual Measurement Method (DMM)*: As is known, LDA measurements are measurements of velocity components. Because of this property, there are sometimes difficulties in accurately resolving a component, say that one in the secondary flow, which is much smaller against the other components. Usually the direct measurement of such a small velocity component requires exact optical alignment, which is, however, often impossible or very complicated and time-consuming. A method which enables the lowest secondary flow to be exactly measured has been developed by Zhang (2005) and is denoted as the Dual Measurement Method (DMM). The method has been successfully applied to resolve the secondary flow structure in a high speed jet flow, see also Zhang and Parkinson (2001, 2002).

(D) *Zero Correlation Method (ZCM)*: In measuring the turbulent flows, it is often required to measure both the mean velocity and the turbulence quantities such as the turbulence intensity and turbulent stresses in the Reynolds stress matrix. For measurements of turbulent shear stresses, usually two-component LDA system should be applied to enable the fluctuations in two velocity components

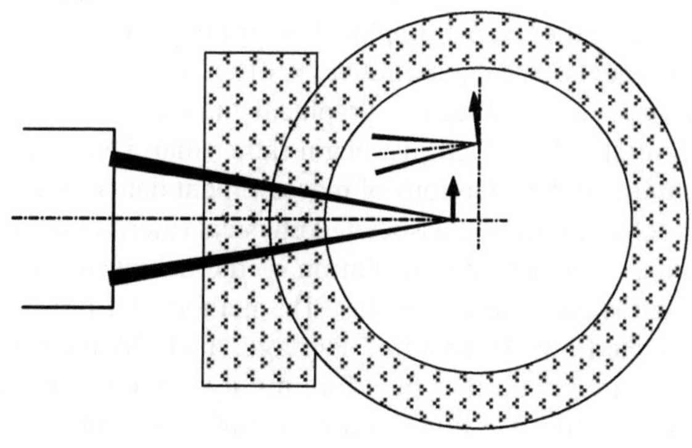

Fig. 1.1 Effective method for measurements of all three velocity components of the flow in a circular pipe (Zhang 2004a)

to be measured simultaneously. The corresponding technique is known as the two-component coincident measurement technique. Although most LDA systems are designed and equipped for doing such measurements, turbulence measurements can be simplified by accounting for the common randomness of velocity fluctuations. For stationary flows, this fluctuation randomness indicates that velocity fluctuations occur symmetrically around the mean flow direction. Based on such an assumption a special method that is called the Zero Correlation Method (ZCM) has been developed by Zhang (1999). The method enables the complete turbulence quantities to be simply measured without requiring the two-component coincident measurement technique.

As outlined above, LDA application methods represent an important category of LDA techniques and play a crucial role in correctly and efficiently carrying out the flow measurements. They form the main subject of this technical book.

1.4 Purposeful Flow Measurements and Rational Measurement Evaluations

Each flow encountered in the engineering applications is arranged to execute the given functions and thus specified by corresponding relevant parameters. It appears to be always important to LDA users to be aware of such parameters prior to starting each measurement. Although the LDA technique nowadays is greatly progressed and the LDA system becomes highly efficient and most convenient for use, it only serves as a useful application tool for flow measurements. In reality, flow investigations will just begin after the measurements have been carried out. For this reason, preliminary studies to specify and clarify the most relevant flow parameters should be made. Such preliminary studies also include how the interested flow quantities can be measured either directly or indirectly, and how accurate the measurement should be. They therefore provide the prerequisite for choosing the available measurement technique and the appropriate measurement method.

On the other hand, it is often a hard task for engineers and researchers to properly evaluate the measurement data. As indicated in Sect. 1.3 referring to the PIV method, the simple graphical mapping of the flow field directly from the PIV measurements, even quantitative, is often no more informative than the qualitative flow visualization. With the graphical mapping of the interested flow field, investigations should indeed just begin rather than just be finished. On one side, more deep studies and evaluations of measurement data require the knowledge of both the flow mechanics and the associated physical (thermal or chemical) processes. On the other side, the poor time-resolution of the achieved measurement data is obviously the significant shortage of the PIV method that prevents the user from further studying the flow. Against this shortage, the LDA method provides highly time-resolved measurements. It is therefore highly suitable for diagnosing the flow for instance by detecting the flow instability, turbulence intensity and exact flow profiles in the region of boundary layers and so on. For this reason, the LDA method is sometimes considered to be comparable with the method of blood test in a clinical laboratory, while the PIV method is comparable with the x-ray method. That is why the LDA method as a diagnostic tool, which provides a lot of useful flow information, is widely applied in practice.

1.5 Purposes of this Book

Since the first LDA application and for a long time thereafter, developments in LDA technology are mainly restricted to LDA fundamentals and instrumentations, as stated in Sect. 1.3. Great advancements in this category of LDA technology enabled the LDA method to become the most favorite technique for flow measurements. The LDA method has thus found its widest applications in experimental flow investigations. As also stated in Sect. 1.3, unique intimate knowledge of LDA fundamentals and instrumentations does not fully ensure correct flow measurements. For LDA users in practical applications, the application methods appear to be as much important and helpful as LDA fundamentals.

The current book therefore tries to completely summarize knowledge that is available in the aspect of LDA application methods. For completeness, the basic fundamentals of LDA measurement techniques and many other relevant aspects like the particle dynamics and velocity transformation algorithm will also be shown. To certain optical aspects, like those in specifying the LDA measurement volume in the circular pipe flows, the mathematical calculations seem to be rather complex. They are nevertheless all crucial for clarifying the background of each optical aspect and for estimating the extent of corresponding influences in measurements. Moreover, they also serve as the basis for further investigations of related optical aspects.

Finally some special applications of LDA methods will be presented. It will also show that the LDA method is just as suitably applicable for solid mechanics as for flow mechanics.

This book with its main subject thus serves as an extended reference for guidance on the LDA applications to users who are attempting to optimize the optical conditions and to gain the maximum results from measurements. Because it is the first book dealing with LDA application methods, it would contribute to the further build-up of LDA technology. Although the book has its weight in the application methods which mainly addresses LDA users, it can also be referred to by developers and manufactures of LDA systems.

Chapter 2
Specifications of Engineering Turbulent Flows

Most flows encountered in the practical applications are turbulent viscous flows. Because of the randomness of flow fluctuations, the flow turbulence probably belongs to the most complex phenomenon in our natural world. It is almost only the flow turbulence that decisively influences the physical (e.g. thermal or chemical) processes in respective fluid flows. On one side, basic research works have been continuously and highly concentrated on the general properties of flow turbulence (Bradshaw 1978, Hinze 1975, Lumley et al. 1996). Efforts have also been made to establish the appropriate turbulence models, especially in the computational fluid dynamics (CFD). On the other side, countless experimental investigations of turbulent flows and their influences on respective flow processes in engineering applications have been carried out. Here the most often used parameter for quantitative characterization of the flow turbulence is simply the turbulence intensity as the outcome directly from statistical evaluation of flow fluctuations. It also represents the most easily obtainable turbulence quantity in experimental flow measurements.

In the aspect of flow dynamics, the significant features of flow turbulence are the time-dependent velocity fluctuations and their statistics regarding the time and spatial extent of fluctuations. With respect to these features of flow turbulence, the most appropriate measurement method is obviously the LDA method that enables the velocity fluctuations in a turbulent flow to be highly resolved and has already been widely applied in experimental flow investigations. For this reason only turbulence quantities will be considered here that are tightly and directly related with LDA measurements in engineering flows.

2.1 Turbulent Flow Properties

2.1.1 Statistical Views of Flow Turbulences

Turbulent flows are known as the flows with irregular fluctuations of fluid particles in motion. To describe the turbulent flow with velocity fluctuations, the flow velocity is usually split into a time-averaged mean and a fluctuation velocity. In regarding

Zh. Zhang, *LDA Application Methods*, Experimental Fluid Mechanics,
DOI 10.1007/978-3-642-13514-9_2,

the velocity component u for instance, the flow velocity with fluctuations is then expressed as

$$u(t) = \bar{u} + u'(t) \tag{2.1}$$

In accordance with this treatment of a turbulent flow, there is $\overline{u'(t)} = 0$ which means that the time-average of velocity fluctuations vanishes. In order to statistically quantify the extent of flow fluctuations, the standard deviation of the mean velocity of a stationary turbulent flow has always been applied, as it is calculated for the given velocity component by

$$\sigma_{\mathrm{u}} = \sqrt{\frac{1}{T}\int_0^{\mathrm{T}} u'^2 \mathrm{d}t} \tag{2.2}$$

Velocity fluctuations around the mean velocity are of different magnitudes and take place to different magnitudes by different probabilities. While velocity events in velocity classes close to the mean velocity happen at high probabilities, velocity events far from the mean appear rather few. Mostly the probability distribution of fluctuation velocities is of a symmetrical bell form as shown in Fig. 2.1 for example from measurements of a stationary turbulent flow. In most cases dealing with stationary turbulent flows, flow fluctuations are stochastic and can be approximated by the Gaussian probability density function as given by

$$\mathrm{pdf_u} = \frac{1}{\sqrt{2\pi}\sigma_{\mathrm{u}}} \mathrm{e}^{-\frac{(u-\bar{u})^2}{2\sigma_{\mathrm{u}}^2}} \tag{2.3}$$

In this expression, σ_{u} in m/s is the standard deviation of the mean velocity $\bar{u}$ and has been specified in Eq. (2.2). It represents the statistical variability of velocity and

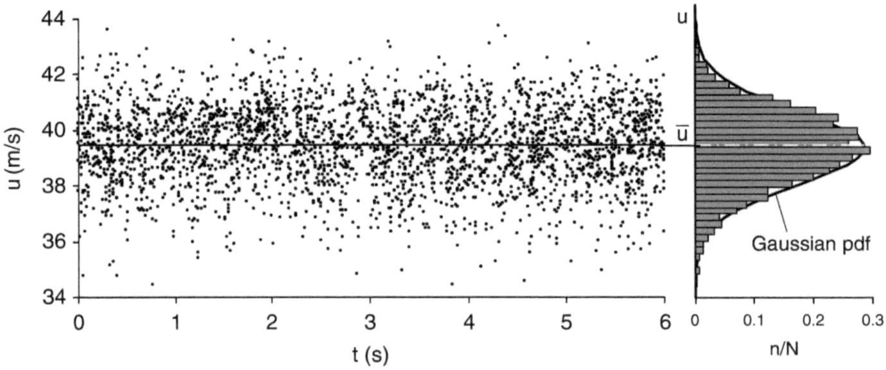

Fig. 2.1 Fluctuation velocity and its probability distribution in a high speed jet flow measured by LDA method

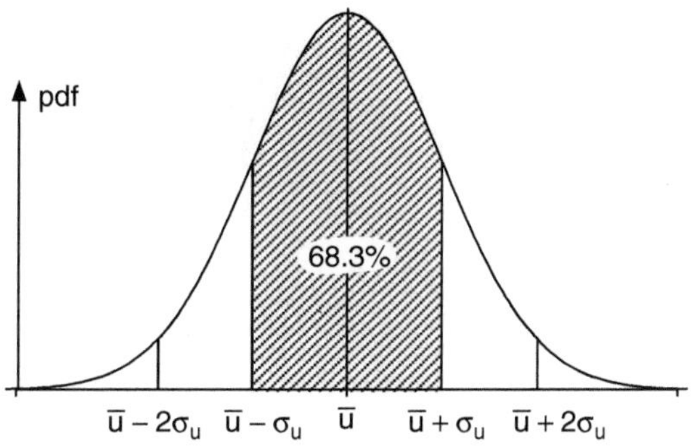

Fig. 2.2 Gaussian probability density function

specifies the most available range of flow fluctuations around the time-average of respective velocity components. In Fig. 2.2 the corresponding probability density function has been shown to approximate the distribution of fluctuation velocities. The probability of random velocities occurring in the range $u = \bar{u} \pm \sigma_\mathrm{u}$ is calculated as

$$P(\sigma_\mathrm{u}) = \int_{\bar{u}-\sigma_\mathrm{u}}^{\bar{u}+\sigma_\mathrm{u}} \mathrm{pdf}_\mathrm{u} \mathrm{d}u = \frac{2}{\sqrt{\pi}} \int_0^{1/\sqrt{2}} \mathrm{e}^{-z^2} \mathrm{d}z = \mathrm{erf}\left(\frac{1}{\sqrt{2}}\right) \approx 68.3\% \tag{2.4}$$

Here the substitution of $z = \frac{u-\bar{u}}{\sqrt{2}\sigma_\mathrm{u}}$ was applied. The term erf(x) is the error function. It vanishes when $x = 0$ and approaches unity as x tends to infinity.

Furthermore, the probability of random velocities occurring in the range $u = \bar{u} \pm 2\sigma_\mathrm{u}$ is given by

$$P(2\sigma_\mathrm{u}) = \int_{\bar{u}-2\sigma_\mathrm{u}}^{\bar{u}+2\sigma_\mathrm{u}} \mathrm{pdf}_\mathrm{u} \mathrm{d}u = \frac{2}{\sqrt{\pi}} \int_0^{\sqrt{2}} \mathrm{e}^{-z^2} \mathrm{d}z = \mathrm{erf}\left(\sqrt{2}\right) \approx 95.4\% \tag{2.5}$$

The standard deviation σ_u generally represents a statistical parameter that depicts the extent of non-uniformity in a data series as in the process fluctuations. In relating to LDA measurements of a stationary turbulent flow, the method of calculating the standard deviation σ_u will be shown in Chap. 5.

2.1.2 Isotropic and Anisotropic Turbulences

Since velocity fluctuations in each turbulent flow are always three-dimensional, it distinguishes between isotropic and anisotropic turbulences, as illustrated in Fig. 2.3 in the general field coordinate system. Local isotropy of turbulence is given at large Reynolds numbers and conditions without remarkable influences of any boundaries.

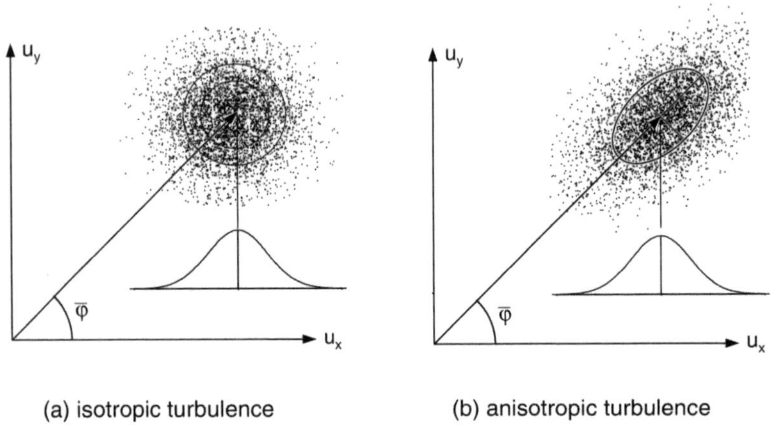

Fig. 2.3 Isotropic and anisotropic turbulences (two-dimensional), shown in the form of scatter diagrams of the fluctuation velocity measured by LDA method

In such a turbulent flow, velocity fluctuations in all spatial directions are of the same statistical level i.e. $\sigma_{\mathrm{x}} = \sigma_{\mathrm{y}} = \sigma_{\mathrm{z}} = \sigma$ in terms of standard deviations. The standard deviation of the mean velocity can then be directly applied to calculate the turbulence intensity of the related turbulent flow:

$$\mathrm{Tu} = \frac{\sigma}{\bar{u}} \tag{2.6}$$

with $\bar{u}$ as the mean velocity in the main flow direction.

As a matter of fact, turbulence encountered in most practical flows is anisotropic, mainly because of reactions of existing boundaries on flow fluctuations. The turbulent flow with anisotropic turbulence as shown in Fig. 2.3b represents the most common case. It is characterised by two distinct aspects: the first aspect is that velocity fluctuations are confirmed to more or less symmetrical distribution on both sides of the mean velocity vector. This property of the flow turbulence simply arises from the randomness of velocity fluctuations. It can be made of use in order to simplify both the specification of the flow turbulence and the measurement of corresponding turbulence quantities, see Chap. 8. The second aspect is that the velocity fluctuation along the mean velocity vector is of maximum magnitude while it is of minimum magnitude perpendicular to the mean velocity vector. The turbulence with minimum fluctuation magnitude along the mean velocity vector is practically uncommon. Its mathematical treatment is equal to that of the common case. For this reason, as well as for simplicities of representing the measurement techniques, only the common case of anisotropic turbulences according to Fig. 2.3b will be further considered.

The asymmetrical distribution of velocity fluctuations around the mean velocity vector has usually been found in the flows where large velocity gradients are present. This type of velocity fluctuations can be confirmed for instance in the shear flow of turbulent boundary layers, as shown by Zhang and Zhang (2002). To some extent, the approximation of symmetrical velocity fluctuations can be made in order to simplify both the measurements and data processing. In principle, it merely depends on the requirement of the measurement accuracy.

2.2 Reynolds Turbulent Stresses

The basic equations of describing the stationary turbulent viscous flows are momentum equations which are also called the Reynolds-averaged Navier–Stokes (RANS) or simply Reynolds equations. In the Cartesian coordinate system with u, v and w as three orthogonal velocity components at a local point in the flow, the Reynolds equations are given as

$$\rho\left(\bar{u}\frac{\partial\bar{u}}{\partial x}+\bar{v}\frac{\partial\bar{u}}{\partial y}+\bar{w}\frac{\partial\bar{u}}{\partial z}\right)=-\frac{\partial\bar{p}}{\partial x}+\mu\nabla^2\bar{u}+\left(\frac{\partial\rho\overline{u'u'}}{\partial x}+\frac{\partial\rho\overline{u'v'}}{\partial y}+\frac{\partial\rho\overline{u'w'}}{\partial z}\right) \tag{2.7}$$

$$\rho\left(\bar{u}\frac{\partial\bar{v}}{\partial x}+\bar{v}\frac{\partial\bar{v}}{\partial y}+\bar{w}\frac{\partial\bar{v}}{\partial z}\right)=-\frac{\partial\bar{p}}{\partial y}+\mu\nabla^2\bar{v}+\left(\frac{\partial\rho\overline{v'u'}}{\partial x}+\frac{\partial\rho\overline{v'v'}}{\partial y}+\frac{\partial\rho\overline{v'w'}}{\partial z}\right) \tag{2.8}$$

$$\rho\left(\bar{u}\frac{\partial\bar{w}}{\partial x}+\bar{v}\frac{\partial\bar{w}}{\partial y}+\bar{w}\frac{\partial\bar{w}}{\partial z}\right)=-\frac{\partial\bar{p}}{\partial z}+\mu\nabla^2\bar{w}+\left(\frac{\partial\rho\overline{w'u'}}{\partial x}+\frac{\partial\rho\overline{w'v'}}{\partial y}+\frac{\partial\rho\overline{w'w'}}{\partial z}\right) \tag{2.9}$$

In these equations ∇^2 is the Laplace operator which operates for instance

$$\nabla^2 u=\frac{\partial^2 u}{\partial x^2}+\frac{\partial^2 u}{\partial y^2}+\frac{\partial^2 u}{\partial z^2} \tag{2.10}$$

Flow fluctuations in the turbulent flows represent the momentum exchange between fluid particles. Their statistical properties are given by local gradients of Reynolds turbulent stresses which are involved in above equations in terms of $\partial\rho\overline{u'u'}/\partial x$ and other similar terms. For simplicity and later convenience, constant density of fluid is assumed and the Reynolds stresses are represented by the variance and the covariance of respective velocity components. The Reynolds turbulent stresses are commonly expressed in the following matrix form

$$\sigma_{\mathrm{mn}}=\begin{vmatrix}\sigma_{xx} & \tau_{xy} & \tau_{xz}\\ \tau_{yx} & \sigma_{yy} & \tau_{yz}\\ \tau_{zx} & \tau_{zy} & \sigma_{zz}\end{vmatrix}=\begin{vmatrix}\overline{u'^2} & \overline{u'v'} & \overline{u'w'}\\ \overline{v'u'} & \overline{v'^2} & \overline{v'w'}\\ \overline{w'u'} & \overline{w'v'} & \overline{w'^2}\end{vmatrix} \tag{2.11}$$

The turbulent stresses σ_{xx}, σ_{yy} and σ_{zz} in $\mathrm{m}^2/\mathrm{s}^2$ are called the Reynolds normal stresses. For flow fluctuations that approximately fulfill the Gaussian probability distribution, each normal stress is equal to the square of the respective standard deviation that has been defined in Eq. (2.2) and shown for application in Eq. (2.3). Correspondingly, turbulent stresses τ_{mn} with $m \neq n$ are called the Reynolds shear stresses. Both turbulent normal and shear stresses are time-averages of velocity fluctuations in corresponding velocity components. According to the definition of Reynolds shear stresses in Eq. (2.11) there are simply

$$\tau_{xy} = \tau_{yx}, \quad \tau_{xz} = \tau_{zx}, \quad \tau_{yz} = \tau_{zy} \tag{2.12}$$

In addition, each turbulent shear stress can be positive and negative, depending on the flow fluctuations themselves and the used coordinate system. This is, however, only of mathematical relevance. It does not represent any difference in flow properties. As can also be recognized from Reynolds equations given above, it is not the Reynolds stresses themselves but their gradients which determine the flow states and the related flow dynamics. For this reason, only the absolute maximum value of the turbulent shear stress behaves as a flow parameter. Other relevant turbulent parameters are the principal normal stresses, as described below.

Turbulences with equal normal stresses i.e. $\sigma_{xx} = \sigma_{yy} = \sigma_{zz}$ are called isotropic turbulence, as already concerned in Fig. 2.3a. It also indicates that all turbulent shear stresses at the considered local point in the flow vanish automatically. This circumstance indeed represents a special case of the general anisotropic turbulence with $\sigma_{xx} \neq \sigma_{yy} \neq \sigma_{zz}$ which has been found in most practical turbulent flows.

The Reynolds stress matrix given in Eq. (2.11) is related to the Cartesian coordinate system $x - y - z$. According to the matrix algebra there exists an orthogonal coordinate system in which all related turbulent shear stresses vanish. The Reynolds stress matrix then takes the form

$$\sigma_{mn} = \begin{vmatrix} \sigma_{11} & 0 & 0 \\ 0 & \sigma_{22} & 0 \\ 0 & 0 & \sigma_{33} \end{vmatrix} \tag{2.13}$$

The remaining normal stresses are known as the principal normal stresses at the considered local point in the turbulent flow. They are ordered with respect to their values to σ_{I}, σ_{II} and σ_{III}, conventionally with $\sigma_{I} > \sigma_{II} > \sigma_{III}$.

The turbulence state in a turbulent flow can thus be described by all three normal stresses, which also determine the maximum of the absolute shear stress (Chap. 6). Another intrinsic property of the turbulence state at a local point in the flow is that the sum of all three normal stresses remains constant as given by $\sigma_{I} + \sigma_{II} + \sigma_{III} = \sigma_{xx} + \sigma_{yy} + \sigma_{zz}$, independent of the applied coordinate system. This constant sum is known as the first invariant of the matrix given by Eq. (2.11). It actually represents the mean value of the turbulent kinetic energy that is involved in the flow fluctuations and usually written as

$$k = \frac{1}{2}\rho\left(\overline{u'^2} + \overline{v'^2} + \overline{w'^2}\right) \tag{2.14}$$

In using the first invariant of the Reynolds stress matrix, the local turbulence intensity of a turbulent flow is calculated by

$$\mathrm{Tu} = \frac{1}{\sqrt{\bar{u}^2 + \bar{v}^2 + \bar{w}^2}} \sqrt{\frac{1}{3}\left(\overline{u'^2} + \overline{v'^2} + \overline{w'^2}\right)} \tag{2.15}$$

For isotropic turbulence with $\overline{u'^2} = \overline{v'^2} = \overline{w'^2}$ Eq. (2.15) reduces to Eq. (2.6).

It should be mentioned that the turbulence intensity actually behaves as a statistical parameter and does not signify any mechanical significance of the flow turbulence. The physical property of the flow turbulence is indeed always related to the turbulent kinetic energy i.e. in the proportional form of Tu^2.

Chapter 3
LDA Principles and Laser Optics

The technique of Laser Doppler Anemometry (LDA), as the name stands for, is a technique of using the laser light and the Doppler effect for velocity measurements. It is an optical method and hence tightly related to both the physical and geometrical optics. In order to depict the functionality of the LDA method, some physical properties of the light and the light wave propagation in medium will firstly be considered.

3.1 Light Wave and Its Propagation

The light is the electromagnetic wave which is specified by its amplitude, the polarization and the wavelength λ. In the LDA technology regarding special laser light properties, the polarization of the laser light is less interesting, at least for LDA users. It will only be considered by splitting the laser beam into parts and transmitting them through single-mode fibers to the LDA head. In addition, it will also be taken into account if the change in the light intensity due to refraction has to be concerned based on Fresnel equations. For laser light refractions at not too large refraction angles, the effect of laser light polarization on the change in the laser light intensity has usually been neglected. By disregarding the light polarization the spatial propagation of a plane light wave of the amplitude E_0 in the positive x-direction can be expressed by

$$E = E_0 \cos(\omega t - kx) \tag{3.1}$$

The parameters $\omega = 2\pi/T$ and $k = 2\pi/\lambda$, respectively, represent the angular frequency and the angular wavenumber, or simply wavenumber of the light wave. They are coupled by the wave propagation speed i.e. the light speed in the medium in which the light propagates. As known, the wave propagation speed is expressed by the phase velocity that is determined by

$$\frac{\mathrm{d}(\omega t - kx)}{\mathrm{d}t} = 0 \tag{3.2}$$

Zh. Zhang, *LDA Application Methods*, Experimental Fluid Mechanics,
DOI 10.1007/978-3-642-13514-9_3, © Springer-Verlag Berlin Heidelberg 2010

The light speed is then obtained as

$$c = \frac{\mathrm{d}x}{\mathrm{d}t} = \frac{\omega}{k} \tag{3.3}$$

Because of the constant light speed in the homogeneous medium, this equation signifies that the light wave can be represented either by the angular frequency ω in the time domain or equivalently by the wavenumber k in the space. The light speed in Eq. (3.3) can also be expressed as $c = \nu\lambda$ with $\nu = 1/T$ as the oscillation frequency of the light wave. In the vacuum, the light speed is measured as $c = 2.99792458 \times 10^8$ m/s. In the dielectric or non-conducting medium, like water for instance, the light speed is less than that in the vacuum. The ratio between them is denoted by n and called the refractive index of the respective dielectric medium. For the case that the light propagates from one medium (n_1) into another (n_2), the light speed then changes from c_1 to c_2 according to the following relation

$$\frac{c_2}{c_1} = \frac{n_1}{n_2} \tag{3.4}$$

Since the light frequency does not change while the light is refracted at the interface between two mediums, Eq. (3.4) can further be written with respect to $c = \nu\lambda$ as

$$\frac{c_2}{c_1} = \frac{n_1}{n_2} = \frac{\lambda_2}{\lambda_1} \tag{3.5}$$

This means that the wavelength of the light in a medium is reciprocally proportional to the medium refractive index n.

The refractive index n is an optical and physical parameter of a dielectric medium. It is not only a function of the medium and the medium temperature, but also a function of the light wavelength (color). The latter phenomenon is known as the dispersion or the chromatic aberration. In LDA measurements, the optical dispersion is generally neglected without causing any significant measurement errors (Chap. 13).

The transmission of light from one medium into another is also related to the change in the direction of light propagation. This is described by the law of refraction (also known as Snell's law) according to Fig. 3.1:

$$n_1 \sin\varepsilon_1 = n_2 \sin\varepsilon_2 \tag{3.6}$$

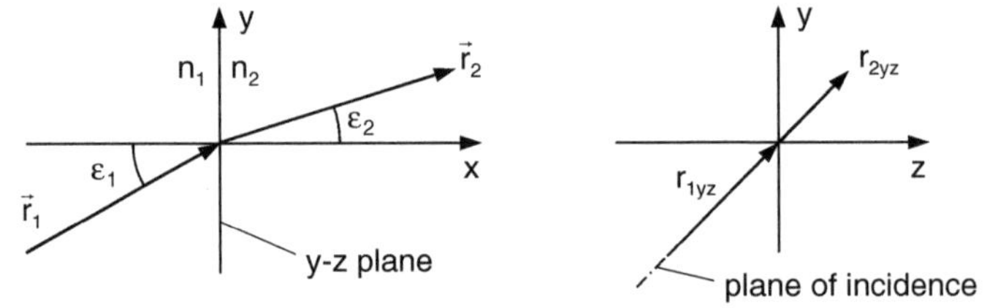

Fig. 3.1 Light ray refraction and the plane of incidence

In this equation, ε_1 and ε_2 represent the angles of incidence and refraction (or transmission), respectively. They are measured from the normal of the medium interface. Corresponding mediums are denoted as the incident and the refractive mediums.

The law of refraction given above can be generalized. According to Fig. 3.1 the x-axis agrees to the normal of the medium interface. The incident ray and the normal of the medium interface determine a plane that is known as the plane of incidence. The law of refraction also specifies that the refracted ray must also lie in the plane of incidence. For the purpose of performing extended calculations, the propagation directions of both the incident and the refracted rays are denoted by unit vectors $\vec{r}_1$ and $\vec{r}_2$, respectively. Their projections onto the medium interface ($y - z$ plane) are also shown in Fig. 3.1. Corresponding vector components are given by r_{1yz} and r_{2yz} in the $y - z$ plane. Clearly, these two vector components are calculated by $\sin\varepsilon_1$ and $\sin\varepsilon_2$, respectively. Equation (3.6) is then expressed as

$$r_{2yz} = \frac{n_1}{n_2} r_{1yz} \tag{3.7}$$

Because two sub-vectors $\vec{r}_{1yz}$ and $\vec{r}_{2yz}$ are parallel vectors in the $y - z$ plane, following expressions can be immediately obtained

$$r_{2y} = \frac{n_1}{n_2} r_{1y} \tag{3.8}$$

$$r_{2z} = \frac{n_1}{n_2} r_{1z} \tag{3.9}$$

These expressions represent the law of refraction in the form of using vectors i.e. vector components. For later convenience of completing some special calculations, Eq. (3.6) is further considered in the form

$$1 - \frac{n_1^2}{n_2^2} = \frac{\sin^2\varepsilon_1 - \sin^2\varepsilon_2}{\sin^2\varepsilon_1} = \frac{\cos^2\varepsilon_2 - \cos^2\varepsilon_1}{\sin^2\varepsilon_1} \tag{3.10}$$

With respect to $r_{1x} = \cos\varepsilon_1$ and $r_{2x} = \cos\varepsilon_2$ it follows

$$1 - \frac{n_1^2}{n_2^2} = \frac{r_{2x}^2 - r_{1x}^2}{\sin^2\varepsilon_1} = \frac{r_{1x}^2}{\sin^2\varepsilon_1}\left(\frac{r_{2x}^2}{r_{1x}^2} - 1\right) = \frac{1}{\tan^2\varepsilon_1}\left(\frac{r_{2x}^2}{r_{1x}^2} - 1\right) \tag{3.11}$$

from which one obtains

$$\frac{r_{2x}^2}{r_{1x}^2} - 1 = \left(1 - \frac{n_1^2}{n_2^2}\right)\tan^2\varepsilon_1 \tag{3.12}$$

The law of refraction is thus interpreted in the form of unit vector components r_{1x} and r_{2x}. This equation will be applied in Chap. 14 to simplify the characterization of astigmatism which, as an optical aberration, is related to the refraction of a focused laser beam.

3.2 The Doppler Effect

The Doppler effect in optics is associated with the light propagation and accounts for the frequency shift when the light source is moving or light is reflected off a moving surface. Because there is no absolute motion according to the special theory of relativity, the Doppler effect must be described by the same mathematical formula, whether it is arising from the moving light source or the moving observer. This statement also relies on the physical reality that the light emitted from a moving light source is independent of the light source motion.

In Fig. 3.2, an optical interaction system with a moving light source and a fixed receiver has been shown. The initial distance between the light source and the receiver is given by s. The time used to transform the light through this distance is t, so that there is $s = ct$. For simplicity it is assumed that the light source emits the monochromatic light of wavelength λ_0. In the first instance, the light source is assumed to be seated in the space. The number of waves on the path s is then given by $s/\lambda_0 = ct/\lambda_0$. In the second instance, the light source is assumed to move at a velocity equal to $\vec{u}_s$. Because the light velocity is independent of the motion of the light source, the time used for wave transmission through the path s is still equal to t. Within this time, the light source itself moves from the plane a to the plane b. The waves which were initially distributed on the path s are now squeezed into the path $ct - (\vec{u}_s \cdot \vec{l})t$. Because it deals with the same number of waves there is

$$\frac{ct}{\lambda_0} = \frac{ct - \left(\vec{u}_s \cdot \vec{l}\right) t}{\lambda_1} \tag{3.13}$$

The wavelength of the light wave perceived at the receiver is then obtained as

$$\lambda_1 = \left(1 - \frac{\vec{u}_s \cdot \vec{l}}{c}\right) \lambda_0 \tag{3.14}$$

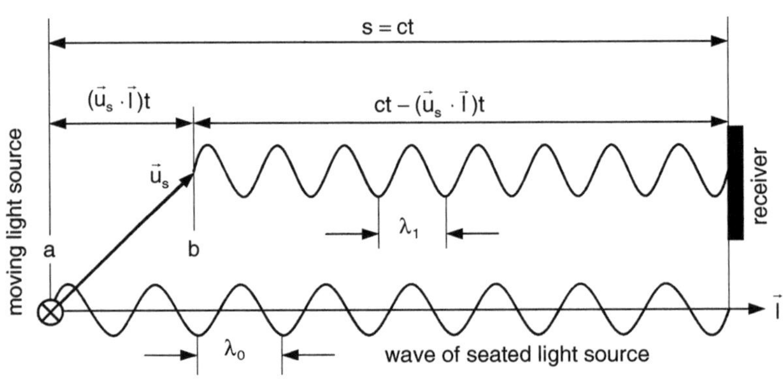

Fig. 3.2 Doppler effect in an optical system with moving light source

With respect to the constant light speed given by $\lambda_1 \nu_1 = \lambda_0 \nu_0 = c$, the frequency of the light wave is calculated as

$$\nu_1 = \frac{\nu_0}{1 - \vec{u}_s \cdot \vec{l}/c} \tag{3.15}$$

This frequency of the light wave perceived at the receiver is shifted against the frequency ν_0 of the light wave that is emitted by the light source. The associated phenomenon is called the Doppler effect. It depends on the relative motion between the light source and the receiver.

Because of $\vec{u}_s \cdot \vec{l}/c << 1$ the above equation is simplified to

$$\nu_1 = \nu_0 \left(1 + \frac{\vec{u}_s \cdot \vec{l}}{c}\right) \tag{3.16}$$

Equation (3.15) i.e. (3.16) is derived by assuming the moving light source and the fixed receiver. According to the special theory of relativity, such a system is totally equivalent to the system with a fixed light source and a moving receiver, when the moving velocity of the receiver is set by $\vec{u}_r = -\vec{u}_s$. Because of this total equivalence the frequency of the light wave received by the moving receiver is obtained directly from Eq. (3.16) as

$$\nu_1 = \nu_0 \left(1 - \frac{\vec{u}_r \cdot \vec{l}}{c}\right) \tag{3.17}$$

In relying on LDA principles, a scattering system is considered to consist of a fixed light source, a moving object (i.e. a small particle) and a fixed observer to receive the light scattered by the moving particle. The corresponding optical arrangement has been illustrated in Fig. 3.3. The particle moves at a velocity equal to $\vec{u}_p$. The light frequency from the light source is ν_0. The frequency which is observed by the moving particle is ν_1 and can be calculated by Eq. (3.17) using substitutions $\vec{u}_p = \vec{u}_r$ and $\vec{l}_1 = \vec{l}$. The particle in its moving system scatters the incident light at the same frequency ν_1. In a certain spatial direction ($\vec{l}_2$), the scattered light is then received as the light of another frequency ν_2 by the fixed receiver because of the Doppler effect.

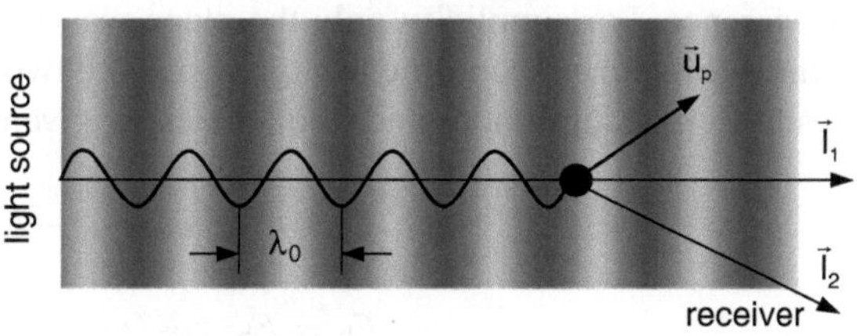

Fig. 3.3 Optical interaction system with a light source, a moving object and a receiver to explain the Doppler effect in LDA measurements

The relationship between the frequencies ν_1 and ν_2 is obtained from Eq. (3.16) by simply changing the indices

$$\nu_2 = \nu_1 \left(1 + \frac{\vec{u}_p \cdot \vec{l}_2}{c}\right) \tag{3.18}$$

Combining Eqs. (3.17) and (3.18) yields

$$\nu_2 = \nu_0 \left(1 - \frac{\vec{u}_p \cdot \vec{l}_1}{c}\right)\left(1 + \frac{\vec{u}_p \cdot \vec{l}_2}{c}\right) \tag{3.19}$$

This equation is found as the basic theory of the laser Doppler anemometry. The shifted frequency according to Eq. (3.19) is then a function of the particle velocity which is considered to be equal to the flow velocity. Because the flow velocity is always negligible against the light speed as given by $\vec{u}_p \cdot \vec{l}_1/c << 1$ and $\vec{u}_p \cdot \vec{l}_2/c << 1$, Eq. (3.19) is further simplified to

$$\nu_2 \approx \nu_0 \left(1 - \frac{\vec{u}_p \cdot \vec{l}_1}{c} + \frac{\vec{u}_p \cdot \vec{l}_2}{c}\right) \tag{3.20}$$

The shifted frequency ν_2 is in the order of ν_0 and therefore still too high to be measured by conventional devices that are found in usual laboratories. In order to make use of the Doppler effect for flow measurements, the dual beam configuration has been confirmed to be highly effective. In fact, the configuration of using two laser beams has been widest applied and comes to be the standard in LDA measurements. The physical background of this configuration is given in Sect. 3.4. The key technique in it is the superposition of two light waves with different frequencies, as described in the next section.

3.3 Superposition of Two Plane Light Waves

The dual beam configuration of LDA optics (see Sect. 3.4) relies upon the superposition of two light waves that are differently shifted by the Doppler effect. As known in general, light is the electromagnetic oscillation in the form of waves. The superposition of two light waves of different frequencies leads to the so-called optical interference. For simplicity only the plane waves are considered, which propagate in the x-direction. According to Fig. 3.4a and b two harmonic waves are assumed to have different amplitudes and frequencies, as given by

$$E_a = E_{a0} \cos(\omega_a t - k_a x) \tag{3.21}$$

$$E_b = E_{b0} \cos(\omega_b t - k_b x) \tag{3.22}$$

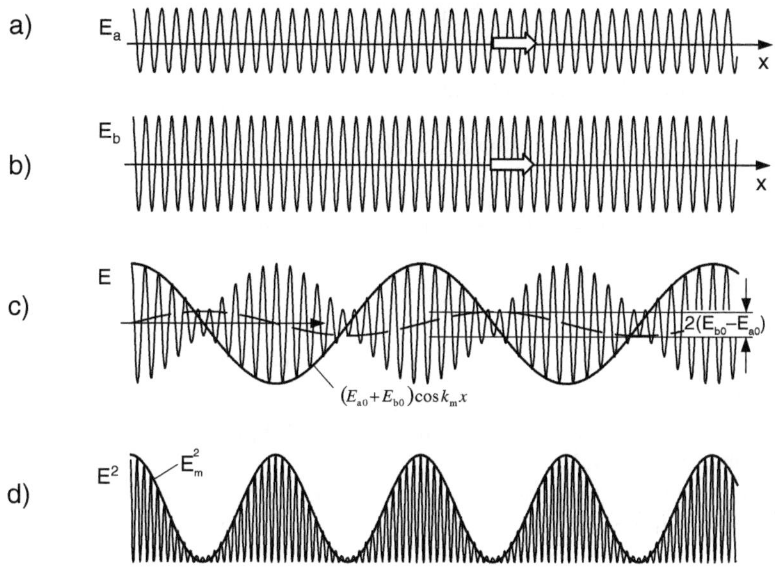

Fig. 3.4 Superposition of two light waves

Different amplitudes of two waves have been assumed, because in LDA measurements with the dual beam configuration the intensities of two chains of laser lights scattered by the particle are always different. This is true even if two laser beams are initially of equal light intensity.

The superposition of these two waves is simply given as

$$E = E_{a0}\cos(\omega_a t - k_a x) + E_{b0}\cos(\omega_b t - k_b x) \tag{3.23}$$

The spatial distribution of this superimposed wave can be obtained at a given time, as shown in Fig. 3.4c for the case $E_{b0} = 1.5E_{a0}$ and $\omega_b = 1.1\omega_a$ i.e. $k_b = 1.1k_a$. Obviously the superimposed wave possesses both a high frequency and a low modulation frequency. For calculating these two frequencies Eq. (3.23) is rearranged as

$$E = E_{a0}\left[\cos(\omega_a t - k_a x) + \cos(\omega_b t - k_b x)\right] + (E_{b0} - E_{a0})\cos(\omega_b t - k_b x) \tag{3.24}$$

By applying the trigonometric identity

$$\cos\alpha + \cos\beta = 2\cos\frac{1}{2}(\alpha+\beta)\cos\frac{1}{2}(\alpha-\beta) \tag{3.25}$$

to the first term on the r.h.s. of Eq. (3.24), then the following equation is obtained:

$$E = 2E_{a0}\cos\left(\frac{\omega_a+\omega_b}{2}t - \frac{k_a+k_b}{2}x\right)\cos\left(\frac{\omega_a-\omega_b}{2}t - \frac{k_a-k_b}{2}x\right) \\ + (E_{b0}-E_{a0})\cos(\omega_b t - k_b x) \tag{3.26}$$

For further calculations the following abbreviations are applied

$$\overline{\omega} = \frac{1}{2}(\omega_a+\omega_b)\,, \quad \omega_m = \frac{1}{2}(\omega_a-\omega_b) \tag{3.27}$$

and

$$\overline{k} = \frac{1}{2}(k_a+k_b)\,, \quad k_m = \frac{1}{2}(k_a-k_b) \tag{3.28}$$

Herein ω_m and k_m represent the modulation frequency and the modulation wavenumber, respectively.

To the cosine function $\cos(\omega_b t - k_b x)$ in Eq. (3.26), $\omega_b = \overline{\omega} - \omega_m$ and $k_b = \overline{k} - k_m$ are substituted so that

$$\cos(\omega_b t - k_b x) = \cos\left(\overline{\omega}t - \overline{k}x\right)\cos(\omega_m t - k_m x) + \sin\left(\overline{\omega}t - \overline{k}x\right)\sin(\omega_m t - k_m x) \tag{3.29}$$

Eq. (3.26) is then converted to

$$E = (E_{a0}+E_{b0})\cos\left(\overline{\omega}t - \overline{k}x\right)\cdot\cos(\omega_m t - k_m x) \\ + (E_{b0}-E_{a0})\sin\left(\overline{\omega}t - \overline{k}x\right)\cdot\sin(\omega_m t - k_m x) \tag{3.30}$$

At the time $t = 0$, the superimposed wave represents a spatial wave distribution given by

$$E = (E_{a0}+E_{b0})\cos\left(\overline{k}x\right)\cos(k_m x) + (E_{b0}-E_{a0})\sin\left(\overline{k}x\right)\sin(k_m x) \tag{3.31}$$

In accordance with Fig. 3.4c, the first term on the r.h.s. of this equation represents the main form of the superimposed wave, whose maximum amplitude is given by $E_{a0}+E_{b0}$. Correspondingly the second term depicts an auxiliary wave with a maximum amplitude equal to $E_{b0}-E_{a0}$. It usually represents a negligible value and only disappears, if two plane waves of equal amplitude are superimposed.

The main form of the superimposed wave comprises the high angular frequency equal to $\overline{\omega}$ (i.e. $\overline{k}$ in the spatial wave distribution) and the low modulation frequency equal to ω_m (k_m). The amplitude of the high frequency oscillation is given by the modulated wave

$$E_m = (E_{a0}+E_{b0})\cos(\omega_m t - k_m x) \tag{3.32}$$

It indeed represents the envelope of the high frequency wave, as shown in Fig. 3.4c for $t=0$. Such a modulated wave form can be imagined to be obtainable by transmitting the wave given in Eq. (3.30) through a low pass filter.

The intensity of a light wave that is sensed by the human eyes or the photonic detectors such as photomultiplier tubes is given by the flux density that is proportional to the wave amplitude squared. From the superposition of two waves, as given by Eq. (3.30) and illustrated in Fig. 3.4c, the amplitude of the main wave oscillation has been confirmed to be the modulated wave and given in Eq. (3.32). Because of this the time and spatial distribution of intensities of the superimposed wave can thus be expressed by

$$E_{\mathrm{m}}^2 = (E_{\mathrm{a0}} + E_{\mathrm{b0}})^2 \cos^2 (\omega_{\mathrm{m}} t - k_{\mathrm{m}} x) \tag{3.33}$$

or equivalently as

$$E_{\mathrm{m}}^2 = \frac{1}{2} (E_{\mathrm{a0}} + E_{\mathrm{b0}})^2 [1 + \cos 2 (\omega_{\mathrm{m}} t - k_{\mathrm{m}} x)] \tag{3.34}$$

The flux density that is proportional to E_{m}^2 oscillates with an angular frequency of $2\omega_{\mathrm{m}} = \omega_{\mathrm{a}} - \omega_{\mathrm{b}}$ which is known as the beat frequency. The corresponding spatial distribution of such an oscillation is shown in Fig. 3.4d for the considered example. It is evident that even in the applied example with large amplitude difference ($E_{\mathrm{b0}} = 1.5E_{\mathrm{a0}}$) the superimposed wave can be well approximated by its main part which is specified by Eq. (3.32) for the amplitude and by Eq. (3.34) for the light intensity.

At this moment it should be mentioned that in LDA measurement techniques the comparable beat frequency of the flux density oscillation is considered and measured to calculate the flow velocities. Because this frequency is many orders smaller than the light frequency (of about $6 \cdot 10^{14}$ Hz), it can be accurately measured by means of usual measurement devices.

3.4 LDA Principle

After the Doppler effect and the superposition of two light waves have been treated in foregoing sections, the optical configuration of an LDA-system and its functionality will be demonstrated. A standard one-component LDA system consists of two laser beams. For simplicity two laser beams (A and B) of equal frequencies (ν_0) are considered to intersect at an angle 2α (Fig. 3.5). The cross area of these two laser beams in the flow is called the measurement volume. A particle that is suspended in the flow is assumed to pass through the measurement volume and so scatters the lights of two laser beams simultaneously. Because of the different spatial layout of two laser beams, the moving particle of velocity $\vec{u}_{\mathrm{p}}$ perceives the different light frequencies resulted from different Doppler effects. A detector is spatially located along $\vec{l}_2$ for receiving the light that is scattered from the measurement volume. The considered system including the laser light source, a moving particle and a detector is completely comparable with the optical interaction system that has already been presented in Fig. 3.3 for explaining the Doppler effect. Hence according to

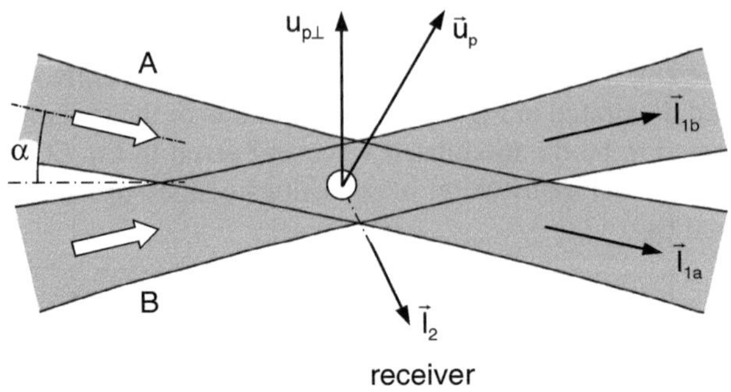

Fig. 3.5 Particle motion through the measurement volume

Eq. (3.20) the frequencies of two light waves received by the detector along $\vec{l}_2$ are given by

$$\nu_{2a} \approx \nu_0 \left(1 - \frac{\vec{u}_p \cdot \vec{l}_{1a}}{c} + \frac{\vec{u}_p \cdot \vec{l}_2}{c}\right) \tag{3.35}$$

and

$$\nu_{2b} \approx \nu_0 \left(1 - \frac{\vec{u}_p \cdot \vec{l}_{1b}}{c} + \frac{\vec{u}_p \cdot \vec{l}_2}{c}\right) \tag{3.36}$$

respectively.

While being received by the photodetector, the two light waves of frequencies ν_{2a} and ν_{2b} get to be superimposed. In accordance with Eq. (3.34), the flux density of the resultant light wave exhibits a low frequency that is known as the beat frequency and equal to twice the modulation frequency ($2\omega_m = \omega_a - \omega_b$). It indeed deals with a low frequency that can be easily measured by means of conventional measurement devices. In the terminology of LDA measurement techniques, this low frequency is called the Doppler frequency. It is calculated from Eqs. (3.35) and (3.36) by

$$\nu_D = |\nu_{2a} - \nu_{2b}| = \frac{\nu_0}{c} \left|\vec{u}_p \cdot \left(\vec{l}_{1b} - \vec{l}_{1a}\right)\right| \tag{3.37}$$

Because of $c/\nu_0 = \lambda_0$ and $\left|\vec{u}_p \cdot \left(\vec{l}_{1b} - \vec{l}_{1a}\right)\right| = 2u_{p\perp} \sin\alpha$ with $u_{p\perp}$ as the component of the particle velocity perpendicular to the bisector of the two laser beams, the above equation becomes

$$\nu_D = 2\frac{u_{p\perp}}{\lambda_0} \sin\alpha \tag{3.38}$$

The Doppler frequency is directly proportional to the velocity component $u_{p\perp}$ of the particle motion, however, independent of the particle motion direction. In assuming the particle velocity to be equal to the velocity of the fluid flow, the corresponding flow velocity component can be obtained by measuring the Doppler frequency. It yields then from Eq. (3.38)

$$u_{\perp} = \frac{\lambda_0}{2 \sin \alpha} \nu_D \tag{3.39}$$

The factor that is multiplied to the Doppler frequency is a physical and geometrical constant. This circumstance implies that the LDA method for flow measurements is a method without system calibration. In addition, as seen from Eq. (3.38), the Doppler frequency detected in the scattered light is independent of the spatial position of the detector. This property enables the detector to be freely positioned for measurements. By using the separate detector, however, it is always time-consuming to align the detector optics to be focused onto the measurement volume. For this reason, most LDA systems are configured as the backscattering system in which the detector unit is integrated into the transmitting unit, see Chap. 4.

Because the Doppler frequency is always positive, independent of the particle flow direction, the velocity component $u_{\perp}$ that is calculated from Eq. (3.39) only corresponds to its absolute value. It is yet impossible to determine the sign of this velocity component i.e. the flow direction. A method to remove the ambiguity of the flow direction is the use of a Bragg cell to shift the frequency in one of two laser beams. This method has come to be a standard in most LDA optics.

Detailed descriptions of an LDA system including photodetectors (photomultiplier) and the use of Bragg cells will be given in Sect. 3.6.

3.5 Fringe Model on the Light Interference

The Doppler frequency in the light scattered by the particle while passing through the measurement volume, as given in Eq. (3.38), can also be calculated by accounting for the interference of laser lights in the measurement volume. The associated calculation method is known as the fringe model. As a matter of fact, the fringe model has most often been applied to explain the principle of LDA method in flow measurements because of its illustrative capability for easy understanding.

Two plane light waves of equal amplitudes (E_0) and frequencies (ω) are postulated to propagate in the directions of $\vec{k}_a$ and $\vec{k}_b$, respectively, and intersect at an angle 2α, as shown in Fig. 3.6. Because of equal angular frequencies and thus equal wavelengths of two plane waves, equal wavenumbers are given as $k = \left|\vec{k}_a\right| = \left|\vec{k}_b\right| = 2\pi/\lambda_0$. For the convenience of analysis, the two-dimensional $z - x$ coordinate system is applied with z as the optical axis which coincides with

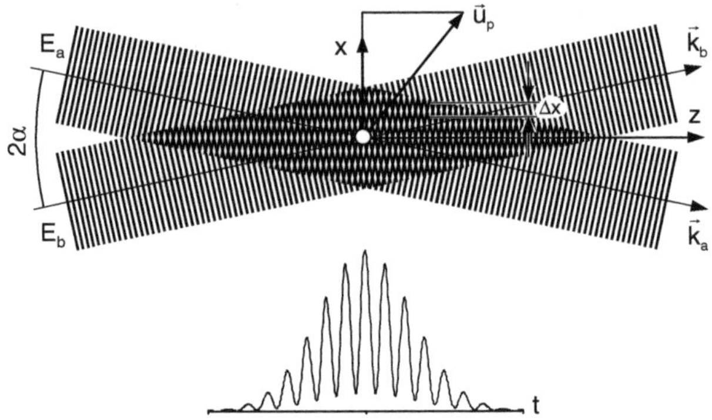

Fig. 3.6 Fringe model of LDA principle and the burst signal

the bisector of two wave vectors $\vec{k}_a$ and $\vec{k}_b$. In the field of the plane wave E_a for instance, the electromagnetic wave at $\vec{r} = (z, x)$ is given by

$$E_a = E_0 \cos\left(\omega t - \vec{k}_a \cdot \vec{r}\right) \tag{3.40}$$

Let us remember, the case of the vector $\vec{r}$ coinciding with the wave vector $\vec{k}$ has already been treated and given in Eq. (3.21).

With respect to the wavenumber $\vec{k}_a = (k\cos\alpha, -k\sin\alpha)$ in the $z - x$ plane Eq. (3.40) becomes

$$E_a = E_0 \cos\left[\omega t - k\left(z\cos\alpha - x\sin\alpha\right)\right] \tag{3.41}$$

Similarly, the electromagnetic wave at the same point, however, in the field of the plane wave E_b can be obtained. The wavenumber in this case is given as $\vec{k}_b = (k\cos\alpha,\ k\sin\alpha)$, so that

$$E_b = E_0 \cos\left[\omega t - k\left(z\cos\alpha + x\sin\alpha\right)\right] \tag{3.42}$$

The superposition of these two plane waves at the point $\vec{r} = (z, x)$ in the $z - x$ plane is obtained by again applying the trigonometric identity according to Eq. (3.25)

$$E = E_a + E_b = 2E_0 \cos\left(kx\sin\alpha\right) \cdot \cos\left(\omega t - kz\cos\alpha\right) \tag{3.43}$$

Obviously the resultant wave at the given point $\vec{r} = (z, x)$ shows the high angular frequency equal to ω. The amplitude of this wave oscillation is $2E_0\cos\left(kx\sin\alpha\right)$ that is a constant and geometrically only a function of the coordinate x. The light intensity that is proportional to the square of the wave amplitude is calculated by

$$E_m^2 = 4E_0^2\cos^2\left(kx\sin\alpha\right) = 2E_0^2\left[1 + \cos 2\left(kx\sin\alpha\right)\right] \tag{3.44}$$

With respect to $k = 2\pi/\lambda_0$ this equation is again written as

$$E_m^2 = 2E_0^2\left[1 + \cos 2\pi\left(\frac{2\sin\alpha}{\lambda_0}x\right)\right] = 2E_0^2\left[1 + \cos\left(2\pi\frac{x}{\Delta x}\right)\right] \tag{3.45}$$

In the direction perpendicular to the optical axis i.e. parallel to the x-axis, the light intensity alternates with a distance equal to

$$\Delta x = \frac{\lambda_0}{2\sin\alpha} \tag{3.46}$$

This distance is known as the fringe spacing in the measurement volume. It is resulting from the interference of two light waves that takes place by overlapping them. To show the dimension of the fringe spacing in the measurement volume, two laser beams of equal wavelength $\lambda_0 = 514.5$ nm are considered to intersect at an half intersection angle $\alpha = 3°$. From the above equation the fringe spacing is calculated as $\Delta x = 5\mu$m. Obviously it deals with a quite small value. By use of a magnified image, the fringe pattern i.e. the light intensity distribution in the measurement volume could be visualized for instance by Miles and Witze (1994, 1996).

In comparing with Eq. (3.39), the velocity component of a particle passing through the measurement volume is calculated by

$$u_\perp = \Delta x \cdot \nu_D \tag{3.47}$$

This equation points out that the Doppler frequency can be considered to be the alternating frequency in the intensity of light that is scattered out by a particle passing through the measurement volume. The corresponding light signal has also been shown in Fig. 3.6. In the terminology of LDA measurements, such a signal is called the Doppler burst. Because the laser beams used in LDA measurements have a Gaussian distribution in the intensity (see Sect. 3.7), the Doppler burst shows its maximum amplitude, as the particle is found in the centre of the measurement volume.

The fringe model presented above is a very useful tool to understand the LDA measurement principle. It also represents a very convenient means to make further studies of diverse optical phenomena influencing the measurement accuracy. To be mentioned here are for instance the change in the fringe spacing at flow measurements in the circular pipes, the fringe distortion caused by either improper optical layout or astigmatism arising from laser beam refractions, and the signal properties in measurements of particle size using Phase Doppler Anemometry (PDA). All of these points excluding that in PDA measurements can be found in the corresponding chapters of this book.

3.6 Frequency Shift Method to Resolve the Flow Direction

The LDA method is based on the evaluation of the burst signals that are generated by particles passing through the measurement volume. An ambiguity to the flow direction of the particle, however, exists because a positive and a negative velocity of the same magnitude will cause the same Doppler frequency. The burst signals thus only involve the magnitude but not the signs of respective velocities. In order to resolve the flow direction from each Doppler burst, the technique of using Bragg cells to slightly shift the frequency in one or both of two laser beams in each laser beam pair has become a standard. The physical principle of the Bragg cell can be found for instance in Albrecht et al. (2003). The purpose of shifting the light frequency is to create the moving fringes in the measurement volume in a predefined direction. The principle of using the frequency shift method to resolve the flow direction is explained here.

According to Fig. 3.7 the light wave frequency of the laser beam A is assumed to be shifted up by ν_{sh}, so that $\nu_{a0} = \nu_0 + \nu_{sh}$. Two laser beams thus show different frequencies. Correspondingly, the light frequencies perceived at the detector in the direction $\vec{l}_2$ can be directly obtained from Eqs. (3.35) and (3.36) as

$$\nu_{2a} = (\nu_0 + \nu_{sh}) \left(1 - \frac{\vec{u} \cdot \vec{l}_{1a}}{c} + \frac{\vec{u} \cdot \vec{l}_2}{c} \right) \tag{3.48}$$

and

$$\nu_{2b} = \nu_0 \left(1 - \frac{\vec{u} \cdot \vec{l}_{1b}}{c} + \frac{\vec{u} \cdot \vec{l}_2}{c} \right) \tag{3.49}$$

respectively. Herein both $\vec{l}_1$ and $\vec{l}_2$ are unit vectors.

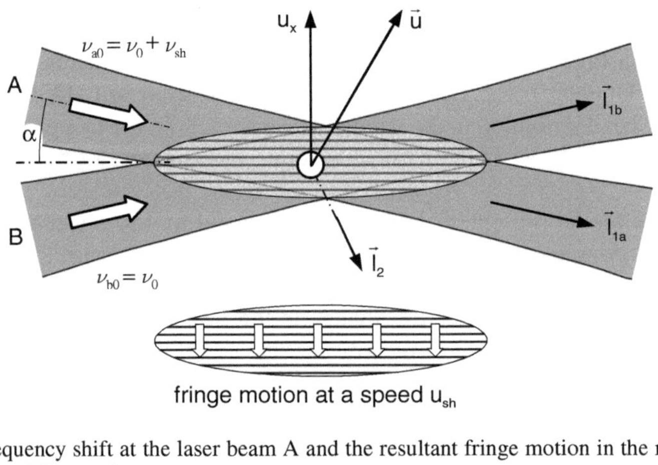

Fig. 3.7 Frequency shift at the laser beam A and the resultant fringe motion in the measurement volume with a shift speed u_{sh}

In the practical application, the shift frequency ν_{sh} is selected in the order of megahertz. It is much lower against the light frequency, however, sufficiently high in comparison to the Doppler frequency caused by the maximum flow velocity. Indeed, the value of the shift frequency should ensure that the difference $\nu_{2a} - \nu_{2b}$ is constantly positive. Under this condition the effective frequency of light signals measured by the receiving unit like the photomultiplier (PM) is simply

$$\nu_{PM} = \nu_{2a} - \nu_{2b} = \nu_{sh} + \frac{\nu_0}{c}\vec{u} \cdot \left(\vec{l}_{1b} - \vec{l}_{1a}\right) + \frac{\nu_{sh}}{c}\vec{u} \cdot \left(\vec{l}_2 - \vec{l}_{1a}\right) \tag{3.50}$$

The third term on the r.h.s. of above equation is negligible against the second term because of $\nu_{sh} << \nu_0$.

With respect to the definition of the positive sign of the velocity component u_x according to Fig. 3.7, the vector difference $\vec{l}_{1b} - \vec{l}_{1a}$ coincides with the positive x-axis. Because of $\vec{u} \cdot \left(\vec{l}_{1b} - \vec{l}_{1a}\right) = 2u_x \sin\alpha$ Eq. (3.50) is then simplified to

$$\nu_{PM} = \nu_{sh} + 2\nu_0 \frac{u_x}{c} \sin\alpha \tag{3.51}$$

The second term on the r.h.s. of this equation represents the Doppler frequency caused by the flow velocity. With respect to $c = \lambda_0 \nu_0$ and in using the fringe model according to Eq. (3.46), Eq. (3.51) is then rewritten as

$$\nu_{PM} = \nu_{sh} + \nu_D = \nu_{sh} + \frac{u_x}{\Delta x} \tag{3.52}$$

The velocity component u_x is then resolved as

$$u_x = \Delta x \left(\nu_{PM} - \nu_{sh}\right) \tag{3.53}$$

Thus from direct comparison between the frequency ν_{PM} which is detected at the photomultiplier and the preset shift frequency ν_{sh} both the value and the sign of the velocity component u_x can be exactly determined. According to Eq. (3.53) there is $u_x > 0$ from $\nu_{PM} > \nu_{sh}$. In particular, the zero velocity of the flow i.e. a particle seated in the measurement volume can also be measured. In this case, there is $\nu_{PM} - \nu_{sh} = 0$.

The technique of using the shift frequency to detect the flow direction is to generate the moving fringe pattern in the measurement volume. According to Eq. (3.52) the detected frequency is simply given as the superposition of the Doppler frequency and the shift frequency. By rewriting Eq. (3.52) to be $\nu_{PM} = (\nu_{sh}\Delta x + u_x)/\Delta x$ it is evident that the detected frequency results from the superposition of the particle velocity and a velocity equal to $\nu_{sh}\Delta x$. This implies that fringes in the measurement volume move in the negative x-direction at the constant speed equal to $u_{sh} = -\nu_{sh}\Delta x$ which is called fringe shift speed. The postulated fringe motion in the measurement volume can be demonstrated in the following way.

3.6.1 Fringe Shift Speed

According to Fig. 3.7 the angular frequency of two laser beams are assumed to be $\omega_a = 2\pi\nu_a$ and $\omega_b = 2\pi\nu_b$, respectively. The corresponding wave equations can be obtained from Eqs. (3.41) and (3.42). For simplicity, the fringe distribution along the x-axis will be considered. This means that $z = 0$ has to be applied, so that

$$E_a = E_0 \cos(\omega_a t + k_a x \sin\alpha) \tag{3.54}$$

$$E_b = E_0 \cos(\omega_b t - k_b x \sin\alpha) \tag{3.55}$$

The superposition of these two plane waves is obtained by again applying the trigonometric identity according to Eq. (3.25)

$$E = E_a + E_b = 2E_0 \cos\left(\omega_m t + \overline{k} x \sin\alpha\right) \cdot \cos(\overline{\omega} t + k_m x \sin\alpha) \tag{3.56}$$

with $\overline{\omega} = \frac{1}{2}(\omega_a + \omega_b)$, $\omega_m = \frac{1}{2}(\omega_a - \omega_b)$, $\overline{k} = \frac{1}{2}(k_a + k_b)$ and $k_m = \frac{1}{2}(k_a - k_b)$.

The resultant light wave at the given x shows the high angular frequency equal to $\overline{\omega}$. The amplitude of this wave oscillation is again the modulated wave $2E_0 \cos\left(\omega_m t + \overline{k} x \sin\alpha\right)$ with the low frequency ω_m. The light intensity that is proportional to the square of the wave amplitude is calculated by

$$E_m^2 = 4E_0^2 \cos^2(kx \sin\alpha) = 2E_0^2\left[1 + \cos 2\left(\omega_m t + \overline{k} x \sin\alpha\right)\right] \tag{3.57}$$

In the current calculation with respect to the shift frequency, there are $\omega_a = \omega_0 + \omega_{sh}$ and $\omega_b = \omega_0$, so that $\omega_m = \omega_{sh}/2 = \pi\nu_{sh}$. Correspondingly there is $\overline{k} = \frac{1}{2}(k_a + k_b) \approx 2\pi/\lambda_0$ because of $\nu_{sh} << \nu_0$. With such simplifications as well as with respect to $\Delta x = \lambda_0/(2\sin\alpha)$ from Eq. (3.46) one obtains from Eq. (3.57)

$$E_m^2 = 2E_0^2\left[1 + \cos 2\pi\left(\nu_{sh} t + \frac{x}{\Delta x}\right)\right] \tag{3.58}$$

It represents an apparent harmonic wave that would move in the negative x-direction. In reality, it only signifies the unsteadiness of the fringe pattern as if this is rolling at a constant speed. It does not indicate any energy transport. For $\nu_{sh} = 0$ one obtains Eq. (3.45).

The speed with which the fringe pattern rolls is determined from the condition

$$\nu_{sh} t + \frac{x}{\Delta x} = \text{const} \tag{3.59}$$

to

$$u_{sh} = \frac{dx}{dt} = -\nu_{sh}\Delta x \tag{3.60}$$

This speed is denoted as the fringe shift speed and has been postulated before based on Eq. (3.52).
The shift frequency, created by means of Bragg cells, is for instance 40 MHz in some LDA applications. With respect to a fringe spacing $\Delta x = 5\mu\text{m}$, see Sect. 3.5, the shift speed of fringes in the measurement volume is calculated to $u_{\text{sh}} = -200$ m/s. Because this speed is sufficiently higher than all possible fluctuation velocities in most flows, the sign of each measured velocity component u_x in a turbulent flow can be determined without ambiguity.

Some LDA users would sometimes like to check the accuracy of the shift frequency preset in the LDA system. A simple and accurate method for performing this task is presented in Chap. 18.

3.7 Gaussian Beam Properties

3.7.1 Geometrical Specifications of the Gaussian Beam

The laser beams that are applied in LDA techniques are generally single-mode laser beams. The intensity distribution in the cross section of such a laser beam can be approximated by the Gaussian distribution as given by

$$I(r) = I_0 \text{e}^{-2(r/w)^2} \tag{3.61}$$

The light intensity on the beam axis is denoted by I_0. The thickness of a Gaussian beam is confirmed to be equal to $2w$. At $r = w$ the light intensity falls down to a level of about $\text{e}^{-2} = 13.5\%$ of the light intensity on the beam axis ($r = 0$), as shown in Fig. 3.8.

The light intensity represents the time rate of flow of radiant energy i.e. the radiant flux density. The total power involved in a light beam is obtained by integrating the light intensity distribution across the light beam section, yielding

$$P = 2\pi I_0 \int_0^\infty \text{e}^{-2(r/w)^2} r\text{d}r = \frac{1}{2}\pi w^2 I_0 \tag{3.62}$$

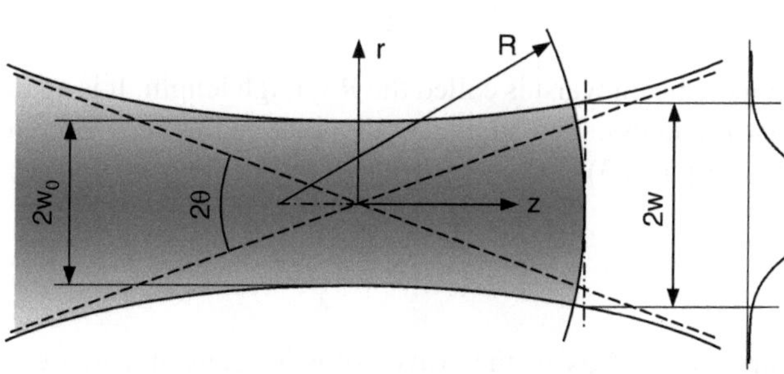

Fig. 3.8 Geometrical and optical specifications of the Gaussian beam

Light beams with the Gaussian distribution always comprise a well-defined beam waist. This property indicates that the Gaussian beam is a focused beam. In fact, such a light beam is geometrically completely describable by merely using the beam waist diameter. According to Fig. 3.8 the geometrical feature of a Gaussian beam mainly includes the curvature radius R of the wave front and the beam thickness $2w$ representing the beam divergence. The wave front is considered as a surface on which the phase is constant. From the wave optics the curvature radius of the front surface of a Gaussian beam and the beam thickness at the distance z from the beam waist are given by

$$R = z\left[1 + \left(\frac{\pi w_0^2}{\lambda z}\right)^2\right] \tag{3.63}$$

and

$$w = w_0\sqrt{1 + \left(\frac{\lambda z}{\pi w_0^2}\right)^2} \tag{3.64}$$

respectively.

For large values of the distance z the thickness of the light beam linearly increases with the distance.

Obviously the beam thickness at the beam waist, given by $2w_0$, is the most essential parameter that determines all geometrical features of a Gaussian beam. According to Eq. (3.63) the curvature radii of the beam front, both at the beam waist ($z = 0$) and at the large distance ($z = \infty$), are infinite. The beam can thus be considered as the plane wave beam. The position that the Gaussian beam possesses the smallest curvature radius ($R_{\min}$) is obtained from Eq. (3.63) under the condition

$$\frac{dR}{dz} = 0 \tag{3.65}$$

to

$$z_R = \frac{\pi \cdot w_0^2}{\lambda} \tag{3.66}$$

This distance from the beam waist is called the Rayleigh length. It is again a function of the beam waist thickness. The smallest curvature radius in a Gaussian beam is then obtained from Eq. (3.63)

$$R_{\min} = 2z_R = \frac{2\pi \cdot w_0^2}{\lambda} \tag{3.67}$$

This smallest curvature radius in the wave front is relevant when evaluating the uniformity of the fringe pattern in the LDA measurement volume. In the case that the measurement volume does not coincide with the waists of two laser beams, non-uniform fringe in the measurement volume will be created. The largest fringe distortion occurs when the measurement volume is formed by laser beam crossing on the Rayleigh length of both laser beams. More about this property and the influence of the fringe distortion on the measurement accuracy will be described in Chap. 16.

In using the Rayleigh length as a characteristic parameter, both the curvature radius of the front surface and the thickness of a Gaussian beam at a distance z from the beam waist are expressed by

$$R = z\left[1 + \left(\frac{z_R}{z}\right)^2\right] \tag{3.68}$$

and

$$w = w_0\sqrt{1 + \left(\frac{z}{z_R}\right)^2} \tag{3.69}$$

respectively.

Especially at the Rayleigh length there is

$$w_R = \sqrt{2}w_0 \tag{3.70}$$

At great distance it yields from Eq. (3.68) with $z >> z_R$

$$R = z \tag{3.71}$$

This last equation signifies that the front surface of a Gaussian beam is a circular surface which has its centre at the beam waist.

The light intensity in the Gaussian beam at the Rayleigh length can be calculated from Eq. (3.62). Because the total power in the beam remains constant, the light intensity in the center of the beam, if compared to the center light intensity at the beam waist, is given by

$$\frac{I_{0R}}{I_{0w}} = \frac{w_0^2}{w_R^2} = \frac{1}{2} \tag{3.72}$$

The divergence of a Gaussian beam can be expressed by the corresponding divergence angle 2θ under the condition $z \to \infty$. This can be obtained by accounting for $\tan\theta = \mathrm{d}w/\mathrm{d}z$ and carrying out the corresponding calculation from Eq. (3.69). Because it usually deals with a very small angle, the approximation $\tan\theta \approx \theta$ can be applied. The half divergence angle of a Gaussian beam is then obtained as

$$\theta = \frac{w_0}{z_R} \tag{3.73}$$

With regard to the Rayleigh length given in Eq. (3.66) this divergence angle of a Gaussian beam is again expressed as

$$\theta = \frac{\lambda}{\pi w_0} \tag{3.74}$$

Obviously laser beams that have large diameters at the beam waist show the negligible divergence angle. For instance for a laser beam of $\lambda = 500$ nm and $w_0 = 1$ mm, the divergence angle is only about 0.009°.

3.7.2 Transmission Performance of the Gaussian Beam

The transmission performance of a Gaussian beam through a lens of the focal length f is considered here according to Fig. 3.9. The Gaussian beam considered is assumed to have a beam waist thickness $2w_0'$. The corresponding Rayleigh length is given by z_R'. Based on lens optics the following geometrical relationships concerning the Gaussian beam prior to and after the lens are available

$$w_0 = \frac{f}{\sqrt{(s'-f)^2 + z_R'^2}} w_0' \tag{3.75}$$

$$s = f + \frac{f^2\,(s'-f)}{(s'-f)^2 + z_R'^2} \tag{3.76}$$

In most cases and for simplicity, a Gaussian beam after passing through a lens can be still considered as a Gaussian beam. All geometrical specifications of a Gaussian beam, as presented in Sect. 3.7.1, remain unchanged.

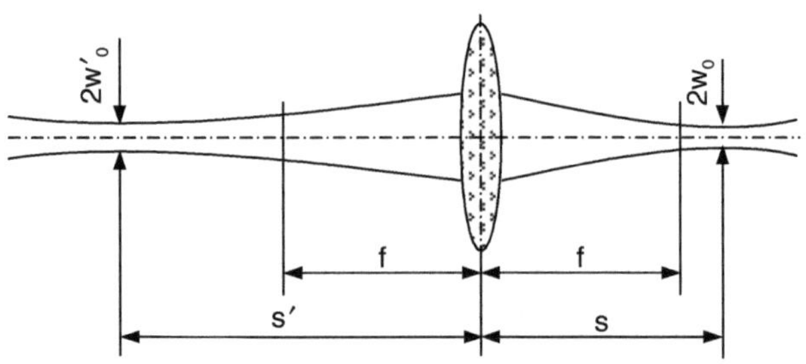

Fig. 3.9 Transmission performance of the Gaussian beam

3.8 Measurement Volume Size

On the optical side of LDA techniques, the measurement volume behaves as the key element in system operations. Both the measurement volume size and the specified optical performance determine the quality of flow measurements. In general, the measurement volume should always be created by arranging the intersection of two laser beams on their waists respectively. On one side, this requirement facilitates the high light intensity in the measurement volume, as this is necessary for detecting small particles passing through the measurement volume. On the other side, the plane wave front at the laser beam waist enables one to create uniform fringes in the measurement volume and hence to enhance the reliability and accuracy of measurements. Otherwise fringe distortion in the measurement volume will occur and lead to measurement errors (see Chap. 16).

For this reason and in general, the measurement volume is created at the waists of two laser beams. The form of the measurement volume can be approximated to be an ellipsoid, as illustrated in Fig. 3.10. The thickness i.e. the diameter of the measurement volume is given by the laser beam thickness at the beam waist

$$d_{\mathrm{mv}} = \frac{2w_0}{\cos\alpha} \tag{3.77}$$

with α as the half intersection angle between two laser beams.

The thickness of the measurement volume is proportional to the laser beam thickness. It depends therefore on the optical arrangement of laser beams regarding the focal length of the used optical lens because of Eq. (3.75). More about this dependence will be presented in Chap. 4 on the optical configuration of a concrete LDA system. In general, the thickness of the measurement volume is in the order of about 0.05–0.1 mm.

With regard to the fringe spacing in the measurement volume, as given in Eq. (3.46), the number of fringes in the measurement volume is then calculated as

$$N = \frac{d_{\mathrm{mv}}}{\Delta x} = 4\frac{w_0}{\lambda_0}\tan\alpha \tag{3.78}$$

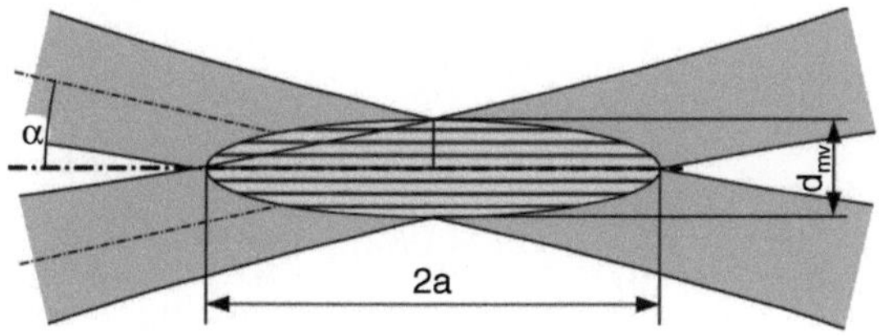

Fig. 3.10 Specification of the measurement volume size

The length of the measurement volume depends, for the same reason, both on the laser beam waist thickness and the crossing angle between two laser beams. According to Fig. 3.10 it is calculated as

$$2a = \frac{d_{\mathrm{mv}}}{\tan\alpha} = \frac{2w_0}{\sin\alpha} \tag{3.79}$$

In comparison with the thickness of the measurement volume, the measurement volume has usually a finite length of about 0.5–3 mm, again depending on the optical arrangement of laser beams.

The geometrical dimension of the measurement volume is defined here independent of the particle size. In reality, the calculated measurement volume size is applicable to small particles of diameters that are comparable to or smaller than the fringe spacing. For large particles that could still scatter the laser light and hence be detected even if the particle center is outside of the measurement volume, the effective detection volume is larger than the independent geometrical measurement volume. Such a dependence of the detection volume size on the particle size is especially crucial in the particle size and mass flux measurements (Zhang et al. 1998, Zhang and Ziada 2000) by means of the Phase Doppler Anemometry (PDA), which is an extended method of LDA, see Albrecht et al. (2003).

Chapter 4
LDA Systems

4.1 Hardware and Optical Components

Based on advanced development of laser and computer technologies as well as on extended requirements in high quality flow measurements, LDA systems have become commercially well available and state of the art products. The hardware of an LDA system consists of transmitting and receiving units. The optical component in the transmitting unit commonly counts the laser, the laser beam transmitter including the Bragg cells and splitters, the fiber and the LDA-head, as shown in Fig. 4.1a for a system of Dantec Dynamics. The laser that is mostly used in LDA measurements is the argon-ion laser that basically provides three selectable wavelengths of 514.5, 488 and 476.5 nm. An LDA system is usually configured to use the laser light of the wavelength 514.5 (green) and 488 (blue) nanometers. After getting into the transmitter, the laser is separated and split into a pair of green and a pair of blue beams. For the purpose of resolving velocity directions, the light frequencies respectively in one green and one blue beams are shifted by Bragg cells, typically for 40 MHz. In some optical configurations, the Bragg cell also serves as the beam splitter. Four laser beams are then conducted into four fibers that are bundled and connected to the LDA head. A two-component LDA head is usually configured such that the plane of two green beams is perpendicular to the plane of two blue beams. This arrangement ensures the measurements of two perpendicular velocity components. The front lens on the LDA head enables all four laser beams to be focused at a unique point for forming the LDA measurement volume. By changing this front lens for different focal lengths the distance of the measurement volume to the LDA head can be changed. Corresponding geometrical and optical properties of the measurement volume in using the lens of different focal lengths will be described in Sect. 4.2.

Because of the use of fiber techniques, LDA optics is sometimes also called the fiber-optic LDA.

The receiving unit of a LDA system commonly includes the components such as the receiving optics, photodetectors like photomultipliers (PM), the signal processor and a computer for both controlling measurements and evaluating measurement data. In the act of using two pair of laser beams, the backward scattered laser light

Zh. Zhang, *LDA Application Methods*, Experimental Fluid Mechanics,
DOI 10.1007/978-3-642-13514-9_4,

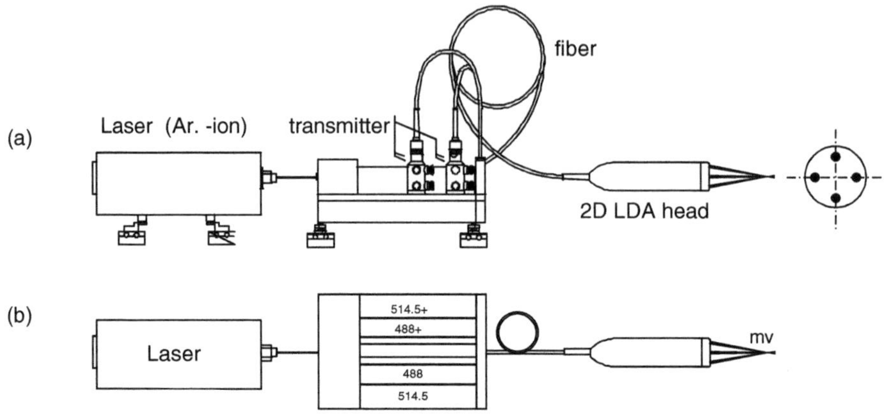

Fig. 4.1 Transmitting optics of an LDA system (Dantec Dynamics)

contains two sequences of light signals and hence is informative of two-component flow velocities. As shown in Fig. 4.2 for the most common case, the LDA head with the corresponding front lens also serves as the receiving unit. The scattered laser light is collected and focused onto the plane end of a supplementary fiber. At another end of the fiber which is usually integrated into the transmitter, the light is firstly separated into two parts of wavelengths 514.5 and 488 nm. These are then guided to respective photomultipliers which convert the light signals into electronic signals. The signal processor and the computer finally work out signals involving the flow velocities and the velocity-time relations.

The LDA system, as shown in Fig. 4.2, is called the backward scattering system. The most relevant advantage of such a system is the consistency of the optical alignment between the transmitting and the receiving units. The LDA head can thus be placed mobile for flow measurements, without realignment each time. In contrast, the forward scattering system necessitates a separate optical receiver for receiving the forward scattered laser light. The background of sometimes configuring such

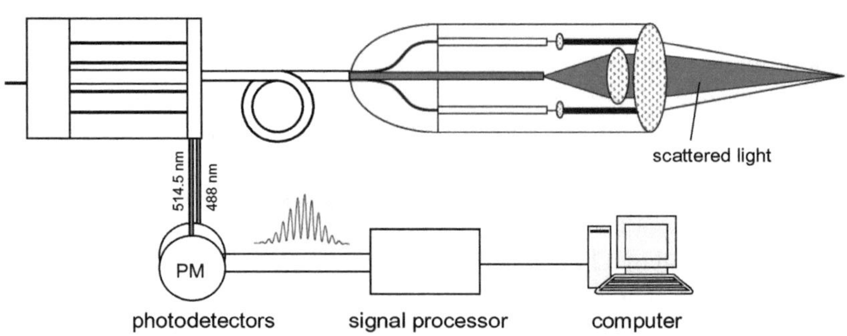

Fig. 4.2 Receiving optics of a backward scattering LDA system

a system is the utilization of high intensity of the forward scattered laser light. Because of the separation between the transmitting and the receiving units an optical realignment is usually always indispensable when the measurement point i.e. the measurement volume in the flow is changed. The realignment of the measurement volume will be very time-consuming, if it deals with measurements of internal flows. All refractions of both the transmitting light beams and the scattering light on diverse medium interfaces (as air-glass and glass-flow) have to be concerned.

LDA systems that use other laser lights (e.g. diode laser) rather than the argon-ion lasers have also been found in the practical applications.

In some few other applications with special optical configurations, coincident measurements of three particular velocity components have been carried out (Hüttmann et al. 2007, Richter and Leder 2006). In these applications, the additional laser light of the wavelength of 476.5 nm has usually been used. As demonstrated in Fig. 4.3 for instance, usually it deals with the measurements of non-orthogonal velocity components. The complete flow information including all three velocity components and the turbulence quantities in the Cartesian coordinate system can be obtained through the appropriate coordinate transformation. More about this technique will be presented in Chap. 6.

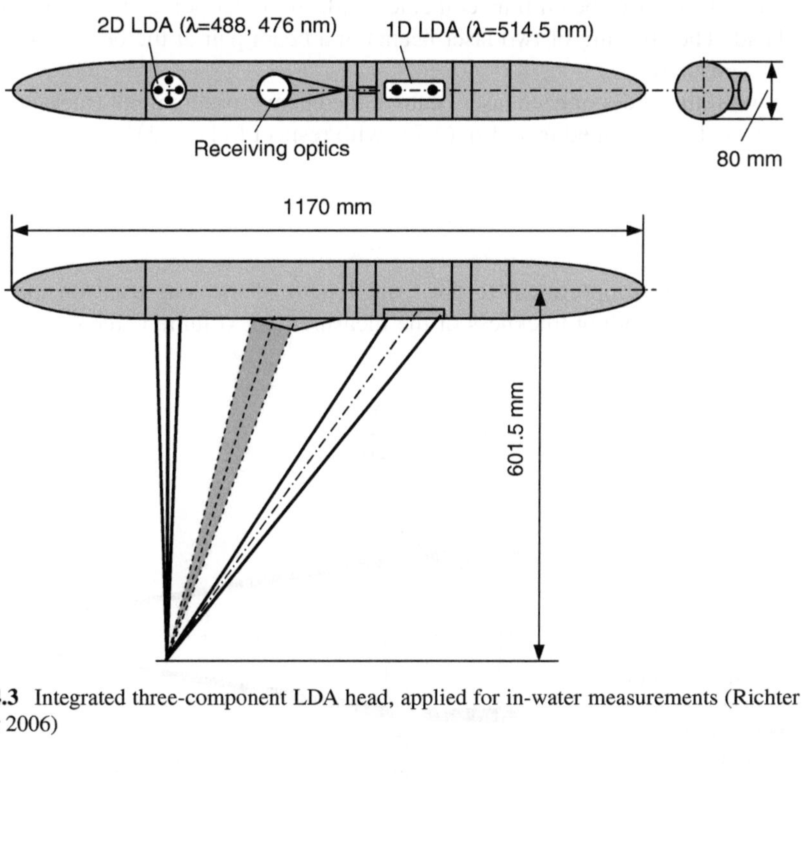

Fig. 4.3 Integrated three-component LDA head, applied for in-water measurements (Richter and Leder 2006)

4.2 Specification of LDA Measurement Volumes

It has been indicated in Sect. 3.8 that the measurement volume size depends on the optical configuration and the used optical lens which determines both the beam crossing angle and the waist thickness of each laser beam. In all LDA configurations, the laser beam crossing is achieved by the lens on the LDA head. Usually each laser beam prior to the lens is configured to be a nearly parallel light bundle. The diameter of such a light bundle is of about $2w'_0 = 1 \sim 2$ mm. By considering this thickness as the waist diameter of a light bundle and according to Eq. (3.74) for the laser light with a wavelength $\lambda = 500$ nm, the divergence angle of the light bundle of a diameter $2w'_0 = 2$ mm is only about 0.009°. Correspondingly the Rayleigh length is calculated from Eq. (3.66) to be $z'_R = 6283$ mm. In all commercial LDA systems, the laser beam thickness $2w'_0$ prior to the lens is configured to be a fixed value. Another fixed value is the distance $2d$ between two laser beams of a beam pair (Fig. 4.4). By changing the lens on the LDA head, both the geometrical dimensions and the optical properties (brightness, fringe spacing etc.) of the measurement volume will get updated. Applying Eq. (3.76) with respect to $f << z'_R$ i.e. $f/z'_R << 1$ yields

$$s = f \tag{4.1}$$

The waist of the laser beam thus coincides with the focal point of the lens on the LDA head. The crossing of two laser beams in a beam pair at the beam waists has been thus ensured.

The waist thickness of each laser beam at the beam crossing point (measurement volume) can be calculated from Eq. (3.75) with respect to $(s' - f)/z'_R << 1$ as

$$w_0 = \frac{f}{z'_R} w'_0 \tag{4.2}$$

It is thus directly proportional to the focal length of the lens used on the LDA head. The corresponding thickness of the measurement volume is then given from Eq. (3.77) as

$$d_{mv} = \frac{2f}{z'_R \cos\alpha} w'_0 \tag{4.3}$$

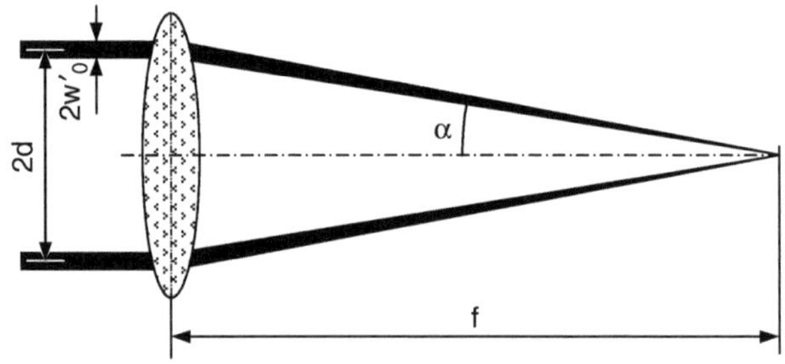

Fig. 4.4 Creation of LDA measurement volume and basic parameters determining the measurement volume size

In using the lens with a small focal length, for instance, small thickness and thus high brightness of the measurement volume can be achieved. This could be very meaningful for LDA measurements at which small natural particles in the flow should be used.

Furthermore, the focal length of the lens simply determines the crossing angle between two laser beams and thus the fringe spacing in the measurement volume. This determination relation is given from Eq. (3.46) as

$$\Delta x = \frac{\lambda_0}{2\sin\alpha} = \frac{\lambda_0}{2}\sqrt{1+\frac{f^2}{d^2}} \tag{4.4}$$

The number of fringes in the measurement volume can be calculated by Eq. (3.78). As it can be confirmed with respect to $\tan\alpha = d/f$ and Eq. (4.2), the number of fringes remains unchanged while changing the lens for another focal length:

$$N = 4\frac{w_0}{\lambda_0}\frac{d}{f} = 4\frac{d}{\lambda_0}\frac{w_0'}{z_R'} \tag{4.5}$$

The length of the measurement volume, according to Eq. (3.79) with respect to Eq. (4.2), is calculated as

$$2a = \frac{2f}{\sin\alpha}\frac{w_0'}{z_R'} = 2f\frac{w_0'}{z_R'}\sqrt{1+\frac{f^2}{d^2}} \tag{4.6}$$

It largely depends on the focal length of used front lens on the LDA head.

In Table 4.1 considering the optical configuration of a standard LDA head manufactured by Dantec Dynamics, calculation examples of geometrical properties of the measurement volume have been shown. While the thickness of the measurement volume is usually in the order of about 0.1 mm, the measurement volume length varies between 0.5 and 5 mm. It should be mentioned that the use of the lens with a long focal length will reduce both the brightness of the measurement volume and the spatial resolution in velocity measurements. In addition, the effective aperture of the receiving optics to the measurement volume decreases too, leading to weakening of signals to be detected.

Table 4.1 Optical configurations and geometrical properties of the measurement volume, example of a standard LDA head ϕ60 of Dantec Dynamics

$\lambda_0 = 514.5$ nm, $2w_0' = 2.2$ mm, 2d = 38 mm				
Focal length of the lens f	mm	160	400	600
Beam intersection angle $2\alpha_0$	deg	13.54	5.44	3.63
Beam waist radius w_0	mm	0.024	0.060	0.089
Measurement volume diameter d_{mv}	mm	0.05	0.12	0.18
Measurement volume length 2a	mm	0.40	2.51	5.65
Fringe spacing Δx	μm	2.18	5.42	8.13
Number of fringes N	–	22	22	22

Chapter 5
Basic Data Processing Methods in LDA Measurements

Most flows encountered in nature and in practical applications are turbulent flows which are specified by high rate velocity fluctuations. In stationary flows, flow fluctuations are merely originated from flow turbulence and hence are completely of randomness. Depending upon the environmental flow conditions, the fluctuation frequency may reach the level of thousands of hertz. For measurements of turbulent flows with such a high velocity fluctuation rate, the LDA method probably provides a most efficient tool. It is non-intrusive, highly accurate and able to highly resolve the flow both in the time and spatial extensions.

LDA measurements basically provide the time series of velocities, to be accurate, the velocity components (see Fig. 2.1). The data processing from LDA measurements finally depends on the specification parameters of respective flow processes and thus on the purpose of conducting flow measurements. The simplest case is probably the measurement of the velocity profile in a channel flow, from which the volumetric flow rate can be calculated. In many other cases, the flow turbulence and its effect on the associated flow processes could be much more relevant. Especially in dealing with non-stationary or periodic flows, corresponding data processing methods have to be worked out, with which all process-relevant flow parameters and dimensionless numbers can be obtained from the time series of measured velocities. In addition, even in the case of dealing with the mean flow velocity and because of the effect of velocity bias, the arithmetic mean from LDA measurements could be insufficient for the required measurement accuracy. Special data processing will then be needed, see Chaps. 11, 12 and 17.

In this chapter, only the general methods of data processing will be described.

5.1 Direct Data Processing for Mean Velocities and Velocity Fluctuations

The stationary turbulent flow is described by the mean velocity and the flow fluctuations, as expressed by Eq. (2.1) with respect to the velocity component u. Based on LDA measurements, the arithmetic mean i.e. the sample mean of the velocity component u, for instance, is calculated as

Zh. Zhang, *LDA Application Methods*, Experimental Fluid Mechanics,

DOI 10.1007/978-3-642-13514-9_5,

$$\bar{u} = \frac{1}{N}\sum_{\mathrm{i}=1}^{\mathrm{N}} u_{\mathrm{i}} \tag{5.1}$$

with N as the total number of velocity samples.

In assuming the known distribution of measurement data like that in Eq. (2.3) or in Fig. 2.1, the sample mean in the above equation can also be expressed by

$$\bar{u} = \int_{-\infty}^{\infty} p_{\mathrm{u}} u \mathrm{d}u \tag{5.2}$$

Herein p_u simply represents the probability density function of velocity distribution from the measurement. It is related to the velocity component u and is not necessarily of the symmetrical form.

Velocity fluctuations involved in the corresponding velocity component can be described, in the statistical specification, by the standard deviation of the mean velocity, as presented in Eq. (2.2). As a statistical measure the so-called root mean square (rms) is calculated by

$$rms_{\mathrm{u}} = \sqrt{\overline{u'^2}} = \sqrt{\frac{1}{N}\sum_{\mathrm{i}=1}^{\mathrm{N}}(u_{\mathrm{i}} - \bar{u})^2} \tag{5.3}$$

Basically, the standard deviation differs from the root mean square in the above equation in that it is calculated from the same statistics as in Eq. (5.3), however, with $1/(N-1)$ in place of $1/N$. Because of large sample size (N) in all LDA measurements, the difference between two statistical quantities is negligibly small, so that there is usually no need to make any distinction between them. The standard deviation σ_u in the mean velocity of a velocity component can thus be represented by the root mean square. In using $\sigma_{\mathrm{u}} = rms_{\mathrm{u}}$ it is further calculated from above equation

$$\sigma_{\mathrm{u}}^2 = \frac{1}{N}\sum_{\mathrm{i}=1}^{\mathrm{N}}(u_{\mathrm{i}} - \bar{u})^2 = \frac{1}{N}\sum_{\mathrm{i}=1}^{\mathrm{N}} u_{\mathrm{i}}^2 - \frac{2}{N}\sum_{\mathrm{i}=1}^{\mathrm{N}} u_{\mathrm{i}}\bar{u} + \frac{1}{N}\sum_{\mathrm{i}=1}^{\mathrm{N}} \bar{u}^2 \tag{5.4}$$

Because of $\frac{1}{N}\sum u_{\mathrm{i}}\bar{u} = \bar{u}^2$ and $\frac{1}{N}\sum \bar{u}^2 = \bar{u}^2$ one obtains

$$\sigma_{\mathrm{u}}^2 = \overline{u^2} - \bar{u}^2 \tag{5.5}$$

This relationship will be used in Chap. 17 for estimating the effect of velocity bias on the measurement accuracy.

The square of the standard deviation, as given in Eq. (5.5), is also called the variance. It represents the normal turbulent stress related to the respective velocity

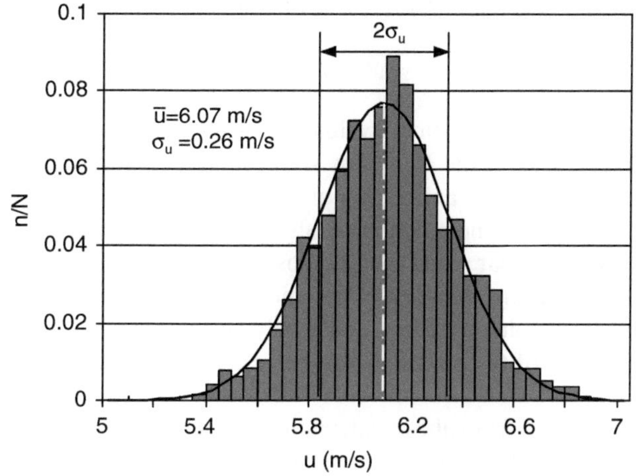

Fig. 5.1 Normalized histogram of velocities and the related Gaussian distribution

component. Both the mean velocity and the standard deviation can be graphically represented by the so-called histogram, as shown in Fig. 5.1 for example of an LDA measurement. This histogram is calculated in such a way that the number of velocity events in each bin is normalized by the total number N of velocity samples. Because of the randomness of velocity fluctuations around the mean velocity, the histogram is generally of symmetrical form and can be well approximated by the Gaussian probability density function as given by Eq. (2.3). In Fig. 5.1 with calculated Gaussian distribution, the mean velocity is confirmed to have the maximum probability in the histogram. The standard deviation measures the variability of velocities. In detail, in the velocity spread of $2\sigma_{\mathrm{u}}$, the fluctuation velocity occurs with a probability of 68.3%, see Fig. 2.2.

It should be mentioned that because of the effect of velocity bias (Chap. 17) the symmetry of the histogram constructed from LDA measurements is always more or less disturbed.

In relying on LDA measurements, from which the root mean square is calculated by Eq. (5.3), the turbulence intensity of the considered turbulent flow with three-dimensional velocity fluctuations is represented by

$$Tu = \frac{1}{\sqrt{\bar{u}^2 + \bar{v}^2 + \bar{w}^2}} \sqrt{\frac{1}{3}(rms_{\mathrm{u}}^2 + rms_{\mathrm{v}}^2 + rms_{\mathrm{w}}^2)} \tag{5.6}$$

Another statistical parameter that can be obtained from measurements is the covariance that is related to two orthogonal velocity components. Like Eq. (5.4) and in considering the velocity components u and v, this is calculated as

$$\overline{u'v'} = \frac{1}{N} \sum_{\mathrm{i=1}}^{\mathrm{N}} (u_{\mathrm{i}} - \bar{u})(v_{\mathrm{i}} - \bar{v}) \tag{5.7}$$

In the similar way of calculating Eq. (5.5), one obtains

$$\overline{u'v'} = \overline{uv} - \overline{u}\,\overline{v} \tag{5.8}$$

This parameter from the statistical evaluation of the flow turbulence actually represents one shear stress in the Reynolds stress matrix that has been shown in Eq. (2.11). It also represents the anisotropy of flow turbulence because it usually does not vanish. According to the calculation algorithm that is shown in Eq. (5.8), the value of the turbulent shear stress may be positive and negative, depending on both the velocity of fluctuations themselves and the choice of the coordinate system. For this reason, an anisotropic turbulence could only be completely described by considering all components of the turbulent shear stress in the respective Reynolds stress matrix.

As can be confirmed from above calculations, determinations of basic turbulence quantities such as $\overline{u'v'}$ clearly require synchronized i.e. coincident measurements of two velocity components. More about the related measurement technique is outlined in Chaps. 6 and 8.

Based on the graphical presentation of turbulent flows in the form of histogram i.e. the probability distribution of flow velocities according to Fig. 5.1, the form of the distribution is sometimes representative for comparing different turbulence properties in internal flows (Durst et al. 1992, 1996). There are two parameters that are often used to describe the form of velocity distributions: the skewness and flatness. The skewness of the probability distribution of the velocity component u is defined by

$$S = \overline{u'^3} = \frac{1}{N}\sum_{\mathrm{i}=1}^{\mathrm{N}}(u_{\mathrm{i}} - \bar{u})^3 \tag{5.9}$$

It depicts the scale of the asymmetry of the probability distribution of the velocity component u around its mean value. In most stationary turbulent flows, random velocity fluctuations occur symmetrically around the respective means. The skewness is then close to zero.

The flatness of the probability distribution of a velocity component is defined by

$$F = \overline{u'^4} = \frac{1}{N}\sum_{\mathrm{i}=1}^{\mathrm{N}}(u_{\mathrm{i}} - \bar{u})^4 \tag{5.10}$$

In reality and like the standard deviation, the flatness as a statistical parameter represents the variability of the velocity that is considered. In order to show that this flatness is equivalent to, as well as related with the standard deviation, the Gaussian probability distribution given at Eq. (2.3) is considered as a reference. Corresponding flatness is calculated as

$$F_{\mathrm{Gauss}} = \int_{-\infty}^{\infty} \mathrm{pdf}_{\mathrm{u}}(u - \bar{u})^4 \mathrm{d}u = 3\sigma_{\mathrm{u}}^4 \tag{5.11}$$

It is simply proportional to the fourth power of the standard deviation. Because of this, the flatness generally does not provide any more about the extent of flow fluctuations than the standard deviation. In addition, it has hardly any physical and mechanical significance, while the standard deviation squared (σ_{u}^2) represents the turbulent kinetic energy.

From the derived equation for the flatness of a Gaussian probability distribution, the so-called flatness factor, also known as the kurtosis factor, is given as

$$K_{\mathrm{Gauss}} = \frac{F}{\sigma_{\mathrm{u}}^4} = 3 \tag{5.12}$$

5.2 Weighting Facilities of Mean Velocity and Fluctuations

Statistical calculations given in Eqs. (5.1), (5.2), (5.3), (5.4), (5.5), (5.6), and (5.7) are simply calculations of respective arithmetical means. Because data acquisitions in all LDA measurements are not time-equidistant, results from above calculations do not exactly represent the corresponding time-averages of respective flow parameters. The apparent deviations of the sample means from those of time-averages are called the bias. Among many sources leading to bias in velocity measurements, velocity fluctuations have been concerned to play a primary role. This type of bias is known as velocity bias. It arises from the mechanism that the high velocities will be more frequently sampled than the low velocities, provided tracer particles in the flow are distributed homogeneously and uniformly. Consequently, the sample mean of velocities from the measurement shifts towards the upper value. This phenomenon was firstly confirmed by McLaughlin and Tiederman (1973) and has been ever since widely investigated with respect to its estimations and corrections. In all traditional considerations, the velocity bias has been uniquely categorized to be an error that is involved in LDA measurements. Because the appearance of velocity bias is directly related to the non-uniformity of flow velocities, the velocity bias is indeed a flow rather than an optical phenomenon. In addition, the biased mean velocity exactly represents a characteristic mean, which should be applied to calculate the momentum flux in the flow. For more details about this concept, see Chap. 17.

With respect to the background of velocity bias, different correction methods have been developed and implemented. The most common method is to use a weighting factor f_{w} in calculating arithmetical means of respective flow parameters. The mean velocity, its standard deviation and the covariance will then be calculated as

$$\bar{u} = \sum_{\mathrm{i}=1}^{\mathrm{N}} f_{\mathrm{w,i}} u_{\mathrm{i}} \tag{5.13}$$

$$rms_{\mathrm{u}} = \sqrt{\sum_{\mathrm{i}=1}^{\mathrm{N}} f_{\mathrm{w,i}}(u_{\mathrm{i}} - \bar{u})^2} \tag{5.14}$$

$$\overline{u'v'} = \sum_{\mathrm{i}=1}^{\mathrm{N}} f_{\mathrm{w,i}}(u_{\mathrm{i}} - \bar{u})(v_{\mathrm{i}} - \bar{v}) \tag{5.15}$$

The weighting factor in dimensionless form is used here for instance to compensate for the effect of irregularity in sampling velocities. For this purpose, McLaughlin and Tiederman (1973) suggested the use of the reciprocal of each individual velocity, related to a total sum, as the weighing factor in the above equations. Theoretically it should be the reciprocal of the magnitude of each velocity vector, which is usually not available because of the lack of three-component LDA measurements.

Another more available weighting factor used to correct the effect of velocity bias is the transit time of each particle while passing through the LDA measurement volume. This correction method is indeed comparable to that of using inverted magnitude of velocity vectors simply because of the proportionality between them by assuming the constant measurement volume thickness. The measurement of the transit time of each particle, however, is in principle much more convenient than that of each velocity vector. Also to be mentioned are measurement uncertainties that might result from both the non-uniformity of fringe spacing in the measurement volume (Chap. 16) and the use of non-monodisperse particles. Large particles, for instance, lead to the existence of long transit time while passing through the measurement volume (i.e. the detection volume in the present aspect).

Besides the use of waiting factors, another method to suppress the effect of velocity bias is to sample flow velocities by controlling LDA detection unit for time-equidistant signals. This method, however, has nowadays not found any practical applications.

Actually, the most important thing concerning the effect of velocity bias is the estimation of the maximal possible extent of influences rather than the correction of it by means of complex data processing or expensive measurement techniques with hard- and software modifications. In the case where the resultant inaccuracy does not exceed the limit that is specified by measurements, no correction of velocity bias is necessary. In addition, velocity bias that is involved in the sample mean of velocities, for instance, does not imply any error, if the momentum flux (instead of the volumetric flux) in the area of measurement volume should be calculated. Based on these viewpoints it appears to be indispensable to quantify the effect of velocity bias first. Because the velocity bias is actually a flow phenomenon and thus can be considered to be related with a turbulence parameter like the turbulence intensity, its accurate quantification can be conducted anyway. Based on this conception, complete calculations for quantifying the effect of velocity bias in turbulent flows with three-dimensional velocity fluctuations have been carried out by Zhang (2002), as this will be shown in Chap. 17 for details.

Chapter 6
Linear Transformation of Velocities and Turbulent Stresses

LDA measurements are known as measurements of velocity components, as has been demonstrated by Eq. (3.39). Because these velocity components are found in the LDA coordinate system, it is always necessary to transform them into the flow field system. This transformation applies not only to the mean velocities, but also to the turbulence quantities. Very often it deals with the two-dimensional orthogonal transformation because usually two orthogonal velocity components are directly obtainable from LDA measurements. Exceptions will be encountered if for instance the third velocity component does not agree to the perpendicular of the other two. In this chapter, both the orthogonal and the non-orthogonal velocity transformations between the LDA and the flow field systems will be presented. It mainly takes account of the two-dimensional coordinate transformation, as this is of great practical relevance. For three-dimensional coordinate transformation concerning the turbulence quantities the readers are referred to Appendix C.

6.1 Orthogonal Linear Transformation

6.1.1 Velocity Transformation

Velocity components in the LDA optical system are denoted by u_1, u_2 and u_3 that are assumed to be perpendicular to each other. The flow field is usually given in the Cartesian coordinate system (x, y, z) that may not agree with the LDA optical system. The velocity transformation between these two coordinate systems is the simplest orthogonal transformation. It is assumed that LDA components u_i ($i = 1$, 2, 3) lie at angles α_i, β_i and γ_i respectively to the flow field coordinates x, y and z, as shown in Fig. 6.1. Then the following velocity transformation is available

$$\begin{bmatrix} u_1 \\ u_2 \\ u_3 \end{bmatrix} = \begin{bmatrix} \cos\alpha_1 & \cos\beta_1 & \cos\gamma_1 \\ \cos\alpha_2 & \cos\beta_2 & \cos\gamma_2 \\ \cos\alpha_3 & \cos\beta_3 & \cos\gamma_3 \end{bmatrix} \begin{bmatrix} u_\mathrm{x} \\ u_\mathrm{y} \\ u_\mathrm{z} \end{bmatrix} = R \begin{bmatrix} u_\mathrm{x} \\ u_\mathrm{y} \\ u_\mathrm{z} \end{bmatrix} \tag{6.1}$$

Zh. Zhang, *LDA Application Methods*, Experimental Fluid Mechanics,
DOI 10.1007/978-3-642-13514-9_6, © Springer-Verlag Berlin Heidelberg 2010

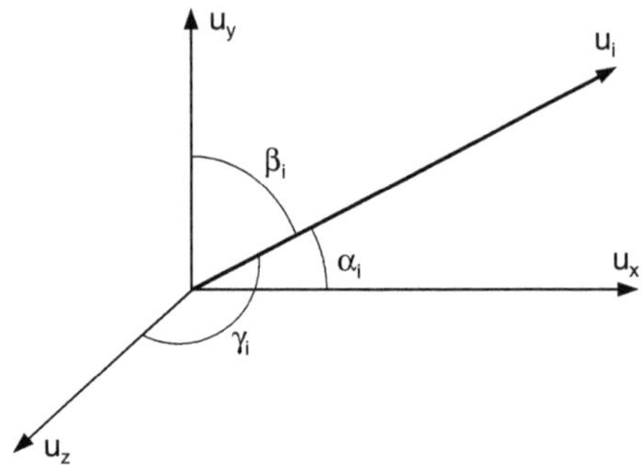

Fig. 6.1 Relationship between velocity components in both the LDA system with orthogonal coordinates and the Cartesian field coordinate system (x, y, z)

In this equation, R represents the orthogonal transformation matrix. Its inverse is simply equal to its transpose, as given by

$$R^{-1} = R' \tag{6.2}$$

Eq. (6.1) is given to generally transform velocities between two coordinate systems. In a two-dimensional $x - y$ plane, as shown in Fig. 6.2 with redefined angles ($\alpha_1 = \varphi$, $\alpha_2 = 90 + \varphi$, $\beta_1 = 90 - \varphi$ and $\beta_2 = \varphi$ in degree), the transformation matrix is simplified to

$$R = \begin{bmatrix} \cos\varphi & \sin\varphi \\ -\sin\varphi & \cos\varphi \end{bmatrix} \tag{6.3}$$

With substitutions of $u_1 = u$ and $u_2 = v$, corresponding velocity transformations are given by

$$u = u_\mathrm{x}\cos\varphi + u_\mathrm{y}\sin\varphi \tag{6.4}$$

$$v = -u_\mathrm{x}\sin\varphi + u_\mathrm{y}\cos\varphi \tag{6.5}$$

as well as in the reverse form

$$u_\mathrm{x} = u\cos\varphi - v\sin\varphi \tag{6.6}$$

$$u_\mathrm{y} = u\sin\varphi + v\cos\varphi \tag{6.7}$$

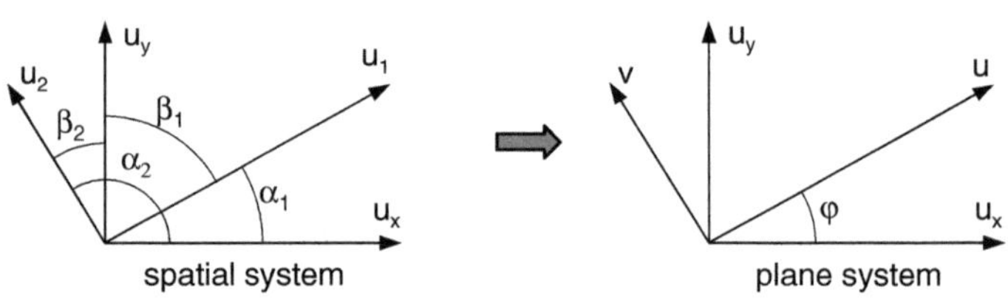

Fig. 6.2 Simplification of expressing velocity components in the plane coordinate system

These transformations of velocity components can be directly applied to both the mean and fluctuation velocities. For the transformation of turbulent stresses which are components of a Reynolds stress matrix, corresponding calculations will be shown in the following section.

6.1.2 Turbulent Stress Transformation

In Chap. 2 with respect to Fig. 2.3, it has been distinguished between isotropic and anisotropic turbulences. Because isotropic turbulence is characterised by turbulent normal stresses that are independent of the coordinate system, it will not be further considered.

The transformation of anisotropic turbulence quantities that are generally represented by the Reynolds stress matrix according to Eq. (2.11) again follows the matrix algebra. It is assumed that in the LDA coordinate system the stress tensor is given by σ_{ij}. The turbulent stresses in the flow system $(x-y-z)$ can then be obtained from the following linear transformation

$$\sigma_{mn} = \begin{bmatrix} \sigma_{xx} & \tau_{xy} & \tau_{xz} \\ \tau_{yx} & \sigma_{yy} & \tau_{yz} \\ \tau_{zx} & \tau_{zy} & \sigma_{zz} \end{bmatrix} = R'\sigma_{ij}R \tag{6.8}$$

In again considering the turbulent stress distribution in the $x-y$ plane (Fig. 6.2), the Reynolds turbulent stresses related to the velocity components u_x and u_y are then derived from Eq. (6.8) as

$$\sigma_{xx} = \sigma_{uu}\cos^2\varphi + \sigma_{vv}\sin^2\varphi - \tau_{uv}\sin 2\varphi \tag{6.9}$$

$$\sigma_{yy} = \sigma_{uu}\sin^2\varphi + \sigma_{vv}\cos^2\varphi + \tau_{uv}\sin 2\varphi \tag{6.10}$$

$$\tau_{xy} = \frac{1}{2}(\sigma_{uu} - \sigma_{vv})\sin 2\varphi + \tau_{uv}\cos 2\varphi \tag{6.11}$$

These equations describe the orthogonal transformation of two-dimensional turbulent stresses between two coordinate systems, which are in the same two-dimensional plane and of a difference φ in the rotation angle. In reality, such a coordinate transformation also represents a way to calculate the plane distribution of respective turbulence quantities, as shown below.

6.1.3 Directional Distribution of Turbulent Stresses

6.1.3.1 On the Basic Parameters σ_{xx}, σ_{yy} and τ_{xy}

The anisotropic turbulence is characterized by the directional dependence of intensities of velocity fluctuations (Fig. 2.3b). In many cases, this directional dependence i.e. the directional distribution of flow fluctuations is interesting because from it the maximum of both the turbulent normal and shear stresses can be determined. In general, the calculation of directional distribution of turbulent stresses is simply the transformation of respective turbulent stresses from given values in a basic coordinate system. This can be demonstrated by again considering Eq. (6.8) for velocity components in the $x - y$ plane (Fig. 6.2). The inverse calculation of Eq. (6.8) is given as follows

$$\sigma_{\mathrm{ij}} = \begin{vmatrix} \sigma_{\mathrm{uu}} & \tau_{\mathrm{uv}} \\ \tau_{\mathrm{vu}} & \sigma_{\mathrm{vv}} \end{vmatrix} = R\sigma_{\mathrm{mn}}R' \tag{6.12}$$

The component σ_{uu}, which is found by φ in the $x - y$ coordinate system (measured from the x-axis counter clockwise, Fig. 6.3) will be denoted by $\sigma_{\varphi\varphi}$. Thus one obtains from the above equation with substitutions of $\sigma_{\varphi\varphi} = \sigma_{\mathrm{uu}}$ and in like manner $\tau_{\varphi,\varphi+90^\circ} = \tau_{\mathrm{uv}}$

$$\sigma_{\varphi\varphi} = \sigma_{\mathrm{xx}} \cos^2 \varphi + \sigma_{\mathrm{yy}} \sin^2 \varphi + \tau_{\mathrm{xy}} \sin 2\varphi \tag{6.13}$$

$$\tau_{\varphi,\varphi+90^\circ} = -\frac{1}{2}(\sigma_{\mathrm{xx}} - \sigma_{\mathrm{yy}}) \sin 2\varphi + \tau_{\mathrm{xy}} \cos 2\varphi \tag{6.14}$$

Eq. (6.13) for instance shows the directional distribution of the turbulent normal stress in the $x - y$ plane based on given quantities $\sigma_{\mathrm{xx}}, \sigma_{\mathrm{yy}}$ and τ_{xy}. As will be shown in Sect. 6.3, the square root of this normal stress and its directional distribution is precisely described by an ellipse function.

From the directional distribution of Reynolds turbulent stresses and according to Eq. (2.13) there exist in the two-dimensional $x - y$ plane two principal normal stresses with corresponding vanishing shear stresses. The orientation of two orthogonal principal normal stresses (σ_{11} and σ_{22}) are found from Eq. (6.13) by setting $\mathrm{d}\sigma_{\varphi\varphi}/\mathrm{d}\varphi = 0$ which leads to

$$\tan 2\varphi_{\mathrm{m}} = \frac{2\tau_{\mathrm{xy}}}{\sigma_{\mathrm{xx}} - \sigma_{\mathrm{yy}}} \tag{6.15}$$

Obviously two angles $\varphi_{\mathrm{m1}} = \varphi_{\mathrm{m}}$ and $\varphi_{\mathrm{m2}} = \varphi_{\mathrm{m}} + 90^\circ$ satisfy this condition. For the common reason $0 \leq \varphi_{\mathrm{m}} < 90$ is agreed on in further calculations. For the anisotropic turbulence as shown in Fig. 2.3b, where the main orientation of velocity fluctuations is given by $\varphi_{\mathrm{m}} \approx \overline{\varphi}$ as in most cases (see Chap. 8), there is $\tau_{\mathrm{xy}} > 0$.

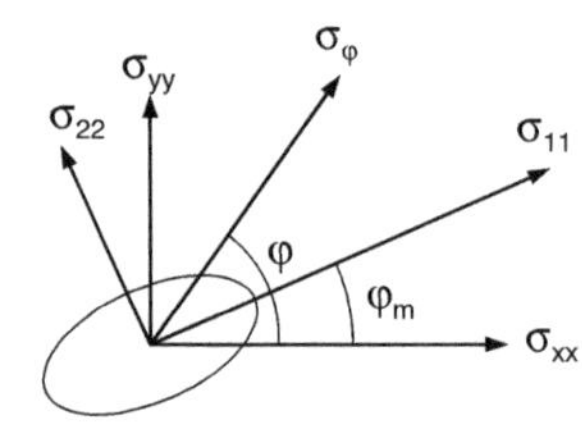

Fig. 6.3 Plane distribution of the turbulent normal stress with the first principal normal stress at φ_{m}

The principal normal stress that is found at φ_m is denoted by σ_{11}. It is calculated from Eq. (6.13) firstly as

$$\sigma_{11} = \sigma_{xx}\frac{1+\cos 2\varphi_m}{2} + \sigma_{yy}\frac{1-\cos 2\varphi_m}{2} + \tau_{xy}\sin 2\varphi_m \tag{6.16}$$

and further to

$$\sigma_{11} = \frac{1}{2}(\sigma_{xx}+\sigma_{yy}) + \frac{1}{2}(\sigma_{xx}-\sigma_{yy})\cos 2\varphi_m + \tau_{xy}\sin 2\varphi_m \tag{6.17}$$

With respect to $\tau_{xy} > 0$ the principal normal stress σ_{11} in Eq. (6.16) represents the first principal of normal stress as $\sigma_I = \sigma_{11}$. Correspondingly, there is $\sigma_{II} = \sigma_{22}$ for the second principal of normal stress. As in the common case, it is $\sigma_I > \sigma_{II}$.

Inserting τ_{xy} from Eq. (6.15) into Eq. (6.17) yields

$$\sigma_{11} = \frac{1}{2}(\sigma_{xx}+\sigma_{yy}) + \frac{1}{2\cos 2\varphi_m}(\sigma_{xx}-\sigma_{yy}) \tag{6.18}$$

To make further calculations, Eq. (6.15) is written, with respect to $\tau_{xy} > 0$, as

$$\cos 2\varphi_m = \frac{\sigma_{xx}-\sigma_{yy}}{\sqrt{(\sigma_{xx}-\sigma_{yy})^2 + 4\tau_{xy}^2}} \tag{6.19}$$

Eq. (6.18) then becomes

$$\sigma_{11} = \frac{1}{2}\left(\sigma_{xx}+\sigma_{yy}\right) + \sqrt{\frac{1}{4}(\sigma_{xx}-\sigma_{yy})^2 + \tau_{xy}^2} \tag{6.20}$$

The second principal of normal stress is found at $\varphi_{m2} = \varphi_m + 90°$, as shown in Fig. 6.3. With the substitution of angle φ_{m2} in Eq. (6.18) one obtains immediately

$$\sigma_{22} = \frac{1}{2}(\sigma_{xx}+\sigma_{yy}) - \sqrt{\frac{1}{4}(\sigma_{xx}-\sigma_{yy})^2 + \tau_{xy}^2} \tag{6.21}$$

The corresponding Reynolds stress matrix is then given by

$$\sigma_{I,II} = \begin{bmatrix} \sigma_{11} & 0 \\ 0 & \sigma_{22} \end{bmatrix} \tag{6.22}$$

One of the most important properties of the stress matrix is that the sum of all three normal stresses is independent of the coordinate system used to represent the turbulent stresses, see Eq. (2.14). In a two-dimensional plane, the so-called first invariant of the matrix is thus expressed by

$$I_1 = \sigma_{11} + \sigma_{22} = \sigma_{xx} + \sigma_{yy} \tag{6.23}$$

As pointed out in Sect. 2.2 in the context of Eq. (2.12), only the absolute maximum value of the turbulent shear stress is representative for the flow state and can be considered as a parameter of the turbulent flow. In the similar way of getting φ_m for the maximum normal stress in the $x - y$ plane, the maximum value of the absolute turbulent shear stress is obtained from Eq. (6.14) to be given at the angle φ_τ which satisfies the condition

$$\tan 2\varphi_\tau = -\frac{\sigma_{xx} - \sigma_{yy}}{2\tau_{xy}} = -\frac{1}{\tan 2\varphi_m} \tag{6.24}$$

The relationship between φ_τ and φ_m in this equation also means $\varphi_\tau = \varphi_m \pm 45°$. Because each turbulent shear stress is always related to the flow fluctuations in two orthogonal velocity components, the two velocity components are commonly assumed to be found counter clockwise at $\varphi_{\tau,1} = \varphi_m - 45$ and $\varphi_{\tau,2} = \varphi_m + 45$, respectively. For simplicity $\varphi_\tau = \varphi_{\tau,1}$ is applied in further calculations, in which $\varphi_\tau < 45$ and thus $2\varphi_\tau < 90$ are available because of $\varphi_m < 90°$.

With respect to Eq. (6.24) the maximum shear stress is then derived from Eq. (6.14) to

$$\tau_{max} = \left[1 + \frac{(\sigma_{xx} - \sigma_{yy})^2}{4\tau_{xy}^2}\right] \tau_{xy} \cos 2\varphi_\tau \tag{6.25}$$

With respect to $\cos 2\varphi_\tau = \sqrt{1/(1 + \tan^2 2\varphi_\tau)}$ one obtains

$$\tau_{max} = \sqrt{\left(\frac{\sigma_{xx} - \sigma_{yy}}{2}\right)^2 + \tau_{xy}^2} \tag{6.26}$$

The maximum turbulent shear stress has been shown to be a function of parameters σ_{xx}, σ_{yy} and τ_{xy}. As seen in Eq. (6.14), these parameters are also used to represent the directional distribution of the turbulent shear stress. In actual fact, the interested distribution can also be shown to be a function of the maximum shear stress τ_{max} only. For this reason the maximum shear stress is calculated from Eq. (6.14) firstly as

$$\tau_{max} = -\frac{1}{2}(\sigma_{xx} - \sigma_{yy}) \sin 2\varphi_\tau + \tau_{xy} \cos 2\varphi_\tau \tag{6.27}$$

and then with respect to $\varphi_\tau = \varphi_m - 45°$ as

$$\tau_{max} = \frac{1}{2}(\sigma_{xx} - \sigma_{yy}) \cos 2\varphi_m + \tau_{xy} \sin 2\varphi_m \tag{6.28}$$

Combining this equation with Eq. (6.14) yields

$$\frac{\tau_{\varphi,\varphi+90^\circ}}{\tau_{max}} = \frac{\sin 2\varphi - \frac{2\tau_{xy}}{\sigma_{xx}-\sigma_{yy}}\cos 2\varphi}{-\cos 2\varphi_m - \frac{2\tau_{xy}}{\sigma_{xx}-\sigma_{yy}}\sin 2\varphi_m} \tag{6.29}$$

The expression $2\tau_{xy}/(\sigma_{xx} - \sigma_{yy})$ in this equation will be replaced by Eq. (6.15). After a short rearrangement one obtains

$$\tau_{\varphi,\varphi+90^\circ} = \tau_{max} \sin 2(\varphi_m - \varphi) \tag{6.30}$$

The directional distribution of the turbulent shear stress has been shown to be a sine function that has an amplitude equal to τ_{max}. At the angle $\varphi = \varphi_m$, at which the maximum normal stress is found, the shear stress vanishes as $\tau(\varphi_m) = 0$.

The turbulent shear stress according to Eq. (6.30) may be positive and negative. This is only of mathematical significance, which must be considered while carrying out the coordinate transformation of the Reynolds stress matrix. By only considering the absolute value of the turbulent shear stress and its spatial distribution, Eq. (6.30) represents a polar rose curve in polar coordinates, as shown in Fig. 6.4 for $\varphi_m = 30$ and $\varphi_m = 0^\circ$, respectively.

Another special case is given for $\varphi = 0$. From Eq. (6.30) one obtains then

$$\tau_{xy} = \tau_{max} \sin 2\varphi_m \tag{6.31}$$

The maximum shear stress clearly behaves as a key parameter with which other dependent turbulence quantities can be shown in the simple form. Combining Eq. (6.31) with Eq. (6.15) for instance yields

$$\sigma_{xx} - \sigma_{yy} = 2\tau_{max} \cos 2\varphi_m \tag{6.32}$$

With respect to Eq. (6.26) for $\tau_{max} > 0$ Eq. (6.20) is written as

$$\sigma_{11} = \frac{1}{2}\left(\sigma_{xx} + \sigma_{yy}\right) + \tau_{max} \tag{6.33}$$

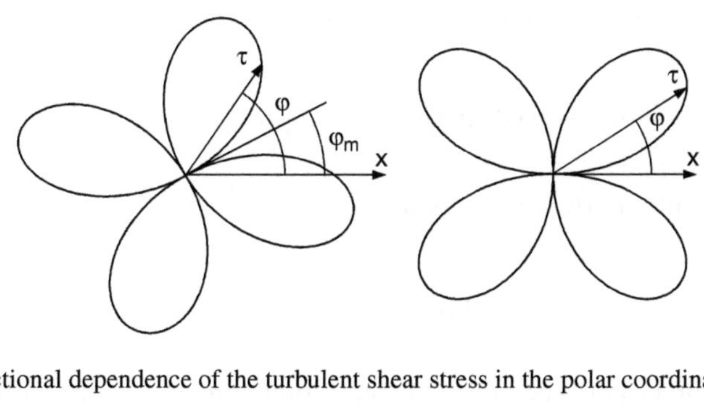

Fig. 6.4 Directional dependence of the turbulent shear stress in the polar coordinate system

This relationship can be well confirmed in the graphical Mohr's stress circle, see Sect. 6.3.2.

There are also some other relations which, however, appear not to be of much relevance and therefore are not shown here.

6.1.3.2 On the Basic Parameters σ_{11} and σ_{22}

In Eqs. (6.13) and (6.14), the directional distributions of both the normal and shear stresses have been shown in the function of given parameters σ_{xx}, σ_{yy} and τ_{xy}. On the other side, these distributions can also be shown as the function of two principal normal stresses σ_{11} and σ_{22}. According to Fig. 6.3 and in accounting for σ_{11} and σ_{22} as independent parameters, the normal stress in Eq. (6.13) is directly written, by changing indices and with respect to $\tau_{12} = 0$ at φ_m, as

$$\sigma_{\varphi\varphi} = \sigma_{11}\cos^2(\varphi - \varphi_m) + \sigma_{22}\sin^2(\varphi - \varphi_m) \tag{6.34}$$

Similarly there is from Eq. (6.14)

$$\tau_{\varphi,\varphi+90°} = -\frac{1}{2}(\sigma_{11} - \sigma_{22})\sin 2(\varphi - \varphi_m) \tag{6.35}$$

From these two general expressions of directional distributions of turbulent stresses some basic relationships can be obtained. For $\varphi = 0$ and $\varphi = 90°$ for instance, one obtains

$$\sigma_{xx} = \sigma_{11}\cos^2\varphi_m + \sigma_{22}\sin^2\varphi_m \tag{6.36}$$

and

$$\sigma_{yy} = \sigma_{11}\sin^2\varphi_m + \sigma_{22}\cos^2\varphi_m \tag{6.37}$$

respectively.

In addition, it results from Eq. (6.35) for $\varphi = 0$

$$\tau_{xy} = -\frac{1}{2}(\sigma_{11} - \sigma_{22})\sin(-2\varphi_m) = \frac{1}{2}(\sigma_{11} - \sigma_{22})\sin 2\varphi_m \tag{6.38}$$

In comparing with Eq. (6.31), one obtains

$$\tau_{max} = \frac{\sigma_{11} - \sigma_{22}}{2} \tag{6.39}$$

This relationship can also be obtained when comparing Eq. (6.35) with Eq. (6.30).

6.1.3.3 Approximation $\varphi_m \approx \overline{\varphi}$ and Simplifications

In above calculations, the angle φ_m for principal normal stress has often been applied to represent the directional distribution of the turbulent normal and shear stresses. An approximation to this angle should be noted. In a great deal of turbulent flows, the principal normal stress σ_{11} which is found at φ_m approximately coincides with the main flow direction $\overline{\varphi}$, as already illustrated in Fig. 2.3b. This approximation enables the angle φ_m to be determined merely from two velocity components in the form $\tan\varphi_m = \bar{u}_y/\bar{u}_x$. If compared with Eq. (6.15) it is evident that the approximation $\varphi_m \approx \overline{\varphi}$ much contributes to the simplification of treating turbulence quantities. In particular, experimental estimations of all related turbulence quantities can be considerably simplified, as will be fully described in Chap. 8.

6.2 Non-orthogonal Transformation

In most LDA applications, the LDA optics is designed for coincident measurements of two orthogonal velocity components in a two-dimensional plane. Based on these measurements, velocity components including the flow fluctuations and turbulence quantities in the flow field coordinate system can be obtained by orthogonal coordinate transformations, as treated in Sect. 6.1. The very different case will be encountered for instance at measurements of all three velocity components according to Fig. 4.3, in which the third velocity component is usually not perpendicular to the other two. Also in more frequent cases dealing with indirect measurements of the third velocity component through an additional one-component measurement, as illustrated in Fig. 6.5, non-orthogonal velocity components are again encountered.

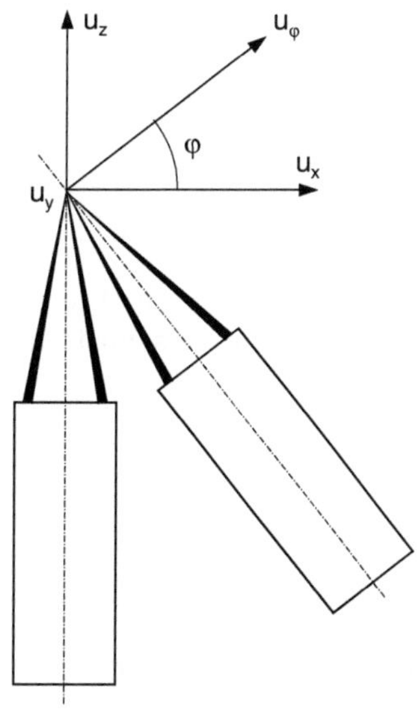

Fig. 6.5 Indirect measurement of the third velocity component u_z

In such situations and with regard to the coordinates given in Fig. 6.5, the third velocity component u_z is obtained from measurements of u_x and u_φ by

$$u_z = \frac{1}{\sin\varphi}\left(u_\varphi - u_x \cos\varphi\right) \tag{6.40}$$

This velocity transformation is usually applicable only to the mean velocity. It could, however, not be supposed to be able to transform each flow fluctuation. That is to say that the fluctuation velocity u'_z could not be determined from the fluctuation velocities u'_x and u'_φ, if these are not in coincidence. This circumstance of non-coincident measurements, however, does not restrict the determination of interested turbulence quantities based on statistical evaluations of flow fluctuations. Corresponding methods will be shown in Sect. 6.2.2.

6.2.1 Velocity Transformation

The case that is shown in Fig. 6.5 can be generalized to that of in Fig. 6.6 where velocity transformation between LDA and flow field system should be conducted. Obviously both velocity components u and v can be written as

$$u = u_x \cos\varphi_u + u_y \sin\varphi_u \tag{6.41}$$

and

$$v = u_x \cos\varphi_v + u_y \sin\varphi_v \tag{6.42}$$

respectively.

For mathematical convenience such a velocity transformation is again described by the matrix form as

$$\begin{bmatrix} u \\ v \end{bmatrix} = \begin{bmatrix} \cos\varphi_u & \sin\varphi_u \\ \cos\varphi_v & \sin\varphi_v \end{bmatrix} \begin{bmatrix} u_x \\ u_y \end{bmatrix} = R \begin{bmatrix} u_x \\ u_y \end{bmatrix} \tag{6.43}$$

It is yet the purpose of the current section to calculate velocity components u_x and u_y in the flow system from velocity components u and v as measured by LDA.

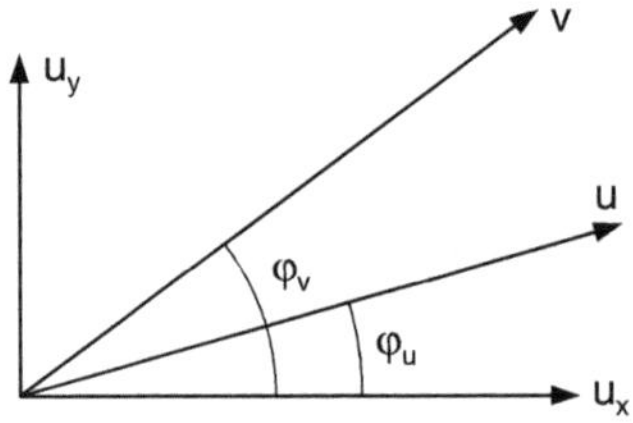

Fig. 6.6 Method of transforming velocity components u and v from LDA measurements into the field velocities u_x and u_y

Thus Eq. (6.43) is written in its inverse form as

$$\begin{bmatrix} u_x \\ u_y \end{bmatrix} = R^{-1} \begin{bmatrix} u \\ v \end{bmatrix} \tag{6.44}$$

with

$$R^{-1} = \frac{1}{\sin(\varphi_v - \varphi_u)} \begin{bmatrix} \sin\varphi_v & -\sin\varphi_u \\ -\cos\varphi_v & \cos\varphi_u \end{bmatrix} \tag{6.45}$$

Both the transformation matrix R and R^{-1} are no longer orthogonal matrixes as that in Eq. (6.3) and therefore there is $R^{-1} \neq R'$. Only for $\varphi_v - \varphi_u = 90°$ the orthogonal transformation is again in effect.

For the special case of $\varphi_u = 0$ and in concerning the mean velocity components, one obtains from Eq. (6.44) besides $\bar{u}_x = \bar{u}$ also

$$\bar{u}_y = \frac{1}{\sin\varphi_v}(\bar{v} - \bar{u}\cos\varphi_v) \tag{6.46}$$

It completely agrees with Eq. (6.40) for the third velocity component.

6.2.2 Turbulent Stress Transformation

The measurement arrangement shown in Fig. 6.6 does not directly provide turbulent stresses as given in Reynolds stress matrix. In order to establish relationships between turbulent stresses in LDA and flow field coordinate systems, it will firstly be assumed that coincident measurements of velocity components u and v enables the covariance $\overline{u'v'}$ to be calculated. This means that the related turbulent shear stress τ_{uv} is assumed to be known. This assumption is only of mathematical reasoning. As soon as all relationships have been established, it will be clear how the parameter τ_{uv} can be eliminated. Based on the rule of matrix calculations, the Reynolds stress matrix in the flow field system with Cartesian coordinates is calculated by

$$\sigma_{mn} = \begin{bmatrix} \sigma_{xx} & \tau_{xy} \\ \tau_{yx} & \sigma_{yy} \end{bmatrix} = R^{-1}\sigma_{ij}\left(R^{-1}\right)' \tag{6.47}$$

Herein the Reynolds stresses relating to velocity components u and v are given by σ_{uu}, σ_{vv} and τ_{uv} with $\tau_{uv} = \tau_{vu}$. The transformation matrix $(R^{-1})'$ is the transpose of the matrix R^{-1} which is given by Eq. (6.45).

From Eq. (6.47) one obtains[1]

[1] Another possibility to transform the turbulent stresses from the non-orthogonal into the orthogonal coordinate system will be presented here based on Fig. 6.6. In applying Eq. (6.41) and (6.42) to velocity fluctuations in respective velocity components, one obtains

$$u' = u'_x\cos\varphi_u + u'_y\sin\varphi_u$$
$$v' = u'_x\cos\varphi_v + u'_y\sin\varphi_v$$

$$\sigma_{xx} = \frac{\sigma_{uu}\sin^2\varphi_v + \sigma_{vv}\sin^2\varphi_u - 2\tau_{uv}\sin\varphi_u\sin\varphi_v}{\sin^2(\varphi_v - \varphi_u)} \tag{6.48}$$

$$\sigma_{yy} = \frac{\sigma_{uu}\cos^2\varphi_v + \sigma_{vv}\cos^2\varphi_u - 2\tau_{uv}\cos\varphi_u\cos\varphi_v}{\sin^2(\varphi_v - \varphi_u)} \tag{6.49}$$

$$\tau_{xy} = -\frac{\sigma_{uu}\sin 2\varphi_v + \sigma_{vv}\sin 2\varphi_u - 2\tau_{uv}\sin(\varphi_u + \varphi_v)}{2\sin^2(\varphi_v - \varphi_u)} \tag{6.50}$$

Like for Eq. (6.46), simplifications for special case of $\varphi_u = 0$ can be made. Corresponding results are then applicable to the case that has been shown in Fig. 6.5. Especially for $\varphi_v - \varphi_u = 90°$ the above equations are simplified to Eqs. (6.9), (6.10), and (6.11).

The above calculations that lead to Eqs. (6.48), (6.49), and (6.50) assumed the known turbulent stresses σ_{uu}, σ_{vv} and τ_{uv}. In effect, the turbulent shear stress τ_{uv} is usually not available because of restrictions in the respective measurement technique and the environmental conditions. To eliminate the parameter τ_{uv}, there are two ways for it to be applied. Firstly, an additional measurement is required, for instance the measurement of velocity component u_x. Together with measurements of velocity components u and v (Fig. 6.6) all turbulence quantities can be obtained theoretically. For the special case of $\varphi_u + \varphi_v = 0$ this can be well realized, as demonstrated below. Secondly, the approximation of $\varphi_m \approx \overline{\varphi}$, whose meaning has been explained at the end of Sect. 6.1, can be applied. This leads to the launch of the so-called Zero-Correlation Method (ZCM) that will be outlined in detail in Chap. 8.

Special case with $\varphi_u + \varphi_v = 0$:

A special arrangement of measurements with $\varphi_u + \varphi_v = 0$ i.e. $\varphi_v = -\varphi_u = -\varphi$ is considered here. It corresponds to two measurements which lie symmetrically to the x-axis, as shown in Fig. 6.7. For this special case Eqs. (6.48), (6.49), and (6.50) are then simplified as

$$\sigma_{xx} = \frac{\sigma_{uu} + \sigma_{vv} + 2\tau_{uv}}{4\cos^2\varphi} \tag{6.51}$$

Consequently there is

$$u'v' = u_x'^2\cos\varphi_u\cos\varphi_v + u_y'^2\sin\varphi_u\sin\varphi_v + u_x'u_y'\sin(\varphi_u + \varphi_v)$$

The sample means of corresponding fluctuations are calculated as

$$\sigma_{uu} = \overline{u'^2} = \sigma_{xx}\cos^2\varphi_u + \sigma_{yy}\sin^2\varphi_u + \tau_{xy}\sin 2\varphi_u$$
$$\sigma_{vv} = \overline{v'^2} = \sigma_{xx}\cos^2\varphi_v + \sigma_{yy}\sin^2\varphi_v + \tau_{xy}\sin 2\varphi_v$$
$$\tau_{uv} = \sigma_{xx}\cos\varphi_u\cos\varphi_v + \sigma_{yy}\sin\varphi_u\sin\varphi_v + \tau_{xy}\sin(\varphi_u + \varphi_v)$$

From these three equations with σ_{uu}, σ_{vv} and τ_{uv} as given quantities, the turbulent stresses σ_{xx}, σ_{yy} and τ_{xy} can be resolved. The same results as given in Eqs. (6.48), (6.49), and (6.50) are obtained.

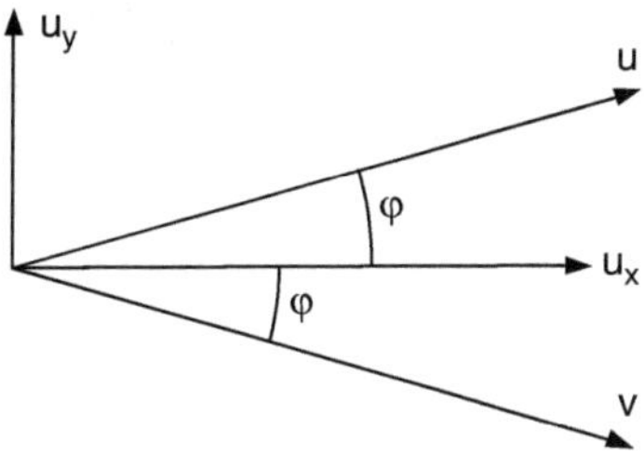

Fig. 6.7 Special arrangement of LDA measurements to resolve the turbulent stress $\overline{u_y'^2}$ through the measurements of velocity components u, v and u_x

$$\sigma_{yy} = \frac{\sigma_{uu} + \sigma_{vv} - 2\tau_{uv}}{4\sin^2\varphi} \tag{6.52}$$

$$\tau_{xy} = \frac{\sigma_{uu} - \sigma_{vv}}{2\sin 2\varphi} \tag{6.53}$$

From Eq. (6.51) and (6.52) one obtains, by eliminating the term τ_{uv}

$$\sigma_{yy} = \frac{\sigma_{uu} + \sigma_{vv} - 2\sigma_{xx}\cos^2\varphi}{2\sin^2\varphi} \tag{6.54}$$

This equation signifies that, from the additional measurement of velocity component u_x with the available normal stress σ_{xx}, the normal stress σ_{yy} can be obtained. Together with the shear stress from Eq. (6.53) the complete two-dimensional state of flow turbulence becomes available. Corresponding applications of Eqs. (6.53) and (6.54) can be found in Tropea (1983).

The measurement technique presented here supposes three independent measurements of σ_{uu}, σ_{vv} and σ_{xx}. It is simply based on statistical evaluation of flow fluctuations without any simplification. Therefore it generally applies not only to LDA but also to other measurement methods. The method, however, appears sometimes too expensive, because for obtaining the mean velocity distribution in the $x-y$ plane basically two single measurements are sufficient, see Eq. (6.46) for instance. An approximation method to simplify the turbulence measurements will be presented in Chap. 8.

6.3 Graphical Presentation of Turbulent Stresses

6.3.1 Ellipse Form of the Turbulence Distribution

It has been confirmed in Sect. 6.1.3 that at φ_m the turbulent shear stress vanishes and the corresponding stress matrix is given by Eq. (6.22). In the function of two principal normal stresses σ_{11} and σ_{22}, the normal stress distribution in the $x-y$ plane has been shown by Eq. (6.34). In order to graphically show the related turbulence quantities in the velocity diagram, substitutions of $\hat{\varphi} = \varphi - \varphi_m$, $r_{\hat{\varphi}}^2 = \sigma_{\hat{\varphi}\hat{\varphi}}$, $a^2 = \sigma_{11}$ and $b^2 = \sigma_{22}$ will be applied. Based on the definition of respective turbulent stresses given in Eq. (2.11), the new parameters $r_{\hat{\varphi}}$, a and b all have the unit m/s. They represent the standard deviations of respective mean velocities i.e. the root mean square (*rms*) of flow fluctuations in relying on Eq. (5.3). Equation (6.34) is then rewritten as

$$r_{\hat{\varphi}}^2 = a^2 \cos^2 \hat{\varphi} + b^2 \sin^2 \hat{\varphi} \tag{6.55}$$

This equation can be considered to be originated from following two equations:

$$r_{\hat{\varphi},x} = a \cos \hat{\varphi} \tag{6.56}$$

$$r_{\hat{\varphi},y} = b \sin \hat{\varphi} \tag{6.57}$$

with $r_{\hat{\varphi},x}$ and $r_{\hat{\varphi},y}$ as two components of $r_{\hat{\varphi}}$. Clearly $r_{\hat{\varphi}}$ precisely represents the radial coordinate (radius) of an ellipse in the polar coordinate system. Both axes of the ellipse are given by a and b, respectively, see Fig. 6.8. The case shown in Fig. 6.8a with corresponding ellipse orientation is often encountered in the practical flows.

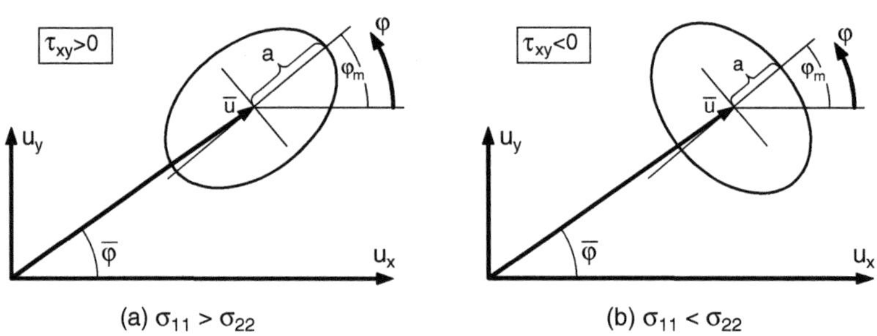

Fig. 6.8 Ellipse form of the turbulence distribution in a two-dimensional plane

6.3.2 Expressions of Turbulent Stresses in Mohr's Stress Circle

Flow turbulences in a two-dimensional $x - y$ plane are again considered here. In using the trigonometric identity $\cos 2\varphi = 2\cos^2 \varphi - 1 = 1 - 2\sin^2 \varphi$, Eq. (6.13) is rewritten as

$$\sigma_{\varphi\varphi} - \frac{1}{2}\left(\sigma_{xx} + \sigma_{yy}\right) = \frac{1}{2}\left(\sigma_{xx} - \sigma_{yy}\right)\cos 2\varphi + \tau_{xy} \sin 2\varphi \tag{6.58}$$

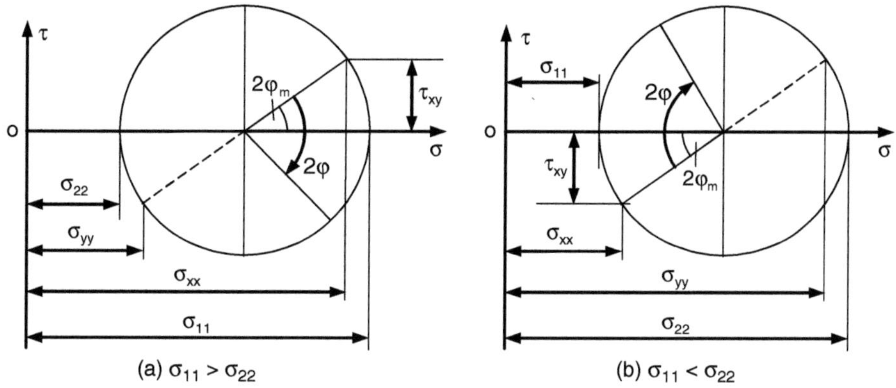

Fig. 6.9 Two-dimensional turbulent stresses expressed in the Mohr's stress circle

Squaring this equation and adding it to the square of Eq. (6.14) yields

$$\left(\sigma_{\varphi\varphi} - \frac{\sigma_{xx} + \sigma_{yy}}{2}\right)^2 + \tau^2_{\varphi,\varphi+90^\circ} = \left(\frac{\sigma_{xx} - \sigma_{yy}}{2}\right)^2 + \tau^2_{xy} \tag{6.59}$$

This equation precisely represents a circle in the coordinate system with normal and shear stresses as two orthogonal coordinates, as shown in Fig. 6.9. The centre of the circle is given by $\sigma = (\sigma_{xx} + \sigma_{yy})/2$ and $\tau = 0$ and the radius by $R = \sqrt{\left(\frac{\sigma_{xx}-\sigma_{yy}}{2}\right)^2 + \tau^2_{xy}}$.

To show the angle φ_m at which the normal stress σ_{11} is present, Fig. 6.9 concerned two cases respectively for $\sigma_{11} > \sigma_{22}$ and $\sigma_{11} < \sigma_{22}$. Correspondingly there are $\tau_{xy} > 0$ in Fig. 6.9a and $\tau_{xy} < 0$ in Fig. 6.9b. In both cases, there is $\sigma_{\varphi\varphi} = \sigma_{11}$ at $2\varphi = 2\varphi_m$.

The graphical presentation of the turbulent stresses is similar to the graphical presentation of the mechanical stresses in the technical mechanics, where the stress circle is called Mohr's stress circle. The difference has to be mentioned: in the flow dynamics, all turbulent normal stresses are always positive, while mechanical stresses in the technical mechanics may be negative, depending on the external loads.

Chapter 7
Tracer Particles and Particle Motion Equations

Flow measurements by means of LDA technique always require particles being suspended in the flow to scatter the laser light. Because the measured particle velocity is used to represent the flow velocity, particles have to be able to follow the flow, especially flow fluctuations. Such a capability of particles is determined by both the particle size and the particle density. In general, a particle in the flow does not behave as a fluid particle because of the difference in their densities. In measuring non-stationary flows or the flows with large velocity fluctuations for instance, particles may have different velocities as the flow. This velocity difference is again the source of diverse forces exerted on the particle. The capability of particles of tracking the flow fluctuations should thus be accounted for with respect to both the particle size and the density of the particle substance.

The motion of a particle in the flow is governed by the dynamical equilibrium between effective forces acting on the particle and the inertial force of the particle. Amongst diverse effective forces, the drag force arising from the viscosity of the flow is usually decisive. Because this force is generally proportional to the particle surface while the particle inertial force is a function of the particle volume and the density of the particle substance, small particles always possess higher capabilities of tracking the flows than large particles. In details, it concerns the particle capabilities of tracking the flow fluctuations and the velocity changes along the streamlines, which may also be curved as a result of the flow arrangement. Since most practical flows are turbulent flows that are specified by the pressure fluctuation and the fluctuation of flow velocities, the particle in the flow also suffers from the pressure force and other force effects. In the straight flow with velocity fluctuations for instance, all force effects on the particle motion are finally found in the amplitude and phase difference of the particle motion in response to the flow fluctuations. This can be well demonstrated, for instance by calculating the particle motion in an oscillation flow.

The capability of particles of tracking the curved streamlines represents a further aspect of greater complexity. As we know, the streamline curvature in the flow is always related to a pressure gradient perpendicular to the streamlines. Such a transverse pressure gradient leads to deviation of the particle motion from streamlines

Zh. Zhang, *LDA Application Methods*, Experimental Fluid Mechanics,
DOI 10.1007/978-3-642-13514-9_7,

of the flow. The possibility of making sure whether a particle is able to follow the streamline curvature depends on how the deviation of the particle motion from the curved streamline can be specified. This is unfortunately not straightforward as that for particles in the flow with straight streamlines.

The effective forces exerted on a moving particle in the practical flows are of different mechanisms and therefore different extent and importance. The most significant forces are the viscous drag force, the pressure force and the forces associated with the particle acceleration.

Although in many cases the natural particles in the flow (e.g. water flow) are sufficiently able to scatter the laser light and thus can be made use of for LDA measurements, as the author of this book has often experienced, it is nevertheless indispensable to make the quantitative reference regarding the capability of seeded particles of tracking the flows. For this reason, diverse flow forces and the respective significances in diverse flows will be considered in this chapter.

7.1 Effective Forces Exerted on the Particle in the Flow

7.1.1 Viscous Drag Force

The drag force exerted on a particle arises from the fluid viscosity and the velocity difference between the particle and the viscous fluid. Since the velocity difference is a vector quantity, the drag force as a vector quantity has its direction coinciding with the difference of two velocity vectors. In assuming the spherical particle of the diameter d_p, the viscous drag force exerted on the particle is calculated by

$$\vec{F}_D = c_\mathrm{D}\frac{1}{4}\pi d_\mathrm{p}^2 \cdot \frac{1}{2}\rho_\mathrm{f}|\vec{u}_\mathrm{f} - \vec{u}_\mathrm{p}|\left(\vec{u}_\mathrm{f} - \vec{u}_\mathrm{p}\right) \tag{7.1}$$

In this determination equation, the drag coefficient c_D is a function of the Reynolds number and can be given, for quasi-stationary flow around the particle, as

$$c_\mathrm{D} = \frac{24}{\mathrm{Re}} \qquad (\mathrm{Re} < 1) \tag{7.2}$$

$$c_\mathrm{D} = \frac{24}{\mathrm{Re}}\left(1 + 0.15\mathrm{Re}^{0.687}\right) \quad (1 < \mathrm{Re} < 1000) \tag{7.3}$$

Equation (7.2) is known as the Stokes law. In concerning the particle motion in the fluid flow, the Reynolds number in above equations is calculated with the relative velocity between the particle and the flow. Because for LDA measurements always small particles are available, the Reynolds number applied to such small particles is usually small. For a particle of diameter $d_\mathrm{p} = 0.02$ mm in the water flow ($\nu = 1.0 \cdot 10^{-6}\mathrm{m}^2/\mathrm{s}$) with a relative velocity of $u_\mathrm{f} - u_\mathrm{p} = 0.1\mathrm{m/s}$, for instance, the Reynolds number is calculated as $\mathrm{Re} = 2$. The drag coefficient, if calculated by

Eq. (7.3), is 15, which is about 25% larger than that calculated by Eq. (7.2). Because of its simplicity the Stokes law given by Eq. (7.2) is often applied to theoretically investigate the dynamical feature of a small particle in the flow. The flow for instance can be assumed to be a sinusoidal function of the time. By choosing high oscillation frequency in the assumed flow, the flow turbulence could be simulated. The fact that the drag coefficient for Re > 1 according to Eq. (7.3) is greater than that from Eq. (7.2) clearly indicates that the real capability of a particle of tracking the flow is higher than that calculated by regarding Eq. (7.2). More about the particle dynamics in diverse types of flows will be shown in Sects. 7.2, 7.3, 7.4, and 7.5.

For particle motion in the same direction as the flow and in using the Stokes flow resistance law, the viscous drag force exerted on the particle is calculated from Eq. (7.1) as

$$\mathrm{F_D} = 3\pi\mu d_\mathrm{p} \cdot (u_\mathrm{f} - u_\mathrm{p}) \tag{7.4}$$

Herein $\mu = \rho_\mathrm{f}\nu$ denotes the dynamical viscosity of the fluid. The drag force is directly proportional to the particle size and the velocity difference. It obviously represents the most significant force exerted on a particle. The drag force is effective, as long as the velocity difference between the particle and the flow exists.

In the general flows with velocity fluctuations or instability, there are also other forces which are proportional to the third power of the particle diameter (such as the pressure force and the force due to added mass, see following sections). The drag force with its linear dependence on the particle diameter clearly dominates in influencing the particle motion in the flow when the particle is sufficiently small.

7.1.2 Gravitational and Lift Forces

The motion of tracer particles in the flow undergoes the influence of the gravity. The associated gravitational force is given by $F_\mathrm{gravity} = m_\mathrm{p}g$ with m_p as the particle mass and g the gravitational acceleration. On the other side, there exists a lift force exerted on the particle in the counter direction of the gravitational force. The difference between them is given by

$$F_\mathrm{gravity} - F_\mathrm{lift} = \frac{1}{6}\pi d_\mathrm{p}^3\rho_\mathrm{p}g - \frac{1}{6}\pi d_\mathrm{p}^3\rho_\mathrm{f}g = \frac{1}{6}\pi d_\mathrm{p}^3 g(\rho_\mathrm{p} - \rho_\mathrm{f}) \tag{7.5}$$

For particles of density that does not much differ from the density of the fluid, the total effect of the gravitational and lift forces can be neglected. Usually the choice of appropriate particles for water flow measurements is not difficult. On the contrary, in dealing with the measurements of air flows that has a density of about 1.2 kg/m^3, there is much restricted possibility in selecting the appropriate particles. Available particles may be the hollow solid particles or liquid particles generated by atomizers. In common cases, the large density difference between the flow and particles can be neglected as long as the particles are sufficiently small. This can be demonstrated by comparing the gravitational force with the drag force arising from the fluid viscosity. The ratio of the gravitational force to the drag force is given by

$$\frac{\text{gravitational force}}{\text{drag force}} \sim \frac{d_\mathrm{p}^3}{d_\mathrm{p}} = d_\mathrm{p}^2 \tag{7.6}$$

Because for small particles the volumetric force including gravitational and lift forces are negligible against the drag force, the corresponding influence on the particle motion can be neglected. For this reason very fine particles such as aerosols of 1 μm in diameter should always be applied for air flow measurements.

7.1.3 Pressure Force

The pressure force exerted on the particle arises from the local pressure gradient in the flow, that may be caused for instance by flow fluctuations in association with both the flow turbulence and the flow periodicity, by the change of velocities along the streamlines as well as by the gravity. For simplicity, only the velocity variation along the straight streamlines i.e. one-dimensional flows and the turbulent flow fluctuations are considered. According to the Euler equation for incompressible flows along the x-direction, the pressure gradient along the streamline is given by

$$-\frac{1}{\rho_\mathrm{f}}\frac{\partial \mathrm{p}}{\partial x} = \frac{\mathrm{d}u_\mathrm{f}}{\mathrm{d}t} = \frac{\partial u_\mathrm{f}}{\partial t} + u_\mathrm{f}\frac{\partial u_\mathrm{f}}{\partial x} \tag{7.7}$$

The first term on the r.h.s. of this equation represents the effect of the temporal flow fluctuations, while the second term accounts for the flow acceleration along the streamlines. For flows without velocity fluctuations ($\partial u_\mathrm{f}/\partial t = 0$) the above equation can also be derived from the Bernoulli equation when applied to a stationary flow.

Within the longitudinal region around a small particle, a linear pressure distribution can be assumed. By integrating all infinitesimal pressure forces exerted normally on the spherical particle surface, the resultant pressure force is calculated as

$$F_\mathrm{p} = -\frac{\pi}{6}d_\mathrm{p}^3\frac{\partial \mathrm{p}}{\partial x} = \frac{\pi}{6}d_\mathrm{p}^3\rho_\mathrm{f}\left(\frac{\partial u_\mathrm{f}}{\partial t} + u_\mathrm{f}\frac{\partial u_\mathrm{f}}{\partial x}\right) \tag{7.8}$$

The calculation of this force is just comparable to that of the lift force of a particle in a flow, see Eq. (7.5). Such a similarity is obvious because the lift force indeed arises in the same way from the pressure gradient $\mathrm{d}p/\mathrm{d}z = \rho_\mathrm{f} g$, which is caused by the gravitation.

In general, the pressure gradient that is related with the velocity acceleration $u_\mathrm{f}\,\mathrm{d}u_\mathrm{f}/\mathrm{d}x$ along the streamlines can be of the same order as or much greater than the pressure gradient caused by the gravity. For the flow through a nozzle of circular section for instance, the pressure gradient can be calculated by applying the Bernoulli equation with respect to the volumetric flow rate $\dot{Q} = Au_\mathrm{f} = \mathrm{const}$ to

$$\frac{\mathrm{d}p}{\mathrm{d}x} = -\frac{1}{2}\rho_\mathrm{f} u_\mathrm{f}^2 \frac{8}{d} \tan\varphi \tag{7.9}$$

Herein d and φ denote the local diameter and the half contraction angle of the nozzle, respectively. For a large flow velocity u_f through a nozzle of small diameter, the pressure gradient $\mathrm{d}p/\mathrm{d}x$ may be many times larger than the pressure gradient $\rho_\mathrm{f} g$ that is caused by the gravity. On the other hand, the resultant pressure force exerted on the particle impels the particle to be accelerated. According to the Newton's second law of motion, the corresponding acceleration of the particle is related with the pressure force by $F_\mathrm{p} = m_\mathrm{p}\,\mathrm{d}u_\mathrm{p}/\mathrm{d}t$ with m_p as the particle mass. Combining this force with the corresponding term in Eq. (7.8) for stationary flow and with respect to $\mathrm{d}u_\mathrm{p}/\mathrm{d}t = u_\mathrm{p}\,\mathrm{d}u_\mathrm{p}/\mathrm{d}x$ yields

$$\frac{\mathrm{d}u_\mathrm{p}}{\mathrm{d}x} = \frac{\rho_\mathrm{f}}{\rho_\mathrm{p}}\frac{u_\mathrm{f}}{u_\mathrm{p}}\frac{\mathrm{d}u_\mathrm{f}}{\mathrm{d}x} \tag{7.10}$$

The particle acceleration, because of the pressure force, is independent of the particle size, but depends on the density of used particles. This result can be well demonstrated by observing air bubbles rising in water. Air bubbles of different sizes (although not spherical) show nearly equal rising velocity. It has to be mentioned that in the real flow the particle undergoes the combined effect of pressure force, viscous drag force, the force arising from the added mass, etc. This will be shown in Sects. 7.4 and 7.5.

7.1.4 Force from Added Mass

When dealing with the motion of a spherical particle at a constant velocity in a fluid at rest, the resistance received by the particle is described by the Stokes flow resistance which accounts for the regular change of the fluid around the particle and thus the constant momentum exchange between the particle and the fluid. In the special case as in the non-viscous potential flow, a spherical particle at the constant velocity does not undergo any resistance from the fluid, which is known as the d'Alembert's paradox. The regular and stationary flow structure around the particle, however, could not be maintained when the free particle in the fluid undergoes the acceleration, for instance, because of the force of gravity or as a result of the Stokes resistance. In association with the acceleration of the particle, more fluid around the particle will be set in motion. This precisely implies that additional momentum exchange between the particle and the fluid takes place. The particle motion thus suffers from an additional resistance, even in the non-viscous fluid. On the other hand, the particle receives an additional force which approves the particle motion, when it is decelerated in the fluid.

The mass that is additionally displaced by the particle because of the particle acceleration is called added mass or the virtual mass. From the theory for potential flow and as concerns the added mass being equally accelerated as the particle, the added mass is equal to the half mass of the fluid that the particle displaces. For the general case of the particle motion in the flow of a velocity u_f and on the same streamline as the flow, the associated force exerted on the particle is given as

$$F_\mathrm{add} = -\frac{1}{2}\frac{\pi}{6} d_\mathrm{p}^3 \rho_\mathrm{f} \frac{\mathrm{d}(u_\mathrm{p} - u_\mathrm{f})}{\mathrm{d}t} \tag{7.11}$$

Because of the added mass the associated force acts as a negative force and counteracts the change of the velocity difference.

7.2 Particle Motion Equation

The motion of a particle in the flow is determined by the equilibrium of diverse forces exerted on the particle. Because this force equilibrium is usually concerned with the flow direction, the vertical gravitational and lift forces will be neglected for simplicity. In fact, the force of gravity and the lift force counteract each other and the remaining effect is negligible if the particle density does not much differ from the fluid density. Some other forces that are associated with the unsteady motion of the particle and are only of less significance will not be considered here.

As effective forces exerted on the particle, the frictional Stokes drag force for $\mathrm{Re} < 1$, the pressure force and the force associated with the added mass will be taken into account. From the Newton's second law of motion the equation of the particle motion in the straight flow is given by

$$\frac{\pi}{6} d_\mathrm{p}^3 \rho_\mathrm{p} \frac{\mathrm{d}u_\mathrm{p}}{\mathrm{d}t} = 3\pi \mu d_\mathrm{p}(u_\mathrm{f} - u_\mathrm{p}) + \frac{\pi}{6} d_\mathrm{p}^3 \rho_\mathrm{f} \frac{\mathrm{d}u_\mathrm{f}}{\mathrm{d}t} - \frac{1}{2}\frac{\pi}{6} d_\mathrm{p}^3 \rho_\mathrm{f} \frac{\mathrm{d}(u_\mathrm{p} - u_\mathrm{f})}{\mathrm{d}t} \tag{7.12}$$

In this equation, the Stokes drag force according to Eq. (7.2) for small Reynolds number i.e. $\mathrm{Re} < 1$ has been applied. The condition of $\mathrm{Re} < 1$ is indeed only satisfied when particles in the flow are sufficiently undersized or the relative velocity between the particle and the flow is sufficiently small. For the particle flow with $Re > 1$, Eq. (7.3) should be applied in which an additional term regarding the viscous drag force is of concern. Because of this feature Eq. (7.12) in assuming $Re < 1$ indeed represent the worst case with respect to the applicability of tracer particles in the flow.

Equation (7.12) will be applied to some special flows to reveal the capability of particles of following the flow variations.

7.3 Particle Motion in the Straight Flow of Constant Velocity

A very common case encountered in the practical flows with particles is the one-dimensional stationary carrying flow. Particles that are seeded into the flow have their initial velocities which are usually not equal to the flow velocity. Under the effect of diverse forces a particle of given diameter and initial velocity will be accelerated or decelerated, until the constant velocity equal to the flow velocity is reached.

For simplicity the particle motion is assumed to be merely influenced by the drag force and the force from the added mass. This means that it deals with a flow field of constant pressure. The particle will be accelerated or slowed down because of the velocity difference between the particle and the flow. From Eq. (7.12) one obtains the particle motion equation rearranged as

$$\left(1+\frac{1}{2}\frac{\rho_{\mathrm{f}}}{\rho_{\mathrm{p}}}\right)\frac{\mathrm{d}u_{\mathrm{p}}}{\mathrm{d}t}=\frac{18\mu}{\rho_{\mathrm{p}}d_{\mathrm{p}}^{2}}(u_{\mathrm{f}}-u_{\mathrm{p}})=\frac{1}{\tau}(u_{\mathrm{f}}-u_{\mathrm{p}}) \tag{7.13}$$

In this equation, the time scale is defined by

$$\tau=\frac{\rho_{\mathrm{p}}d_{\mathrm{p}}^{2}}{18\mu} \tag{7.14}$$

It is called the relaxation time and is a constant representing the particle and flow properties. For a particle of diameter $d_{\mathrm{p}} = 0.01$mm and density $\rho_{\mathrm{p}} = 1.5\,\mathrm{gr/cm^3}$ in the water flow ($\mu = 1.0\cdot 10^{-3}$Pas) for instance, the relaxation time is calculated to $\tau = 8.3\cdot 10^{-6}$s.

To the carrying flow of constant velocity u_{f} = const, the initial velocity difference between the particle and the flow is assumed to be $u_{\mathrm{f}} - u_{\mathrm{p0}}$. Integrating Eq. (7.13) yields

$$\frac{u_{\mathrm{p}}-u_{\mathrm{p0}}}{u_{\mathrm{f}}-u_{\mathrm{p0}}}=1-\mathrm{e}^{-t/A\tau} \tag{7.15}$$

Herein $A = 1+\frac{1}{2}\rho_{\mathrm{f}}/\rho_{\mathrm{p}}$ is a constant that regards the density ratio between the particle substance and the carrying flow.

Figure 7.1, as a calculation example for a particle ($d_{\mathrm{p}} = 0.02$mm) of different density ratios, represents the development of the particle velocity relative to the constant flow velocity. Regarding the particle with $\rho_{\mathrm{p}}/\rho_{\mathrm{f}} = 2$ (i.e. $A = 1.25$) for instance, the particle will reach a velocity of at least 86.5% ($1-\mathrm{e}^{-2} = 0.865$) of the stationary flow velocity after a period of $t = 2A\tau = 1.1\cdot 10^{-4}$s. Because it generally deals with a very short time, the particle of the given diameter could be considered to be able to well follow the flow, despite initially large velocity difference.

From Eq. (7.15) the relaxation time τ can be used as a useful parameter to evaluate the capability of a particle following the flow. It is only a function of the particle and fluid properties but not of the flow properties (e.g. Reynolds number). Whether

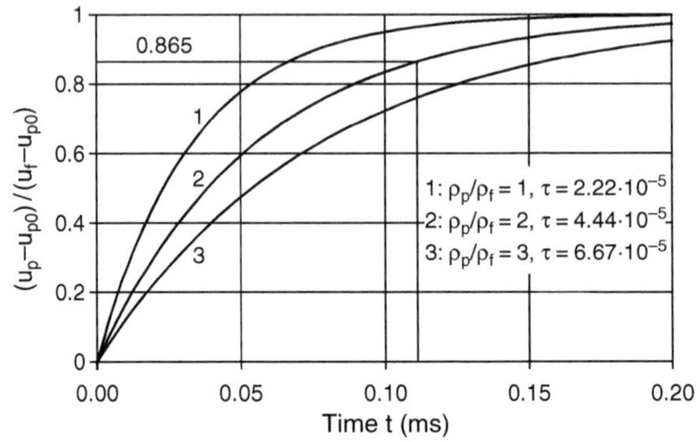

Fig. 7.1 Development of the particle motion velocity relative to the constant flow velocity for $d_p = 0.02$ mm, $\rho_f = 1000\text{kg/m}^3$ and $\mu = 0.001$ Pas

a particle of a certain relaxation time can be applied as the tracer particle for LDA measurements depends on each individual flow and the requirement of the measurement accuracy. In any case, the relaxation time has been widely used to evaluate the applicability of particles for measurements, also in other types of flows. To be mentioned is that the result shown in Eq. (7.15) is derived for Re < 1. This is in any case reasonable, when the particle velocity reaches a value which is close to the velocity of the carrying flow.

7.4 Particle Motion in Nozzle and Diffuser Flows

The particle motion in the flow depends on the state of the fluid flow. Basically the particle motion can be considered to be along the streamlines, when the flow does not rapidly change its direction. Because the velocity of fluid along a streamline is usually not constant, the particle motion will generally differ from the fluid flow. As a result of the difference in the density, the particle will respond to the velocity change along the streamline always with a time delay. The simplest flow with changeable velocity along the streamlines is probably the flow in the nozzle or the diffuser, where the flow along the streamlines is either accelerated or decelerated, with resultant pressure gradients. As already indicated in Sect. 7.1.3, the contribution of the pressure force exerted on the particle to the particle acceleration is independent of the particle size. Additional forces are obviously the viscous drag force and the force because of the added mass. It is interesting to determine the particle motion in such practical flows. For simplicity, the fluid flow is assumed to linearly change along the axis of the nozzle or diffuser, as given by

$$u_f = ax + u_{f0} \tag{7.16}$$

For the flow in the nozzle there is $a > 0$, while in the diffuser flow it is $a < 0$. At the entrance of the nozzle and the diffuser, both the particle and the flow have equal velocities $u_{p0} = u_{f0} = u_0$. With respect to $\mathrm{d}u_f/\mathrm{d}t = u_p \mathrm{d}u_f/\mathrm{d}x \approx u_f \mathrm{d}u_f/\mathrm{d}x$ and $\mathrm{d}u_p/\mathrm{d}t \approx u_f \mathrm{d}u_p/\mathrm{d}x$ as well as the relaxation time according to Eq. (7.14), one obtains from Eq. (7.12)

$$\frac{\mathrm{d}u_p}{\mathrm{d}x} = \frac{1}{\tau}\left(1 - \frac{u_p}{u_f}\right) + \frac{\rho_f}{\rho_p}\frac{\mathrm{d}u_f}{\mathrm{d}x} - \frac{1}{2}\frac{\rho_f}{\rho_p}\frac{\mathrm{d}(u_p - u_f)}{\mathrm{d}x} \tag{7.17}$$

In accounting for $\mathrm{d}u_f/\mathrm{d}x = a$ and thus $\mathrm{d}x = \mathrm{d}u_f/a$, the above equation becomes

$$\left(1 + \frac{1}{2}\frac{\rho_f}{\rho_p}\right)\frac{\mathrm{d}u_p}{\mathrm{d}u_f} = -\frac{1}{a\tau}\frac{u_p}{u_f} + \left(\frac{3}{2}\frac{\rho_f}{\rho_p} + \frac{1}{a\tau}\right) \tag{7.18}$$

The combined influence of the particle size and the density, the dynamical viscosity and the flow acceleration in the nozzle on the particle motion is confirmed to be given only by the product $a\tau$, which is dimensionless.

For further calculations the following abbreviations are applied:

$$A = 1 + \frac{1}{2}\frac{\rho_f}{\rho_p}, \quad B = \frac{1}{a\tau}, \quad C = \frac{3}{2}\frac{\rho_f}{\rho_p} + \frac{1}{a\tau} \tag{7.19}$$

In using the parameter substitution of $\frac{u_p}{u_f} = \overline{u}_p$ from which $\frac{\mathrm{d}u_p}{\mathrm{d}u_f} = u_f \frac{\mathrm{d}\overline{u}_p}{\mathrm{d}u_f} + \overline{u}_p$ is obtained, Eq. (7.18) is further written as

$$\frac{\mathrm{d}\overline{u}_p}{\mathrm{d}u_f} = \frac{C - (A + B)\overline{u}_p}{A u_f} \tag{7.20}$$

This equation with A, B and C as constants represents the particle motion in responding to the velocity change in the straight nozzle or diffuser flows. Because the parameters $\overline{u}_p$ and u_f have been separated, Eq. (7.20) can be integrated, firstly leading to

$$\ln \frac{C - (A + B)\,\overline{u}_p}{C - (A + B)\,\overline{u}_{p0}} = \frac{A + B}{A} \ln \frac{u_{f0}}{u_f} \tag{7.21}$$

With respect to $\overline{u}_{p0} = 1$ at $t = 0$ i.e. $x = 0$ the particle velocity can then be resolved as

$$\overline{u}_p = \left(1 - \frac{C}{A + B}\right)\left(\frac{u_{f0}}{u_f}\right)^{1+B/A} + \frac{C}{A + B} \tag{7.22}$$

Of concern is that Eq. (7.22) is applicable to both the straight nozzle and diffuser flows. For nozzle flows there is $a > 0$ according to Eq. (7.16), while for diffuser flows it is $a < 0$. It should be mentioned that the purpose of current calculations is to evaluate the particle capability of responding to the change in the flow velocity. The assumption of linear dependence of flow velocity on the flow path in both the nozzle and the diffuser contributes to the simplification of calculations. In the practice, this linearity in the flow velocity requires the non-linear design of nozzle or diffuser with respect to the constant volumetric flow rate $\dot{Q} = u_f A_f$, with A_f as the flow section area.

7.4.1 Nozzle Flow

The linear change of the flow velocity in the straight nozzle is given by Eq. (7.16) with $a > 0$. Based on the assumption that at the nozzle entrance the particle and the fluid flow have equal velocities, the particle velocity along the particle streamline has been derived and shown in Eq. (7.22). To a certain nozzle flow with $a =$ 20 (m/s)/m for example, Fig. 7.2 shows the ratio of the particle velocity to the flow velocity along the flow path for different particle densities. At the inlet of the nozzle, every particle suffers from a delay in the acceleration against the flow acceleration, so that there is $\overline{u}_p < 1$. The particle acceleration relative to the flow acceleration just at the nozzle inlet, at which $u_{f0} = u_0$ and $\overline{u}_{p0} = 1$ are available, can be directly obtained from Eq. (7.20) as

$$\left(\frac{d\overline{u}_p}{du_f}\right)_{x=0} = \frac{1}{u_{f0}} \frac{C - (A + B)}{A} \tag{7.23}$$

Further, because of $\overline{u}_p = u_p/u_f$ the above equation is rewritten as

$$\left(\frac{du_p}{du_f}\right)_{x=0} - \frac{u_{p0}}{u_{f0}} = \frac{C - (A + B)}{A} \tag{7.24}$$

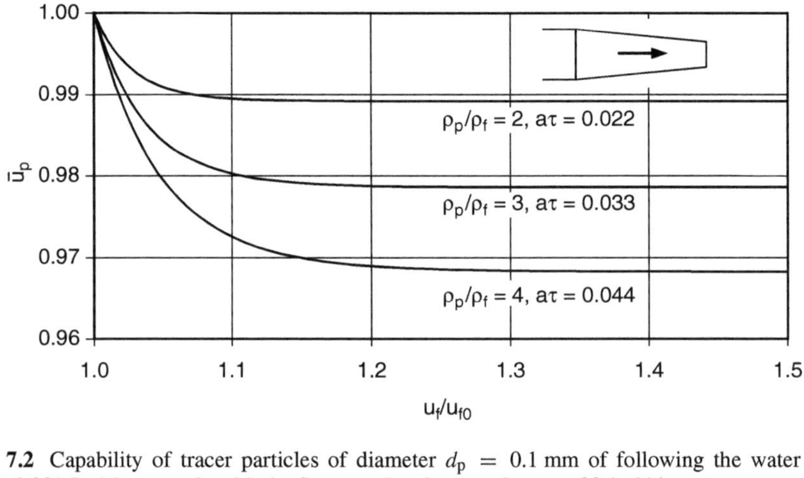

Fig. 7.2 Capability of tracer particles of diameter $d_p = 0.1$ mm of following the water flow ($\mu = 0.001$ Pas) in a nozzle with the flow acceleration equal to $a = 20$ (m/s)/m

With respect to $u_{p0}/u_{f0} = 1$ at the nozzle inlet one obtains

$$\left(\frac{du_p}{du_f}\right)_{x=0} = \frac{C-B}{A} = \frac{3}{1+2\rho_p/\rho_f} \tag{7.25}$$

It is a function of the relative density of the particle only and is independent of the particle size and the viscosity of the flow. Such a result indeed can also be expected from Eq. (7.17). Because at the nozzle entrance the velocity difference between the particle and the flow is zero, the viscous drag force vanishes. Thus with the reject of the first term on the r.h.s. of Eq. (7.17), the particle motion is determined merely by the density ratio ρ_p/ρ_f.

For tracer particles with the density ratio $\rho_p/\rho_f<2$ in the considered example as shown in Fig. 7.2, the final particle velocity is merely about 1% less than the flow velocity. This difference seems to be really negligible for most applications.

As it can also be further confirmed from Fig. 7.2 for the given example, the ratio of particle to flow velocities tends to be constant after a short time at which the flow velocity reaches a value of about $u_f/u_{f0} \approx 1.1 \sim 1.2$. The related constant velocity ratio can be obtained from Eq. (7.22) by setting $u_{f,\infty} = \infty$ to be given as

$$\bar{u}_{p,\infty} = \frac{C}{A+B} \tag{7.26}$$

This final particle velocity can be considered as a measure of the applicability of tracer particles used in LDA measurements. Because of Eq. (7.19), the final particle velocity is obviously a function of the density ratio ρ_p/ρ_f and the product $a\tau$. Corresponding relationships have been show in Fig. 7.3. The almost linear dependence of the final particle velocity on the parameter $a\tau$ can be made use of to

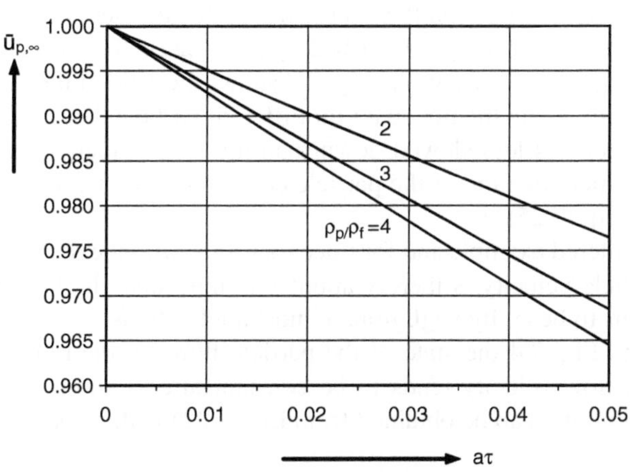

Fig. 7.3 Final reachable ratio of particle to flow velocities in the nozzle flow, in the function of the particle density and the combined parameter $a\tau$

explicitly simplify Eq. (7.26). Because for $a\tau = 0$ there is $\overline{u}_{p,\infty} = 1$, the linearization of Eq. (7.26) will be performed at $a\tau = 0$. With respect to Eq. (7.19) it yields from Eq. (7.26)

$$\left.\frac{\mathrm{d}\overline{u}_{p,\infty}}{\mathrm{d}\,(a\tau)}\right|_{a\tau=0} = -\left(1 - \frac{\rho_f}{\rho_p}\right) \tag{7.27}$$

The linearization of Eq. (7.26) is then obtained as

$$\overline{u}_{p,\infty} = 1 - \left(1 - \frac{\rho_f}{\rho_p}\right) a\tau \tag{7.28}$$

This relationship serves for rapidly evaluating the applicability of interested tracer particles to be used in LDA measurements of nozzle flows. The relaxation time τ of the particle is calculated by Eq. (7.14). The flow acceleration a in the nozzle flow depends not only on the nozzle geometry but also on the flow velocity itself. From the constant volumetric flow rate $\dot{Q} = u_f A_f$ through the nozzle the flow acceleration in the nozzle is calculated as

$$a = \frac{\mathrm{d}u_f}{\mathrm{d}x} = -\frac{u_f}{A_f}\frac{\mathrm{d}A_f}{\mathrm{d}x} \tag{7.29}$$

In general, the flow in a nozzle of small dimension undergoes the greater acceleration than the flow in a large geometrically analogous nozzle.

7.4.2 Diffuser Flow

The diffuser flow in the present context is given by Eq. (7.16) with $a < 0$. Under the assumption of equal particle and flow velocities at the diffuser entrance, the particle motion in such a diffuser flow is described by Eq. (7.22). To a certain diffuser flow with $a = -20(\mathrm{m/s})/\mathrm{m}$ for example, Fig. 7.4 shows the ratio of the particle velocity to the flow velocity along the flow path for different particle density ratios. Because the particle velocity is less slowed down than the flow velocity, there is $\overline{u}_p > 1$. At the diffuser inlet, the ratio of the particle deceleration to the flow deceleration is again given by Eq. (7.25).

For the considered example and for tracer particles with the density ratio $\rho_p/\rho_f < 2$ the final particle velocity is merely about 1% more than the flow velocity. This difference seems to be really negligible in most applications.

According to Fig. 7.4 the state of the particle flow i.e. the ratio of the particle velocity to the flow velocity tends to be constant after a short time. The related constant velocity ratio can be obtained from Eq. (7.22) with respect to $1 + B/A < 0$ and $u_{f,\infty} = 0$

$$\overline{u}_{p,\infty} = \frac{C}{A + B} \tag{7.30}$$

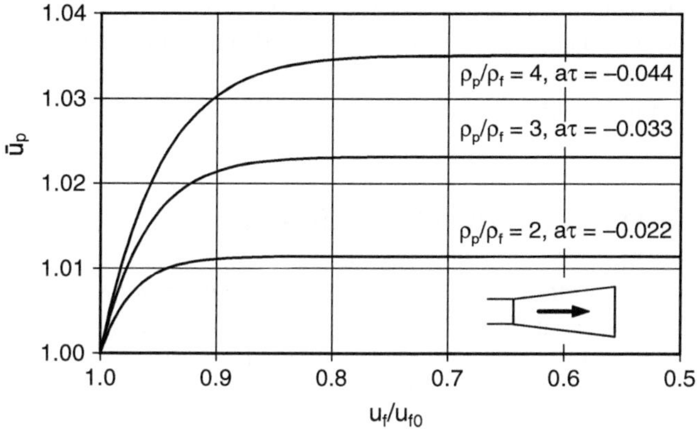

Fig. 7.4 Capability of tracer particles of diameter $d_p = 0.1$ mm of following the water flow ($\mu = 0.001$ Pas) in a diffuser with the flow deceleration equal to $a = -20$ (m/s)/m

It is formally equal to Eq. (7.26). The constant A, B and C are again calculated from Eq. (7.19), however, with $a\tau < 0$ for diffuser flows. Figure 7.5 shows corresponding relationships calculated from the above equation. The almost linear dependence of the final particle velocity on the parameter $a\tau$ can be again utilized to simplify Eq. (7.30). Like calculations leading to Eq. (7.28) for $a\tau << 1$, corresponding calculations are summarized as follows

$$\left.\frac{d\overline{u}_{p,\infty}}{d(-a\tau)}\right|_{a\tau=0} = 1 - \frac{\rho_f}{\rho_p} \tag{7.31}$$

$$\overline{u}_{p,\infty} = 1 - \left(1 - \frac{\rho_f}{\rho_p}\right) a\tau \tag{7.32}$$

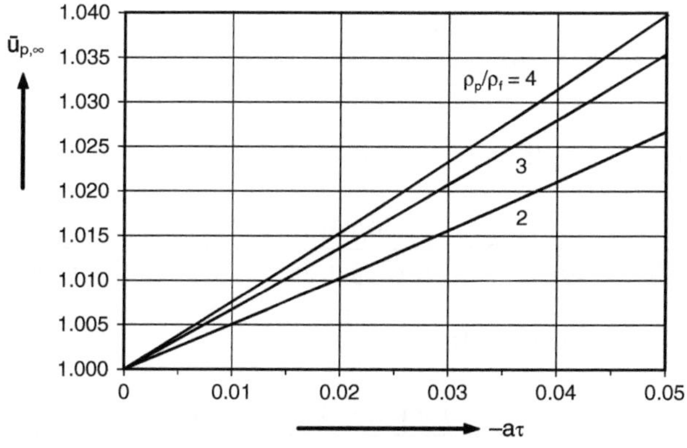

Fig. 7.5 Final reachable ratio of particle to flow velocities in the diffuser flow, in the function of the particle density and the combined parameter $a\tau$

This last relationship has the same form as Eq. (7.28). Because of $a\tau < 0$ for diffuser flows there is $\overline{u}_{p,\infty} > 1$. Thus a rapid and direct method of evaluating the applicability of interested tracer particles in diffuser flow measurements by means of LDA method is available.

7.5 Particle Motion in the Oscillation Flow

One of the most important aspects dealing with tracer particles for LDA measurements is the particle capability of following velocity fluctuations in a turbulent flow. Because of the high frequency of velocity fluctuations commonly up to thousands hertz it is reasonable to make sure what type of tracer particles should be applied, in order to resolve the true velocity fluctuations in each interested flow. The reaction behavior of a particle to the flow fluctuations can be estimated basically by accounting for a spatially uniform fluctuation flow. The most simple case is to assume the flow fluctuation to be a sinusoidal function with the mean velocity u_0, the angular frequency ω and the amplitude A_f as follows

$$u_f = u_0 + A_f \sin \omega t \tag{7.33}$$

The assumption of spatially uniform flow is for the present consideration necessary, because then the spatial pressure distribution in function of $\partial u_f/\partial x$ in Eq. (7.8) disappears. The introduction of the sinusoidal flow model aims to simulate the non-steadiness of the flow in function of $\partial u_f/\partial t \neq 0$, to which the pressure force exerted on the particle is directly related. It should be mentioned that around the particle the local pressure gradient $\partial p/\partial x$ arising from the fluctuation velocity $\partial u_f/\partial t$ according to Eq. (7.8) can be assumed to be constant, as long as the particle size is sufficiently small against the turbulent structures.

By inserting Eq. (7.33) into Eq. (7.12) one obtains the rearranged particle motion equation as

$$\frac{\mathrm{d}\left(u_p - u_0\right)}{\mathrm{d}t} + a\left(u_p - u_0\right) = b \sin \omega t + c \cos \omega t \tag{7.34}$$

with corresponding particle and fluid constants

$$a = \frac{36\mu}{d_p^2\left(2\rho_p + \rho_f\right)}, \quad b = \frac{36\mu A_f}{d_p^2\left(2\rho_p + \rho_f\right)} \quad \textit{and} \quad c = \frac{3\rho_f \omega A_f}{2\rho_p + \rho_f} \tag{7.35}$$

Integrating Eq. (7.34) then yields

$$u_p - u_0 = \frac{(ab + c\omega)\sin \omega t + (ac - b\omega)\cos \omega t}{a^2 + \omega^2} + C \cdot \mathrm{e}^{-at} \tag{7.36}$$

Herein C is the integration constant.

It is again especially interesting to know the final state of the particle oscillation in the assumed oscillation flow according to Eq. (7.33). This is obtained from Eq. (7.36) by setting $t \to \infty$, so that the last term on the r.h.s. of the equation disappears. The particle oscillation in the flow is then independent of the initial velocity difference between the flow and the particle and is described by

$$u_\mathrm{p} - u_0 = \frac{(ab + c\omega)\sin\omega t + (ac - b\omega)\cos\omega t}{a^2 + \omega^2} \tag{7.37}$$

Obviously the particle motion has been shown to be a pure oscillation. To further simplify this equation, the relationship $(ab + c\omega)^2 + (ac - b\omega)^2 = (a^2 + \omega^2)(b^2 + c^2)$ is applied, so that the following substitutions can be applied

$$\frac{ab + c\omega}{\sqrt{(a^2 + \omega^2)(b^2 + c^2)}} = \cos\varphi_\mathrm{p}, \qquad \frac{b\omega - ac}{\sqrt{(a^2 + \omega^2)(b^2 + c^2)}} = \sin\varphi_\mathrm{p} \tag{7.38}$$

Eq. (7.37) then becomes

$$u_\mathrm{p} - u_0 = \sqrt{\frac{b^2 + c^2}{a^2 + \omega^2}}\left(\sin\omega t\cos\varphi_\mathrm{p} - \cos\omega t\sin\varphi_\mathrm{p}\right) = A_\mathrm{p}\sin\left(\omega t - \varphi_\mathrm{p}\right) \tag{7.39}$$

As can be seen, it deals with an oscillation of the particle at the same oscillation frequency as the fluid flow, however with an amplitude modification and a phase difference. The amplitude A_p of the particle oscillation is calculated, with respect to Eq. (7.35), as

$$A_\mathrm{p} = \sqrt{\frac{b^2 + c^2}{a^2 + \omega^2}} = \sqrt{\frac{1 + \frac{1}{12^2}\left(\frac{\omega d_\mathrm{p}^2}{\nu}\right)^2}{1 + \frac{1}{36^2}\left(1 + 2\frac{\rho_\mathrm{p}}{\rho_\mathrm{f}}\right)^2\left(\frac{\omega d_\mathrm{p}^2}{\nu}\right)^2}} \cdot A_\mathrm{f} \tag{7.40}$$

or

$$\frac{A_\mathrm{p}}{A_\mathrm{f}} = \sqrt{\frac{1 + \frac{1}{12^2}N_\mathrm{S}^4}{1 + \frac{1}{36^2}\left(1 + 2\frac{\rho_\mathrm{p}}{\rho_\mathrm{f}}\right)^2 N_\mathrm{S}^4}} \tag{7.41}$$

Herein the Stokes number is applied which is defined by

$$N_\mathrm{S} = \sqrt{\frac{\omega d_\mathrm{p}^2}{\nu}} \tag{7.42}$$

with ν as the kinematic viscosity of the flow.

Correspondingly, the phase difference in Eq. (7.39) is calculated as

$$\tan\varphi_{\mathrm{p}} = \frac{b\omega - ac}{ab + c\omega} = \frac{24\left(\rho_{\mathrm{p}}/\rho_{\mathrm{f}} - 1\right)N_{\mathrm{S}}^2}{432 + \left(2\rho_{\mathrm{p}}/\rho_{\mathrm{f}} + 1\right)N_{\mathrm{S}}^4} \tag{7.43}$$

The Stokes number combines the influences of the flow fluctuation frequency, the particle size and the kinematic viscosity of the fluid on the particle motion in the flow. Its values in usual flows with tracer particles can be estimated. The flow fluctuation rate is assumed for instance to be 1000 hertz. For a particle of diameter $d_{\mathrm{p}} = 0.1$ mm and the flow of kinematic viscosity $\nu = 1.0 \cdot 10^{-6}$ m^2/s (water at p = 1 bar and $T = 20$°C), the Stokes number is calculated to $N_{\mathrm{S}} \approx 8$. This value can be considered to be the upper value of tracer particles usually applied in LDA flow measurements. Table 7.1 shows the variation range of the Stokes numbers in diverse situations in the water flow ($\nu = 1.0 \cdot 10^{-6}$).

In using the Stokes number, as defined in Eq. (7.42), the influences of the particle size and the particle density on the oscillation amplitude of the particle in the flow has been separated. This clearly shows the great advantage against the use of the relaxation time, which is defined in Eq. (7.14) and is basically available for stationary rather than the oscillation carrying flows.

The amplitude ratio given in Eq. (7.41) as a function of the density ratio $\rho_{\mathrm{p}}/\rho_{\mathrm{f}}$ and the Stokes number is shown in Fig. 7.6 for a wide range of the Stokes number. In general, particles with the density ratio $\rho_{\mathrm{p}}/\rho_{\mathrm{f}}<1.2$ could well follow the flow oscillation, even at high oscillation frequency and for large particles. The maximum diminishment of the oscillation amplitude is about 10%. For $\rho_{\mathrm{p}} = \rho_{\mathrm{f}}$ there are both $A_{\mathrm{p}} = A_{\mathrm{f}}$ and $\varphi_{\mathrm{p}} = 0$. The particle will follow the flow exactly regardless of size. It indeed deals with a fluid particle.

As the Stokes number increases, the amplitude ratio $A_{\mathrm{p}}/A_{\mathrm{f}}$ tends to be a constant that is only a function of the density ratio. This can be easily confirmed, as for very large Stokes numbers it results from Eq. (7.41)

$$\frac{A_{\mathrm{p}}}{A_{\mathrm{f}}} = \frac{3}{1 + 2\rho_{\mathrm{p}}/\rho_{\mathrm{f}}} \tag{7.44}$$

For particles with $\rho_{\mathrm{p}}/\rho_{\mathrm{f}} = 1.2$ for instance there is $A_{\mathrm{p}}/A_{\mathrm{f}} = 0.88$.

Table 7.1 Stokes number in the water flow in function of the particle size and the fluctuation rate (frequency) of the flow

	Particle diameters (mm)				
Frequency (Hz)	$d_{\mathrm{p}} = 0.02$	$d_{\mathrm{p}} = 0.04$	$d_{\mathrm{p}} = 0.06$	$d_{\mathrm{p}} = 0.08$	$d_{\mathrm{p}} = 0.1$
100	0.5	1.0	1.5	2.0	2.5
200	0.7	1.4	2.1	2.8	3.5
500	1.1	2.2	3.4	4.5	5.6
1000	1.6	3.2	4.8	6.3	7.9

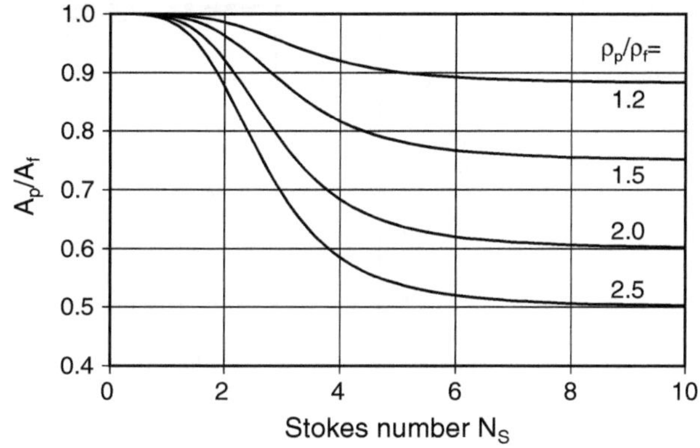

Fig. 7.6 Amplitude of the particle oscillation in the oscillation carrying flow in the function of the Stokes number for different particle density ratios

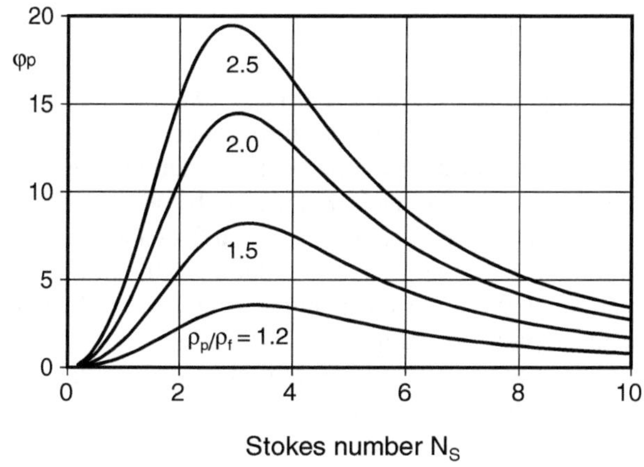

Fig. 7.7 Phase difference of the particle oscillation in the oscillation carrying flow in the function of the Stokes number for different particle density ratio

In Fig. 7.7, the phase difference between the particle and the flow oscillations has been shown. For particles with a density ratio $\rho_p/\rho_f < 1.2$ the phase difference could be kept within 3.5°. It should be mentioned that for measurements of turbulent flows the phase difference is actually irrelevant. Also the existence of a maximum phase difference for each given particle density ratio ρ_p/ρ_f appears to be less interesting.

From above considerations with an oscillation flow model both the capability of particles of following the flow fluctuations and its limitation could be well demonstrated. The calculation results can be applied to simulate the particle behaviors in turbulent flows by considering Eq. (7.33). On one hand, the turbulence intensity of a flow, that is given by $|u'|/\overline{u}$, can be simulated in terms of the amplitude ratio A_f/u_0 of a sinusoidal fluctuation. On the other hand, the frequency of the turbulent flows

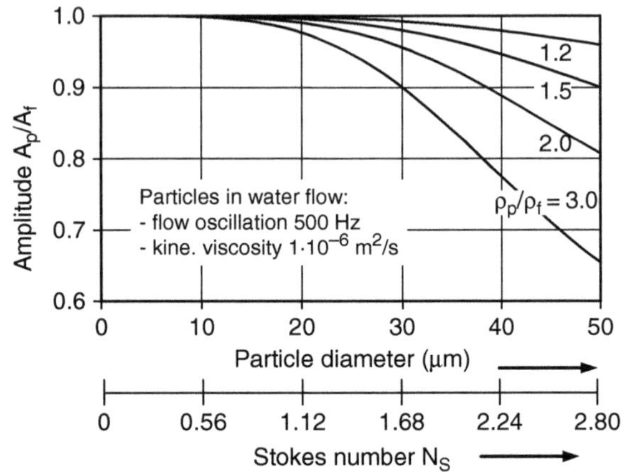

Fig. 7.8 Amplitude ratio of the particle to the water flow oscillations in the function of the particle size for different particle density ratios

that commonly takes hundreds to thousands of hertz, can be simulated by the angular frequency ω. For each particle of a given size, the related Stokes number can be calculated according to Eq. (7.42). From Eq. (7.41) then the amplitude ratio of the particle to the flow fluctuations can be further determined. Fig. 7.8 represents such a calculation example for different particles in a turbulent water flow, in which the frequency of velocity fluctuations is assumed to be 500 Hz. For particles with a density ratio below 2.0 (for glass sphere it is about 2.4) and diameters of less than 30 μm, the amplitude ratio of the particle oscillation to the flow fluctuations is calculated to be greater than 95%. For most engineering flows and the flow evaluation, such a level of measurement accuracy is well acceptable.

With respect to the calculated particle motion in the oscillation flow there are two special aspects that should be accounted for. Firstly, for very small particles the particle motion in the flow is governed by the viscous drag force, as this has already been indicated in Sect. 7.1. Secondly, for large particles i.e. at large Stokes numbers the volumetric forces dominate in the particle dynamics. These two special cases will be treated below.

7.5.1 *Particle Flows of Small Stokes Numbers*

For LDA measurements, small particles are in favor to be used because of their high capabilities of following the flows and the flow fluctuations. In general, small particles are specified by small Stokes numbers. For $N_S < 1$ Eq. (7.41) is simplified by neglecting the high order term as N_S^8 in the Taylor series expansion to

$$\frac{A_p}{A_f} = 1 - \frac{1}{288}\left[\frac{1}{9}\left(1 + 2\frac{\rho_p}{\rho_f}\right)^2 - 1\right] N_S^4 \tag{7.45}$$

On the other side, as mentioned before, the viscous drag force dominates in influencing the particle motion when particles are sufficiently small. In order to quantify this circumstance, only the viscous drag force in Eq. (7.12) is considered. By again assuming the same sinusoidal flow oscillation as Eq. (7.33), the corresponding amplitude ratio can be calculated from the same calculation procedure, leading to

$$\frac{A_p}{A_f} = \frac{1}{\sqrt{1+\frac{1}{18^2}\frac{\rho_p^2}{\rho_f^2}N_S^4}} \approx 1 - \frac{1}{648}\frac{\rho_p^2}{\rho_f^2}N_S^4 \tag{7.46}$$

Also to be mentioned is that in using the relaxation time according to Eq. (7.14), the expression at square root in above equation can also be written as $\sqrt{1+\omega^2\tau^2}$. The corresponding expression for A_p/A_f can be found in some earlier publications as cited in Carter et al. (2001).

The comparison between Eqs. (7.45) and (7.46) has been shown in Fig. 7.9. It clearly demonstrates that for small particles the viscous drag force dominates. Since in deriving Eq. (7.46) both the pressure force arising from the flow fluctuation $\partial u_f/\partial t$ according to Eq. (7.7) and the force owing to the added mass were neglected, the amplitude ratio A_p/A_f will not be in unity, even for the density ratio $\rho_p/\rho_f = 1$. For this reason Eq. (7.45) is more precise than Eq. (7.46) and could be considered to be resulted only from the viscous drag force.

Correspondingly the phase difference that is given in Eq. (7.43) is simplified for $N_S < 1$ as

$$\tan\varphi_p \approx \frac{1}{18}\left(\frac{\rho_p}{\rho_f} - 1\right)N_S^2 \tag{7.47}$$

For $N_S << 1$ there are $A_p = A_f$ and $\varphi_p = 0$, as expected.

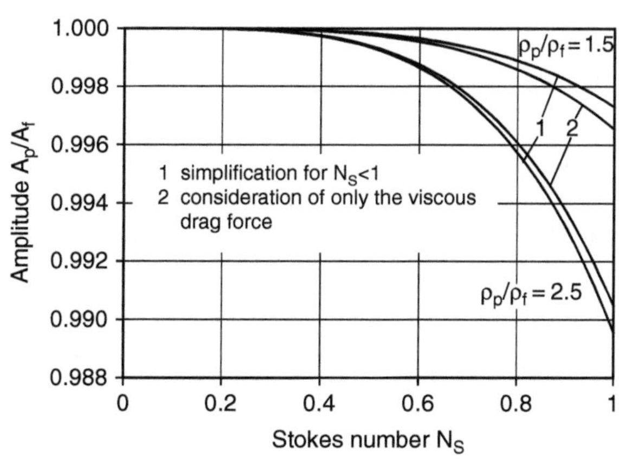

Fig. 7.9 Comparison of calculation results respectively from the simplification for small Stokes numbers and from the consideration of only the viscous drag force

7.5.2 Particle Flows of Large Stokes Numbers

In addition to the viscous drag force, the other effective forces exerted on the particle in the spatially uniform flow are the pressure force arising from the local flow fluctuation rate $\partial u_{\mathrm{f}}/\partial t$ and the force because of the added mass. As it has been shown in Sect. 7.1, these last two forces are proportional to the third power of the particle diameter. Such a particle dynamical property implies that at large particles both volume forces dominate in influencing the particle motion against the viscous drag force which becomes negligible. This circumstance has already been confirmed in Fig. 7.6, as for large Stokes numbers ($N_{\mathrm{S}} > 10$) the particle oscillation amplitude is independent of the Stokes number i.e., according to Eq. (7.42), independent of the fluid viscosity. The corresponding calculation result has been given in Eq. (7.44). In effect, the same result can also be obtained from Eq. (7.34) by directly neglecting the viscous drag force, that means to set $a = 0$ and $b = 0$. Because the particle motion is independent of the particle size, Eq. (7.44) can also be used to explain why the rising velocity of air bubbles in water is independent of the bubble size, as mentioned in Sect. 7.1.3.

Chapter 8
Zero Correlation Method (ZCM)

8.1 Shear Stress Measurements with Non-coincident LDA

The most significant turbulence parameters of a turbulent flow are specified by corresponding Reynolds stresses that are given in the matrix form at Eq. (2.11). The knowledge about the turbulent shear stresses for instance τ_{uv} basically demands, according to Eq. (5.7), coincident measurements of two velocity components u and v. For such measurements a two-component LDA-system has usually to be applied and optically arranged for direct measurements of u and v (Chap. 4). This is well available because most LDA systems are designed so. The different case, however, will be encountered in measurements of all three velocity components. In such a case, the third i.e. on-axis velocity component has usually to be separately measured by off-axis alignment of the LDA head. This can be achieved either directly by the method according to Fig. 8.1a, or indirectly according to Fig. 8.1b. In the case of indirect measurements, the on-axis velocity component is calculated by transforming the velocity components u_{x} and u_{φ}

$$u_{\mathrm{y}} = \frac{u_{\varphi} - u_{\mathrm{x}} \cos \varphi}{\sin \varphi} \tag{8.1}$$

This equation has already been given by Eq. (6.40).

Because it deals here with non-coincident measurements of velocity components u_{x} and u_{y}, the turbulent shear stress $\overline{u'_{\mathrm{x}} u'_{\mathrm{y}}}$ could not be directly obtained by regarding Eq. (5.7). For indirectly obtaining any turbulent shear stress, the Reynolds stress matrix shown in Eq. (2.11) is considered again. Because of equalities $\tau_{\mathrm{xy}} = \tau_{\mathrm{yx}}, \tau_{\mathrm{xz}} = \tau_{\mathrm{zx}}$ and $\tau_{\mathrm{yz}} = \tau_{\mathrm{zy}}$ there are indeed altogether six independent Reynolds turbulent stresses. According to the statement given by Durst et al. (1981) and for the most general case of non-coincident LDA measurements, six single measurements are needed for resolving all six Reynolds stresses in a three-dimensional space. In the case of considering the flow in a two-dimensional plane, then three single measurements are necessary. The corresponding measurement method has already been shown in Sect. 6.2 with respect to Fig. 6.7.

Measurements of complete turbulent stresses in a two-dimensional plane within a flow field, however, can be simplified by making use of a highly reasonable

Zh. Zhang, *LDA Application Methods*, Experimental Fluid Mechanics,
DOI 10.1007/978-3-642-13514-9_8, © Springer-Verlag Berlin Heidelberg 2010

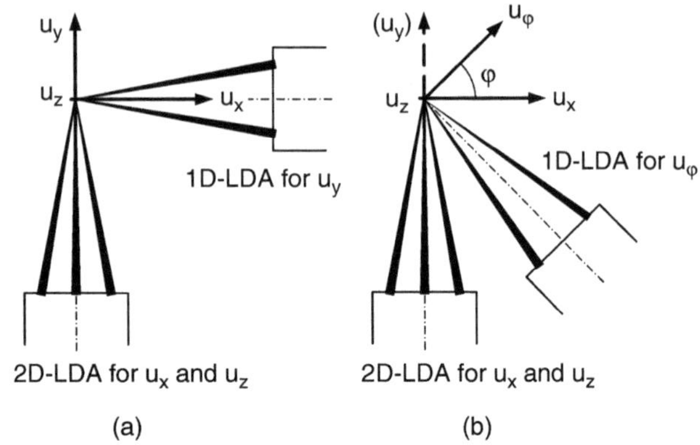

Fig. 8.1 Separate measurement of the on-axis velocity component u_y by relocating the LDA head; The velocity component u_z is perpendicular to the drawing plane (**a**) direct measurement of the velocity component u_y; (**b**) indirect measurement of the velocity component u_y

assumption to the turbulent flow. The so-called Zero Correlation Method (ZCM) (Zhang and Eisele 1998a, Zhang 1999) enables the complete stress components in a two-dimensional plane to be determined by just carrying out two single measurements. The method thus does not provide any more expenses, because two measurements are necessary anyway for obtaining both the magnitude and the direction of the mean velocity. This method will be presented in below.

8.2 Basics of ZCM

The significant attribute of a turbulent flow is the stochastic velocity fluctuations. In dealing with a stationary turbulent flow, flow fluctuations basically comprise the fluctuations of both the magnitude and the direction of the velocity vector, as illustrated in Fig. 8.2 for a turbulent flow which was measured by the LDA method. Because of the stochastic feature of flow fluctuations, the fluctuation of the velocity direction in the $x - y$ plane can be considered to symmetrically lie on both sides of the mean velocity vector that has its direction at $\overline{\varphi}$. For further consideration, the velocity component along the mean velocity vector is denoted by u_1.

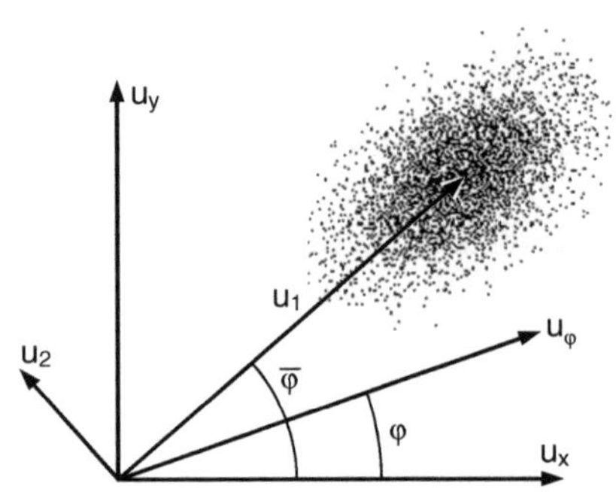

Fig. 8.2 Turbulent flow and velocity fluctuations measured by LDA

In its perpendicular, the velocity component is u_2. Clearly there is $\overline{u}_2 = 0$. The symmetrical distribution of velocity fluctuations around the mean velocity vector is mathematically expressed by the zero correlation between flow fluctuations in velocity components u_1 and, u_2 as given by

$$C = \overline{u_1' u_2'} = 0 \tag{8.2}$$

From velocity components u_1 and u_2 the velocity components u_x and u_y can be obtained from the orthogonal coordinate transformation that has already been outlined in Sect. 6.1. The same transformation is also applicable to the flow fluctuations. In effect, Eqs. (6.6) and (6.7) can be directly applied to represent the fluctuation velocities in the present case as

$$u_\mathrm{x}' = u_1' \cos\overline{\varphi} - u_2' \sin\overline{\varphi} \tag{8.3}$$

$$u_\mathrm{y}' = u_1' \sin\overline{\varphi} + u_2' \cos\overline{\varphi} \tag{8.4}$$

In these equations, the mean flow direction is denoted by angle $\overline{\varphi}$ that is calculated by

$$\tan\overline{\varphi} = \frac{\overline{u}_\mathrm{y}}{\overline{u}_\mathrm{x}} \tag{8.5}$$

For later convenience this equation is also represented as

$$\tan 2\overline{\varphi} = \frac{2\tan\overline{\varphi}}{1-\tan^2\overline{\varphi}} = \frac{2\overline{u}_\mathrm{x}\overline{u}_\mathrm{y}}{\overline{u}_\mathrm{x}^2 - \overline{u}_\mathrm{y}^2} \tag{8.6}$$

The statistical turbulence properties including the Reynolds normal and shear stresses can be calculated from the velocity fluctuations u_x' and u_y' given above. With respect to the zero correlation condition given by Eq. (8.2), the following relationships are obtained

$$\overline{u_\mathrm{x}'^2} = \overline{u_1'^2}\cos^2\overline{\varphi} + \overline{u_2'^2}\sin^2\overline{\varphi} \tag{8.7}$$

$$\overline{u_\mathrm{y}'^2} = \overline{u_1'^2}\sin^2\overline{\varphi} + \overline{u_2'^2}\cos^2\overline{\varphi} \tag{8.8}$$

$$\overline{u_\mathrm{x}' u_\mathrm{y}'} = \frac{1}{2}\left(\overline{u_1'^2} - \overline{u_2'^2}\right)\sin 2\overline{\varphi} \tag{8.9}$$

Also to be mentioned is that these three equations can also be directly obtained from Eqs. (6.9), (6.10), and (6.11). The velocity components u and v there need to be considered to be the components on the main flow direction $\overline{\varphi}$ and its perpendicular, respectively. Because of predefined zero correlation condition by Eq. (8.2) $\tau_\mathrm{uv} = 0$ should be applied.

Based on similar calculations, the turbulence properties related to any other components u_φ, according to Fig. 8.2, are given by

$$\overline{u'^2_\varphi} = \overline{u'^2_1}\cos^2(\varphi - \overline{\varphi}) + \overline{u'^2_2}\sin^2(\varphi - \overline{\varphi}) \tag{8.10}$$

$$\overline{u'_\varphi u'_{\varphi+90^\circ}} = -\frac{1}{2}\left(\overline{u'^2_1} - \overline{u'^2_2}\right)\sin 2(\varphi - \overline{\varphi}) \tag{8.11}$$

Because both normal stresses $\overline{u'^2_1}$ and $\overline{u'^2_2}$ are positive, the normal stress $\overline{u'^2_\varphi}$ is positive in all directions. In fact, $\overline{u'^2_1}$ and $\overline{u'^2_2}$ represent two principal normal stresses and $\overline{u'^2_\varphi}$ lies between them. From Eqs. (8.7) and (8.8) two principal normal stresses can be resolved as

$$\overline{u'^2_1} = \frac{1}{2}\left(\overline{u'^2_x} + \overline{u'^2_y}\right) + \frac{1}{2\cos 2\overline{\varphi}}\left(\overline{u'^2_x} - \overline{u'^2_y}\right) \tag{8.12}$$

$$\overline{u'^2_2} = \frac{1}{2}\left(\overline{u'^2_x} + \overline{u'^2_y}\right) - \frac{1}{2\cos 2\overline{\varphi}}\left(\overline{u'^2_x} - \overline{u'^2_y}\right) \tag{8.13}$$

Eq. (8.12) is indeed equal to Eq. (6.18) that represents the principal normal stress at the angle φ_m in the $x - y$ plane, see also Fig. 6.3. Obviously the zero correlation condition given by Eq. (8.2) assumes the angle φ_m for the principal normal stress to be equal to the main flow direction.

The above two equations are subsequently inserted into Eqs. (8.10) and (8.11), respectively. The following expressions are then obtained

$$\overline{u'^2_\varphi} = \frac{1}{2}\left(\overline{u'^2_x} + \overline{u'^2_y}\right) + \frac{\cos 2(\varphi - \overline{\varphi})}{2\cos 2\overline{\varphi}}(\overline{u'^2_x} - \overline{u'^2_y}) \tag{8.14}$$

$$\overline{u'_\varphi u'_{\varphi+90^\circ}} = -\frac{\sin 2(\varphi - \overline{\varphi})}{2\cos 2\overline{\varphi}}\left(\overline{u'^2_x} - \overline{u'^2_y}\right) \tag{8.15}$$

These last two equations indicate that the complete Reynolds stresses in the $x - y$ plane can be well resolved from two independent i.e. non-coincident measurements ($\overline{u'^2_x}$ and $\overline{u'^2_y}$). In comparison with the method of three measurements, as presented in Sect. 6.2 with respect to Fig. 6.7, the method presented here clearly shows advantages in the simplification of turbulence measurements. This accessibility is simply based on the assumption of zero correlation condition $\overline{u'_1 u'_2} = 0$ as specified in Eq. (8.2). For this reason the method shown above is called Zero Correlation Method (ZCM).

The turbulent shear stress $\overline{u'_x u'_y}$ is obtained by setting $\varphi = 0$ in Eq. (8.15), which results in

$$\overline{u'_x u'_y} = \frac{1}{2}\tan 2\overline{\varphi}\left(\overline{u'^2_x} - \overline{u'^2_y}\right) \tag{8.16}$$

and because of Eq. (8.6)

$$\overline{u'_x u'_y} = \frac{\overline{u}_x \overline{u}_y}{\overline{u}_x^2 - \overline{u}_y^2} \left(\overline{u'^2_x} - \overline{u'^2_y} \right) \tag{8.17}$$

Eq. (8.16) is comparable to Eq. (6.15). It can be directly applied to Fig. 8.1a.

8.3 Extension of ZCM

8.3.1 Non-orthogonal Velocity Components

The Zero Correlation Method (ZCM) introduced in the last section applies to the orthogonal velocity components like u_x and u_y in a two-dimensional plane, as shown in Fig. 8.1a. Because LDA arrangement like that in Fig. 8.1b is also often available, the ZCM must be modified to extend its applications.

From Eq. (8.14) the turbulent normal stress $\overline{u'^2_y}$ is resolved as

$$\overline{u'^2_y} = \frac{\cos\varphi \cos(2\overline{\varphi} - \varphi)\overline{u'^2_x} - \cos 2\overline{\varphi}\,\overline{u'^2_\varphi}}{\sin\varphi \sin(2\overline{\varphi} - \varphi)} \tag{8.18}$$

This equation signifies that from measurements of two non-orthogonal turbulence components $\overline{u'^2_x}$ and $\overline{u'^2_\varphi}$ in a two-dimensional plane the turbulence component $\overline{u'^2_y}$ in the same plane can be determined as well. One has again the case of orthogonal normal stresses. All results achieved in Sect. 8.2, especially Eq. (8.16), can be applied. The mean flow angle $\overline{\varphi}$ is calculated by Eq. (8.5), where the mean velocity component $\overline{u}_y$ is obtained from Eq. (8.1).

8.3.2 Three-Dimensional Flow Turbulence

Calculations presented above are performed for turbulence properties in a two-dimensional plane. For purposes of applying the ZCM to three-dimensional turbulent flows in the Cartesian coordinate system, Eq. (8.17) is taken into account as the reference for other components. The completeness of turbulent stresses can be expressed in a matrix as given by

$$\overline{u'_i u'_j} = \begin{vmatrix} \overline{u'^2_x} & \frac{\bar{u}_x \bar{u}_y}{\bar{u}_x^2 - \bar{u}_y^2}\left(\overline{u'^2_x} - \overline{u'^2_y}\right) & \frac{\bar{u}_x \bar{u}_z}{\bar{u}_x^2 - \bar{u}_z^2}\left(\overline{u'^2_x} - \overline{u'^2_z}\right) \\ \frac{\bar{u}_x \bar{u}_y}{\bar{u}_x^2 - \bar{u}_y^2}\left(\overline{u'^2_x} - \overline{u'^2_y}\right) & \overline{u'^2_y} & \frac{\bar{u}_y \bar{u}_z}{\bar{u}_y^2 - \bar{u}_z^2}\left(\overline{u'^2_y} - \overline{u'^2_z}\right) \\ \frac{\bar{u}_x \bar{u}_z}{\bar{u}_x^2 - \bar{u}_z^2}\left(\overline{u'^2_x} - \overline{u'^2_z}\right) & \frac{\bar{u}_y \bar{u}_z}{\bar{u}_y^2 - \bar{u}_z^2}\left(\overline{u'^2_y} - \overline{u'^2_z}\right) & \overline{u'^2_z} \end{vmatrix} \tag{8.19}$$

Obviously three non-coincident measurements, from which both the mean velocities ($\overline{u}_x$, $\overline{u}_y$ and $\overline{u}_z$) and the turbulence components ($\overline{u_x'^2}$, $\overline{u_y'^2}$ and $\overline{u_z'^2}$) at a fixed point in the flow are obtainable, are sufficient to complete the Reynolds stress matrix with nine elements. The advantage of the ZCM has been thus again demonstrated.

8.4 Restriction and Validation of ZCM

The zero correlation method is indeed an approximation method that is introduced to simplify the turbulence measurements. Respective application restrictions and accuracies must be considered. From Eqs. (8.14) and (8.15), it is evident that the main flow angle $\overline{\varphi}$ in the used coordinate system should not be equal or very close to 45°. This limitation could also be confirmed from Eq. (8.16) when measurements of u_x and u_y should be accomplished. It is therefore recommended to arrange the appropriate $x - y$ coordinate system for measurements. At best the x-axis is set to closely agree with the main flow direction.

The extent of errors resulting from the ZCM depends on the homogeneity of turbulence in the respective flow to be measured. In actual fact, the method does work accurately as long as the zero correlation condition according to Eq. (8.2) is highly satisfied. This is always the case when the local flow does not sensitively affected by the rigid surface or boundaries in the flow. An experimental verification of the method was conducted once through measurements in a turbulent channel flow (Zhang and Eisele 1998a, Zhang 1999). In this validation measurement, a two-component coincident LDA system was applied, so that turbulent stresses $\overline{u_x'^2}$, $\overline{u_y'^2}$ and $\overline{u_x'u_y'}$ were directly obtained. Based on such measurements the validation of the ZCM could be well accomplished by following calculations:

(1) From the coincident measurements of velocity components u_x and u_y other velocity components u_φ and $u_{\varphi+90°}$ for each given angle φ can be calculated via coordinate transformation according to Eq. (6.4) as

$$u_\varphi = u_x \cos\varphi + u_y \sin\varphi \tag{8.20}$$

$$u_{\varphi+90°} = -u_x \sin\varphi + u_y \cos\varphi \tag{8.21}$$

Subsequently the covariance i.e. the correlation between velocity fluctuations u'_φ and $u'_{\varphi+90°}$ is calculated by

$$\overline{u'_\varphi u'_{\varphi+90°}} = \sum_{i=1}^{N} \left(u_\varphi - \overline{u}_\varphi\right)\left(u_{\varphi+90°} - \overline{u}_{\varphi+90°}\right) \tag{8.22}$$

This covariance indeed represents the turbulent shear stress and is a function of angle φ.

(2) On the other side the covariance $\overline{u'_{\varphi}u'_{\varphi+90^\circ}}$ for each given angle φ can also be directly calculated from two normal stresses $\overline{u'^2_x}$ and $\overline{u'^2_y}$ by using Eq. (8.15). Because of the use of this equation as a result of the ZCM, the calculation results must be considered to be approximate.
(3) Calculation results from calculations in (1) and (2), respectively, will be compared.

This validation procedure was applied to the validation measurements (Zhang and Eisele 1998a, Zhang 1999). Figure 8.3a shows the comparison of respective calculations in function of each given angle φ. Obviously the ZCM provides satisfactory measurement results.

The same validation can be completed by concerning the turbulence component $\overline{u'^2_{\varphi}}$. On one side, this component can be calculated from velocity data that are obtained by Eq. (8.20), leading to accurate values of variance. On the other side, it can also be calculated directly from $\overline{u'^2_x}$ and $\overline{u'^2_y}$ by means of Eq. (8.14), leading to approximated values. The comparison between two calculations in relying on the mentioned validation measurement is shown in Fig. 8.3b. The same satisfactory results were obtained.

Strictly, the zero correlation condition according to Eq. (8.2) is not satisfied if applied to the turbulent boundary layer. With respect to the main flow direction $u_1 = u_x$ and to the perpendicular $u_2 = u_y$, the covariance $\overline{u'_1u'_2}$ i.e. $\overline{u'_xu'_y}$ in the turbulent boundary layer is indeed a function of the distance to the wall surface and therefore does not disappear as it is assumed in the ZCM. The error arising from

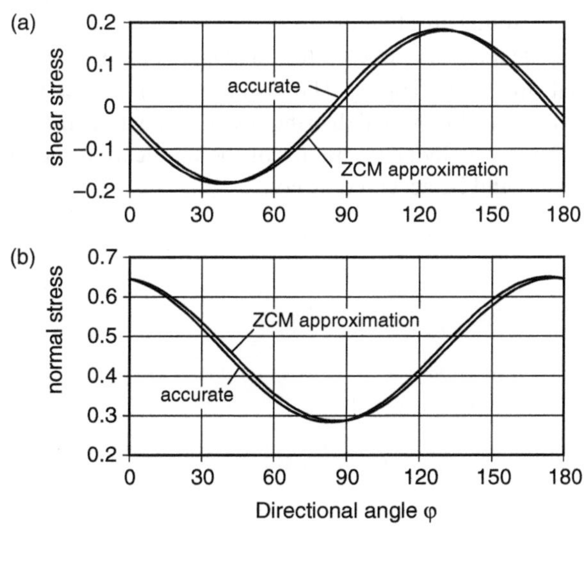

Fig. 8.3 Experimental validations of the ZCM by comparing the accurate turbulent stress in directional distribution with that calculated by the ZCM (Zhang and Eisele 1998a)

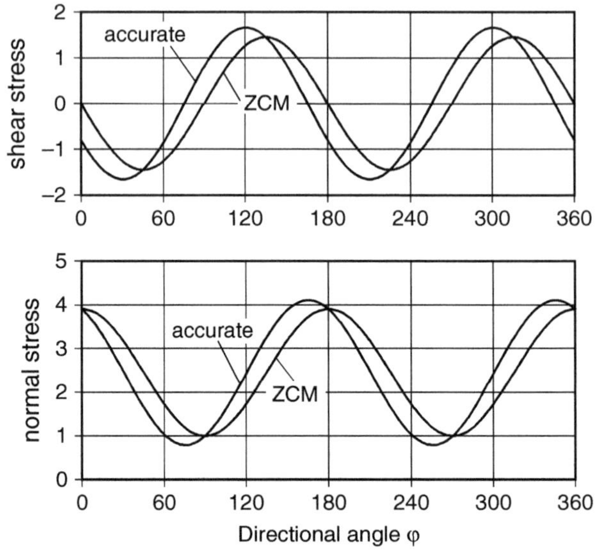

Fig. 8.4 Comparison of accurate turbulent stress in directional distribution with that calculated by ZCM for the flow in the turbulent boundary layer at $y^+ = 90$ (Zhang and Zhang 2002)

the assumption $\overline{u'_x u'_y} = 0$, when applied to the turbulent boundary layer, has been analyzed by Zhang and Zhang (2002). With known turbulent stresses $\overline{u'^2_x}, \overline{u'^2_y}$ and $\overline{u'_x u'_y}$ at a certain wall distance y^+ the directional distributions of the normal and shear stresses can be calculated by Eqs. (6.13) and (6.14), respectively. On the other hand, these distributions can also be calculated by assuming $\overline{u'_x u'_y} = 0$ which in effect simulates the application of ZCM. Figure 8.4 represents the comparison of respective calculations to the turbulent state at $y^+ = 90$ in the turbulent boundary layer. With certain reservation the ZCM in this application provides quite satisfactory result. This obviously arises from the fact that the amplitudes of both the normal and shear stresses in their trigonometric functions are mainly determined by the normal stresses $\overline{u'^2_x}$ and $\overline{u'^2_y}$.The assumption of $\overline{u'_x u'_y} = 0$ in the ZCM merely results in a shift in the directional distribution of respective turbulent stresses. In practical applications, it is indeed a matter of the application requirement, whether the resultant error could be accepted or not.

Chapter 9
Dual Measurement Method (DMM)

9.1 Possibility of Resolving the Secondary Flow

LDA measurements are component measurements. This means that the measured velocity components are always referred to as coordinates in the LDA optical system. For direct measurements of velocity components in the flow field coordinate system, LDA coordinates have to be arranged to agree with the flow field coordinates. In other cases, all interested velocity components in the flow field can be obtained from coordinate transformation, as described in Chap. 6. Correspondingly, LDA users in practical applications always try to make sure either the accurate coincidence between two coordinate systems or an accurate rotation by a given angle. The alignment error has mostly been assumed to be small and hence is often neglected. This is generally allowed in most flow measurements at which the flow field could still be truly represented from the measurements despite the measurement errors. In contrast, however, there are cases at which a small error in LDA alignment to the flow field coordinates could lead to total misinterpretation of the actual flow field. Such a case has been for instance encountered by Zhang and Parkinson (2001, 2002) while trying to measure the very weak but very important secondary flow structure in a high speed water jet of a Pelton turbine (Fig. 9.1) In principle, it is generally impossible to align LDA velocity components u and v to be exactly coincident with velocity components u_x (axial) and u_t (tangential) of the jet flow. An inevitable small alignment error $\tau \neq 0$ simply means that the measured velocity component v additionally comprises a part of the axial velocity component, as given by

$$v = u_t \cos\tau + u_x \sin\tau \tag{9.1}$$

Although the bias angle τ as an error parameter is usually very small, the term $u_x \sin\tau$ in the above equation, however, could be still very large because of high values of the axial velocity component u_x in the high speed jet flow. In the case that this term is comparable to or larger than the term $u_t \cos\tau \approx u_t$, the secondary flow pattern represented by u_t would be sensitively and totally misinterpreted from measurements. For this reason the mentioned secondary flow in a cross section of the high speed jet flow could be measured neither directly nor indirectly through coordinate transformation.

Zh. Zhang, *LDA Application Methods*, Experimental Fluid Mechanics,
DOI 10.1007/978-3-642-13514-9_9, © Springer-Verlag Berlin Heidelberg 2010

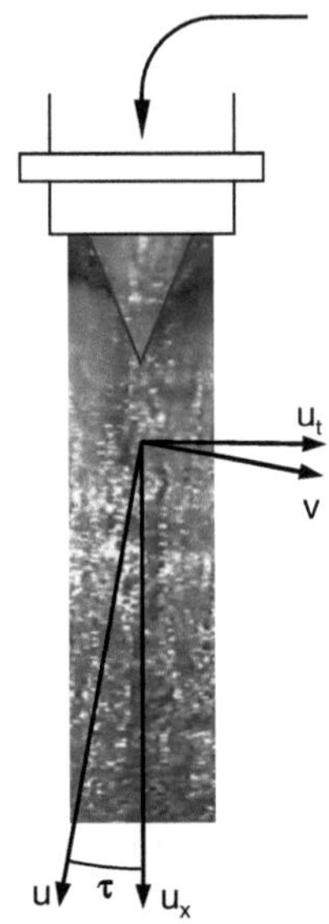

Fig. 9.1 Possible inaccurate LDA alignment to a high speed jet flow and the associated problem of resolving the secondary flow structure in the jet cross section

Such a problem arising from inaccurate LDA alignment could be basically solved, when the alignment error i.e. the bias angle τ could be accurately identified and then the appropriate correction calculation of measurement date is implemented. In the mentioned measurement of the secondary flow structure in the high speed water jet, a method that is known as the Dual Measurement Method (DMM) was developed and applied to accurately identify the geometrical deviation in the LDA alignment to the jet flow. After finding out the alignment error, measurement data could be correctly evaluated. It dealt with a quite interesting flow phenomenon which could be detected by DMM.

The background of the dual measurement method is the two-measurement principle which is sometimes also called two-step method. The most famous example of using this principle is the Michelson-Morley experiment (Hecht 1990) that was constructed, by subsequently rotating the apparatus horizontally for 90°, to be able to measure the difference of light speeds in different directions, provided that this difference would exist. Another example that is directly related to LDA techniques is to exactly check the constant shift frequency generated in one of two laser beams of a laser beam pair, as it will be shown in Chap. 18.

In this chapter, the dual measurement method will be firstly presented in the version of its initial application to the high speed jet flow with complex secondary flow structures (Zhang and Parkinson 2001, 2002). Then the extended form of DMM (Zhang 2005) will be shown that can be applied to other special cases.

The readers would probably be interested in how the laser beams could enter into the jet that has a turbulent and hence opaque surface. The corresponding measurement technique of using an optical wedge can be found in Chap. 18 showing application examples.

9.2 DMM in Basic Form

The dual measurement method was initially developed to accurately resolve the secondary flow structure in the high speed jet flow of a Pelton turbine. The jet was generated by an injector which was connected to a bend, as shown in Fig. 9.2a Because of the bend effect and the resultant change in the flow state, the flow after passing through the bend is characterised by the existence of a secondary flow structure across the pipe section, as shown in Fig. 9.2b based on LDA measurements of tangential velocity components. It deals with a typical secondary flow pattern that clearly demonstrates two identical areas with flow rotations. This flow structure remains while passing through the injector, as it was measured and has been shown in Fig. 9.2c, provisionally without mentioning how this secondary flow structure could be measured. Although it deals with a small scale secondary flow structure, it represents the main reason for the jet instability. One of the most serious disturbances on the jet because of related secondary flows is the generation of a chain of droplets on the jet surface, as indicated in Fig. 9.2a. The measurement of the secondary flow structure in the mentioned high speed water jet represents a highly difficult task and could only be conducted by means of DMM.

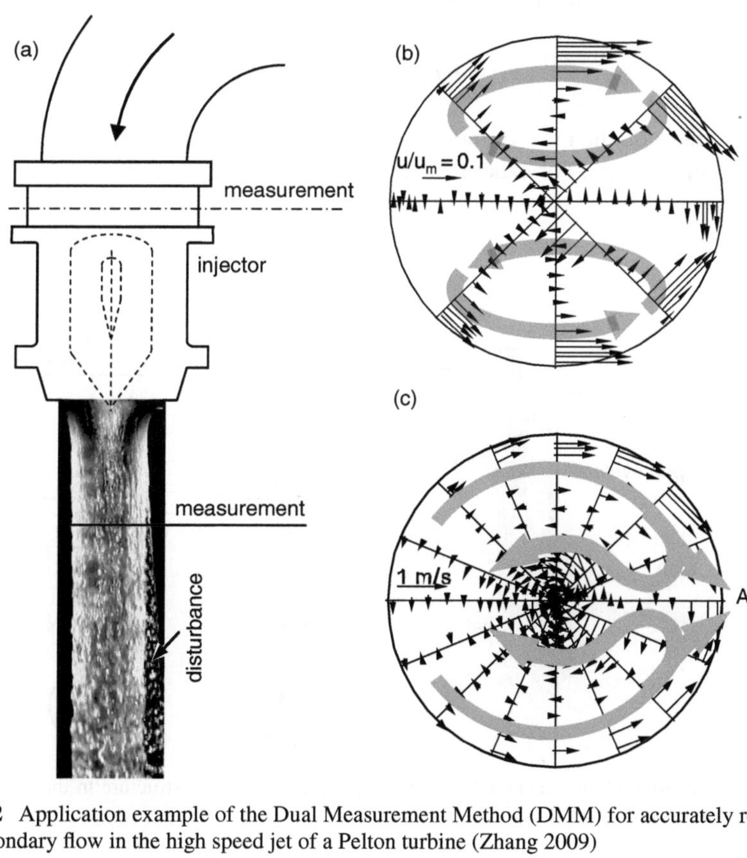

Fig. 9.2 Application example of the Dual Measurement Method (DMM) for accurately resolving the secondary flow in the high speed jet of a Pelton turbine (Zhang 2009)

The jet flow was made accessible for LDA measurements from all directions around the jet axis, as shown in Fig. 9.3a where a support platform for LDA head was installed. The LDA head is mounted on the traversing board of the support device that enables a velocity profile in the jet to be measured. For the measurement technique of getting the laser beams into the jet with turbulent and rough surface the readers are referred to Chap. 18.

The LDA optics was aligned for direct measurements of the tangential velocity component without using coordinate transformation. In assuming an alignment error i.e. a bias angle $\tau \neq 0$ as shown in Fig. 9.3b, the measured velocity component v takes

$$v_0 = u_t \cos\tau + u_x \sin\tau \tag{9.2}$$

In this equation the positive bias angle is defined as it is shown in Fig. 9.3.

The dual measurement method is constructed so as to arrange an additional flow measurement by simply rotating the LDA head around the jet axis by 180° (Fig. 9.3c). While rotating the measurement system the alignment error i.e. the bias angle τ can be assumed to be constant. The measured velocity component v this time at the same point in the jet flow is then given by

$$v_1 = -u_t \cos\tau + u_x \sin\tau \tag{9.3}$$

The second term on the r.h.s. of the above equation is the part arising from the axial velocity component u_x. For exact LDA arrangement with $\tau = 0$ there is

$$v_1 = -v_0 \tag{9.4}$$

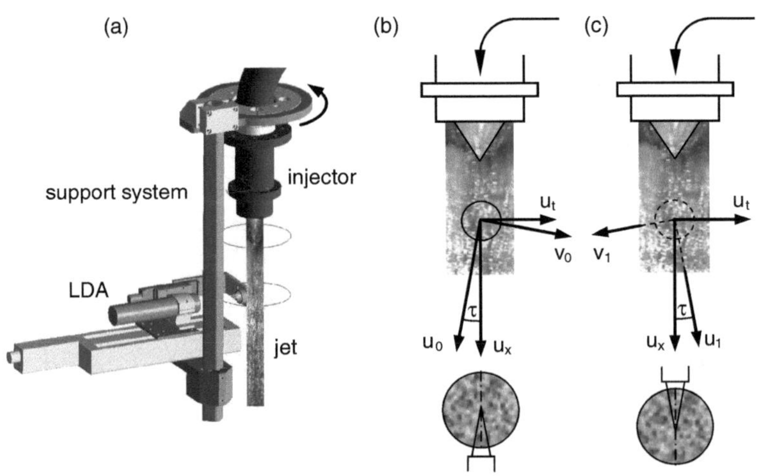

Fig. 9.3 DMM principle for accurately resolving the secondary flow structure in the high speed jet of a Pelton turbine

This equation represents a criterion of the faultless LDA alignment for direct measurement of velocity component u_t. Both values (v_1 and v_0) act as mirrored about $v = 0$. This ideal "mirrored view" of two velocity components v_1 and v_0 will be more or less disturbed by errors with $\tau \neq 0$ in LDA alignment. It can be concluded that any deviation from Eq. (9.4) is quantitatively related to the bias angle τ Since this bias angle τ is actually a geometrical or mechanical arrangement error, it basically causes a systematic error which, according to Eqs. (9.2) and (9.3), takes $u_x \sin\tau$. This systematic error in the velocity measurement is called velocity shift because it acts as an additive quantity in both Eqs. (9.2) and (9.3).

From Eqs. (9.2) and (9.3) one obtains

$$v_{sh} = u_x \sin\tau = \frac{v_0 + v_1}{2} \tag{9.5}$$

as well as for $\tau << 1$

$$u_t = \frac{v_0 - v_1}{2} = v_0 - v_{sh} \tag{9.6}$$

It is now clear that through twice measurements of the same flow the velocity shift as the outcome of the LDA alignment error can be exactly identified. This dual deal of the jet flow measurement leads to direct determination of the tangential velocity component that otherwise could not be accurately measured. Because of this, the applied method is called the Dual Measurement Method (DMM).

With regard to the assumption that the velocity shift v_{sh} is a kind of systematic error involved in the LDA arrangement, it needs only to be determined one time at a fixed point in the flow by means of DMM. It can then be directly applied, according to Eq. (9.6), to correct the measurement results that are achieved at other points in the flow.

In the above mentioned example, the dual measurement method was applied to identify the bias angle τ and the associated velocity shift. The completed two measurements of the same velocity profile across the jet have been shown in Fig. 9.4a Because it deals with the same velocity profile, the "mirrored view" of two-time measurements would have been expected. That is to say that two velocity profiles should have symmetrically lain on both sides of the neutral line with $v = 0$, if the bias angle τ would be zero. The measured deviation of the symmetry line from $v = 0$, as shown in Fig. 9.4a, just corresponds to the velocity shift calculated by Eq. (9.5). In this measurement example, the velocity shift reads at 0.4 m/s. It corresponds to a bias angle of $\tau = 0.92°$ (for $u_x = 25$ m/s). Obviously this is a quite small angle. The associated velocity shift, however, is of the same order as the existing velocity component itself or even higher. With respect to the velocity shift determined from Fig. 9.4a measurement results could be immediately corrected, as shown in Fig. 9.4b. Both measurements after correction then behave as mirrored at $v = 0$, as expected.

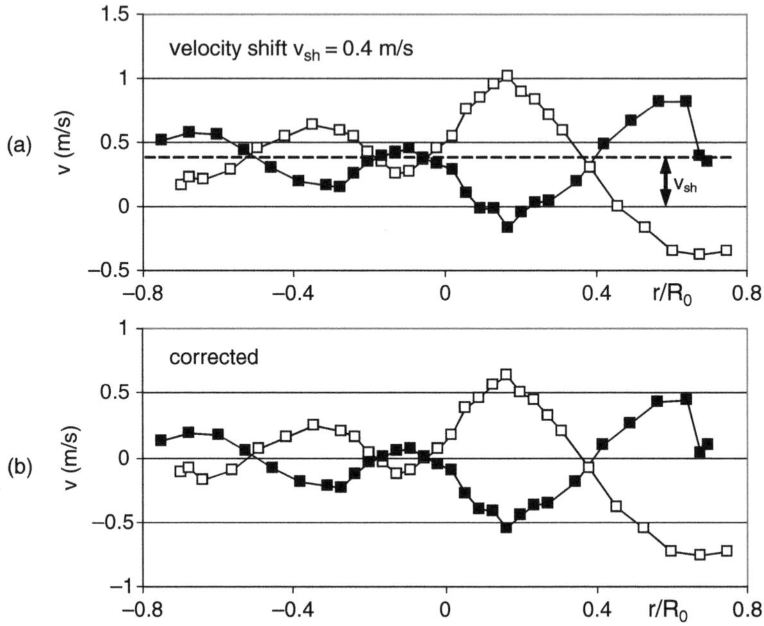

Fig. 9.4 Determination of the velocity shift by means of DMM and correction of the measurement results across the jet flow

With respect to the constant systematic error of $v_{\text{sh}} = 0.4$ in the installed system the complete secondary flow structure in a jet section has been measured and corrected, as already shown in Fig. 9.2c. Obviously the secondary flow that contains two counter-rotating vortices exhibits the similar flow structure as in the pipe ahead of the injector. At the point A both streams from two flow areas come together. Because of the absence of any rigid boundaries that could guide the jet flow, the fluid tends to escape from the jet. As the consequence a chain of water droplets comes about, as this has often been observed in hydro power plants with Pelton turbines since more than half century (Zhang 2009).

The comparison between Fig. 9.4a and Fig. 9.4b points out that a small bias angle τ could lead to total misinterpretation of the flow. In fact, none of the two measurements in Fig. 9.4a represents the real flow. While the real flow pattern in the jet section involves the swirling flow structures (Fig. 9.2c), each uncorrected measurement (Fig. 9.4a) simply shows a transversal motion of the fluid almost with $v > 0$.

Basically it is sufficient to apply the DMM to a single point in the flow to determine the bias angle τ and hence the systematic error in form of the velocity shift v_{sh} In the presented example as shown in Fig. 9.4, the dual measurement method has been applied to a survey across the jet. This could be well realized by the constructed measurement system according to Fig. 9.3. The constant velocity shift across the jet straightforwardly demonstrates the reliability of DMM. The velocity shift as the systematic error involved in measurements has been thus verified.

It should be emphasized that the velocity shift could only be assumed as a constant systematic error, when the bias angle τ is kept constant while rotating the LDA support system. The consistency of the bias angle τ thus represents the prerequisite for applying the DMM.

9.3 DMM with Coordinate Transformation

In the last section, the direct dual measurement method has been presented. It is known as the direct method because in the applied example the LDA coordinates were initially set to be coincident with the flow coordinates by $u = u_x$, $v = u_y$ and $w = u_z$. Sometimes it is advantageous to rotate the LDA coordinate system by an angle α against the flow system, so that a coordinate transformation matrix $R(\alpha)$ generally exists between two coordinate systems (Chap. 6).

Because the angle of rotation of one coordinate system against another is by no means exact, its real value has to be assumed to involve a small LDA alignment error i.e. bias angle τ (<<1) and hence takes $\alpha + \tau$ according to Fig. 9.5 As for the general case, the velocity component to be measured in the secondary flow is denoted by u_y, instead of u_t that is used for the example in Sect. 9.2.

The alignment of a two-component LDA head according to Fig. 9.5a enables the measurements of velocity components u_0 and v_0. The velocity component u_x and u_y in the flow system are then simply obtained by

$$u_x = u_0 \cos(\alpha + \tau) + v_0 \sin(\alpha + \tau) \tag{9.7}$$

$$u_y = -u_0 \sin(\alpha + \tau) + v_0 \cos(\alpha + \tau) \tag{9.8}$$

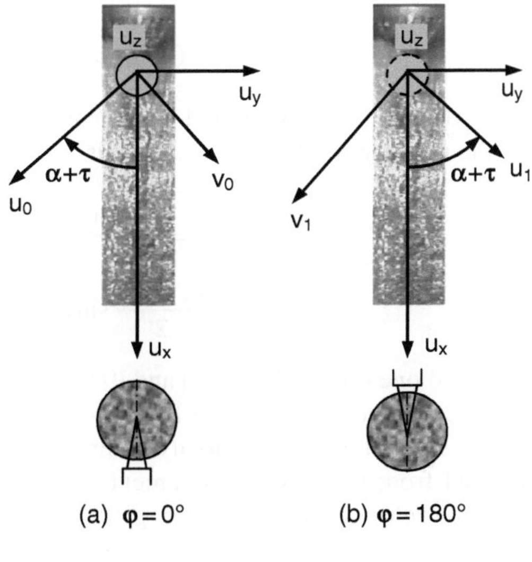

Fig. 9.5 DMM principle for accurately resolving the secondary flow structure in a high speed jet flow

Since the small alignment error τ only causes the negligible relative change in the axial velocity component u_x, so $\tau = 0$ can be applied to Eq. (9.7). Thus u_x acts as a component that is not affected by the alignment error τ. To Eq. (9.8) the approximations $\sin(\alpha + \tau) \approx \sin\alpha + \tau\cos\alpha$ and $\cos(\alpha + \tau) \approx \cos\alpha - \tau\sin\alpha$ because of $\tau << 1$ will be applied. This leads to

$$u_y = (-u_0 \sin\alpha + v_0 \cos\alpha) - u_x\tau \tag{9.9}$$

and further with the velocity shift $v_{sh} = u_x\tau$

$$u_y = (-u_0 \sin\alpha + v_0 \cos\alpha) - v_{sh} \tag{9.10}$$

Like in Eq. (9.5) the velocity shift appears again in the form of $v_{sh} = u_x \sin\tau$ i.e.$v_{sh} \approx u_x\tau$ because of $\tau << 1$. In addition, it has been again confirmed as an error that shifts the measurement results. Basically the velocity component u_y is intended to be determined by the first term on the r.h.s. of Eq. (9.10). This will be only true if $\tau = 0$ is true. The existence of the velocity shift v_{sh} as the consequence of the alignment error τ could significantly influence the determination of velocity component u_y, as already shown in the last section.

In order to examine the velocity shift existing in Eq. (9.10), the dual measurement method is again applied. This means that the LDA set-up needs to be turned around the x-axis by 180°, as shown in Fig. 9.5b. Attention has to be paid so that the bias angle τ as a systematic error has to remain constant.

From Fig. 9.5b and in analogy to Eq. (9.8), the velocity component u_y is directly written as

$$u_y = u_1 \sin(\alpha + \tau) - v_1 \cos(\alpha + \tau) \tag{9.11}$$

Based on similar calculation as that leading to Eq. (9.10) and newly with $u_x = u_1 \cos\alpha + v_1 \sin\alpha$ one obtains from the above equation

$$u_y = (u_1 \sin\alpha - v_1 \cos\alpha) + v_{sh} \tag{9.12}$$

Eq. (9.10) and (9.12) are found as basic equations to resolve both the velocity shift and the velocity component u_y. It yields

$$v_{sh} = u_x\tau = \frac{v_0 + v_1}{2}\cos\alpha - \frac{u_0 + u_1}{2}\sin\alpha \tag{9.13}$$

$$u_y = \frac{u_1 - u_0}{2}\sin\alpha + \frac{v_0 - v_1}{2}\cos\alpha \tag{9.14}$$

For $\alpha = 0$ one obtains Eq. (9.5) and (9.6), respectively.

Basically, the application of DMM in this case with coordinate transformation is completed. Although the velocity component u_y in Eq. (9.14) seems to be directly obtained from the dual measurements without via the calculation of the velocity shift, the velocity shift, however, gives a clear indication about the inaccuracy in the LDA alignment. It will also be used, according to Eq. (9.10), to directly correct measurements at other measurement points in the flow with the same LDA setup. In practical applications, one can estimate the possible velocity shift according to $\nu_{sh} = u_x\tau$ by assuming the possible alignment error τ before carrying out the LDA measurement. When the estimated velocity shift is negligible against the velocity component u_y to be measured, then the dual measurement would not be necessary.

Sometimes it may be convenient to conduct calculations by using the velocity component u_x which is often assumed to be known. For this purpose, two cases in Fig. 9.5 are again considered. The velocity component ν in LDA system can be expressed by

$$\nu_0 = u_x \sin(\alpha + \tau) + u_y \cos(\alpha + \tau) \tag{9.15}$$

$$\nu_1 = u_x \sin(\alpha + \tau) - u_y \cos(\alpha + \tau) \tag{9.16}$$

From these two equations and with respect to $\sin(\alpha + \tau) \approx \sin\alpha + \tau\cos\alpha$ it follows

$$\frac{\nu_0 + \nu_1}{2} = u_x \sin(\alpha + \tau) = u_x \sin\alpha + u_x\tau\cos\alpha \tag{9.17}$$

The velocity shift can thus be resolved as

$$\nu_{sh} = u_x\tau = \frac{\nu_0 + \nu_1}{2\cos\alpha} - u_x \tan\alpha \tag{9.18}$$

It is completely equivalent to Eq. (9.13).

Especially it follows from (9.15) and (9.16) directly

$$u_y = \frac{\nu_0 - \nu_1}{2\cos(\alpha + \tau)} \approx \frac{\nu_0 - \nu_1}{2\cos\alpha} \tag{9.19}$$

This equation is completely equivalent to Eq. (9.14). Further for $\alpha = 0$, Eq. (9.6) is again obtained.

9.4 Extension of DMM

The Dual Measurement Method (DMM) has been constructed based on the "mirrored view" of two measurements. Through the direct comparison between two measurements of the same velocity component, the alignment error in LDA optics and the associated velocity shift can be exactly determined. However, it often comes about that the positioning of the LDA head at 180° for the second measurement is impossible. This will limit the application of DMM in such a way that by positioning the LDA head at another angle the measurement cannot be compared with the first measurement, because it deals with the measurements of two different velocity components. This means that two measurements are generally not sufficient to identify the alignment error and the associated velocity shift. As an extension of DMM, a method of carrying out three arbitrary measurements at a fixed point in the flow from three different azimuth angles φ has been developed and also experimentally validated by Zhang (2005). The concept was that at a fixed point in the flow, any three velocity components measured at three different azimuth angles have to lie in a same plane and thus are related to a unique plane vector, when the LDA optics is aligned to the flow without any alignment error. Otherwise a unique plane vector would not exist, when the LDA system is aligned to the flow system with an error $\tau \neq 0$. Indeed, this concept has been clearly shown in Fig. 9.3. Because of the alignment error $\tau \neq 0$ the two measured velocity components v_0 and v_1 do not lie in the plane perpendicular to the jet axis.

9.4.1 Direct Component Measurements

The flow is given in the flow coordinate system by $\vec{u}_{\text{flow}} = (u_x, u_y, u_z)$. Herein u_x represents the component in the main flow direction. As an example, the jet flow with the axial component u_x has been shown in Fig. 9.3 The velocity components related to LDA are involved in the velocity vector $\vec{u}_{\text{LDA}} = (u, v, w)$. Usually a two-component LDA system is applied and the measured velocity components are denoted by u and v, respectively. The theoretically exact coincidence between the LDA and the flow systems is assumed to be given by $u = u_x$ and $v = u_y$ The corresponding arrangement of the LDA head to the flow (Fig. 9.6a, b) is denoted as the basic arrangement. Deviations from this basic arrangement are confirmed by

(1) bias angle $\tau \neq 0$. It is unknown and should be identified through measurements. Here τ is like in Fig. 9.3 the angle between velocity components u and u_x;
(2) rotation of the LDA head about the x-axis for further carrying out two measurements at two other azimuth angles (Fig. 9.6c, $\varphi = \varphi_1$ and $\varphi = \varphi_2$).

With respect to these two deviations, specified by angles τ and φ, the general relationship between velocity components in two coordinate systems should be established. The velocity shift is expected to appear again in the form $v_{\text{sh}} = u_x\tau$. For the reason of mathematical simplicity the LDA alignment with τ and φ as parameters is assumed to be achieved by successively getting φ and τ from the basic arrangement with $u = u_x$ and $v = u_y$ (Fig. 9.6a). In the first step, when the LDA head has rotated about the x-axis by φ (Fig. 9.6c), the relationship between velocity components in LDA and flow systems is given by

$$u' = u_x \tag{9.20}$$

$$v' = u_y \cos\varphi - u_z \sin\varphi \tag{9.21}$$

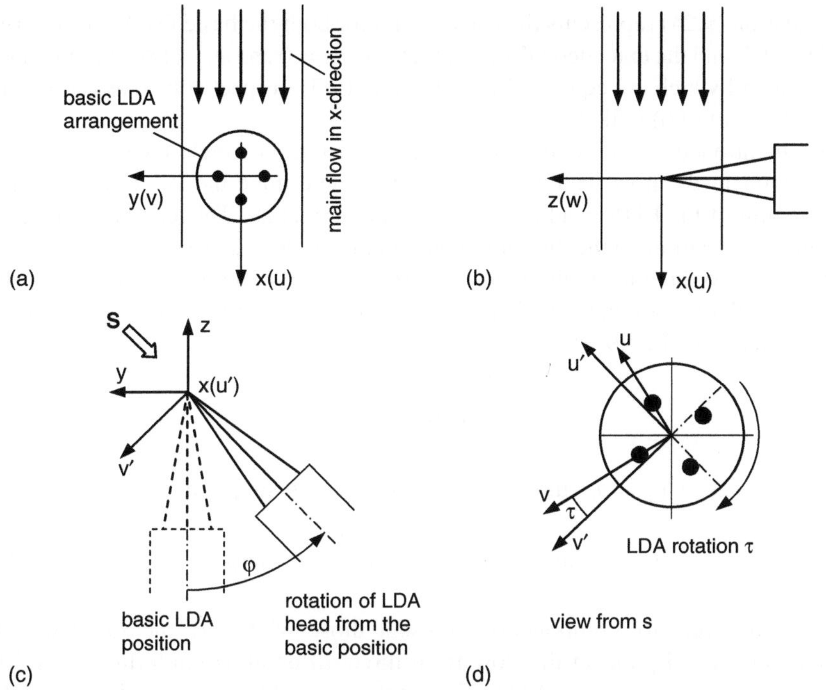

Fig. 9.6 Explanation of extended Dual Measurement Method

or in the form of matrix

$$\begin{bmatrix} u' \\ v' \end{bmatrix} = R' \begin{bmatrix} u_x \\ u_y \\ u_z \end{bmatrix} = \begin{bmatrix} 1 & 0 & 0 \\ 0 & \cos\varphi & -\sin\varphi \end{bmatrix} \begin{bmatrix} u_x \\ u_y \\ u_z \end{bmatrix} \tag{9.22}$$

The LDA head is subsequently turned about its axis by τ (Fig. 9.6d). Corresponding relationship between velocity components is given by

$$\begin{bmatrix} u \\ v \end{bmatrix} = R'' \begin{bmatrix} u' \\ v' \end{bmatrix} = \begin{bmatrix} \cos\tau & -\sin\tau \\ \sin\tau & \cos\tau \end{bmatrix} \begin{bmatrix} u' \\ v' \end{bmatrix} \tag{9.23}$$

From Eqs. (9.22) and (9.23), the general relationship between velocity components in two coordinate systems with τ and φ as parameters is obtained as

$$\begin{bmatrix} u \\ v \end{bmatrix} = R \begin{bmatrix} u_x \\ u_y \\ u_z \end{bmatrix} = \begin{bmatrix} \cos\tau & -\cos\varphi\sin\tau & \sin\varphi\sin\tau \\ \sin\tau & \cos\varphi\cos\tau & -\sin\varphi\cos\tau \end{bmatrix} \begin{bmatrix} u_x \\ u_y \\ u_z \end{bmatrix} \tag{9.24}$$

with $R = R''R'$

Equation (9.24) represents the background for both the basic DMM, as described in Sect. 9.2, and the extended DMM. In fact, for $\varphi_0 = 0$ and $\varphi_1 = 180°$ both v_0 and v_1 are obtained which are equal to Eqs. (9.2) and (9.3), respectively. Correspondingly Eqs. (9.5) and (9.6) will also be obtained.

As mentioned at the beginning of this section, three single measurements at a fixed point in the flow would be necessary to determine the velocity shift, if the positioning of the LDA head at $\varphi_1 = 180°$ is ineffectual. This can be achieved for instance by positioning the measurement volume on the x-axis ($y = 0, z = 0$) and the LDA optics to three azimuth angles $\varphi = 0$, $\varphi_1$and φ_2. The following relationships between velocity components from measurements and in the actual flow field can be obtained from Eq. (9.24):

$$v_0 = u_x \sin\tau + u_y \cos\tau \tag{9.25}$$

$$v_1 = u_x \sin\tau + u_y \cos\varphi_1 \cos\tau - u_z \sin\varphi_1 \cos\tau \tag{9.26}$$

$$v_2 = u_x \sin\tau + u_y \cos\varphi_2 \cos\tau - u_z \sin\varphi_2 \cos\tau \tag{9.27}$$

These three velocity components represent those which are measured by LDA method at a fixed point in the flow from three different azimuth directions. They will be used to determine the LDA alignment error and the associated velocity shift, in order to finally accurately determine the velocity component u_y.

By eliminating $u_z \cos\tau$ in Eqs. (9.26) and (9.27) one obtains

$$v_1 \sin\varphi_2 - v_2 \sin\varphi_1 = u_x \sin\tau(\sin\varphi_2 - \sin\varphi_1) + u_y \cos\tau \sin(\varphi_2 - \varphi_1) \tag{9.28}$$

The term $u_y \cos\tau$ in this equation will be replaced by that from Eq. (9.25). The velocity shift is finally resolved as

$$v_{sh} = u_x \sin\tau = \frac{(v_1 \sin\varphi_2 - v_2 \sin\varphi_1) - v_0 \sin(\varphi_2 - \varphi_1)}{(\sin\varphi_2 - \sin\varphi_1) - \sin(\varphi_2 - \varphi_1)} \tag{9.29}$$

The actual velocity component u_y then results from Eq. (9.25) with $\cos\tau \approx 1$

$$u_y = v_0 - \sin\tau \cdot u_x = v_0 - v_{sh} \tag{9.30}$$

Eq. (9.29) represents the extended DMM in determining the velocity shift by three single measurements. As long as the velocity shift is obtained through this way, it can be applied to Eq. (9.30) to determine the velocity component u_y, which is simply the same as that in Eq. (9.6) and represents the velocity component in the secondary flow. The prerequisite of the applicability of the method is to maintain the bias angle τ as the constant systematic error in the measurement system. This can be ensured for instance by an appropriate mechanical system like that is shown in Fig. 9.3a.

For recognising the LDA arrangement error and the associated velocity shift in such an application, it is recommended that one locates the LDA measurement volume on the x-axis. This enables the LDA users to merely rotate the LDA support system, without having to realign the measurement volume.

The reliability of Eq. (9.29) could be successively verified by making use of the measurement results which have been shown in Fig. 9.2c. As shown, the flow on the jet axis was measured eight times. According to Eq. (9.29) any combination of three initial measurements is sufficient for calculating the velocity shift as a systematic error. Corresponding calculations (altogether 56 combinations) showed the satisfactory consistency of the velocity shift with a maximum uncertainty of about 14% around $v_{sh} = 0.4$ m/s.

In addition, Eq. (9.29) also points out that for the purpose of determining the velocity shift only measurements of the v-component are required. In other words, the one-component LDA instrument can be applied.

Finally, the velocity component u_z at the measuring point can also be derived. From Eq. (9.26) for instance this velocity component is calculated as

$$u_z = \frac{u_x \sin\tau + u_y \cos\varphi_1 \cos\tau - v_1}{\sin\varphi_1 \cos\tau} \tag{9.31}$$

With respect to $v_{sh} = u_x \sin\tau$ and u_y from Eq. (9.30) as well as $\tau << 1$ leading to $\cos\tau \approx 1$, the above equation becomes

$$u_z = \frac{(v_0 - v_{sh})\cos\varphi_1 - (v_1 - v_{sh})}{\sin\varphi_1} \tag{9.32}$$

Correspondingly this velocity component can also be calculated from Eq. (9.27) as

$$u_z = \frac{(v_0 - v_{sh})\cos\varphi_2 - (v_2 - v_{sh})}{\sin\varphi_2} \tag{9.33}$$

Combining Eq. (9.32) and (9.33) to eliminate v_0 yields

$$u_z = \frac{(v_1 - v_{sh})\cos\varphi_2 - (v_2 - v_{sh})\cos\varphi_1}{\sin(\varphi_2 - \varphi_1)} \tag{9.34}$$

These last three equations are fully equivalent.

9.4.2 Method of Using Coordinate Transformation

In Sect. 9.2, DMM was applied to the case at which the LDA coordinate system is rotated against the flow system by the angle $\alpha + \tau$ (see Fig. 9.5) inclusive the unknown alignment error τ. Based on the "mirrored view" of two velocities and the direct comparison between them, the velocity shift as a systematic error can be identified. Restriction of the "mirrored view" may be encountered if the positioning of the LDA head at $\varphi = 180°$ is impossible. Like in the case treated in the last section, three measurements at a single point in the flow are required in order to determine the velocity shift that is resulted from the alignment error. Clearly, the LDA head with rotation by an angle $\alpha + \tau$ (Fig. 9.5a) has to be positioned at $\varphi = 0$, $\varphi_1$and φ_2, successively. To establish the relationship between velocity components in LDA and flow field systems, Eq. (9.24) that is available for $\alpha = 0$ can be made of use. Because the angle α always appears in the form $\alpha + \tau$, Eq. (9.24) is directly taken over for the present use by substituting the bias angle τ through $\alpha + \tau$. In accordance with Eqs. (9.25) to (9.27) the following relationships between measurements and the actual flow are obtained:

$$v_0 = u_x \sin(\alpha + \tau) + u_y \cos(\alpha + \tau) \tag{9.35}$$

$$v_1 = u_x \sin(\alpha + \tau) + u_y \cos\varphi_1 \cos(\alpha + \tau) - u_z \sin\varphi_1 \cos(\alpha + \tau) \tag{9.36}$$

$$v_2 = u_x \sin(\alpha + \tau) + u_y \cos\varphi_2 \cos(\alpha + \tau) - u_z \sin\varphi_2 \cos(\alpha + \tau) \tag{9.37}$$

They are velocity components which are measured by LDA method at a fixed point in the flow from three different azimuth directions.

Equations (9.36) and (9.37) are combined to eliminate $u_z \cos(\alpha + \tau)$. This results in, similar to Eq. (9.28)

$$v_1 \sin\varphi_2 - v_2 \sin\varphi_1 = u_x \sin(\alpha + \tau)(\sin\varphi_2 - \sin\varphi_1) + u_y \cos(\alpha + \tau)\sin(\varphi_2 - \varphi_1) \tag{9.38}$$

The term $u_y \cos(\alpha + \tau)$ in this equation will be replaced by that from Eq. (9.35), leading to

$$u_x \sin(\alpha + \tau) = \frac{v_1 \sin\varphi_2 - v_2 \sin\varphi_1 - v_0 \sin(\varphi_2 - \varphi_1)}{\sin\varphi_2 - \sin\varphi_1 - \sin(\varphi_2 - \varphi_1)} \tag{9.39}$$

With respect to $\tau << 1$ and thus $\sin(\alpha + \tau) \approx \sin\alpha + \tau\cos\alpha$ the velocity shift is finally resolved as

$$v_{sh} = u_x\tau = \frac{1}{\cos\alpha} \cdot \frac{(v_1 \sin\varphi_2 - v_2 \sin\varphi_1) - v_0 \sin(\varphi_2 - \varphi_1)}{(\sin\varphi_2 - \sin\varphi_1) - \sin(\varphi_2 - \varphi_1)} - u_x \tan\alpha \tag{9.40}$$

For $\alpha = 0$ this equation is simplified to Eq. (9.29). Further for the positioning of the LDA head at $\varphi_1 = 180°$ the above equation becomes the same as Eq. (9.5) which represents the basic form of DMM.

After the velocity shift has been determined through Eq. (9.40), the velocity component u_y can then be determined from Eq. (9.35). With respect to $\tau << 1$ and thus $\sin(\alpha + \tau) \approx \sin\alpha + \tau\cos\alpha$ and $\cos(\alpha + \tau) \approx \cos\alpha$ one obtains

$$u_{\text{y}} = \frac{1}{\cos\alpha}(v_0 - u_{\text{x}} \sin\alpha) - v_{\text{sh}} \tag{9.41}$$

For $\alpha = 0$ this equation is simplified to Eq. (9.30).

Because the velocity component u_{x} appears in both Eqs. (9.40) and (9.41), the two-component LDA system is required for the current case with $\alpha \neq 0$.

Chapter 10
Symmetrical Method of 3D-Velocity Measurements

Most LDA measurement systems are two-component systems. In the case where all three velocity components need to be measured, then two separate measurements must be carried out with the necessary realignment of LDA optics between the measurement series. Such measurements with the optical realignment always lead to more expenses and are also time-consuming. They are especially not applicable when the flow measurements should be completed within a short time as demanded by special flow processes. Basically, it seems to be unreliable to directly measure all three velocity components by using a two-component LDA system without realignment. There are cases, however, in which the symmetrical flow distribution can be utilized to cleverly succeed such a flow measurement. It actually deals with a method of simply carrying out the two-component measurements and afterwards making an appropriate evaluation of measurement data.

The simplest case is obviously the axial symmetrical flow like the swirling flow out of a burner for instance. According to Fig. 10.1 measurements of all three velocity components by using a two-component LDA optics can be accomplished by positioning the LDA head firstly along the z-axis for measurements of both the axial and radial velocity components and then along the y-axis for the tangential velocity component. The entire measurements have been thus completed without the optical realignment. As a special case for direct measurements of all three velocity components in this example there is no need to reprocess measurement data.

In reality, such a possibility exists in any type of symmetrical flows that can be encountered in lots of practical applications. Figure 10.2 shows a practical water atomization process flow, at which the flow in the atomization region is clearly three-dimensional, however, symmetrical about the $x - z$ plane. In this example (Zhang et al. 1998, Zhang and Eisele 1999, Zhang and Ziada 2000), it dealt with the measurements of water droplet distribution in the atomization region by means of the Phase Doppler Anemometry (PDA) which is an extension of the LDA method. For quantitative evaluation of the related atomization process the local mass flux of droplets is an extraordinarily important quantity. Its measurements, however, presumes the measurements of all three velocity components at each point in the flow. The method of using two-component LDA system to measure all three velocity

Zh. Zhang, *LDA Application Methods*, Experimental Fluid Mechanics,
DOI 10.1007/978-3-642-13514-9_10, © Springer-Verlag Berlin Heidelberg 2010

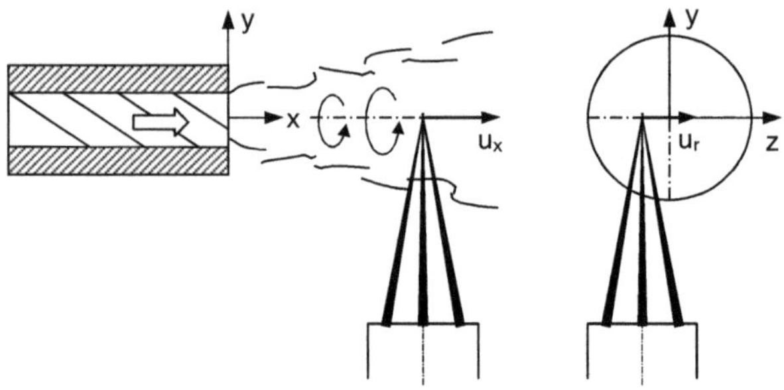

Fig. 10.1 Axial symmetrical flow and the method for measurements of all three velocity components

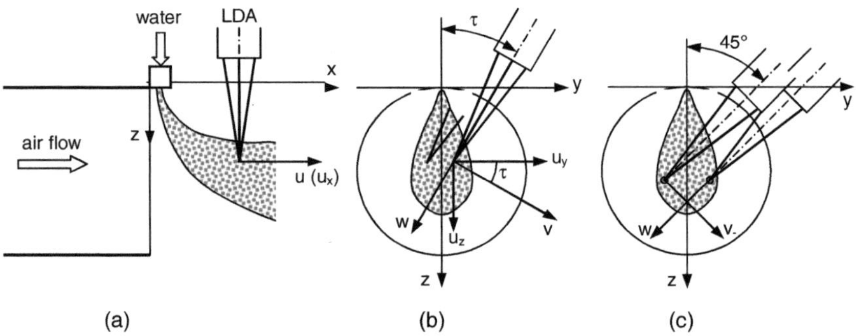

Fig. 10.2 Symmetrical atomization flow and the method for measurements of all three velocity components

components without rearranging the LDA optics was applied there. The main thing to do in this special application is the arrangement of measurements and the measurement data processing. This appears somewhat more complex than in the case that has been shown in Fig. 10.1.

In order to represent the universal applicability of this measurement method, the measurement example shown in Fig. 10.2 will be considered. For other cases and at other optical arrangements the LDA users have to probably find out respective relations for post-processing the measurement data.

According to Fig. 10.2 the flow is symmetrical about the $x - z$ plane. The two-component LDA head is aligned so that the optical axis is inclined by an angle τ in the $y - z$ plane. The two laser beam pairs are arranged to measure the velocity component u which is parallel to the x-axis and the component v which lies in the plane parallel to the $y - z$ plane. Two coordinate systems have been shown in Fig. 10.2: In the LDA-coordinate system, the velocity components are represented by u, v and w; And in the flow coordinate system, they are symbolized by u_x, u_y and u_z. Obviously there is $u_x = u$. In the flow coordinate system, there are additionally $u_x(+y) = u_x(-y)$, $u_y(+y) = -u_y(-y)$ and $u_z(+y) = u_z(-y)$ because of the symmetry of the flow distribution about the $x - z$ plane.

The measurements are arranged by positioning the measurement volume sequentially and symmetrically in flow areas with positive and negative y-values

(Fig. 10.2b). This can be easily achieved by means of a traversing system on which the LDA head is mounted.

While the velocity component u_x in the flow coordinate system is directly measured because of $u_x = u$, the other two components u_y and u_z should be resolved from the measurements of velocity component v alone. In general, the velocity component v in the LDA coordinate system can be expressed by velocity components in the flow coordinate system. Depending on whether this v-component is measured in the positive (+) or negative (−) region of y, the velocity transformation is given by

$$v_+ = u_y\,(+y) \cdot \cos\tau + u_z \sin\tau \tag{10.1}$$

and

$$v_- = u_y\,(-y) \cdot \cos\tau + u_z \sin\tau \tag{10.2}$$

respectively.

With regard to the symmetry condition $u_y\,(+y) = -u_y\,(-y) = u_y$ these are two equations from which both the velocity component u_y and the component u_z can be resolved. By subtracting Eq. (10.2) from Eq. (10.1) as well as adding Eq. (10.1) to Eq. (10.2) one obtains

$$u_y = \frac{v_+ - v_-}{2\cos\tau} \tag{10.3}$$

$$u_z = \frac{v_+ + v_-}{2\sin\tau} \tag{10.4}$$

Together with $u_x = u$, these equations fully describe the three-dimensional flow state at the measurement point in the flow region $y > 0$. It refers to the flow region $y > 0$ because of the specification $u_y = u_y\,(+y)$. The velocity component v_- in above equations refers to the conjugate measurement point in the flow region $y < 0$. As a result of the symmetry of the flow distribution about the $x - z$ plane, corresponding three-dimensional flow states in the flow region $y < 0$ can be obtained immediately.

In the LDA coordinate system according to Fig. 10.2b, the third velocity component can be calculated by (for $y > 0$)

$$w = u_z \cdot \cos\tau - u_y \cdot \sin\tau = \frac{v_+ + v_-}{2\tan\tau} - \frac{v_+ - v_-}{2}\tan\tau \tag{10.5}$$

It is indeed the on-axis velocity component. The restriction of the method is that the inclination angle of the LDA head could not be set at $\tau = 0$ or $\tau = 90°$. The ideal inclination angle is obviously around $\tau = 45°$ (Fig. 10.2c). At $\tau = 45°$ it yields from Eq. (10.5)

$$w = v_- \tag{10.6}$$

It signifies that the velocity component v which is measured in the flow region $y < 0$ is equal to the velocity component w i.e. the on-axis component in the flow region $y > 0$. This results by all means because of the symmetry of the flow field and the LDA inclination at $\tau = 45°$.

The method outlined above demonstrates that at least in the presented LDA alignment, measurements of all three velocity components are very well possible. Obviously data processing and calculation algorithms depend on used coordinate systems for both LDA optics and the flow. The calculation will be complex when for instance the u component in LDA-system does not coincide with the u_x component in the flow system. The calculation made above is obviously the simplest one. For other cases dealing with measurements of symmetrical flows, LDA users have to make their own calculations according to their LDA configurations.

Chapter 11
Non-stationary Turbulent Flows

11.1 Non-stationary Turbulent Flows in the Practice

Non-stationary turbulent flows are often encountered while starting or stopping a flow process. There are also designed non-stationary flow processes like the periodic flows in a reciprocating engine (also known as the piston engine) and the pulsatile flow through an prosthetic heart valve (Hirt et al.1994). Even in a stationary flow system, the local flow unsteadiness could occur. This can be encountered for instance in the alternating vortex flow from the trailing edge of a wing or in the flow at the exit of the impeller of a centrifugal pump (Fig. 11.1). The non-stationary flows are thus expressed either as the time- or phase-dependent flows. As in most cases, the non-stationary flows are usually also turbulent flows. Hence they are usually considered to be composed of the built-in i.e. enforced flow instability and the flow instability because of the flow turbulence. Mostly, as in the case of starting and stopping a flow process and even in the flow with local instabilities (Fig. 11.1), the flow turbulence can be considered as the small-scale flow fluctuations against the enforced flow state changes. The latter and their dependence on the time have been usually of much interest while evaluating a non-stationary flow process. For a general non-stationary turbulent flow, as presented in Fig. 11.2 for instance, the flow at a local point in the flow field can be expressed by the corresponding velocity component:

$$u(t) = \hat{u}(t) + u'(t) \tag{11.1}$$

The enforced non-stationary flow velocity $\hat{u}(t)$ is thus superimposed by the random flow fluctuation that is quantified by deviations of velocities from the enforced velocity profile. From measurements by means of LDA method that enables the rapid change of velocities in the non-stationary flow to be highly resolved and provides a time series of flow velocities, the enforced non-stationary flow $\hat{u}(t)$ can be well reconstructed i.e. regressed by the method of least squares fitting. This method has been widely applied in data analyses commonly based on experimental measurements.

Zh. Zhang, *LDA Application Methods*, Experimental Fluid Mechanics,
DOI 10.1007/978-3-642-13514-9_11,

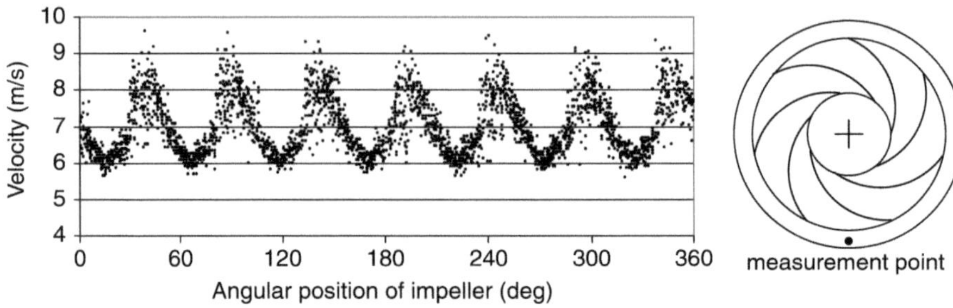

Fig. 11.1 Non-stationary flow at the exit of the impeller of a centrifugal pump, measured by LDA and rearranged to be the function of the angular position of the impeller

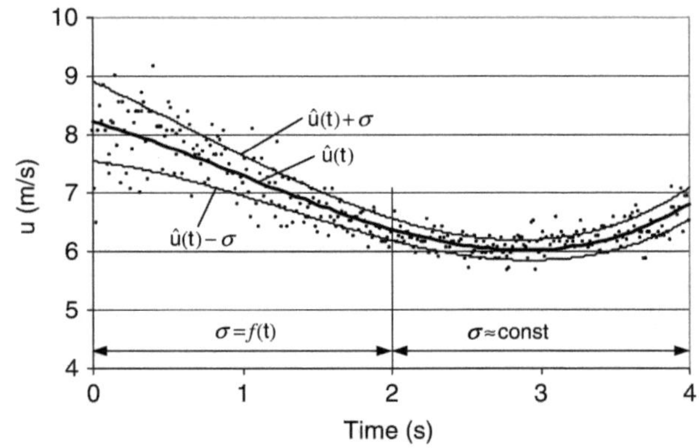

Fig. 11.2 Non-stationary turbulent flow and the method of data processing

Measurement data that do not lie on the fitted curve correspond to velocity fluctuations $u'(t)$ and can be considered, because of randomness, as resulted from the flow turbulence. In dealing with non-stationary flows, it is of most interest how the flow turbulence can be statistically evaluated. With regard to the time-dependent behavior of flow fluctuations, it is to distinguish between two fluctuation forms:

– nearly constant intensity of velocity fluctuations and
– variable intensity of velocity fluctuations in the function of time.

The first form of velocity fluctuations is characterized by the fact that the bandwidth of the flow fluctuation around the regressed curve is approximately constant. This can be confirmed for instance at Fig. 11.2 by considering the flow in the time period from 2 to 4 s. A unique value representing this bandwidth is representative for this form of flow turbulence. As the unique value the so-called standard deviation σ of the mean velocity (Chap. 2) has often been used. The second form of the turbulence distribution can be confirmed at Fig. 11.2 for instance by considering the flow in the time period from 0 to 2 s. A unique value representing the statistics of flow turbulence, however, does not exist in this case. Neither the turbulent normal nor the shear stresses can be clearly represented by use of respective mean values. On the other hand, it is usually not clear what does the flow turbulence in such non-stationary flows actually mean for the related flow process (could be thermal or chemical for instance). Without knowing this, any complicated data processing would generally be of less value and could not be made of use. It is therefore always necessary to carry out the preliminary study and to carefully clarify the significance of respective turbulence quantities. In the case of dealing with turbulence models in computational fluid dynamics (CFD), for instance, the turbulent kinetic energy and hence the complete Reynolds turbulent stresses are known as the most important turbulence quantities. In the example shown in Fig. 11.2 with non-constant turbulence intensities, one can basically calculate the bandwidth of standard deviations of velocities in the function of time. The calculation indeed merely provides the precondition for further evaluations of measurement data with regard to the related physical and engineering flow processes.

The non-stationary flows encountered in the engineering practices can be divided into two categories. The first category is related to the start or stop of the flow process, so that the flow state is purely a function of the time. The second category of non-stationary flows is related to the enforced periodic flow that can be rearranged in function of the phase angle, like that as already shown in Fig. 11.1, for example.

11.2 Time-Resolved Non-stationary Turbulent Flows

11.2.1 Method of Linear Least Squares Fitting

In accounting for a non-stationary turbulent flow, it is in the first place interesting to know the time dependence of the enforced flow part. Based on LDA measurements, such a time-dependent flow can be well reconstructed by the curve fitting methods. Amongst lots of available mathematical functions used to approach the measurement data, the method of linear least squares fitting behaves as the basic form. As long as the linear trend of flow velocities measured in the interested time range is evident (e.g. at Fig. 11.2 in the time range from 0 to 2 s), the method can be applied to approximate the non-stationary turbulent flow. The linear regression function of the enforced flow is expressed as

$$\hat{u}(t) = at + b \tag{11.2}$$

Eq. (11.1) then becomes

$$u(t) = (at + b) + u'(t) \tag{11.3}$$

In order to perform the linear least squares fitting of measurement data, the sample means of both the time as the independent variable and the velocity as the dependent variable will be firstly calculated as

$$\bar{t} = \frac{1}{N}\sum_{\mathrm{i}=1}^{\mathrm{N}} t_{\mathrm{i}}, \quad \bar{u} = \frac{1}{N}\sum_{\mathrm{i}=1}^{\mathrm{N}} u_{\mathrm{i}} \tag{11.4}$$

Other quantities that are required are

$$s_{\mathrm{tt}} = \frac{1}{N}\sum_{\mathrm{i}=1}^{\mathrm{N}} \left(t_{\mathrm{i}} - \bar{t}\right)^2, \quad s_{\mathrm{uu}} = \frac{1}{N}\sum_{\mathrm{i}=1}^{\mathrm{N}} \left(u_{\mathrm{i}} - \bar{u}\right)^2, \quad s_{\mathrm{ut}} = \frac{1}{N}\sum_{\mathrm{i}=1}^{\mathrm{N}} \left(u_{\mathrm{i}} - \bar{u}\right)\left(t_{\mathrm{i}} - \bar{t}\right) \tag{11.5}$$

In all these calculations, no weighting factor has been used. The calculations hence suffer from certain inaccuracies which for instance are arising from the effect of velocity bias (Chap. 17). In particular, there is strictly $\bar{t} \neq (t_1 + t_{\mathrm{N}})/2$ because of velocity bias or other reasons and consecutively of non-constant data sampling rate during LDA measurements. For simplicity and also because of its negligible effect, the velocity bias is not considered here.

In LDA measurements, the time that is related to each velocity event is the independent variable and should be considered as accurate. The method of least squares fitting is formulated as

$$\sum_{\mathrm{i}=1}^{\mathrm{N}} (u_{\mathrm{i}} - \hat{u}_{\mathrm{i}})^2 \rightarrow \min \tag{11.6}$$

In performing such a calculation with respect to $\hat{u}$ as defined in Eq. (11.2), both constants a and b are determined as

$$a = \frac{s_{\mathrm{ut}}}{s_{\mathrm{tt}}}, \quad b = \bar{u} - a\bar{t} \tag{11.7}$$

The enforced non-stationary flow has been thus calculated based on the linear least squares fitting in the period $t_{\mathrm{N}} - t_1$ containing N velocity events. The pure mathematical calculation additionally provides the knowledge about the degree of linearity between dependent and independent variables, given by the correlation coefficient

$$R = \frac{s_{\mathrm{ut}}}{\sqrt{s_{\mathrm{uu}} s_{\mathrm{tt}}}} \tag{11.8}$$

The special case of $R = 1$ is obtained in that all measurement data lie on the fitted curve $\hat{u}(t) = at + b$. This implies the negligible random fluctuations of flow velocities. As will be shown in the next section, the correlation coefficient that is always less than unity actually involves the information about the mean flow turbulence in the time period $t_{\mathrm{N}} - t_1$.

Because the velocity gradient that is given by a may be positive and negative, the correlation coefficient R in Eq. (11.8) may be so. In some applications like those of using a spreadsheet, the correlation coefficient is simply given in the form R^2.

11.2.2 Linear Trend of the Velocity and the Calculation Method

The random velocity fluctuation in the non-stationary flow is denoted by $u'(t)$ for the given velocity component. The associated kinetic energy is represented by the turbulent normal stress or equivalently by the variance $\overline{u'^2}$ as a statistical parameter. In considering the measurement within the time period $t_N - t_1$, it is then calculated according to Eq. (11.3) and with respect to $b = \bar{u} - a\bar{t}$ as

$$\overline{u'^2} = \frac{1}{N} \sum_{i=1}^{N} \left[(u_i - \bar{u}) - a (t_i - \bar{t}) \right]^2 \tag{11.9}$$

The calculation is given here simply as the arithmetic mean. Because of the use of $b = \bar{u} - a\bar{t}$ it is actually only applicable to the flow process with linear trend of velocity in the time period $t_N - t_1$.

Eq. (11.9) is then further written as

$$\overline{u'^2} = \frac{1}{N} \sum_{i=1}^{N} (u_i - \bar{u})^2 - \frac{2a}{N} \sum_{i=1}^{N} (u_i - \bar{u}) (t - \bar{t}) + \frac{a^2}{N} \sum_{i=1}^{N} (t_i - \bar{t})^2 \tag{11.10}$$

and because of Eq. (11.5) also as

$$\overline{u'^2} = s_{uu} - 2as_{ut} + a^2 s_{tt} \tag{11.11}$$

For the second term on the r.h.s. of this equation there is $as_{ut} = a^2 s_{tt}$ because of Eq. (11.7). Equation (11.11) then becomes

$$\overline{u'^2} = s_{uu} - a^2 s_{tt} \tag{11.12}$$

The regular data distribution i.e. the regular distribution of velocity events in the period $t_N - t_1$ will be assumed. This simply means $\bar{t} \approx (t_N + t_1)/2$. Then s_{tt} can be further calculated by converting the summation into the integral calculation as

$$s_{tt} = \frac{1}{N} \sum_{i=1}^{N} (t_i - \bar{t})^2 = \frac{1}{t_N - t_1} \int_{t_1}^{t_N} (t - \bar{t})^2 \, dt = \frac{(t_N - \bar{t})^3 - (t_1 - \bar{t})^3}{3(t_N - t_1)} \tag{11.13}$$

Because of $\bar{t} \approx (t_N + t_1)/2$ there is

$$s_{tt} = \frac{1}{12} (t_N - t_1)^2 \tag{11.14}$$

and from Eq. (11.7)

$$s_{ut} = as_{tt} = \frac{a}{12} (t_N - t_1)^2 \tag{11.15}$$

Eq. (11.12) is then rewritten as

$$\overline{u'^2} = s_{\mathrm{uu}} - a^2 \frac{1}{12}(t_{\mathrm{N}} - t_1)^2 \tag{11.16}$$

with s_{uu} as given in Eq. (11.5).

Furthermore, because of $\Delta\hat{u} = \hat{u}_{\mathrm{N}} - \hat{u}_1 = a(t_{\mathrm{N}} - t_1)$ from Eq. (11.2) the above equation is also written as

$$\overline{u'^2} = \frac{1}{N}\sum_{\mathrm{i=1}}^{\mathrm{N}}(u_{\mathrm{i}} - \bar{u})^2 - \frac{1}{12}(\hat{u}_{\mathrm{N}} - \hat{u}_1)^2 \tag{11.17}$$

The second term on the r.h.s. of this equation behaves as the correction term to the first term in calculating the turbulent normal stress in a non-stationary turbulent flow. Also to be mentioned is that the relationship given in Eq. (11.17) is obtained based on the assumption of linear trend of velocity in the time period $t_{\mathrm{N}} - t_1$.

Following the similar calculations based on linear least squares fitting, the covariance of two orthogonal velocity components (u and v), that represents the turbulent shear stress, can be calculated as

$$\overline{u'v'} = s_{\mathrm{uv}} - \frac{a_{\mathrm{u}}a_{\mathrm{v}}}{12}(t_{\mathrm{N}} - t_1)^2 \tag{11.18}$$

Herein s_{uv} is calculated by the corresponding arithmetic mean as given by

$$s_{\mathrm{uv}} = \frac{1}{N}\sum_{\mathrm{i=1}}^{\mathrm{N}}(u_{\mathrm{i}} - \bar{u})(v_{\mathrm{i}} - \bar{v}) \tag{11.19}$$

The second term on the r.h.s. of Eq. (11.18) is again the correction term. Both the constant a_{u} and a_{v} are equivalent to the constant a in Eq. (11.2). They are related to the velocity components u and v, respectively, and are calculated as

$$a_{\mathrm{u}} = \frac{s_{\mathrm{ut}}}{s_{\mathrm{tt}}} \text{ and } a_{\mathrm{v}} = \frac{s_{\mathrm{vt}}}{s_{\mathrm{tt}}} \tag{11.20}$$

Because of relationships $\hat{u}_{\mathrm{N}} - \hat{u}_1 = a_{\mathrm{u}}(t_{\mathrm{N}} - t_1)$ and $\hat{v}_{\mathrm{N}} - \hat{v}_1 = a_{\mathrm{v}}(t_{\mathrm{N}} - t_1)$ Eq. (11.18) can then be written as

$$\overline{u'v'} = s_{\mathrm{uv}} - \frac{1}{12}(\hat{u}_{\mathrm{N}} - \hat{u}_1)(\hat{v}_{\mathrm{N}} - \hat{v}_1) \tag{11.21}$$

In the last section, the correlation coefficient has been calculated and given in Eq. (11.8). Usually this correlation coefficient is directly accessible while carrying out the linear least squares fitting, for instance by applying a graphical tool with spreadsheet functions. As a matter of fact, the obtained correlation coefficient also contains information about the variance $\overline{u'^2}$ within the time interval $t_{\mathrm{N}} - t_1$ and therefore can be applied to determine the turbulent normal stress. With respect to s_{tt} from Eq. (11.14) and s_{uu} from Eq. (11.16) as well as $s_{ut} = as_{tt}$ from Eq. (11.7), Eq. (11.8) is expressed as

$$R^2 = \frac{s_{ut}^2}{s_{uu}s_{tt}} = \frac{a^2}{12} \cdot \frac{(t_N - t_1)^2}{\overline{u'^2} + \frac{a^2}{12}(t_N - t_1)^2} \tag{11.22}$$

The variance is then resolved as

$$\overline{u'^2} = \frac{a^2}{12}\left(\frac{1}{R^2} - 1\right)(t_N - t_1)^2 \tag{11.23}$$

Because for linear data fitting there is $\hat{u}_N - \hat{u}_1 = a(t_N - t_1)$ Eq. (11.23) is also written as

$$\overline{u'^2} = \frac{1}{12}\left(\frac{1}{R^2} - 1\right)(\hat{u}_N - \hat{u}_1)^2 \tag{11.24}$$

These last two equations can be applied to estimate the variance of each interested velocity component directly from the correlation coefficient R which is obtained from linear data fitting. Obviously the correlation coefficient R represents its mechanical significance always in the form R^2. Also to be mentioned is that it cannot be concluded from Eq. (11.23) that for $a = 0$ there would be $\overline{u'^2} = 0$. This is because for $a = 0$ there is also $s_{ut} = 0$ and further $R = 0$.

Similarly the covariance of two orthogonal velocity components which are measured in the same time series can be obtained immediately when Eq. (8.16) is concerned:

$$\overline{u'v'} = \frac{1}{24}\tan 2\overline{\varphi}\left[\left(\frac{1}{R_u^2} - 1\right)(\hat{u}_N - \hat{u}_1)^2 - \left(\frac{1}{R_v^2} - 1\right)(\hat{v}_N - \hat{v}_1)^2\right] \tag{11.25}$$

This equation is actually only applicable, if the enforced flow direction i.e. the flow angle $\overline{\varphi}$ in the plane of velocity components u and v remains constant despite of the flow instability. Otherwise the above equation is not valid.

Calculations given above based on the method of linear least squares fitting. In many other cases dealing with non-stationary flows, the enforced non-stationary flow can be, however, better approximated by other mathematical functions. Calculations presented in this section are then not applicable.

11.2.3 Time-Dependent Flow Turbulences

In the last section, the linear regression method in data processing has been presented. The method is applicable to each interested non-stationary flow in that in the time interval $t_N - t_1$, it can be approximated by the linear velocity distribution. Within this time interval, each turbulence quantity can be represented by its average, like the turbulent normal stress by Eq. (11.9). This treatment has commonly found its wide practical applications.

In association with the determination of turbulence quantities in non-stationary turbulent flows there are generally two cases that limit the application of the presented method:

- The non-stationary flow could not be reconstructed through the linear or other approximations of measurement data within the interested time interval (Fig. 11.2).
- The turbulence is highly time-dependent and appreciably changes along the regressed curve in function of the time (Fig. 11.2 in the time interval 0–2 s). The flow turbulence in this case should be resolved in time and not be simply approximated by an average.

For both cases dealing with time-dependent turbulence quantities, an appropriate data processing method has to be worked out. Usually the data processing of non-stationary turbulent flows requires the classification of measurement data in the time series with predefined time steps which are also called time-averaging windows (Fig. 11.3). Inside each small sized time window, the flow could be roughly considered to be quasi-stationary. This treatment, however, is only valid if no appreciable change in the flow state within the time window is confirmed. This condition can be generally ensured by setting the time window Δt to be sufficiently small. Otherwise the variable flow within the time window has to be concerned. Because it deals now with a small sized window, the variable flow within it can be well approximated by the linear velocity distribution. In any case, either of quasi-steady or of variable flow state in each time window Δt, one has to make sure that sufficient measurement data are included for reliable statistical calculations of interested flow parameters.

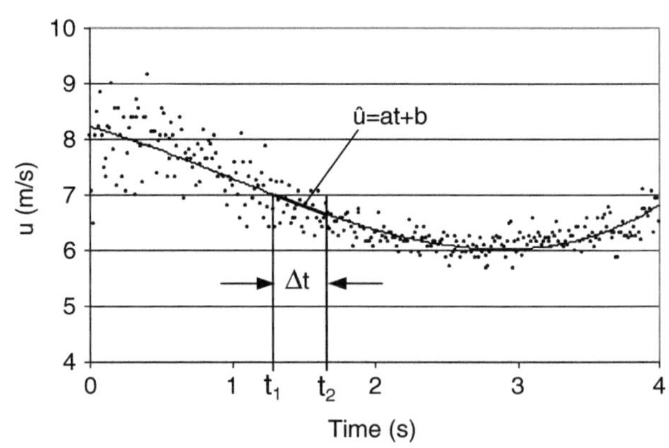

Fig. 11.3 Method of evaluating the non-stationary turbulent flows by specifying the appropriate time-averaging window size

The mean velocity inside the time window Δt in which n velocity events are included is obtained by the sample mean as

$$\bar{u} = \frac{1}{n}\sum_{\mathrm{i}=1}^{\mathrm{n}} u_{\mathrm{i}} \tag{11.26}$$

As a matter of fact, this mean velocity is approximately equal to the middle value of $\hat{u}$ in the time window Δt. It is thus independent of the slope of linear velocity distributions and further almost non-sensitive to the window size Δt. In contrast, the determination of turbulence quantities in the respective time window strongly depends on both the flow distribution in it and the window size itself. For simplicity, the general case of linear velocity distribution with regular velocity fluctuations in the time window Δt is considered. Calculations executed in Sect. 11.2.2 using the linear least squares fitting method in the time interval $t_{\mathrm{N}} - t_1$ can then be directly applied to the present case with small sized time window $\Delta t = t_{\mathrm{n}} - t_1$. From Eqs. (11.16) and (11.18) one obtains by substituting $\Delta t = t_{\mathrm{n}} - t_1$ immediately the Reynolds normal and shear stresses being averaged in the time window

$$\overline{u'^2} = \frac{1}{n}\sum_{\mathrm{i}=1}^{\mathrm{n}}(u_{\mathrm{i}} - \bar{u})^2 - \frac{a_{\mathrm{u}}^2}{12}(\Delta t)^2 \tag{11.27}$$

$$\overline{u'v'} = \frac{1}{n}\sum_{\mathrm{i}=1}^{\mathrm{n}}(u_{\mathrm{i}} - \bar{u})(v_{\mathrm{i}} - \bar{v}) - \frac{a_{\mathrm{u}}a_{\mathrm{v}}}{12}(\Delta t)^2 \tag{11.28}$$

The temporal gradients of velocities inside the time window are denoted by a_{u} and a_{v}, respectively for velocity components u and v. In both equations, the second terms on the r.h.s. behave as the correction terms. Obviously the actual Reynolds stresses within the time window Δt are not simply equal to those from respective arithmetic average in the summation form of using $(u_{\mathrm{i}} - \bar{u})$ and $(v_{\mathrm{i}} - \bar{v})$ in above equations. Both first terms on the r.h.s. of equations given above are therefore called pseudo or apparent turbulent stresses. The apparent turbulent normal stress plainly implies the overestimation of flow turbulence. Hence turbulent stresses from arithmetic average should generally be corrected by respective terms like $a_{\mathrm{u}}^2/12 \cdot \Delta t^2$ for the normal stress. It is, however, evident that for sufficiently small window all correction terms disappear. On the other hand, the time window of small size usually also implies few or insufficient data that are included in each time window for reliable statistical calculations. For this reason, the time window should be sufficiently large so that the real turbulence quantities can be accurately determined by accounting for respective correction terms in above equations. For stationary turbulent flow there is $a_{\mathrm{u}} = 0$ and $a_{\mathrm{v}} = 0$.

It should be mentioned that the necessity of accounting for the respective correction terms in Eqs. (11.27) and (11.28) depends on the requirement of calculation accuracies and therefore on the engineering application purposes of respective turbulence quantities. With regard to Eq. (11.28) for instance, it is possible that the apparent shear stress (the first term on the r.h.s. of the equation) in the used coordinate system is equal or close to zero and hence the correction term dominates. The correction term, however, can still be neglected if it is sufficiently small against the normal stress in Eq. (11.27). This is grounded in that the relevant turbulence behavior in a turbulent flow is the directional dependence of respective turbulence quantities rather than that in a given spatial direction, see Figs. 8.3 and 8.4. In other words, the directional dependence of the turbulent stresses is mainly determined by the turbulent normal stresses.

11.3 Phase-Resolved Non-stationary Turbulent Flows

The most often encountered non-stationary flows in the engineering applications are periodic flows. As an example of this type of non-stationary flows the fluctuation velocity at the impeller exit of a centrifugal pump has been already shown in Fig. 11.1. The velocity data used in the diagram were achieved by LDA measurements and rearranged to be phase-resolved i.e. in the function of the angular

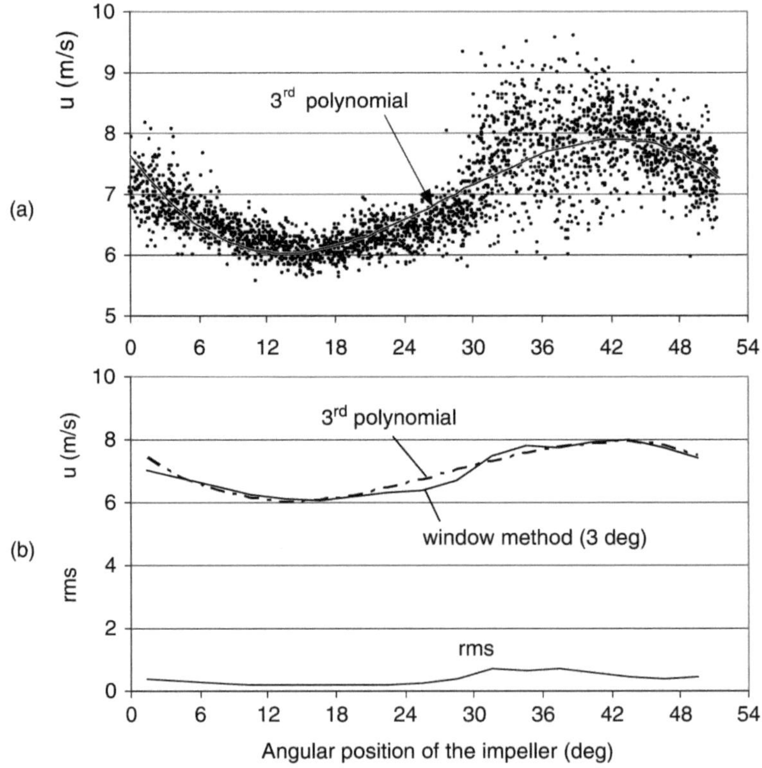

Fig. 11.4 Method of data processing by means of the small-sized phase averaging window (3°) without overlap, (**a**) Rearrangement of data from Fig. 11.1 to a sub-period of 51.4° (360/7), (**b**) Curve fitting and window method

position of the impeller. Obviously the enforced periodic flow is superimposed by the random flow fluctuations. The main objective of evaluating and processing such measurement data is to resolve both the periodicity of the flow and the turbulent flow fluctuations. In the example shown in Fig. 11.1, the flow repeated seven times in each pump rotation, corresponding to the flows in seven channels of the pump impeller. Because flows in all seven channels are primitively identical, the measurement data could also be rearranged to a sub-period which is specified by the phase angle range from 0 to 51.4° see Fig. 11.4a. The data rearrangement in this way greatly contributes to the simplification of further data processing. In order to resolve the phase-dependent mean velocity, usually appropriate data regression that is not necessarily linear can be calculated. In the example shown in Fig. 11.4a, an approach with a polynomial of 3rd order has been applied.

In principle, any deviation of velocities from the fitted velocity distribution should be considered to be from the random flow fluctuations. However, this deviation may actually also come from errors in the flow repeatability from circle to circle and thus is not of the randomness of velocity fluctuations as a result of the flow turbulence. The outcome of this situation, however, should only be considered at flows with low turbulence intensities, when the errors in the flow repeatability could not be simply accepted as the turbulent flow fluctuations.

On the side of measurement techniques, each periodic flow like that in the rotating machinery has to be resolved by the phase angle. This can be easily achieved by applying a rotary encoder that is mounted on the rotor shaft for instance and connected to the data acquisition unit.

11.3.1 Method of Linear Least Squares Fitting

As stated before with respect to Fig. 11.4a, the enforced velocity distribution of a non-stationary turbulent flow within an entire period can be well approximated by the regression of measurement data in accordance with an appropriate polynomial of at least second order. In order to utilize the advantages of linear least squares fitting to evaluate the phase-dependent flow turbulence, the periodic flow of interest can be approached by dividing a period into partial lengths and subsequently by assuming the linear distribution of velocities in form $\hat{u}(\varphi) = a\varphi + b$ in each of them. This approach is comparable with that described in Sect. 11.2.1 and applied in Sect. 11.2.2. For the purpose of direct application, similar calculations are presented here, as shown below.

In assuming the linear distribution of the enforced flow in the partial length $\varphi_N - \varphi_1$, the velocity component u for instance with fluctuations can be expressed as

$$u(\varphi) = (a\varphi + b) + u'(\varphi) \tag{11.29}$$

Like that shown in Sect. 11.2.1 both constant a and b are calculated respectively by

$$a = \frac{s_{u\varphi}}{s_{\varphi\varphi}} \text{ and } b = \bar{u} - a\overline{\varphi} \tag{11.30}$$

In the second equation, $\bar{u}$ and $\overline{\varphi}$ are sample means of N velocity events that are involved in the considered phase interval i.e. partial length $\varphi_N-\varphi_1$ and calculated by

$$\bar{u}=\frac{1}{N}\sum_{i=1}^{N}u_i \text{ and } \overline{\varphi}=\frac{1}{N}\sum_{i=1}^{N}\varphi_i \tag{11.31}$$

Other available arithmetic means are

$$s_{\varphi\varphi}=\frac{1}{N}\sum_{i=1}^{N}(\varphi_i-\overline{\varphi})^2,\quad s_{uu}=\frac{1}{N}\sum_{i=1}^{N}(u_i-\bar{u})^2,\quad s_{u\varphi}=\frac{1}{N}\sum_{i=1}^{N}(u_i-\bar{u})(\varphi_i-\overline{\varphi}) \tag{11.32}$$

While performing the data regression, the correlation coefficient that represents the degree of linearity between dependent and independent variables is given as

$$R=\frac{s_{u\varphi}}{\sqrt{s_{uu}s_{\varphi\varphi}}} \tag{11.33}$$

It specifies a statistical quantity that most computational tools like the spreadsheet can provide with, usually in the form R^2. It also represents a quantity that involves the information about the averaged flow turbulence within the phase interval $\varphi_N-\varphi_1$, see Sect. 11.3.2 below.

All the above equations can be further simplified when regular data distribution within the phase interval $\varphi_N-\varphi_1$ is assumed. The assumption simply implies the approximation $\overline{\varphi}\approx(\varphi_N+\varphi_1)/2$. Following the similar calculations as those in Sect. 11.2.2, one obtains

$$s_{\varphi\varphi}=\frac{1}{12}(\varphi_N-\varphi_1)^2 \tag{11.34}$$

$$s_{uu}=\overline{u'^2}+\frac{a^2}{12}(\varphi_N-\varphi_1)^2 \tag{11.35}$$

$$s_{u\varphi}=as_{\varphi\varphi}=\frac{a}{12}(\varphi_N-\varphi_1)^2 \tag{11.36}$$

In all these calculations, accurate angle measurements (φ) have been assumed. Also to be mentioned is that Eq. (11.35) actually represents the determination equation of the turbulent normal stress $\overline{u'^2}$.

11.3.2 Linear Trend of the Velocity and the Calculation Method

In order to calculate respective turbulence quantities in the interested phase interval $\varphi_N-\varphi_1$, the random flow fluctuation $u'(\varphi)$ should be basically resolved from Eq. (11.29). In likeness to Sect. 11.2.2 or directly from Eq. (11.35), the turbulent normal stress in the phase interval $\varphi_N-\varphi_1$ is calculated as

$$\overline{u'^2} = \frac{1}{N}\sum_{\mathrm{i}=1}^{\mathrm{N}} (u_\mathrm{i} - \bar{u})^2 - \frac{a^2}{12}(\varphi_\mathrm{N} - \varphi_1)^2 \tag{11.37}$$

And because of $\hat{u}_\mathrm{N} - \hat{u}_1 = a\,(\varphi_\mathrm{N} - \varphi_1)$ it is also written as

$$\overline{u'^2} = \frac{1}{N}\sum_{\mathrm{i}=1}^{\mathrm{N}} \left(u_\mathrm{i} - \bar{u}\right)^2 - \frac{1}{12}\left(\hat{u}_\mathrm{N} - \hat{u}_1\right)^2 \tag{11.38}$$

Herein $\hat{u}_1$ and $\hat{u}_\mathrm{N}$ represent the linearized velocities at $\varphi = \varphi_1$ and $\varphi = \varphi_\mathrm{N}$, respectively.

Correspondingly the turbulent shear stress that is given by the covariance of two orthogonal velocity components is calculated as

$$\overline{u'v'} = s_\mathrm{uv} - \frac{a_\mathrm{u} a_\mathrm{v}}{12}(\varphi_\mathrm{N} - \varphi_1)^2 \tag{11.39}$$

with

$$s_\mathrm{uv} = \frac{1}{N}\sum_{\mathrm{i}=1}^{\mathrm{N}} (u_\mathrm{i} - \bar{u})\,(v_\mathrm{i} - \bar{v}) \tag{11.40}$$

Both a_u and a_v are constants that are obtained from calculating linear least squares fitting for velocity components u and v, respectively, see Eq. (11.30).

In an analogy to Sect. 11.2.2, the correlation coefficient that is calculated by Eq. (11.33) can be utilized to directly calculate the variance i.e. the turbulent normal stress. From Eqs. (11.23) and (11.24) regarding the velocity component u one obtains immediately

$$\overline{u'^2} = \frac{a^2}{12}\left(\frac{1}{R^2} - 1\right)(\varphi_\mathrm{N} - \varphi_1)^2 \tag{11.41}$$

$$\overline{u'^2} = \frac{1}{12}\left(\frac{1}{R^2} - 1\right)\left(\hat{u}_\mathrm{N} - \hat{u}_1\right)^2 \tag{11.42}$$

It should be mentioned again that in all above calculations the repeatability errors of mean velocity distribution from circle to circle in the periodic flow have been neglected. In addition, all calculations assumed the regular data distribution within the phase interval $\varphi_\mathrm{N} - \varphi_1$, leading to the use of $\bar{\varphi} \approx (\varphi_\mathrm{N} + \varphi_1)/2$. This condition could be more or less distorted when the measurement data is rearranged from the time to the phase domain under the condition of non-linearity between the time and the phase angle. This should only be considered when the associated effect is significantly large.

11.3.3 Phase-Dependent Flow Turbulences

More complex data processing is encountered when accounting for the periodic flows with complex velocity distribution and non-regular velocity fluctuations. Not only the mean velocities but also the turbulence quantities usually need to be resolved in the function of the phase angle. The method of linear least squares fitting presented in the last section appears to be not applicable, because it only applies to the phase interval $\varphi_{\rm N} - \varphi_1$ with linear velocity distribution and almost consistent flow turbulence.

As in Sect. 11.2.3 of processing measurement data in the time domain by means of small sized time-averaging windows, periodic flows with complex velocity distribution and non-regular velocity fluctuations can be treated by specifying appropriate small sized phase-averaging windows. With sufficiently small phase angle window, both velocities and turbulence quantities within each window can be assumed to be constant and calculated by arithmetic average. In doing this, overlapping of the phase angle window can be arranged in order to smooth calculation results.

An example of calculating non-stationary turbulent flows by means of the window method named above has been shown in Fig. 11.4b, relying on the rearranged measurement data that had been shown in Fig. 11.4a. The mean velocity distribution was calculated by specifying a window size equal to $\Delta\varphi = 3^\circ$ without overlapping. The extent of velocity fluctuation has been shown as the time-dependent standard deviation in form of the root mean squares (*rms*).

In principle, calculations of mean velocity distributions will usually be less affected by the predefined phase-averaging window size. However, calculation results of the flow turbulence notably depend on the applied window size, if velocity gradients in these windows are significantly large. This is comparable with the data processing of time-dependent flow velocities (Sect. 11.2.3). The effect of velocity gradients on the data processing can only be neglected, when the window size is sufficiently small. Otherwise, the linear velocity distribution $\hat{u} = a\varphi + b$ within the phase angle window $\Delta\varphi$ should be assumed, so that

$$u(\varphi) = (a\varphi + b) + u'(\varphi) \tag{11.43}$$

Calculations performed in the last section based on linear least squares fitting in the phase interval $\varphi_{\rm N} - \varphi_1$ can then be directly applied to the present case of using the small sized phase angle window $\Delta\varphi = \varphi_{\rm n} - \varphi_1$, in which n velocity events are involved. From Eqs. (11.37) and (11.39) and by substituting $\Delta\varphi = \varphi_{\rm n} - \varphi_1$ one immediately obtains the respective turbulence quantities representing Reynolds normal and shear stresses in the considered phase angle window as

$$\overline{u'^2} = \frac{1}{n}\sum_{\rm i=1}^{\rm n}(u_{\rm i} - \bar{u})^2 - \frac{a_{\rm u}^2}{12}\Delta\varphi^2 \tag{11.44}$$

$$\overline{u'v'} = \frac{1}{n}\sum_{\rm i=1}^{\rm n}(u_{\rm i} - \bar{u})(v_{\rm i} - \bar{v}) - \frac{a_{\rm u}a_{\rm v}}{12}\Delta\varphi^2 \tag{11.45}$$

They are comparable to Eqs. (11.27) and (11.28), respectively (Zhang et al. 1996, 1997). The temporal gradients of velocities inside the phase angle window are denoted by a_u and a_v, corresponding to velocity components u and v. In both equations, the second terms on the r.h.s. again behave as the correction terms. Correspondingly the arithmetic averages in the summation form of using $(u_i - \bar{u})$ and $(v_i - \bar{v})$ in above equations are called pseudo or apparent turbulent stresses. It is, however, evident that for sufficiently small window size both correction terms become negligible. The respective arithmetic average can then be directly applied to represent the actual turbulent stresses ($\overline{u'^2}$ and $\overline{u'v'}$). On the other hand, the small window size usually also implies few or insufficient data that are included in each window for reliable statistical calculations. For this reason, each specified phase angle window should be sufficiently large and the related turbulence quantities should be determined by accounting for respective correction terms given in above equations. For a quasi-stationary turbulent flow inside the phase-averaging window there are $a_u = 0$ and $a_v = 0$.

As in Sect. 11.3.2, the repeatability errors of mean velocity distribution from circle to circle in the periodic flow have been neglected, while carrying out the above calculations.

Chapter 12
Turbulent Flow with Spatial Velocity Gradient

In most applications, LDA measurements are considered as measurements of local flow velocities because of the small dimension of the measurement volume. Typically the LDA measurement volume possesses a thickness of about 0.1 mm and a finite length of about 0.3–3 mm, depending on the geometrical layout of LDA optics, see Sects. 3.8 and 4.2. This geometrical property of the measurement volume permits one to presume the uniform flow across the LDA measurement volume. On the side of instrumentations, the LDA unit is also commonly designed without being able to resolve the possible velocity distribution within the measurement volume. The assumption of the uniform flow within the LDA measurement volume is available, as long as the measurement volume is positioned sufficiently far away from the boundaries or in the flow where the spatial velocity gradient arising from eddy motions, for instance, is statistically negligible. In all these cases, no special care should be taken with regard to the measurement volume size. Because of the small dimension of the measurement volume, the LDA method has been acknowledged as being a method with high spatial resolutions in flow field measurements.

The high spatial resolution property of the LDA method can be made of use for special flow investigations. The well-known application is the measurement of the flow distribution in the near-wall region for instance in the turbulent boundary layers. Because of the large velocity gradient within the thin boundary layer the LDA measurement volume should be arranged so that the LDA optical plane is fairly parallel to the plane of the boundary layer (Fig. 12.1). By traversing the LDA head along the normal of the wall (z-axis) the velocity profile can be well measured. Such a measurement arrangement is generally able to resolve the velocity distribution within the near-wall region. Nevertheless, the measurement of velocity profiles in the viscous sub-layer region of a turbulent boundary layer could still be difficult because this viscous sub-layer has a thickness which is comparable to or even less than the thickness of the LDA measurement volume. In the practical application of flows with non-uniform velocity distribution and thus non-negligible influence on the measurement accuracy, the constant velocity gradient within the thickness of the thin measurement volume can be assumed. Naturally the arithmetic average of measurement data (velocity) has to be considered as existing at the centre of the measurement volume. While this can be sometimes accepted, the evaluation

Zh. Zhang, *LDA Application Methods*, Experimental Fluid Mechanics,
DOI 10.1007/978-3-642-13514-9_12,

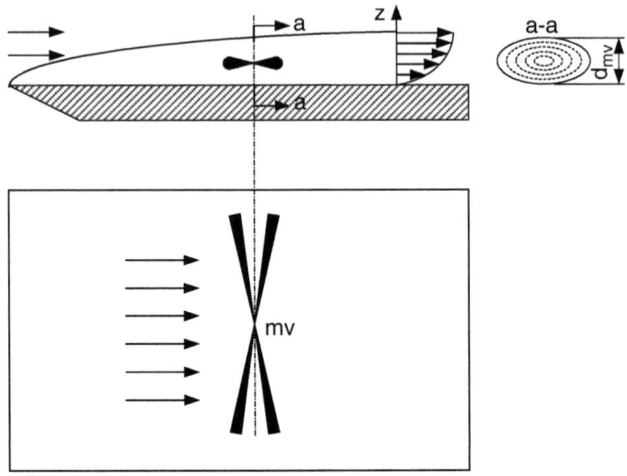

Fig. 12.1 Optical arrangement of LDA for measurements of the velocity distribution in the near-wall region of a boundary layer

of measurement data to estimate the turbulence quantities requires the special process. This is comparable with the apparent turbulence intensity in the measurement of non-stationary turbulent flows, which has been treated in Chap. 11. In particular, the same remains true that the turbulence intensity is always overestimated. The associated broadening effect in turbulence measurements becomes significant, if the non-uniform velocity distribution is along the length of the measurement volume. Corresponding descriptions of the associated phenomenon and the quantitative specifications can be found for instance by Albrecht et al. (2003) and Durst et al. (1998).

Although LDA method can be satisfactorily applied to flow measurements in the near-wall region according to the optical arrangement as shown in Fig. 12.1, this LDA arrangement in the flow with velocity gradient, however, is not always available. This may be encountered for instance in the measurement of the flow in a circular pipe, at which the length scale of LDA measurement volume is usually aligned perpendicular to the pipe wall. Because the velocity gradient in this case lies along the length of the measurement volume, the apparent turbulence intensity becomes significant. For LDA measurements that are not restricted to the near-wall region of the turbulent boundary layer, however, the linear velocity distribution i.e. the constant velocity gradient along and within the length of the measurement volume can still be assumed in the first approximation. This applies to most complex engineering flows with velocity gradients and greatly contributes to the simplification of resolving the real from the apparent turbulence intensity.

Another phenomenon that is associated with the non-uniform velocity distribution in the measurement volume and affects the measurement accuracy is that the velocity sampling rate is proportional to the velocity itself, when homogeneous particle distribution in the flow is assumed. This phenomenon leads to some special features in calculating both the mean and turbulence velocities from the corresponding arithmetic average of measurement data. Because it is purely related to flow

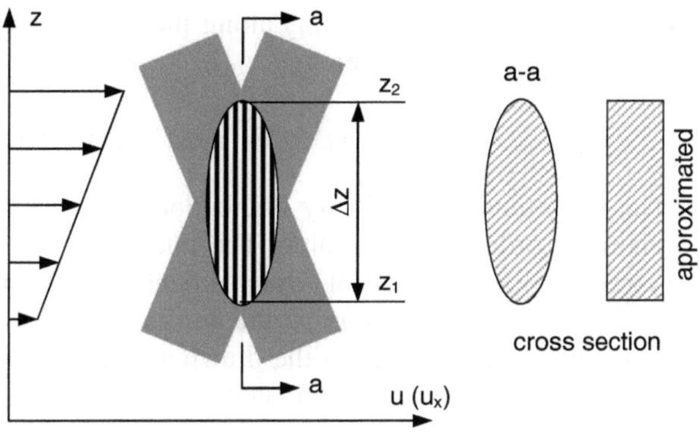

Fig. 12.2 Optical arrangement for measurements of flows with velocity gradient along the LDA optical axis

velocities and is of the same mechanism as the flow instability in influencing the measurements, this phenomenon is also called velocity bias, which has been well known in measurements of non-stationary flows and the flows with high turbulence intensity (Chap. 17). In the case of large velocity gradient being present along the length of the measurement volume, the bias effect could be significant. It should be mentioned that the velocity bias, against the traditional viewpoint, could indeed not be simply categorized as being the measurement error. For details about this statement see Sect. 12.2 and Chap. 17.

It can be concluded that in dealing with flows that exhibit the velocity gradient within the area of LDA measurement volume, both the measurement and the related data processing suffer from two effects:

– the effect that leads to *apparent* turbulence intensity and
– the effect that leads to *velocity bias* in all flow quantities.

Two expressions have been used here to specify the two different effects. For LDA users it is helpful to know the quantitative outcomes of each and the combined effect. For simplicity and with respect to the practical engineering flows the linear velocity distribution in the LDA measurement volume will be supposed in the following analyses. In the related graphical illustrations, the non-uniform velocity distribution is assumed to be present along the length of the measurement volume (Fig. 12.2). The results are applicable to the velocity distribution along the thickness of the measurement volume. One needs only to make substitution for the respective length scale. For the purpose of comprehension, the case that results in the apparent turbulence intensity without the bias effect, will be considered first.

12.1 Apparent Turbulence Intensity and Related Quantities

The arrangement of LDA optics to the flow with velocity gradient has been illustrated in Fig. 12.2. In the applied coordinate system, the turbulent flow to be measured is considered as the superposition of the time-averaged linear velocity distribution $\hat{u} = \overline{u}_1 + a(z - z_1)$ along the measurement volume and the random velocity fluctuations, as given by

$$u(z) = \overline{u}_1 + a(z - z_1) + u'(z) \tag{12.1}$$

The randomness of velocity fluctuations also implies that each velocity event is randomly ordered to a z-value. The linear velocity distribution is presented by the constant velocity gradient which is denoted by a and can be expressed by $a = (\overline{u}_2 - \overline{u}_1)/(z_2 - z_1)$. In assuming the unit thickness of the measurement volume in the direction perpendicular to the drawing plane, the volumetric mean velocity in the area of $\Delta z = z_2 - z_1$ is calculated as

$$\overline{u} = \frac{1}{z_2 - z_1} \int_{z_1}^{z_2} u \mathrm{d}z = \frac{1}{2}(\overline{u}_1 + \overline{u}_2) \tag{12.2}$$

It is equal to the mean velocity that is found at the centre of the measurement volume. In this calculation, $u' \mathrm{d}z = 0$ has been applied because of randomness of velocity fluctuations and their random distribution along the measurement volume length.

Combining the above equation with $a = (\overline{u}_2 - \overline{u}_1)/(z_2 - z_1)$ to eliminate $\overline{u}_2$ yields

$$\overline{u}_1 = \overline{u} - \frac{1}{2} a(z_2 - z_1) \tag{12.3}$$

Then Eq. (12.1) can be expressed as

$$u(z) = \overline{u} - \frac{1}{2} a(z_1 + z_2 - 2z) + u'(z) \tag{12.4}$$

To further calculate the turbulent normal stress, the fluctuation velocity in the above equation is resolved and then squared as

$$u'^2 = (u - \overline{u})^2 + a(u - \overline{u})(z_1 + z_2 - 2z) + \frac{1}{4} a^2 (z_1 + z_2 - 2z)^2 \tag{12.5}$$

The velocity difference $u - \overline{u}$ in the second term on the r.h.s. of above equation will be interpreted by Eq. (12.4) in function of z, so that it yields, after a rearrangement,

$$u'^2 = (u - \overline{u})^2 - \frac{1}{4} a^2 (z_1 + z_2 - 2z)^2 + a(z_1 + z_2 - 2z) u' \tag{12.6}$$

The averaged turbulent normal stress within the length of the LDA measurement volume can then be obtained by entirely averaging all three terms on the r.h.s. of the above equation. In reality, it deals with the arithmetic average of measurement data. To be expected is that the third term vanishes because of the randomness of velocity fluctuations and their random distribution along the z-axis. The corresponding calculation of the second term can be transformed into an integral under the assumption of uniform distribution of velocity events along the LDA measurement volume (The case of non-uniform distribution of velocity events will be considered in the next section). The averaged turbulent normal stress in the area of the measurement volume is then calculated as

$$\overline{u'^2} = \frac{1}{N}\sum_{\mathrm{i}=1}^{\mathrm{N}}(u_\mathrm{i} - \overline{u})^2 - \frac{1}{4}\frac{a^2}{z_2 - z_1}\int_{z_1}^{z_2}(z_1 + z_2 - 2z)^2 \mathrm{d}z \tag{12.7}$$

The integral calculation can be performed easily. With $\Delta z = z_2 - z_1$ as the measurement volume length one obtains

$$\overline{u'^2} = \frac{1}{N}\sum_{\mathrm{i}=1}^{\mathrm{N}}(u_\mathrm{i} - \overline{u})^2 - \frac{1}{12}a^2(\Delta z)^2 \tag{12.8}$$

The first term on the r.h.s. of the above equation corresponds to the direct arithmetic average (sample mean). Because it does not represent the actual value of the related turbulence quantity it is called the pseudo or apparent normal stress. Using $a\Delta z = \Delta\overline{u}$, Eq. (12.8) is further written as

$$\overline{u'^2} = \overline{u'^2_{\mathrm{app}}} - \frac{1}{12}(\Delta\overline{u})^2 \tag{12.9}$$

For the turbulent flow currently treated, the normal stress, which is calculated by the arithmetic average and thus is of apparent nature, needs to be corrected through a part that considers the spatial velocity difference $\Delta\overline{u} = \overline{u}_2 - \overline{u}_1$ i.e. the linear velocity distribution within the LDA measurement volume. The algorithm is similar to and comparable with that in Chap. 11 considering the non-stationary turbulent flows, see Eqs. (11.27) and (11.44). Only at flows with small spatial velocity gradient and comparably high turbulent velocities or at LDA with short measurement volumes, does the correction term become insufficient. The arithmetic average like in Eq. (12.8) can then be used, for convenience, to represent the actual value of the related turbulent normal stress.

Based on the same calculation procedure the turbulent shear stress is obtained in the form of the covariance of two orthogonal velocity components as

$$\overline{u'v'} = \frac{1}{N}\sum_{\mathrm{i}=1}^{\mathrm{N}}(u_\mathrm{i} - \overline{u})(v_\mathrm{i} - \overline{v}) - \frac{a_\mathrm{u}a_\mathrm{v}}{12}\Delta z^2 \tag{12.10}$$

It is comparable with Eq. (11.28) as well as Eq. (11.45).

The calculations given above have been performed with respect to the flow distribution along the LDA measurement volume that has a length equal to $\Delta z = z_2 - z_1$. The results can also be applied to the case of the flow distribution which is confirmed to be along the cross section of the measurement volume (Fig. 12.1). One needs only to replace the length scale Δz in all above equations through the thickness d_{mv} of the LDA measurement volume. From Eq. (12.8) for instance, one obtains immediately

$$\overline{u'^2} = \overline{u'^2_{\mathrm{app}}} - \frac{1}{12}a^2 d^2_{\mathrm{mv}} \tag{12.11}$$

Because the thickness d_{mv} of the LDA measurement volume is usually very thin, the respective correction terms in both the normal and the shear stresses do behave as negligible.

It should be remembered that above calculations are based on the assumption of uniform distribution of velocity events along the LDA measurement volume. In reality, there are two factors that counteract this assumption. The first one is the velocity bias effect which implies that the high velocities will be more frequently detected by LDA optics than the low velocities. This effect will be considered in detail in the next section of this chapter. The second factor is related to the cross-section form of the LDA measurement volume. Because of the ellipsoidal form of the LDA measurement volume and hence non-constant width in its cross-section along the z-axis (Fig. 12.2), the assumption of uniform distribution of velocity events is not exact. The maximum probability of velocity events is obviously expected in the centre area of the measurement volume. This circumstance also applies to the case of LDA alignment as in Fig. 12.1, where along the thickness of LDA measurement volume the velocity gradient is present. With respect to the elliptic detection area of LDA measurement volume and in assuming the linear velocity distribution across the measurement volume thickness (d_{mv}), the averaged turbulent normal stress is calculated, according to Albrecht et al. (2003), as

$$\overline{u'^2} = \frac{1}{N}\sum_{\mathrm{i}=1}^{\mathrm{N}}(u_{\mathrm{i}} - \overline{u})^2 - \frac{1}{16}a^2 d^2_{\mathrm{mv}} \tag{12.12}$$

In comparison with Eq. (12.11), only a small difference in the correction term is present.

At the end of this section, both the momentum and energy flow rates across the measurement volume should be considered. The mean velocity that is calculated in Eq. (12.2) is indeed the volumetric mean. In the practical application dealing with the flow with non-uniform velocity distribution, the mean velocities that are related to both the momentum and energy flow rates are all relevant flow quantities. The momentum flow rate as a vector quantity is calculated by the product of the velocity vector and the mass flow rate as $\dot{m} \cdot \vec{u}$. Its x-component is given by $\dot{m} \cdot u_{\mathrm{x}}$. The corresponding component of the so-called momentum flux (momentum flow rate per unit area) is then given by $\dot{J}_{\mathrm{x}} = (\rho u_{\mathrm{x}})u_{\mathrm{x}} = \rho u^2_{\mathrm{x}}$. Herein ρu_{x} represents the x-component of the mass flux. The related volumetric flux u_{x} in the unit of $\mathrm{m}^3/(\mathrm{m}^2\mathrm{s})$ is measured by LDA. In the case of Fig. 12.2, it is $u_{\mathrm{x}} = u$. Furthermore the simplified rectangular cross-section with unit width is assumed. The averaged momentum flux component through the measurement volume is then calculated by

$$\bar{J}_x = \rho \frac{1}{\Delta z} \int_{z_1}^{z_2} u^2 dz \tag{12.13}$$

To the term u^2 in the integral, the linear velocity distribution according to Eq. (12.4) is applied. While calculating the integral, all terms with uneven power of velocity fluctuations (u') disappear. One obtains

$$\frac{\bar{J}_x}{\rho} = \bar{u}^2 + \frac{1}{12} a^2 (z_2 - z_1)^2 + \frac{1}{\Delta z} \int_{z_1}^{z_2} u'^2 dz \tag{12.14}$$

On the other side the interested momentum flux can also be written as $\bar{J}_x = \rho \bar{u}_J \bar{u}$, with $\bar{u}$ as the mean of volumetric flux and $\bar{u}_J$ as the mean velocity that is relevant for calculating the momentum flux. Eq. (12.14) then becomes

$$\bar{u}_J \bar{u} = \bar{u}^2 + \frac{1}{12} a^2 (\Delta z)^2 + \overline{u'^2} \tag{12.15}$$

Because of $a\Delta z = \bar{u}_2 - \bar{u}_1 = \Delta \bar{u}$ it is further written as

$$\bar{u}_J \bar{u} = \bar{u}^2 + \frac{1}{12} (\Delta \bar{u})^2 + \overline{u'^2} \tag{12.16}$$

As it will be shown in the next section, the mean velocity $\bar{u}_J$ exactly corresponds to the sample mean undergoing the effect of velocity bias.

In practical applications, the momentum flux correction factor β has been used to represent the mean velocity $\bar{u}_J$ through $\bar{u}_J = \beta \bar{u}$. From Eq. (12.16) it follows

$$\beta = \frac{\bar{u}_J}{\bar{u}} = 1 + \frac{1}{12} \frac{(\Delta \bar{u})^2}{\bar{u}^2} + \frac{\overline{u'^2}}{\bar{u}^2} \tag{12.17}$$

It represents a correction factor that is greater than unity.

For completeness, the mean velocity which is relevant for representing the energy flux is calculated by

$$\overline{u_E^2} \cdot \bar{u} = \frac{1}{\Delta z} \int_{z_1}^{z_2} u^3 dz \tag{12.18}$$

Following the similar calculations and by neglecting all terms with uneven power of velocity fluctuations (u') one finally obtains

$$\overline{u_E^2} = \bar{u}^2 + \frac{1}{4} (\Delta \bar{u})^2 + 3\overline{u'^2} \tag{12.19}$$

Correspondingly, the kinetic energy flux correction factor is calculated as

$$\alpha = \frac{\overline{u_{\mathrm{E}}^2}}{\bar{u}^2} = 1 + \frac{1}{4}\frac{(\Delta\bar{u})^2}{\bar{u}^2} + 3\frac{\overline{u'^2}}{\bar{u}^2} \tag{12.20}$$

It represents a value that is greater than unity.

12.2 Combined Velocity Bias Effect

It has been indicated that the velocity bias as a flow phenomenon is ascribed as the dependence of velocity sampling rate in LDA measurements on the velocity magnitude. Traditionally, the velocity bias has been acknowledged as being related to the non-stationary flows or the flows with velocity fluctuations (McLaughlin and Tiederman 1973). It is confirmed by that high velocities will be more frequently sampled than low velocities, if the homogeneous particle distribution in the flow is assumed. A great number of investigations aiming to estimate and correct the effect of velocity bias have been carried out. Detailed description and quantification of velocity bias in this traditional sense will be shown in Chap. 17.

As a matter of fact, the velocity bias in the similar form and of equal mechanism also exists in the flow with spatial velocity gradient. In this case, the non-uniform velocity distribution in the flow leads to non-uniform particle arrival rate and thus to non-uniform distribution of velocity events along the LDA measurement volume. To be expected is that more velocity events will be detected where the flow velocity is high. The arithmetic mean of velocities will be slightly shifted, if compared with the volumetric mean velocity, towards the upper value of the velocity.

The existence of velocity bias that is currently associated with the finite length of the LDA measurement volume and the spatially non-uniform velocity distribution will influence experimental determinations of both the mean velocity and all turbulent stresses. It is therefore reasonable to estimate the associated relationships in respective calculations. For simplicity, linear velocity distribution according to Eq. (12.4) as well as Fig. 12.2 is again assumed to be present along the measurement volume. In addition, further assumptions are specified as follows:

(1) homogeneous particle distribution in the flow;
(2) constant flow direction along the length of the measurement volume;
(3) proportional dependence of velocity sampling rate on the magnitude of the measured velocity component.

Because of the second assumption the velocity bias which is hypothetically a function of the absolute velocity can be expressed in the proportional function of a velocity component.

Starting from the third assumption, the probability distribution of velocity sampling rate along the length of the measurement volume can be expressed by the probability density function

$$p_u = \frac{1}{N}\frac{dN}{dz} = ku \tag{12.21}$$

In this equation, the constant k must be determined from the condition that the probability of total velocity events sampled during a measurement is equal to unity. With the linear velocity distribution according to Eq. (12.4) one then obtains

$$\int_{z_1}^{z_2} p_u dz = k \int_{z_1}^{z_2} \left[\bar{u} - \frac{1}{2} a\,(z_1 + z_2 - 2z) + u'\right] dz = 1 \tag{12.22}$$

Without any difficulty the constant k can be resolved from this last equation. With $\Delta z = z_2 - z_1$ one obtains

$$k = \frac{1}{\bar{u}\Delta z} \tag{12.23}$$

12.2.1 Mean Velocity

From direct data processing of LDA measurements without any weighting form, the arithmetic mean velocity specifies the biased mean and is given as

$$\bar{u}_{bias} = \frac{1}{N}\sum_{i=1}^{N} u_i \tag{12.24}$$

On the other hand, the mean velocity that undergoes the bias effect is calculated with respect to the probability density function of velocity events as

$$\bar{u}_{bias} = \int_{z_1}^{z_2} p_u u dz = \frac{1}{\bar{u}\Delta z}\int_{z_1}^{z_2} u^2 dz \tag{12.25}$$

The integral in this equation has been already encountered at Eq. (12.13) and calculated at Eq. (12.14) in the last section. Thus, one immediately obtains

$$\bar{u}_{bias}\bar{u} = \bar{u}^2 + \frac{1}{12}(\Delta\bar{u})^2 + \overline{u'^2} \tag{12.26}$$

Obviously the mean velocity is shifted towards the upper value of the velocity ($\bar{u}_{bias} > \bar{u}$). The bias effect will disappear only when the second and the third terms on the r.h.s. of the above equation are negligible. For this reason it deals with a flow phenomenon and is therefore called velocity bias. The traditional aspect of velocity bias is related to the third term i.e. $\overline{u'^2}$.

As can be seen by comparing the above equation with Eq. (12.16), the biased mean velocity precisely represents the mean velocity that is used to calculate the mean momentum flow rate or the mean momentum flux across the measurement volume. In the field of flow dynamics as well as in related investigations, it is usual to distinguish between the volumetric mean velocity and mean velocities that are relevant for the momentum and the energy flux, respectively. While in applying the law of mass conservation the volumetric mean velocity is considered, the corresponding mean velocity should be accounted for when dealing with the momentum equations like Euler, Navier-Stokes or Reynolds equations (see Sect. 2.2). For this reason, the velocity bias represents an error only when the arithmetic mean velocity $\bar{u}_{\text{bias}}$ according to Eq. (12.24) is used as the volumetric mean.

Correspondingly, the ratio of the biased to the volumetric mean velocities is equal to the momentum flux correction factor as $\beta = \bar{u}_{\text{bias}}/\bar{u}$.

For flows without velocity gradient ($a = 0$) as the special case, there is $\Delta\bar{u} = 0$ across the measurement volume. Eq. (12.26) is then simplified as

$$\bar{u}_{\text{bias}} = \bar{u}\left(1 + \frac{\overline{u'^2}}{\bar{u}^2}\right) = \bar{u}\left(1 + \text{Tu}^2\right) \tag{12.27}$$

Here the use of turbulence intensity in the form $\text{Tu}^2 = \overline{u'^2}/\bar{u}^2$ is valid only when the velocity component u approximately represents the main flow. The fact that the biased velocity is simply related to the turbulence intensity agrees well with the approximation that was applied to estimate the bias effect in the traditional aspect of velocity bias (see Chap. 17). Because of this agreement it can be concluded that Eq. (12.26) concerns the effects of velocity bias in both the traditional aspect regarding the flow turbulence and the aspect that regards both the finite extent of LDA measurement volume and the non-uniform velocity distribution along the measurement volume.

As can be seen, Eq. (12.26) would basically also behave as the determination equation (polynomial of second order) for the volumetric mean velocity $\bar{u}$. While the biased mean velocity $\bar{u}_{\text{bias}}$ is obtained by arithmetic mean of velocities according to Eq. (12.24), the actual turbulence quantity $\overline{u'^2}$, at the moment, remains unknown. Its determination will be presented in the next section, where it will be shown that the volumetric mean velocity $\bar{u}$ can be rather more simply calculated than by using Eq. (12.26). The readers are kindly referred to Eq. (12.32).

12.2.2 Turbulent Normal Stress

More complex outcomes of non-uniform velocity distribution and thus non-uniform distribution of velocity events along the length of LDA measurement volume are found in determining actual values of turbulence quantities such as the turbulent normal stress. Firstly, the non-uniform velocity distribution results in the broadening of the turbulence scale, so that an *apparent* turbulent normal stress comes about. This is comparable with the *apparent* turbulence that is resulted from the non-stationary flow and the related data processing. Secondly, both the non- uniformity of velocity events and the flow turbulence lead to *combined velocity bias*. To be noted is that the effect of this combined velocity bias on the mean velocity has been already treated in the last section.

With respect to these aspects the *biased apparent* turbulent normal stress in the velocity component u is calculated by direct data processing of LDA measurements as

$$\overline{u'^2_{\text{app,bias}}} = \frac{1}{N}\sum_{\text{i}=1}^{\text{N}}(u_{\text{i}} - \bar{u}_{\text{bias}})^2 \tag{12.28}$$

It is called the apparent turbulent normal stress just because the velocity difference $u_{\text{i}} - \bar{u}_{\text{bias}}$ of each velocity event stands for the apparent velocity fluctuation. Like calculations of the biased mean velocity in the last section the *biased apparent* turbulent normal stress can be calculated by converting its summation form into the integral form. For this purpose the non-uniformity of velocity events along the measurement volume should be taken into account by again concerning the probability density function that is given in Eq. (12.21) with $k = 1/(\bar{u}\Delta z)$, so that

$$\overline{u'^2_{\text{app,bias}}} = \int_{z_1}^{z_2} p_{\text{u}}\,(u - \bar{u}_{\text{bias}})^2\,\text{d}z = \frac{1}{\bar{u}\Delta z}\int_{z_1}^{z_2} u\,(u - \bar{u}_{\text{bias}})^2\,\text{d}z \tag{12.29}$$

The velocity component u in this equation is related to Eq. (12.4). It linearly changes along the measurement volume and further involves the flow fluctuations. In performing the integral calculation, all terms containing u' and u'^3 disappear because of the randomness of flow fluctuations. One obtains then from above equation

$$\overline{u'^2_{\text{app,bias}}} = (\bar{u} - \bar{u}_{\text{bias}})^2 + \frac{1}{12}\left(3 - 2\frac{\bar{u}_{\text{bias}}}{\bar{u}}\right)(\Delta\bar{u})^2 + \left(3 - 2\frac{\bar{u}_{\text{bias}}}{\bar{u}}\right)\overline{u'^2} \tag{12.30}$$

Combining this equation with Eq. (12.26) to eliminate $\overline{u'^2}$ yields

$$\overline{u'^2_{\text{app,bias}}} = (2\bar{u} - \bar{u}_{\text{bias}})(\bar{u}_{\text{bias}} - \bar{u}) \tag{12.31}$$

The volumetric mean velocity is then resolved as

$$\bar{u} = \frac{1}{4}\left(3\bar{u}_{\text{bias}} + \sqrt{\bar{u}^2_{\text{bias}} - 8\overline{u'^2_{\text{app,bias}}}}\right) \tag{12.32}$$

This equation represents a simple method of determining the volumetric mean velocity $\bar{u}$ directly from two arithmetic means. According to the context it actually belongs to the last section.

Equation (12.31) is further considered. The expression $(\bar{u}_{\text{bias}} - \bar{u})$ in it will be replaced by that from Eq. (12.26). It yields then

$$\overline{u'^2_{\text{app,bias}}} = \left(2 - \frac{\bar{u}_{\text{bias}}}{\bar{u}}\right)\left(\frac{1}{12}(\Delta\bar{u})^2 + \overline{u'^2}\right) \tag{12.33}$$

or because of Eq. (12.9)

$$\overline{u'^2_{\text{app,bias}}} = \left(2 - \frac{\bar{u}_{\text{bias}}}{\bar{u}}\right) \cdot \overline{u'^2_{\text{app}}} \tag{12.34}$$

The biased apparent turbulent normal stress has been thus shown to be the apparent normal stress corrected by a factor which accounts for the bias effect. This relationship exactly represents the combination between the effect of velocity bias and the effect of non-uniform velocity distribution along the LDA measurement volume. In addition, comparing Eqs. (12.31) and (12.34) yields

$$\overline{u'^2_{\text{app}}} = \bar{u}\,(\bar{u}_{\text{bias}} - \bar{u}) \tag{12.35}$$

For vanishing bias effect that is given only at laminar flows ($u' = 0$) with uniform velocity distribution along the LDA measurement volume, there is $\bar{u}_{\text{bias}} \approx \bar{u}$. Accordingly, there is $\overline{u'^2_{\text{app,bias}}} \approx \overline{u'^2_{\text{app}}} \approx \overline{u'^2} \approx 0$, as expected.

The actual turbulent normal stress can be calculated from Eq. (12.26) as

$$\overline{u'^2} = \left(\frac{\bar{u}_{\text{bias}}}{\bar{u}} - 1\right)\bar{u}^2 - \frac{1}{12}(\Delta\bar{u})^2 \tag{12.36}$$

and further with respect to $\Delta\bar{u} = a\Delta z$ as

$$\overline{u'^2} = \left(\frac{\bar{u}_{\text{bias}}}{\bar{u}} - 1\right)\bar{u}^2 - \frac{1}{12}a^2(\Delta z)^2 \tag{12.37}$$

In this equation, the biased mean velocity $\bar{u}_{\text{bias}}$ acts as known according to Eq. (12.24). The volumetric mean velocity has been already calculated in Eq. (12.32).

Based on all above calculations, additional simplifications are considered as follows:

12.2.2.1 Uniform Velocity Distribution

Because of $\Delta\bar{u} = 0$ along the LDA measurement volume, there is no apparent turbulence to be concerned, so that it comes about $\overline{u'^2_{\text{app,bias}}} = \overline{u'^2_{\text{bias}}}$ and $\overline{u'^2_{\text{app}}} = \overline{u'^2}$. From Eq. (12.34) one obtains

$$\overline{u'^2_{\text{bias}}} = \left(2 - \frac{\bar{u}_{\text{bias}}}{\bar{u}}\right) \cdot \overline{u'^2} \tag{12.38}$$

As can be confirmed from this relationship, the biased turbulent normal stress is less than its actual value because of $\bar{u}_{\text{bias}} > \bar{u}$ and hence $(2 - \bar{u}_{\text{bias}}/\bar{u}) < 1$. This circumstance has already been confirmed in earlier studies (Nobach 1998, Zhang 2002), see Chap. 17.

12.2.2.2 Negligible Turbulent Flow Fluctuations $\overline{u'^2} \approx 0$

This case can be considered to be comparable with measurements of laminar flows. From Eq. (12.33) one obtains

$$\overline{u'^2_{\text{app,bias}}} = \frac{1}{12}\left(2 - \frac{\bar{u}_{\text{bias}}}{\bar{u}}\right)(\Delta\bar{u})^2 \tag{12.39}$$

The term $\bar{u}_{\text{bias}} / \bar{u}$ in this equation is substituted by that from Eq. (12.26). It follows then with $\Delta\bar{u} = a\Delta z$

$$\overline{u'^2_{\text{app,bias}}} = \frac{1}{12}a^2(\Delta z)^2 - \frac{a^4}{12^2}\frac{(\Delta z)^4}{\bar{u}^2} \tag{12.40}$$

Although the flow does not show any velocity fluctuations, a non-vanishing normal stress has been confirmed when using Eq. (12.28) to process LDA measurement data. The first term on the r.h.s. of above equation is the apparent part, as can be confirmed by comparing it with Eq. (12.9), and the second term of the part arising from the velocity bias, however, in negative effect. Obviously this second term represents a negligible part against the first term, especially at high velocity flow and the short measurement volume length. This indicates that for all practical flows with LDA measurements, when considering the turbulent normal stress, the velocity bias associated with non-uniform velocity distribution along the LDA measurement volume can be neglected. To be noted is that this conclusion is not applicable to the mean velocity according to Eq. (12.26).

12.3 Method of Resolving the Non-uniform Velocity Distribution

As shown above, the non-uniform velocity distribution for instance in the turbulent boundary layer complicates its accurate measurements. Neither the mean velocity nor the averaged turbulence quantities can be accurately located within the finite length of LDA measurement volume. Only when the LDA optics is arranged as in Fig. 12.1, the spatial distribution of the flow can be more accurately measured by a spatial resolution of about 0.1 mm.

In order to resolve both the mean velocity profile and the turbulence quantities within the finite length of the LDA measurement volume, separate receiving optics can be used, as shown in Fig. 12.3a for measurements of the velocity profile beneath a thin film of falling water. The transmitting optics is arranged to be perpendicular to the wall. The separate receiving optics is aligned with its focus to the measurement

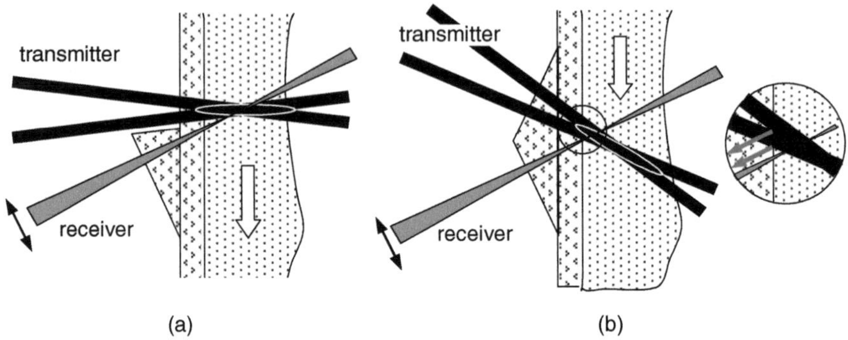

Fig. 12.3 LDA configuration to resolve the velocity profile beneath a thin water film

volume. In this case, a transparent wedge is applied to suppress all possible optical aberrations and hence to enhance the optical performance. Because of the small aperture of the receiving optics, only velocities at a local point along the length of the measurement volume will be measured each time. By traversing the receiving optics as shown in Fig. 12.3a, the velocity profile in the height of the thin water film can then be accurately resolved. Such an optical arrangement with the separate receiving optics has also been found by Wittig et al. (1996).

Another possible optical configuration is illustrated in Fig. 12.3b. This option, however, has a significant deficiency. Because of the symmetrical arrangement of transmitting and receiving optics the partial laser beams that are reflected on the window-water interface, will be directly guided to the receiving unit. This will lead to the rapid overloading of used photodetectors (e.g. photomultiplier) and to high noise in the opto-electronic signals. In addition, because of the non-regular refractions of two laser beams on the window-water interface, the optical performance of the measurement volume such as the fringe spacing and the fringe orientation will be affected.

Chapter 13
Flow Measurements Behind the Plane Window: On-axis

The non-intrusive property of the LDA method represents the greatest advantage against other methods using mechanical sensors for flow measurements. The LDA method thus has found its wide applications in measurements of internal flows such as the flows in ducts and machines. In such applications, both the ducts and the machines have to be configured to posses a window so that the flow is made optically accessible. It is well known by LDA users that the plane windows with parallel surfaces are always the first choice in practical applications, because the use of plane windows will significantly simplify calculations of laser beam transmissions through both the window and the test fluid. Another important advantage in using plane windows is that the fringe spacing in the LDA measurement volume remains constant i.e. independent of the optical properties both of the window and the test fluid, if the LDA optical axis is perpendicularly i.e. on-axis aligned to the window surface. This implies that in using plane windows no remarkable optical aberrations will occur. Refractions of laser beams do not have any remarkable influence on the measurements. These properties will be briefly revealed in this chapter.

13.1 Fringe Spacing

The refraction of light waves on a medium interface follows the law of refraction according to Eq. (3.6). At the on-axis alignment of the LDA head to the plane of the medium interface, two laser beams of each laser beam pair in an LDA system undergo the symmetrical refraction. The half intersection angle between two refracted laser beams in the medium 2 is thus calculated according to Fig. 13.1

$$\sin\alpha_2 = \sin\varepsilon_2 = \frac{n_1}{n_2}\sin\varepsilon_1 \tag{13.1}$$

The speed of light in mediums 1 and 2, respectively, is given by $c_1 = \lambda_1\nu_1$ and $c_2 = \lambda_2\nu_2$ with λ as the wavelength and ν the frequency of the light wave. Firstly, the ratio of two light speeds is equal to the reciprocal of the corresponding refractive index ratio ($c_2/c_1 = n_1/n_2$). Secondly, the frequency of the light wave does not

Zh. Zhang, *LDA Application Methods*, Experimental Fluid Mechanics,
DOI 10.1007/978-3-642-13514-9_13, © Springer-Verlag Berlin Heidelberg 2010

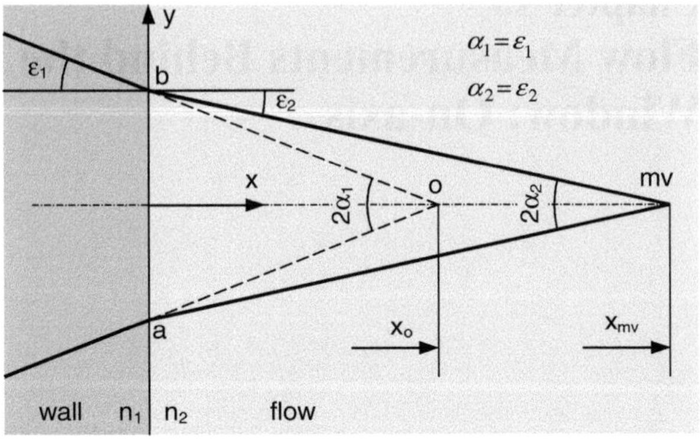

Fig. 13.1 On-axis LDA alignment to the medium interface and the laser beam transmissions in the flow

change with the light ray refraction on the medium interface ($\nu_2 = \nu_1$). From these two conditions the wavelength of the light wave in the medium 2 is given by

$$\lambda_2 = \frac{n_1}{n_2}\lambda_1 \tag{13.2}$$

According to Eq. (3.46) for fringe model of LDA optics, the fringe spacing in the measurement volume, which is present in medium 2, should be calculated using the light wavelength λ_2 and the half intersection angle between two laser beams α_2. With respect to Eqs. (13.1) and (13.2) the fringe spacing in the measurement volume is calculated as ($\alpha_1 = \varepsilon_1$)

$$\Delta x = \frac{\lambda_2}{2\sin\alpha_2} = \frac{\lambda_1}{2\sin\alpha_1} \tag{13.3}$$

This equation signifies that at the on-axis i.e. perpendicular alignment of the LDA head to the medium interface, the fringe spacing in the measurement volume is independent of the fluid properties of the test flow. This is true even though both laser beams pass through a series of windows of different optical properties before reaching the flow.

13.2 Shift of the Measurement Volume

As a result of laser beam refractions on the medium interface, the intersection point of two laser beams, i.e. the measurement volume in the flow, shifts away from the virtual beam intersection point o (Fig. 13.1). The measurement volume as the actual intersection point of two laser beams is denoted by mv. Obviously the location of the measurement volume is a function of the location of the virtual beam intersection point. According to Fig. 13.1 this function can be easily established by considering the distance $y_b - y_a$ on the y-axis. By starting both from the virtual beam intersection point and the measurement volume, respectively, this distance is given by

$$y_b - y_a = 2x_{mv} \tan\alpha_2 = 2x_o \tan\alpha_1 \quad (13.4)$$

Obviously the distance of the measurement volume from the medium interface is proportional to that distance of the virtual beam crossing point. The proportional constant is simply $\tan\alpha_1/\tan\alpha_2$, as can be obtained from the above equation:

$$k_{mv} = \frac{dx_{mv}}{dx_o} = \frac{\tan\alpha_1}{\tan\alpha_2} \quad (13.5)$$

It also represents the ratio of the shift of the actual measurement volume to the shift of the virtual beam crossing point. The latter is equal to the shift of the LDA head, if the LDA head is positioned in the medium 1.

Although the calculation is achieved by considering two mediums, the ratio given in Eq. (13.5) also applies to the case of laser beam transmissions from air through a glass window into the test flow. It corresponds to the case of most available optical arrangement. The application of Eq. (13.5) is independent of the window used in between.

Because of small angles α_1 and α_2 in most LDA configurations, both $\tan\alpha_1$ and $\tan\alpha_2$ can be approximated by $\sin\alpha_1$ and $\sin\alpha_2$, respectively. With regard to Eq. (13.1), that represents the law of refraction, Eq. (13.5) then becomes

$$k_{mv} \approx \frac{\sin\alpha_1}{\sin\alpha_2} = \frac{n_2}{n_1} \quad (13.6)$$

The shift ratio of LDA measurement volume to the LDA head has been shown to be equal to the ratio of refractive indices of two mediums. It can be used to accurately position the LDA measurement volume in the flow by starting from a reference point of the measurement volume that lies, for instance, on the medium interface ($x = 0$).

Because of the simplest geometrical and optical behaviours in on-axis alignment of LDA optics, measurements of all types of internal flows behind a plane window can be well accomplished without having to pay any special attentions. In contrast to this, the optical condition and the measurement facilities will become much complex, when the LDA optical axis is aligned off-axis to the plane wall. This will be thoroughly described in Chap. 14.

13.3 Optical Dispersion and its Negligible Effect

A special phenomenon in the LDA optics with laser beams of different wavelengths should be mentioned. It is the optical dispersion which states that the refractive index of a dielectric medium fundamentally depends on the light wavelength. This phenomenon can be well demonstrated by using a dispersing prism to separate the white light that is composed of numerous wavelengths. The occurrence of the optical dispersion in LDA measurements is confirmed by that the laser beams of different wavelengths, for instance $\lambda = 488$ and $\lambda = 514.5$ nm in a two-component LDA unit, will be differently refracted on the medium interface. As a result, both measurement volumes in the flow will no longer be found at the same point. The displacement between two measurement volumes, however, is usually negligible. This can be demonstrated by considering the laser beam refraction in water based on the following calculation example.

A two-component LDA system applies the laser light of wavelengths $\lambda = 488$ and $\lambda = 514.5$ nm. The half intersection angle between two laser beams in each laser beam pair is assumed to be equal to $\alpha_1 = 6.77°$. The LDA head is assumed to be found in open air ($n_1 = 1$). Corresponding to both wavelengths of the laser light, the refractive index of water at a temperature of 20°C is given by $n_2 = 1.337$ for $\lambda = 488$ and $n_2 = 1.336$ for $\lambda = 514.5$. According to Eq. (13.6) the positions of both measurement volumes in the flow, corresponding to both laser beam pairs, are given by

$$x_{\mathrm{mv},488} = n_2 x_{\mathrm{o}} = 1.337 x_{\mathrm{o}} \tag{13.7}$$

and

$$x_{\mathrm{mv},514} = n_2 x_{\mathrm{o}} = 1.336 x_{\mathrm{o}} \tag{13.8}$$

respectively.

For example, the virtual beam crossing point is assumed to be given at $x_{\mathrm{o}} = 100$ mm, to which the corresponding measurement volumes are then found at about $x_{\mathrm{mv}} = 134$ mm. The displacement between two measurement volumes is calculated to be

$$x_{\mathrm{mv},488} - x_{\mathrm{mv},514} = 0.001 x_{\mathrm{o}} = 0.1 \ \ (\mathrm{mm}) \tag{13.9}$$

In comparison to the measurement volume length of about 0.40 mm (Table 4.1) this displacement can usually be neglected.

Chapter 14
Flow Measurements Behind the Plane Window: Off-axis

The simplest case for internal flow measurements is the on-axis alignment of LDA optics, as this has been shown in the last chapter. There are other cases in which the LDA optics has to be aligned off-axis, i.e. the optical axis is no more perpendicular to the window plane. This situation indeed occurs very frequently, when for instance according to Fig. 14.1 the normal velocity component in the flow behind a plane window should be measured. The off-axis of LDA optics in this case is arranged within the plane containing two laser beams. Because of the asymmetrical beam refraction all geometrical specifications in both forming and shifting the measurement volume will change. This undesired and partly complex circumstance makes the flow measurements difficult. Every measurement result needs to be corrected.

In contrast to the on-axis alignment of LDA optics, the asymmetry of laser beam refractions in the off-axis case becomes a very serious matter when considered further in the aspect of entire optical performances. This is confirmed by the occurrence of optical aberrations in both the transmitting and the receiving optics. The situation with such optical aberrations may be so serious that no LDA measurements can be achieved. This corresponds to the case where both laser beams after the refractions do not intersect at all. The associated optical phenomenon has been recognized to be astigmatism (Zhang 1995, Zhang and Eisele 1995a, b). This phenomenon sensitively affects both the optical features of the LDA measurement volume and the signal quality in the receiving optics. Some LDA users may have probably encountered great difficulties of getting satisfactory optical signals while performing measurements by the LDA head at off-axis position. The cause is nothing other than the effect of astigmatism. In addition, the asymmetry of laser beam refractions leads to different deformation of beam waists on each individual laser beam and hence to the distortion of the LDA measurement volume. This again leads to fringe distortion in the measurement volume and further to measurement errors. Clearly a great deal of troubles in performing LDA measurements is related with the off-axis LDA alignment.

In the past, parallel to identifying all the influences of optical aberrations named above on LDA measurements, diverse measures have also been worked out to enhance the optical performances in the described complex situations. It should be mentioned that the passive method of matching refractive indices of test fluids does help reducing the optical aberration in LDA measurements. The method, however,

Zh. Zhang, *LDA Application Methods*, Experimental Fluid Mechanics,
DOI 10.1007/978-3-642-13514-9_14,

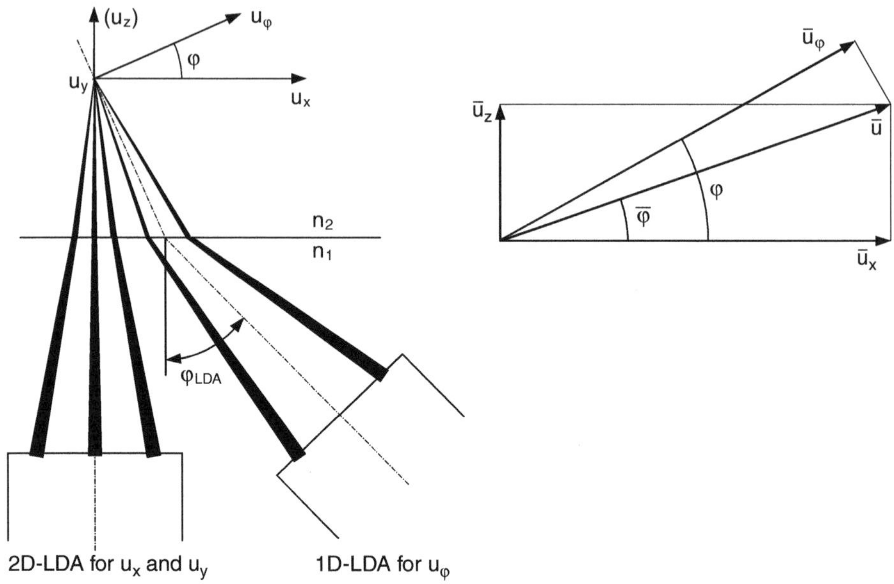

Fig. 14.1 Off-axis alignment of LDA optics for indirect measurements of velocity component u_z

is often not applicable where, for instance, field measurements out of the laboratory should be conducted.

A great deal of optical aspects related with the off-axis alignment of LDA optics will be presented in this chapter.

14.1 Off-axis Measurements and Velocity Transformation

As mentioned before, the off-axis alignment of the LDA head to the normal of the plane window usually aims to measure the velocity component out of the window plane. The optical arrangement according to Fig. 14.1 has found its broad applications, despite the measurement of the third velocity component not being coincident with the measurements of other two in-plane velocity components. In order to obtain the third velocity component u_z i.e. the component along the normal of the window surface, velocity transformation as described in Chap. 6 can be applied. The mean value of this velocity component is then obtained from Eq. (6.40)

$$\bar{u}_z = \frac{1}{\sin\varphi}\left(\bar{u}_\varphi - \bar{u}_x\cos\varphi\right) \tag{14.1}$$

The main flow direction in the $x - z$ plane is calculated by

$$\tan\bar{\varphi} = \frac{\bar{u}_z}{\bar{u}_x} \tag{14.2}$$

In order to determine the related turbulence quantities, the Zero-Correlation Method (ZCM) that is described in Chap. 8 can be applied. From Eq. (8.18) one obtains

$$\overline{u_z'^2} = \frac{\cos\varphi\cos(2\overline{\varphi}-\varphi)\cdot\overline{u_x'^2} - \cos 2\overline{\varphi}\cdot\overline{u_\varphi'^2}}{\sin\varphi\sin(2\overline{\varphi}-\varphi)} \tag{14.3}$$

Further according to Eq. (8.16) the turbulence quantity representing the Reynolds shear stress is obtained as

$$\overline{u_x' u_z'} = \frac{1}{2}\tan 2\overline{\varphi}\left(\overline{u_x'^2} - \overline{u_z'^2}\right) \tag{14.4}$$

Basically, when the off-axis alignment of the LDA head only aims to indirectly measure the on-axis velocity component according to Eq. (14.1), the measurement could be achieved at each off-axis position excluding $\varphi = 0$. Nevertheless, some special aspects associated with this LDA configuration have to be considered for high quality measurements and the maximum achievement from the measurements.

- Only one component measurement is possible. Otherwise two measurement volumes would separate from each other, see Sect. 14.5 of this chapter.
- The angle φ used in above equations is the effective off-set angle of the LDA optical axis in the flow. This angle can be obtained approximately from φ_{LDA} by applying the law of refraction to the LDA optical axis. The velocity component which is measured is then approximately perpendicular to the refracted optical axis (Sect. 14.3).
- Because of different refractions of two laser beams the crossing angle between two laser beams in the fluid flow changes with the off-axis angle of the LDA head. This means that the fringe spacing in the measurement volume also depends on this angle. The measured velocity has thus to be corrected based on the fringe model of LDA optics, see Sect. 14.2.
- The traversing path of the LDA measurement volume is generally two-dimensional, even though the LDA head is traversing along the normal of the window surface, see Sect. 14.4
- The LDA signal quality strongly depends on the off-axis angle of the LDA head, the focal length of used LDA optics and the depth of the measurement volume in the flow. Usually the deterioration of signal qualities is significant so that in worst cases no measurements could be accomplished at all. More to deteriorated signal qualities see Sect. 14.8.

14.2 Fringe Spacing in Measurement Volume and Velocity Corrections

Depending on the off-axis alignment angle of the LDA head, the beam crossing angle in the flow changes. The fringe spacing in the measurement volume is then calculated by $\Delta x_{\text{off}} = \frac{1}{2}\lambda_{\text{n}}/\sin\alpha_{\text{off}}$ according to Eq. (3.46), with λ_{n} as the wavelength of the laser light in the test medium (refractive index n) and α_{off} as the half intersection angle of two laser beams in the flow. Obviously this fringe spacing differs from that at the on-axis configuration $\Delta x_{on} = \frac{1}{2}\lambda_n/\sin\alpha_{on}$. A systematic error in velocity measurements then takes place as a result of this change in the fringe spacing. The ratio of the actual to the initial fringe spacing is obtained as

$$k_{vel} = \frac{\Delta x_{off}}{\Delta x_{on}} = \frac{\sin\alpha_{on}}{\sin\alpha_{off}} \tag{14.5}$$

To each off-axis angle φ_{LDA} in the LDA alignment according to Fig. 14.1, the half intersection angle of two laser beams in the flow α_{off} can be calculated by applying the law of refraction, given by Eq. (3.6), to each laser beam. Thus there is $\alpha_{off} = f(\varphi_{LDA})$ and naturally $\alpha_{on} = f(\varphi_{LDA} = 0)$.

Equation (14.5) represents, because of Eq. (3.47), a correction factor for each velocity that is measured at the off-axis LDA alignment:

$$u_\varphi = \Delta x_{off} \cdot \nu_D = k_{vel}\Delta x_{on}\nu_D = k_{vel}u_{\varphi,\text{measured}} \tag{14.6}$$

The correction of velocities is necessary because the specification of both the optical and geometrical parameters in a LDA-system is usually referred to the open air case. It is equivalent to the case of the on-axis alignment of the LDA head to the plane window (fringe spacing Δx_{on}), as has been shown in Chap. 13.

Figure 14.2 shows an example that represents the correction factor according to Eq. (14.5) as a function of the LDA off-axis angle φ_{LDA} for measurements of a water flow. As can be seen, velocity corrections up to 10% or even 15% are indispensable. The optical configuration at the LDA head, say the half intersection angle α_0 of two laser beams, has indeed the negligible influence on the velocity correction factor.

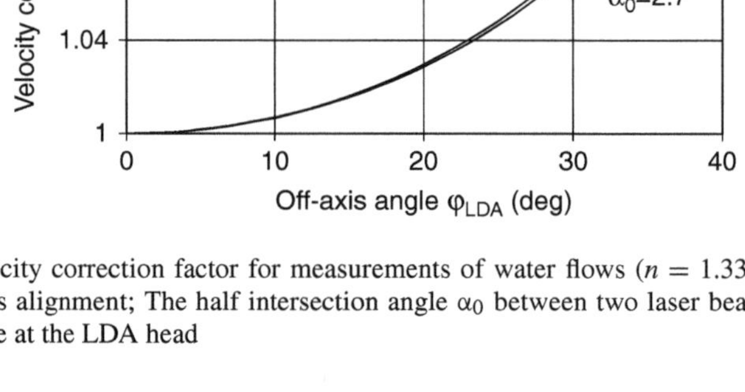

Fig. 14.2 Velocity correction factor for measurements of water flows ($n = 1.333$) by means of LDA in off-axis alignment; The half intersection angle α_0 between two laser beams is related to the design angle at the LDA head

14.3 Refraction of Optical Axis and Orientation of the Measurement Volume

As in common applications of LDA method, the velocity component that is measured with the off-axis LDA is spatially perpendicular to the bisector of two laser beams in the test flow. As a result of optical aberrations the bisector of two refracted laser beams neither coincides with nor is parallel to the refracted optical axis. Such a disagreement is called comatic aberration. This phenomenon, however, does not have significant influence on LDA measurements. This can be demonstrated according to Fig. 14.3a with two laser beams that are refracted in a test medium. In applying the law of refraction according to Eq. (3.6) and because of $\varepsilon_{\mathrm{A1}} = \varphi_{\mathrm{LDA}} + \alpha_0$ and $\varepsilon_{\mathrm{B1}} = \varphi_{\mathrm{LDA}} - \alpha_0$, the refraction angles at two laser beams are calculated, respectively, as

$$\varepsilon_{\mathrm{A2}} = \arcsin\left(\frac{n_1}{n_2}\sin\left(\varphi_{\mathrm{LDA}} + \alpha_0\right)\right) \tag{14.7}$$

$$\varepsilon_{\mathrm{B2}} = \arcsin\left(\frac{n_1}{n_2}\sin\left(\varphi_{\mathrm{LDA}} - \alpha_0\right)\right) \tag{14.8}$$

The inclination angle of the bisector of two laser beams in the test medium is then given by

$$\varepsilon_{\mathrm{b}} = \frac{1}{2}\left(\varepsilon_{\mathrm{A2}} + \varepsilon_{\mathrm{B2}}\right) \tag{14.9}$$

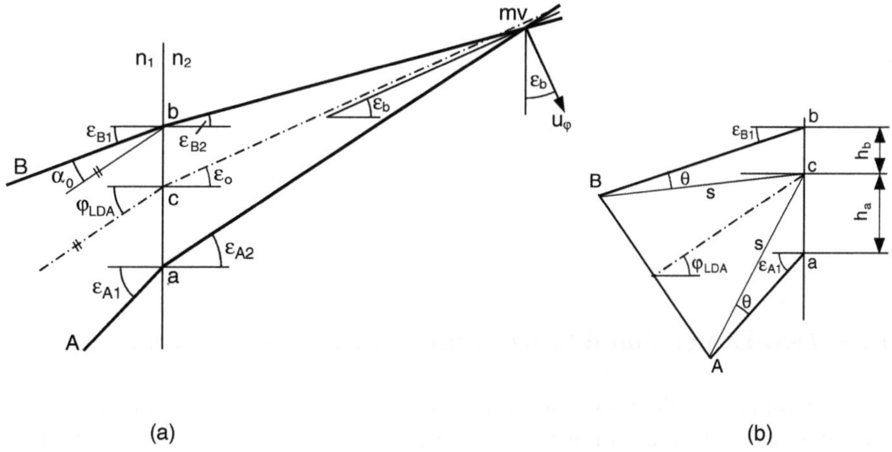

Fig. 14.3 Orientation of LDA measurement volume and the intersection of the LDA optical axis on the medium interface

Because the half intersection angle α_0 is usually small, this last equation can be approximated by considering the linear terms in the Taylor series of ε_{A2} and ε_{B2}, so that it becomes

$$\varepsilon_b = \arcsin\left(\frac{n_1}{n_2}\sin\varphi_{LDA}\right) \tag{14.10}$$

It is equal to the refraction angle of the optical axis ε_o of the LDA head. For this reason, the refracted optical axis in the flow can be used to represent the measurement volume orientation. For water flow ($n_2/n_1 = 1.333$) and the LDA configuration with $\alpha_0 = 3°$ and $\varphi_{LDA} = 30°$, for instance, the uncertainty in ε_b arising from the above simplification is less than 0.02°. The comatic aberration is thus irrelevant. The velocity component measured by the off-axis LDA alignment corresponds to the component that is perpendicular to the refracted optical axis in the flow.

The intersection point of the optical axis on the medium interface is denoted by c, as shown in Fig. 14.3b. Sometimes one needs to know this intersection point, in order to further calculate and to track the laser beams with respect to the optical axis and its refraction on the interface. According to Fig. 14.3b the law of sines will be applied to triangles Aac and Bbc, respectively. Because of $\sin(90 - \varepsilon_{B1}) = \cos\varepsilon_{B1}$ and $\sin(90 + \varepsilon_{A1}) = \cos\varepsilon_{A1}$ there are accordingly

$$\frac{h_a}{\sin\theta} = \frac{s}{\cos\varepsilon_{A1}} \tag{14.11}$$

$$\frac{h_b}{\sin\theta} = \frac{s}{\cos\varepsilon_{B1}} \tag{14.12}$$

Eliminating s and $\sin\theta$ yields

$$\frac{h_a}{h_b} = \frac{\cos\varepsilon_{B1}}{\cos\varepsilon_{A1}} \tag{14.13}$$

With $h = h_a + h_b$ there are further

$$\frac{h_a}{h} = \frac{\cos\varepsilon_{B1}}{\cos\varepsilon_{A1} + \cos\varepsilon_{B1}} \tag{14.14}$$

$$\frac{h_b}{h} = \frac{\cos\varepsilon_{A1}}{\cos\varepsilon_{A1} + \cos\varepsilon_{B1}} \tag{14.15}$$

14.4 Two-Dimensional Shift of the Measurement Volume

In order to measure the flow distribution in an internal flow that is found behind an optical window, the measurement volume needs to be shifted through the flow field. An outstanding behaviour of the off-axis LDA setup is the two-dimensional shift of the measurement volume when the LDA head moves one-dimensionally parallel to the normal of the medium interface. This feature of the measurement volume

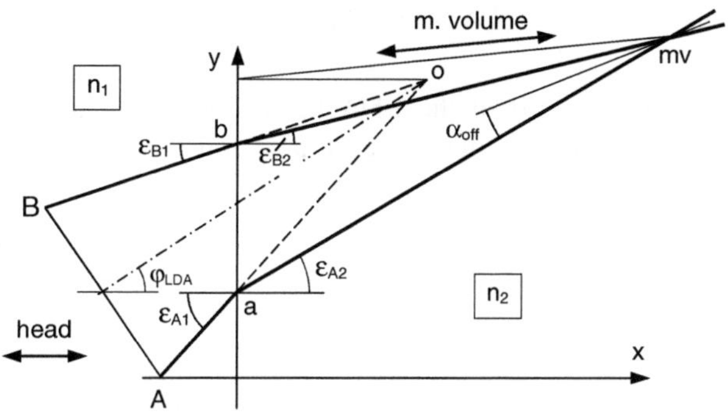

Fig. 14.4 Two-dimensional shift of the LDA measurement volume

shift can be specified by accounting for the geometrical relations between two laser beams, as shown in Fig. 14.4. The virtual and actual intersection points of two laser beams (A and B) are marked by *o* and *mv*, respectively. Herein *mv* stands for the measurement volume. Corresponding incident and refraction angles of both laser beams are denoted by ε_{A1}, ε_{B1}, ε_{A2} and ε_{B2}. Both laser beams A and B intersect the y-axis at y_a and y_b, respectively. The distance between intersection points *a* and *b* can be calculated respectively form the virtual and actual intersection points of two laser beams, as given by

$$x_{mv}(\tan\varepsilon_{A2} - \tan\varepsilon_{B2}) = x_o(\tan\varepsilon_{A1} - \tan\varepsilon_{B1}) = y_b - y_a \tag{14.16}$$

From this equation the following differentiation is obtained

$$k_{mv} = \frac{dx_{mv}}{dx_o} = \frac{\tan\varepsilon_{A1} - \tan\varepsilon_{B1}}{\tan\varepsilon_{A2} - \tan\varepsilon_{B2}} \tag{14.17}$$

This equation represents the x-component of the shift ratio between the actual and virtual crossing points of two laser beams. Because the virtual crossing point moves with the LDA head, the ratio k_{mv} given in above equation simply represents the ratio of the measurement volume shift to the shift of the LDA head. Like in the case of on-axis alignment that has been treated in Chap. 13, the shift ratio in the present case is also independent of the glass window which is found between the LDA head and the flow.

In Fig. 14.4, it has also shown that the shift of the measurement volumes *mv* does not follow a path that is parallel to the x-axis, even though the LDA head is merely shifting in the x-direction. Such a two-dimensional feature of the measurement volume shift can be well quantified. According to Fig. 14.4 the y-coordinate of the measurement volume is given as

$$y_{mv} = y_a + x_{mv}\tan\varepsilon_{A2} \tag{14.18}$$

and because of $y_a = y_o - x_o\tan\varepsilon_{A1}$ as

$$y_{mv} = y_o - x_o \tan \varepsilon_{A1} + x_{mv} \tan \varepsilon_{A2} \tag{14.19}$$

It is of concern that the shift of the LDA head along the x-axis means that the y-coordinate of the virtual crossing point remains constant. Thus differentiation of above equation with respect to $y_o =$ const and $k_{mv} = \mathrm{d}x_{mv} / \mathrm{d}x_o$ yields

$$\frac{\mathrm{d}y_{mv}}{\mathrm{d}x_{mv}} = \tan \varepsilon_{A2} - \frac{\mathrm{d}x_o}{\mathrm{d}x_{mv}} \tan \varepsilon_{A1} = \tan \varepsilon_{A2} - \frac{1}{k_{mv}} \tan \varepsilon_{A1} \tag{14.20}$$

This equation describes the path that the measurement volume follows when the LDA head is moving parallel to the x-direction. In general, the lateral shift of the measurement volume $\mathrm{d}y_{mv} / \mathrm{d}x_{mv}$ does not disappear, except for the special cases of the on-axis alignment, the off-axis alignment with one laser beam perpendicular to the plane of the medium interface, or the flow medium that is the same as the medium in which the LDA head is present. Figure 14.5 as an example shows the two-dimensional shift rate of the measurement volume in water, calculated from Eqs. (14.17) and (14.20) for an LDA optics with $\alpha_0 = 2.75°$.

The results presented here are of special importance for cases where simultaneous three-component velocity measurements would be carried out by using two LDA heads. When trying to shift measurement volumes in the flow, all three measurement volumes will separate from each other, so that both LDA heads need to be realigned, see also Thiele and Eckelmann (1994).

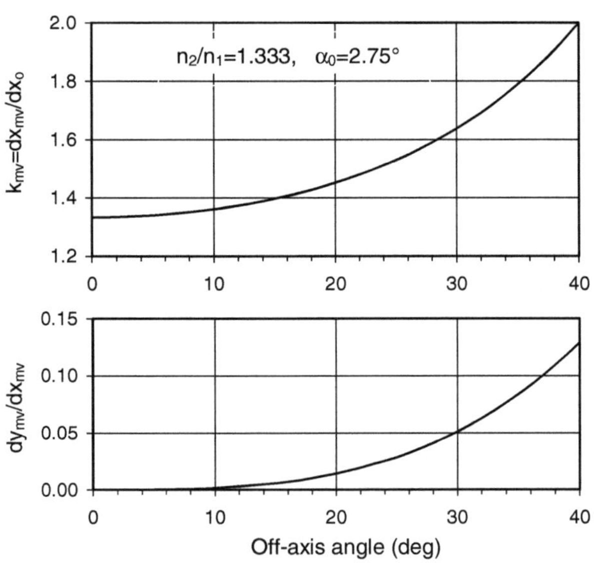

Fig. 14.5 Two-dimensional shift rate of LDA measurement volume in water

14.5 Astigmatism and its Presence in Transmitting Optics

Astigmatism is an optical aberration which refers to the loss of the unique focal point of a light bundle after the refraction through a non-perpendicular interface (Fig. 14.6). While the meridian plane (also called the tangential plane) of the light bundle focuses on the point m after the refraction, the sagittal plane focuses on the point s. The distance between these two focal points is known as the astigmatic difference. Obviously this astigmatic difference depends on both the off-axis angle φ of the light bundle against the normal of the interface and the focusing angle i.e. the thickness of the incident light bundle. In addition, it also depends on the distance of the virtual focal point from the plane interface.

In the context of LDA applications, the focused light bundle that is shown in Fig. 14.6 can be considered as being constructed by four laser beams of a two-component LDA system. One pair of laser beams forms the meridian plane and another pair the sagittal plane. Correspondingly two measurement volumes are obtained, respectively, at the meridian and the sagittal focal point. Because of the remarkable separation between two measurement volumes, two-component coincident LDA measurements become impossible. Also other optical features such as signal qualities and the fringe distortion in the measurement volume are related to the effect of astigmatism aberration. This will be separately shown in following sections.

The occurrence of astigmatism at off-axis LDA alignment and the related effect on measurements are always inevitable, even if only one-component measurements by using two laser beams in the meridian plane will be carried out. Besides the fringe distortion in the measurement volume the astigmatism effect is additionally confirmed in the backward propagation of scattered laser lights from the measurement volume to the receiving optics, which is integrated in the LDA head. As a result,

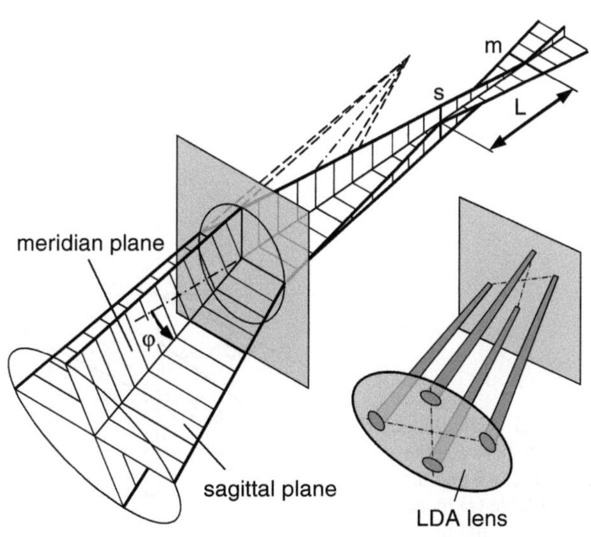

Fig. 14.6 Astigmatism

the existence of astigmatism in the receiving optics directly influences the signal strength and qualities. To quantify such an influence, the astigmatic difference as a quantitative measure of the phenomenon should be considered. It is represented in the present case with two pairs of laser beams (Fig. 14.6) by the distance between two focal points i.e. the available (m) and unavailable (s) measurement volumes.

For practical LDA applications in internal flow measurements, astigmatism takes place twice: in the transparent optical window of thickness d and in the fluid flow, as illustrated in Fig. 14.7 showing the laser beam refractions in the meridian plane. The crossing point of two in-plane laser beams (A and B) is denoted by m. Because of symmetrical refraction of two other laser beams in the sagittal plane, the corresponding crossing point s is also found in the meridian plane. According to the detailed calculations, as shown in Appendix A using the theory of ray optics (Zhang 1995), the displacement between two measurement volumes (m and s) along the x-axis is given by

$$\Delta x_{\mathrm{m,s}} = \frac{1}{T_{20}} \left(\Psi_1 d + \Psi_2 d_{\mathrm{s}} \right) \tag{14.21}$$

Herein

$$\Psi_1 = \frac{\cos \alpha_0 \cos \varphi_{\mathrm{LDA}}}{\sqrt{\frac{n_1^2}{n_0^2} - \left(1 - \cos^2 \alpha_0 \cos^2 \varphi_{\mathrm{LDA}}\right)}} - T_{10} \tag{14.22}$$

$$\Psi_2 = \frac{\cos \alpha_0 \cos \varphi_{\mathrm{LDA}}}{\sqrt{\frac{n_2^2}{n_0^2} - \left(1 - \cos^2 \alpha_0 \cos^2 \varphi_{\mathrm{LDA}}\right)}} - T_{20} \tag{14.23}$$

$$T_{10} = \frac{\tan \varepsilon_{\mathrm{A1}} - \tan \varepsilon_{\mathrm{B1}}}{\tan \varepsilon_{\mathrm{A0}} - \tan \varepsilon_{\mathrm{B0}}} \tag{14.24}$$

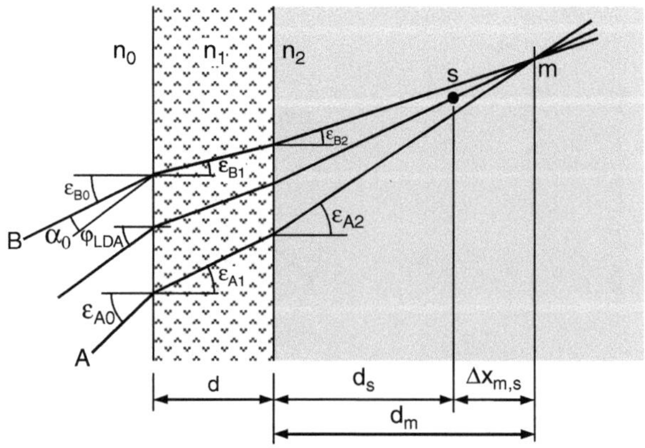

Fig. 14.7 Calculation of the distance between the available (m) and the unavailable (s) measurement volumes in the test flow

$$T_{20} = \frac{\tan \varepsilon_{A2} - \tan \varepsilon_{B2}}{\tan \varepsilon_{A0} - \tan \varepsilon_{B0}} \quad (14.25)$$

with ε_{A1}, ε_{A2}, ε_{B1} and ε_{B2} as the refraction angles of laser beams (A and B) in the medium 1 and 2, respectively. They can be calculated from corresponding incident angles in the form of $\varepsilon_{A0} = \varphi_{LDA} + \alpha_0$ and $\varepsilon_{B0} = \varphi_{LDA} - \alpha_0$ by applying the law of refraction. The half intersection angle between two laser beams in air is designated by α_0. Also worth mentioning is that at small off-axis angles there may be $\varepsilon_{B0} < 0$ according to Fig. 14.7 with given parameter notations. The parameters Ψ_1 and Ψ_2 refer to the window and the test fluid, respectively.

The above equations are obtained based on the assumption that in the medium 0 all four laser beams intersect at a unique point. This medium is denoted as the reference medium. Usually it is the air in that the LDA head is found. The special case with water as the reference medium will be shown in Sect. 14.10.

Figure 14.8 represents an example of calculated displacements between two measurement volumes (m and s) at the off-axis LDA alignment. The calculations are referred to an air-glass-water system with a glass thickness of $d = 20\,\text{mm}$. The displacement i.e. the distance between two measurement volumes is enormous, especially at large off-axis angles. It linearly increases with the depth of the measurement volume in the test flow and is also dominated by this depth. The effect of the glass window on the extent of astigmatism can be read out at $d_s = 0$ in Fig. 14.8. Usually it deals with a relatively small effect because of the small thickness of the used window.

Because of the great distance between two measurement volumes the two-component LDA measurements become impossible. In the practical applications, the LDA off-axis alignment is indeed initially required for separate one-component velocity measurement, as already shown in Fig. 14.1. The two laser beams used for

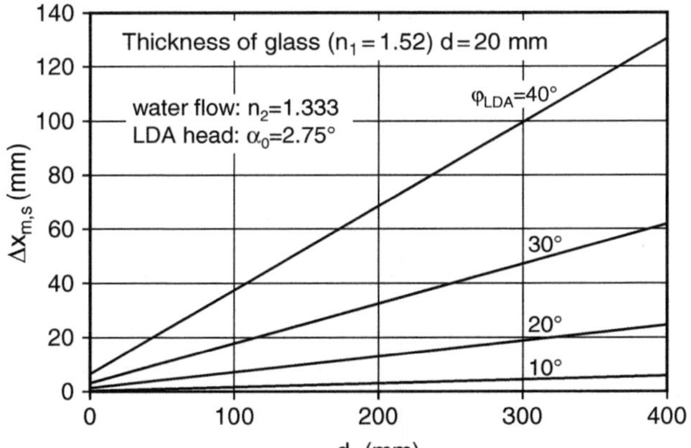

Fig. 14.8 Distance between the available (m) and the unavailable (s) measurement volumes in water as a result of the off-axis alignment of the LDA head

measurements lie in the meridian plane and form the measurement volume at d_m. This measurement volume is therefore called available measurement volume, while the measurement volume at d_s is unavailable. For using d_m as the parameter to express the distance between two measurement volumes, Eq. (14.21) is rearranged by substituting $d_s = d_m - \Delta x_{m,s}$ as

$$\Delta x_{m,s} = \frac{1}{T_{20} + \Psi_2} (\Psi_1 d + \Psi_2 d_m) \tag{14.26}$$

For $d = 0$ there is simply

$$\frac{\Delta x_{m,s}}{d_m} = \frac{\Psi_2}{T_{20} + \Psi_2} \tag{14.27}$$

These two expressions are sometimes preferred because of the use of the available and hence visible measurement volume at the distance d_m. Also worth mentioning is that for small values of the distance d_m the unavailable measurement volume may be found virtually within the window because of $d_s < 0$.

Based on Eq. (14.27) for negligible window thickness, Fig. 14.9 shows the related displacement between the available and unavailable measurement volumes in function of the LDA off-axis angle, as for the medium with $n = 1.333$.

A special case will be considered in which the test medium is equal to the reference medium. For measurements of air flows behind the window, which implies an arrangement of air-glass-air, all related equations can be simplified. Because of $n_2 = n_0 = 1$ there is $T_{20} = 1$. From Eq. (14.23) one obtains

$$\Psi_2 = 0 \tag{14.28}$$

Eq. (14.26) then becomes

$$\Delta x_{m,s} = \Psi_1 d \tag{14.29}$$

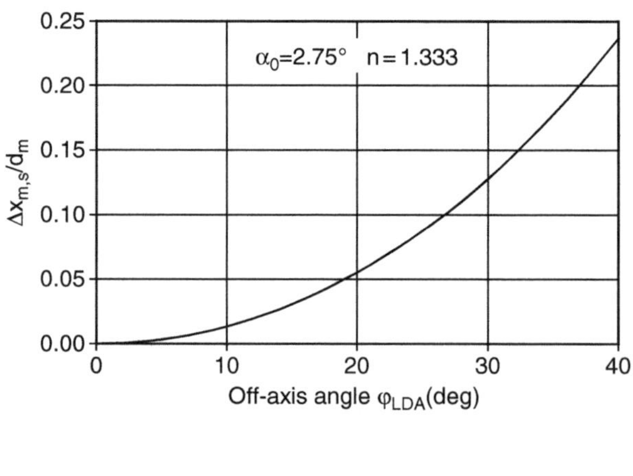

Fig. 14.9 Related displacement between the available and the unavailable measurement volumes in water in function of the LDA off-axis angle (window thickness $d = 0$)

The astigmatism and the associated displacement between two measurement volumes have been thus related to the thickness and the refractive index of the used glass window. They are independent of the depth of the measurement volumes in the air flow. In reality, this is always the case as long as the test medium is equal to the reference medium. See also Sect. 14.10 dealing with the method of suppressing the astigmatism effect.

At the end of this section, it should be emphasized that the LDA off-axis alignment should be performed strictly in the optical plane i.e. the plane containing two laser beams. Deviations from this requirement will lead to separation of two laser beams and thus to failure in forming the measurement volume. More about this aspect will be given in Sect. 14.9.

In the following sections, some other related optical aspects in LDA measurements will be further accounted for.

14.6 Astigmatism at the Focused Laser Beam Bundle

As is known, astigmatism represents an optical aberration and refers in the present context to the loss of the unique focal point of a light bundle after the refraction through a non-perpendicular interface (Fig. 14.6). Since each individual laser beam in an LDA-system is precisely a focused light bundle, its refraction on the medium interface then leads to the occurrence of astigmatism. As a result each refracted laser beam exhibits two characteristic focal points: one in the meridian plane and another in the sagittal plane. Because these two focal points in a laser beam do not coincide, the wave front of the laser beam is thus nowhere of the plane form. The intersection of two such laser beams obviously leads to fringe distortion in the measurement volume. This type of the fringe distortion, however, cannot be simply described as the well-known fringe distortions caused by the improper crossing of two regular laser beams (Chap. 16). The attempt by Li and Tieu (1998) to get the insight into the measurement volume that is distorted by reasons of astigmatism did not refer to the astigmatism associated with the off-axis LDA as presented here.

The extent of the fringe distortion caused by astigmatism in the present case clearly depends on the relative positions of the measurement volume and the respective meridian and sagittal focal points of two laser beams. It depends therefore on the off-axis angle in the individual LDA arrangement. For this reason, the laser beams that suffer from astigmatism should be characterized in function of LDA off-axis angle. Corresponding detailed investigations have been performed by Zhang and Eisele (1996b).

14.6.1 One-time Refraction of a Focused Beam Bundle

A spatially focused beam bundle is considered which is refracted at the interface of mediums 1 and 2 (Fig. 14.10). For this configuration, a coordinate system is chosen

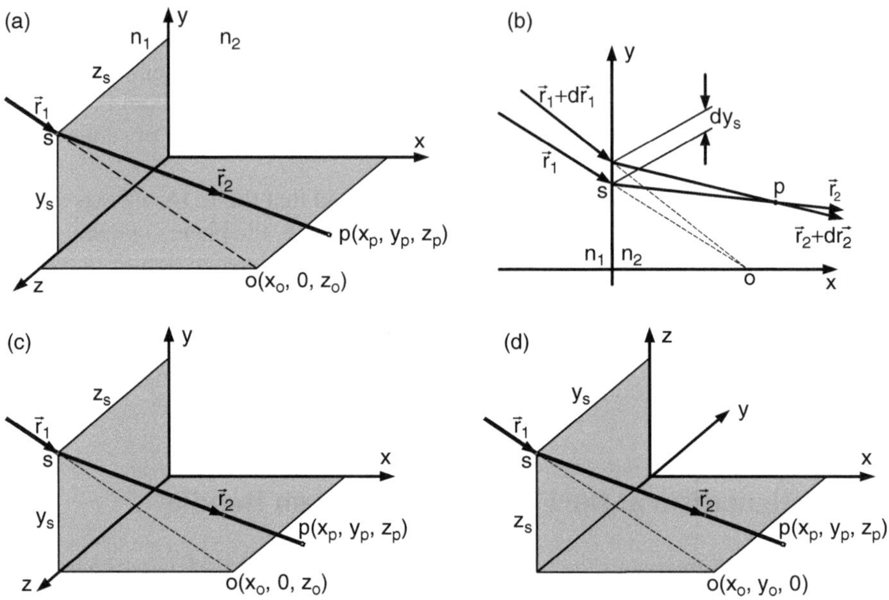

Fig. 14.10 Ray optics to calculate the astigmatism in a focused light bundle

with the $y - z$ plane on the interface between two mediums. The x-axis is then directed from medium 1 to medium 2. The virtual focal point of the incident beam bundle lies on the $x - z$ plane and is thus given at $o = (x_o, 0, z_o)$. In accordance with definitions of astigmatism parameters, two characteristic planes related to the incident beam should be noted: the plane containing the beam axis and the normal of the interface is called the meridian plane; the plane perpendicular to this and through the beam axis is called the sagittal plane.

To simplify the geometric optical calculations the beam bundle will be firstly considered as a single ray which may be described by a straight line with the parallel unit vector $\vec{r}_1 \left(r_{1x}, r_{1y}, r_{1z}\right)$ as follows

$$\frac{x-0}{r_{1x}} = \frac{y-y_s}{r_{1y}} = \frac{z-z_s}{r_{1z}} \tag{14.30}$$

with $s\,(0, y_s, z_s)$ as the intersection point of the straight line on the interface.

The coordinates of the intersection point s can be determined by inserting the coordinates of the known virtual focal point of the incident beam $o = (x_o, 0, z_o)$ into Eq. (14.30). For y_s for instance one obtains

$$y_s = -\frac{r_{1y}}{r_{1x}} x_o \tag{14.31}$$

Analogous to Eq. (14.30) the refracted ray that departs from the intersection point s can be represented by a straight line with $\vec{r}_2\left(r_{2x},\ r_{2y},\ r_{2z}\right)$ as a parallel unit vector

$$\frac{x-0}{r_{2x}}=\frac{y-y_s}{r_{2y}}=\frac{z-z_s}{r_{2z}} \tag{14.32}$$

The unit vector $\vec{r}_2$ depends on the unit vector $\vec{r}_1$ and can be determined by use of the law of refraction according to Eqs. (3.8) and (3.9):

$$r_{2y}=\frac{n_1}{n_2}\cdot r_{1y} \tag{14.33}$$

$$r_{2z}=\frac{n_1}{n_2}\cdot r_{1z} \tag{14.34}$$

with n_1 and n_2 as the refractive index of respective mediums.

The objective of the following calculations is to find out the focal points of the refracted beam for a given incident beam bundle that is focused at $o=(x_o,0,z_o)$. The form of the considered focused beam bundle can be imagined to be created by infinitesimally turning the incident ray $\vec{r}_1$ around the virtual focal point o, so that in the plane of the medium interface ($y-z$ plane) the intersection point s shifts. A certain infinitesimal change $d\vec{r}_1$ in the incident ray will thus cause an infinitesimal change $d\vec{r}_2$ in the refracted ray $\vec{r}_2$ and an infinitesimal shift of the intersection point s in the plane of the medium interface. Usually both the ray $\vec{r}_2$ and $\vec{r}_2+d\vec{r}_2$ propagate spatially and thus do not meet each other. By projecting the rays onto the $x-y$ plane, the intersection between two refracted rays can be confirmed, as shown in Fig. 14.10b. The related infinitesimal shift of the intersection point s in the plane of the medium interface is identified by the component dy_s.

In reality, the intersection of the projected rays $\vec{r}_2$ and $\vec{r}_2+d\vec{r}_2$ just represents a focal point of the simulated beam bundle after the refraction, however, conditionally for the given change $d\vec{r}_1$ in the incident ray and in the section plane that is parallel to the $x-y$ plane. Because the rays $\vec{r}_2$ and $\vec{r}_2+d\vec{r}_2$ generally do not spatially intersect, the observed intersection between them obviously depends on the orientation of the projection plane. This means that no unique focal point could be expected. This phenomenon with the loss of the unique focal point in the refracted beam bundle, as outlined, is known as astigmatism. In order to mathematically specify this type of optical aberrations, the intersection point of $\vec{r}_2$ and $\vec{r}_2+d\vec{r}_2$ is mathematically interpreted as to be found at the coordinate x_p, at which the refracted ray $\vec{r}_2$ does not show any change in the y-coordinate for instance in the projected $x-y$ plane (Fig. 14.10b), despite the infinitesimal change $d\vec{r}_1$ in the incident ray. For simplicity, the change in the incident ray is firstly assumed to be dr_{1y} i.e. in the y-component of the unit vector $\vec{r}_1$. The corresponding focal point at x_p on the refracted beam, considered in the projected $x-y$ plane, then has to fulfil the condition

$$\frac{\partial y_p}{\partial r_{1y}}=0 \tag{14.35}$$

In further derivations, the z-component of the unit vector $\vec{r}_1$ will be assumed to be constant. The change in r_{1y} will then cause a simultaneous change in r_{1x}. On the side of the refracted ray which is described by Eq. (14.32) there will be both a change in the unit vector $\vec{r}_2$ and a displacement of the ray because of the displacement of the intersection point s in the plane of the medium interface. With regard to this behaviour of ray optics the corresponding relation $y = f(x)$ from Eq. (14.32) is inserted into Eq. (14.35), which leads to

$$\frac{\partial y_s}{\partial r_{1y}} + x_p \frac{\partial}{\partial r_{1y}} \left(\frac{r_{2y}}{r_{2x}} \right) = 0 \tag{14.36}$$

The differentiation $\partial y_s / \partial r_{1y}$ can be calculated from Eq. (14.31) which applies to the incident ray. With regard to $r_{1z} = \text{const}$ it follows then

$$\frac{\partial y_s}{\partial r_{1y}} = -x_o \frac{\partial}{\partial r_{1y}} \left(\frac{r_{1y}}{\sqrt{1 - r_{1y}^2 - r_{1z}^2}} \right) = -x_o \frac{1 - r_{1z}^2}{r_{1x}^3} \tag{14.37}$$

Regarding the second term in Eq. (14.36) the following expression with regard to Eq. (14.33) is considered first

$$\frac{r_{2y}}{r_{2x}} = \frac{n_1}{n_2} \frac{r_{1y}}{\sqrt{1 - (n_1 / n_2)^2 r_{1y}^2 - (n_1 / n_2)^2 r_{1z}^2}} \tag{14.38}$$

Its differentiation leads to

$$\frac{\partial}{\partial r_{1y}} \left(\frac{r_{2y}}{r_{2x}} \right) = \frac{n_1}{n_2} \frac{1 - r_{2z}^2}{r_{2x}^3} \tag{14.39}$$

Both Eqs. (14.37) and (14.39) are inserted into Eq. (14.36). Then the x-coordinate, at which the geometrical condition given by Eq. (14.35) is fulfilled, is resolved as

$$x_p = x_o \frac{n_2}{n_1} \frac{r_{2x}^3}{r_{1x}^3} \frac{1 - r_{1z}^2}{1 - r_{2z}^2} \tag{14.40}$$

This is the position on the refracted ray, within which there is no change in y-coordinate despite the given change in the incident ray. The related point at x_p can thus be seen as a focal point of the refracted laser beam bundle, however, conditionally in the section plane which is parallel to $x - y$ plane because of Eq. (14.35).

As can also be seen from Eq. (14.40), x_p obviously depends on the spatial orientation of the incident beam bundle, or in other words, in the specification of used coordinate system. For the case where the meridian plane of the incident beam i.e. the plane containing the unit vector $\vec{r}_1$ and a parallel of x-axis, coincides with or is parallel to the $x - y$ plane (Fig. 14.10c), then there is simply $r_{1z} = r_{2z} = 0$.

Equation (14.35) represents the condition of a focal point in the meridian plane of the refracted beam bundle. From Eq. (14.40) the corresponding focal point on the refracted beam bundle is then given as

$$x_{\mathrm{m}} = x_{\mathrm{o}} \frac{n_2}{n_1} \frac{r_{2\mathrm{x}}^3}{r_{1\mathrm{x}}^3} \tag{14.41}$$

A second case will be considered when departing from the first case the coordinate system is rotated around the x-axis by 90° (Fig. 14.10d). Then Eq. (14.35) represents a focusing condition in the sagittal plane of the refracted beam bundle. Calculations given above are hence available for the ray optics in the sagittal plane. Because of $r_{1\mathrm{y}} = r_{2\mathrm{y}} = 0$ i.e. $1 - r_{1\mathrm{z}}^2 = r_{1\mathrm{x}}^2$ and $1 - r_{2\mathrm{z}}^2 = r_{2\mathrm{x}}^2$ one obtains from Eq. (14.40) the corresponding sagittal focal point on the refracted beam bundle as

$$x_{\mathrm{s}} = x_{\mathrm{o}} \frac{n_2}{n_1} \frac{r_{2\mathrm{x}}}{r_{1\mathrm{x}}} \tag{14.42}$$

Obviously the meridian and sagittal focal points are found at different positions on the refracted beam bundle. The distance between them is given by

$$\Delta x_{\mathrm{m,s}} = x_{\mathrm{m}} - x_{\mathrm{s}} = x_{\mathrm{o}} \frac{n_2}{n_1} \left(\frac{r_{2\mathrm{x}}^2}{r_{1\mathrm{x}}^2} - 1 \right) \frac{r_{2\mathrm{x}}}{r_{1\mathrm{x}}} \tag{14.43}$$

In the conventional specification of astigmatism, the spatial distance between both focal points is called astigmatic difference. With $r_{2\mathrm{x}}$ as the projection of the unit vector $\vec{r}_2$ on the x-axis the astigmatic difference is calculated as

$$\Delta s_{\mathrm{m,s}} = \frac{x_{\mathrm{m}} - x_{\mathrm{s}}}{r_{2\mathrm{x}}} = x_{\mathrm{o}} \frac{n_2}{n_1} \frac{r_{2\mathrm{x}}^2 - r_{1\mathrm{x}}^2}{r_{1\mathrm{x}}^3} \tag{14.44}$$

The unit vector components $r_{1\mathrm{x}}$ and $r_{2\mathrm{x}}$ in this equation can be expressed by $r_{1\mathrm{x}} = \cos\varepsilon_1$ and $r_{2\mathrm{x}} = \cos\varepsilon_2$, respectively, with ε_1 and ε_2 as the incident and the refraction angles of the beam bundle. Equation (14.43) then becomes

$$\Delta x_{\mathrm{m,s}} = x_{\mathrm{o}} \frac{n_2}{n_1} \left(\frac{\cos^2\varepsilon_2}{\cos^2\varepsilon_1} - 1 \right) \frac{\cos\varepsilon_2}{\cos\varepsilon_1} \tag{14.45}$$

Because the two focal points respectively on the meridian and sagittal planes no longer occur at the same place, there exists at each focal point indeed a focal line. This focal line changes its orientation as observed from one focal point (x_{m}) to another (x_{s}). For a finite beam bundle like a laser beam in an LDA-system, the elliptical cross section of the beam between x_{m} and x_{s} should be observed. The existence of an elliptical cross section, instead of the single focal line, arises from the coma effect (see Sect. 14.3) and other third-order aberrations. The comatic aberration, however, is much smaller than that caused by astigmatism.

The results presented above are relevant for LDA measurements and measurement accuracies. The change in the cross section of each refracted laser beam from one to another focal point influences the uniformity of the fringe spacing in the measurement volume that is created by intersection of two laser beams. The extent of the fringe distortion in the measurement volume clearly depends on deviations of all focal points (x_m and x_s) from the measurement volume. Detailed calculations of such deviations will be presented in Sect. 14.7.

Equation (14.43) will be further considered. Substituting x_o from Eq. (14.42) yields

$$\Delta x_{m,s} = x_s \left(\frac{r_{2x}^2}{r_{1x}^2} - 1 \right) \tag{14.46}$$

Because of Eq. (3.12) this can also be written as

$$\Delta x_{m,s} = x_s \left(1 - \frac{n_1^2}{n_2^2} \right) \tan^2 \varepsilon_1 \tag{14.47}$$

The astigmatic difference has thus been shown to depend on both the incident angle of the light bundle and the position of the sagittal focal point in the medium 2. Sometimes it would be necessary to relate the astigmatic difference to the position of the meridian focal point. For this reason $x_s = x_m - \Delta x_{m,s}$ is inserted in above equation:

$$\Delta x_{m,s} = \left(x_m - \Delta x_{m,s} \right) \left(1 - \frac{n_1^2}{n_2^2} \right) \tan^2 \varepsilon_1 \tag{14.48}$$

Then the astigmatic difference, related to x_m, is resolved as

$$\frac{\Delta x_{m,s}}{x_m} = \frac{\left(1 - \frac{n_1^2}{n_2^2} \right) \tan^2 \varepsilon_1}{1 + \left(1 - \frac{n_1^2}{n_2^2} \right) \tan^2 \varepsilon_1} \tag{14.49}$$

For the given two mediums, the related astigmatic difference is thus only a function of the incident angle of the light bundle, as this dependence has been shown in Fig. 14.11 for $n_2/n_1 = 1.333$. At large incident angles of the beam bundle, the astigmatic difference becomes significant. If compared with Fig. 14.9, which stands for a macro light bundle comprising four laser beams and being aligned off-axis at angle φ_{LDA}, it is evident that the displacement or the astigmatic difference does not much depend on the light bundle thickness.

It should be mentioned that Eq. (14.47) and hence Eq. (14.49) can also be obtained from Eq. (14.21) through corresponding simplifications. In the present case with two mediums, one needs to set $d = 0$. In addition, it deals here with a thin beam

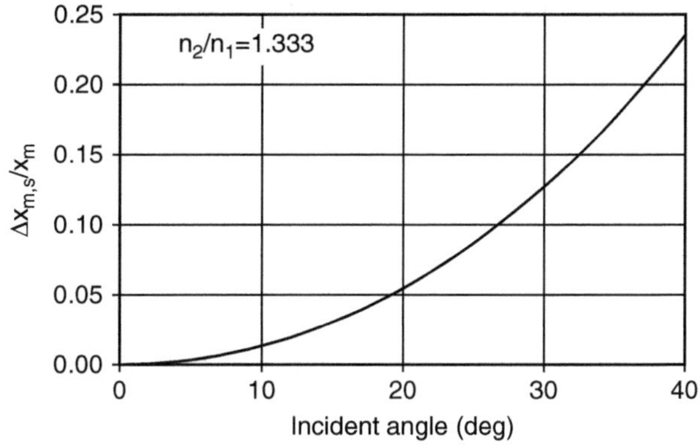

Fig. 14.11 Astigmatism difference in function of the incident angle of a focused light bundle

bundle, so that to the beam crossing angle in Eq. (14.21) the condition $\alpha_0 << 1$ has to be put to use. Corresponding verification calculations have been performed by Zhang (1995).

14.6.2 Multiple Refraction of a Focused Beam Bundle

Multiple refraction of a focused beam bundle is often encountered in LDA measurements of internal flows, at which all laser beams have always to pass through at least one glass window of thickness *d*. In this case, both focal points corresponding to the meridian and the sagittal planes of a beam bundle will be calculated in line with Fig. 14.12. In order to simplify the presentation, only the meridian plane of the beam bundle is sketched in Fig. 14.12. The initial i.e. virtual focal point of the incident beam is again found at $o\,(x_o, y_o)$. To calculate the focal point o_2 of the refracted beam bundle in the medium 2, the focal point o_1 of the refracted beam bundle in medium 1 has to be calculated first.

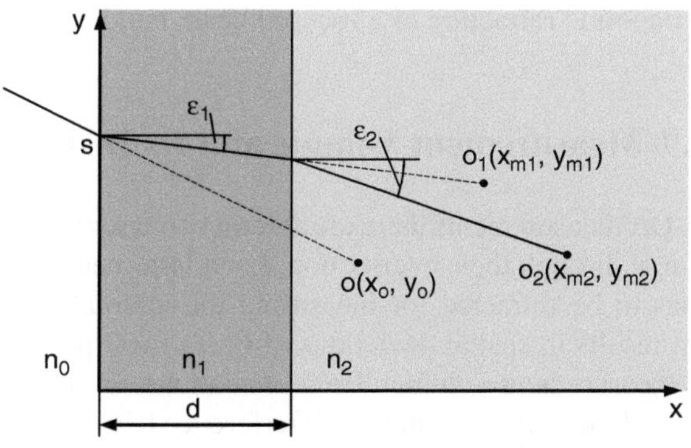

Fig. 14.12 Meridian plane of a focused light bundle with refractions in the medium 1 and 2

The first beam refraction takes place on the interface between mediums 0 and 1. From Eqs. (14.41) and (14.42) the meridian and sagittal focal points of the beam bundle in medium 1 are given at

$$x_{m1} = x_o \frac{n_1}{n_0} \frac{r_{1x}^3}{r_{0x}^3} \tag{14.50}$$

$$x_{s1} = x_o \frac{n_1}{n_0} \frac{r_{1x}}{r_{0x}} \tag{14.51}$$

In Fig. 14.12, only the focal point in the meridian plane has been shown by o_1. Both this point and the related sagittal focal point are virtual. They will be utilized to further calculate the focal points of the refracted beam bundle in the medium 2.

Equations (14.41) and (14.42) will again be applied to the beam that is refracted on the interface between mediums 1 and 2. With regard to the thickness d of the medium 1 one obtains

$$x_{m2} = (x_{m1} - d) \frac{n_2}{n_1} \frac{r_{2x}^3}{r_{1x}^3} + d \tag{14.52}$$

$$x_{s2} = (x_{s1} - d) \frac{n_2}{n_1} \frac{r_{2x}}{r_{1x}} + d \tag{14.53}$$

The displacement between two focal points in medium 2 is then given by

$$\begin{aligned} \Delta x_{m,s} &= x_{m2} - x_{s2} = (x_{m1} - d) \frac{n_2}{n_1} \frac{r_{2x}^3}{r_{1x}^3} - (x_{s1} - d) \frac{n_2}{n_1} \frac{r_{2x}}{r_{1x}} \\ &= x_o \frac{n_2}{n_0} \left(\frac{r_{2x}^2}{r_{0x}^2} - 1 \right) \frac{r_{2x}}{r_{0x}} - d \frac{n_2}{n_1} \left(\frac{r_{2x}^2}{r_{1x}^2} - 1 \right) \frac{r_{2x}}{r_{1x}} \end{aligned} \tag{14.54}$$

Corresponding components of unit vectors in this equation can be substituted by respective refraction angles for instance $r_{2x} = \cos \varepsilon_2$.

The same calculation procedure will be given, if a beam bundle is refracted more than two times. In particular, if $d = 0$, Eq. (14.54) takes the form of Eq. (14.43) for the one-time refraction of a focused beam bundle.

14.7 Measurement Volume and Its Distortion

In LDA measurements there are at least two laser beams that create the measurement volume through their intersection. Each laser beam will suffer from astigmatism, if it has to be refracted for measuring the internal flow. The existence of astigmatism results in spatial deviations of beam focal points (meridian and sagittal) from the measurement volume. These deviations act as the main reason for the distortion and the non-uniformity of interference fringes in the measurement volume. The possibility of calculating these deviations is therefore of great relevance for further investigations of optical properties of the measurement volume. The fringe distortion in the measurement volume in association with astigmatism indeed represents a new type that has not been widely investigated. Conventional fringe distortions in LDA measurement volume have been concerned to be merely caused by improper intersection of two normal laser beams (see Chap. 16).

In order to quantify the separations of diverse focal points as well as deviations of them from the measurement volume, calculation results in the last section dealing with the characterizations of astigmatism will be needed. For practical reasons a one-component LDA system with two laser beams at off-axis alignment will be considered. In order to ensure the intersection of two laser beams after refractions, the off-axis alignment of the LDA head has to be taken exactly in the plane containing both laser beams, as shown in Fig. 14.13. For the general case two-time refractions of laser beams will be firstly considered. The refracted laser beams in the first medium (often a glass window) are denoted by A_1 and B_1, while they are denoted in the test medium by A_2 and B_2, respectively.

Deviations of diverse focal points on the refracted laser beams from the measurement volume can be specified by firstly determining the measurement volume in the test medium. From the first interface (0–1) to the measurement volume in the medium 2, both laser beams (A and B) converge themselves by a lateral distance equal to t (Fig. 14.13a). With respect to this distance the following relation can be established

$$\begin{aligned} & x_{\mathrm{o}}\left(\tan\varepsilon_{\mathrm{A0}} - \tan\varepsilon_{\mathrm{B0}}\right) \\ &= \left[d\tan\varepsilon_{\mathrm{A1}} + (x_{\mathrm{mv}} - d)\tan\varepsilon_{\mathrm{A2}}\right] - \left[d\tan\varepsilon_{\mathrm{B1}} + (x_{\mathrm{mv}} - d)\tan\varepsilon_{\mathrm{B2}}\right] = t \end{aligned} \tag{14.55}$$

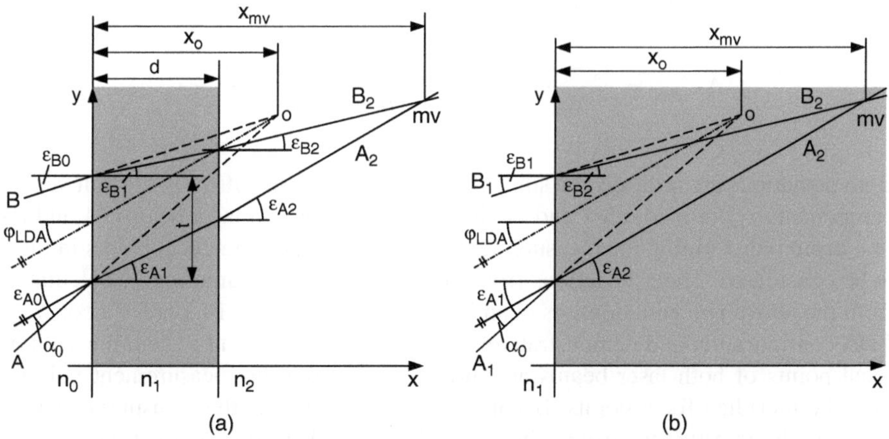

Fig. 14.13 Laser beam transmissions and the formation of the measurement volume (**a**) Two-time refractions as the general case (medium indices 0, 1, 2) (**b**) One-time refraction as a special case (medium indices 1, 2)

In relying on this basic equation, calculations will be given by taking into account the single and the multiple refractions of laser beams, respectively.

14.7.1 Single Refraction of Laser Beams

The occurrence of single or one-time refraction of laser beams, as shown in Fig. 14.13b, is encountered for instance in measuring an open channel flow or an internal flow whose refractive index is matched to that of the optical window. In such cases, Eq. (14.55) is simplified for $d = 0$ as

$$\frac{x_{\mathrm{mv}}}{x_{\mathrm{o}}} = \frac{\tan\varepsilon_{\mathrm{A1}} - \tan\varepsilon_{\mathrm{B1}}}{\tan\varepsilon_{\mathrm{A2}} - \tan\varepsilon_{\mathrm{B2}}} \tag{14.56}$$

Herein the order of two mediums is rearranged by using the subscripts 1 and 2.

For a given off-axis LDA alignment, the distance ratio $x_{\mathrm{mv}}/x_{\mathrm{o}}$ in the above equation is a constant. This implies that it also represents the ratio of the measurement volume shift to the shift of the virtual beam intersection, given by $\mathrm{d}x_{\mathrm{mv}}/\mathrm{d}x_{\mathrm{o}}$. This is equal to Eq. (14.17), as expected.

The above equation is, to eliminate x_{o}, combined with Eqs. (14.41) and (14.42), respectively. This leads to respective deviations of the meridian and sagittal focal points of a laser beam from the measurement volume as

$$\Delta x^*_{\mathrm{m,mv}} = \frac{x_{\mathrm{m}} - x_{\mathrm{mv}}}{x_{\mathrm{mv}}} = \frac{n_2}{n_1}\frac{r^3_{\mathrm{2x}}}{r^3_{\mathrm{1x}}}\frac{\tan\varepsilon_{\mathrm{A2}} - \tan\varepsilon_{\mathrm{B2}}}{\tan\varepsilon_{\mathrm{A1}} - \tan\varepsilon_{\mathrm{B1}}} - 1 \tag{14.57}$$

$$\Delta x^*_{\mathrm{s,mv}} = \frac{x_{\mathrm{s}} - x_{\mathrm{mv}}}{x_{\mathrm{mv}}} = \frac{n_2}{n_1}\frac{r_{\mathrm{2x}}}{r_{\mathrm{1x}}}\frac{\tan\varepsilon_{\mathrm{A2}} - \tan\varepsilon_{\mathrm{B2}}}{\tan\varepsilon_{\mathrm{A1}} - \tan\varepsilon_{\mathrm{B1}}} - 1 \tag{14.58}$$

Both deviations are referred to a single laser beam that is specified by the unit vector components r_{1x} in medium 1 and r_{2x} in medium 2. For two laser beams (*A* and *B*) of a beam pair forming the measurement volume there are four focal points in total to be considered. Their deviations from the measurement volume can be calculated from the above two equations, as quantitatively shown in Fig. 14.14a for a concrete LDA configuration and demonstratively in Fig. 14.14b. As can be seen, the sagittal focal points of both laser beams are further away from the measurement volume than the meridian focal points. Because of such deviations, the measurement volume and the interference fringes in it will clearly be distorted. At least the form of the measurement volume is no longer an exact ellipsoid. It remains, however, still unknown, how the fringes in the measurement volume will be distorted and how the measurement accuracies will be affected.

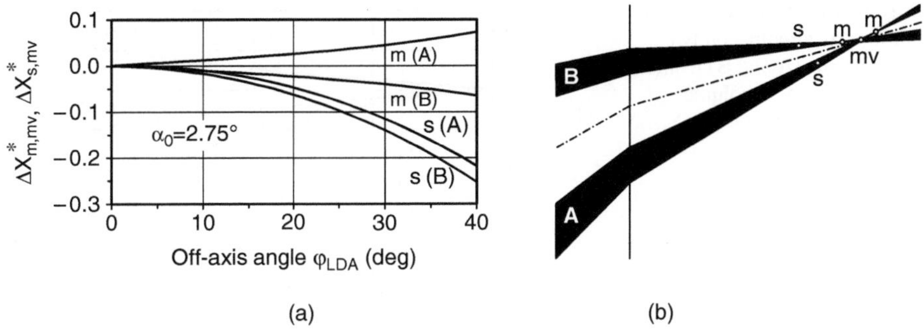

Fig. 14.14 Displacements of all four focal points on two laser beams (A and B) from the measurement volume (one-time refraction) (**a**) Quantitative calculations (**b**) Illustrations

14.7.2 Multiple Refractions of Laser Beams

Multiple refractions of laser beams are found in measuring internal flows where at least one glass window is present, as shown in Fig. 14.13a. Following the similar calculation procedure as that in Sect. 14.7.1, however, with the window thickness $d \neq 0$, Eq. (14.55) is combined with Eqs. (14.50) and (14.51) aiming to eliminate the initial focal point x_0. This leads to

$$x_{m1} = \frac{n_1}{n_0}\frac{r_{1x}^3}{r_{0x}^3}\frac{(\tan\varepsilon_{A1} - \tan\varepsilon_{B1})\,d - (\tan\varepsilon_{B2} - \tan\varepsilon_{A2})\,(x_{mv} - d)}{\tan\varepsilon_{A0} - \tan\varepsilon_{B0}} \tag{14.59}$$

$$x_{s1} = \frac{n_1}{n_0}\frac{r_{1x}}{r_{0x}}\frac{(\tan\varepsilon_{A1} - \tan\varepsilon_{B1})\,d - (\tan\varepsilon_{B2} - \tan\varepsilon_{A2})\,(x_{mv} - d)}{\tan\varepsilon_{A0} - \tan\varepsilon_{B0}} \tag{14.60}$$

Then from Eq. (14.52) as well as Eq. (14.53) the relative deviations of respective focal points from the measurement volume can be derived to be

$$\Delta x^*_{m,mv} = \frac{x_{m2} - x_{mv}}{x_{mv} - d}$$
$$= \frac{n_2}{n_0}\frac{r_{2x}^3}{r_{0x}^3}\frac{\tan\varepsilon_{A2} - \tan\varepsilon_{B2}}{\tan\varepsilon_{A0} - \tan\varepsilon_{B0}} - 1 + \left(\frac{n_2}{n_0}\frac{r_{2x}^3}{r_{0x}^3}\frac{\tan\varepsilon_{A1} - \tan\varepsilon_{B1}}{\tan\varepsilon_{A0} - \tan\varepsilon_{B0}} - \frac{n_2}{n_1}\frac{r_{2x}^3}{r_{1x}^3}\right)\frac{d}{x_{mv} - d} \tag{14.61}$$

$$\Delta x^*_{s,mv} = \frac{x_{s2} - x_{mv}}{x_{mv} - d}$$
$$= \frac{n_2}{n_0}\frac{r_{2x}}{r_{0x}}\frac{\tan\varepsilon_{A2} - \tan\varepsilon_{B2}}{\tan\varepsilon_{A0} - \tan\varepsilon_{B0}} - 1 + \left[\frac{n_2}{n_0}\frac{r_{2x}}{r_{0x}}\frac{\tan\varepsilon_{A1} - \tan\varepsilon_{B1}}{\tan\varepsilon_{A0} - \tan\varepsilon_{B0}} - \frac{n_2}{n_1}\frac{r_{2x}}{r_{1x}}\right]\frac{d}{x_{mv} - d} \tag{14.62}$$

with $x_{mv} - d$ as the depth of the measurement volume in the test medium.

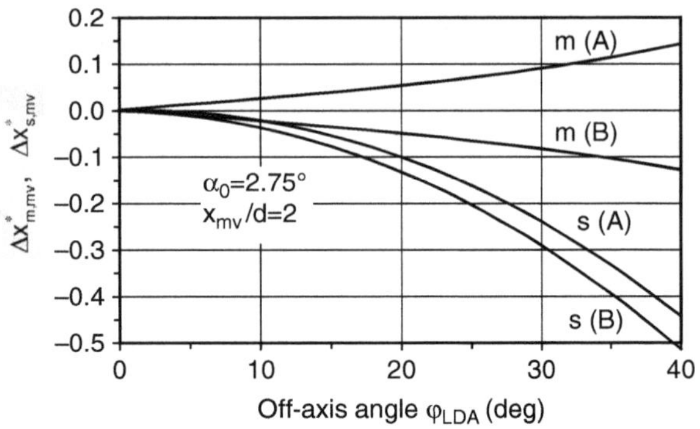

Fig. 14.15 Displacements of all four focal points on two laser beams (A and B) from the measurement volume (two-time refractions)

All deviations calculated in the current case of multiple beam refractions depend on the off-axis angle of the LDA head as well as on the depth of the measurement volume in the test fluid. In the case of very small window thickness ($d/x_{mv} << 1$), Eqs. (14.61) and (14.62) can be simplified, leading to Eqs. (14.57) and (14.58), respectively, for one-time refractions.

Taking into account both laser beams forming the measurement volume, the deviations of all four focal points from the measurement volume are presented in Fig. 14.15 for a concrete LDA configuration with $\alpha_0 = 2.75°$ and $x_{mv}/d = 2$. They are similar to those in Fig. 14.14. In particular, it has been again confirmed that the sagittal focal points of both laser beams are further away from the measurement volume than the meridian focal points. For the same reason the distortion of the measurement volume will be expected.

14.7.3 Astigmatism at the On-axis LDA Alignment

In most cases of practical applications, LDA measurements of channel flows are carried out with on-axis alignment of the LDA head ($\varphi_{LDA} = 0$). Both laser beams suffers from the same effect of astigmatism, however, only in relation to the small beam crossing angle ($2\alpha_0$). Corresponding deviations of laser focal points from the measurement volume can be calculated for instance for the case of $d/x_{mv} << 1$ by Eqs. (14.57) and (14.58) with respective incident and refraction angles of two laser beams. In assuming $\alpha_0 = 2.75°$ and the refractive index ratio $n_1/n_0 = 1.333$, one obtains

$$\Delta x^*_{m,mv} = 0.001 \tag{14.63}$$

$$\Delta x^*_{s,mv} = 0 \tag{14.64}$$

This result signifies that no distortion of the measurement volume takes place. The best optical condition for LDA measurements is reached in this way (see Chap. 13).

14.8 Signal Qualities and the Lens Dependence

14.8.1 Deterioration of Signal Qualities and Strengths

It can be inferred from Fig. 14.6 that in the case of occurring astigmatism, each elementary segment on the front lens of the LDA head has its individual focal point in the medium. The measurement volume that is formed by two laser beams in the meridian plane obviously coincides with the focal points of only a few such elementary segments and thus cannot be seen by others. The scattered laser light from the measurement volume therefore cannot be efficiently detected by the receiving unit. The direct consequence is that the velocity signal rate rapidly decreases with the LDA off-axis angle. In this respect, the influence of astigmatism on the signal quality is rooted in the receiving optics rather than in the transmitting optics, as this is illustrated in Fig. 14.16. The light scattered from the measurement volume backward propagates along the optical axis through the medium interface. It is initially a beam bundle with a unique focal point at *mv* i.e. the measurement volume. After passing through the medium interface the light in the meridian plane has its virtual focal point at o_m, which is the same as that of the transmitting optics. The light in the sagittal plane, however, shows its virtual focal point o_s which does not coincide with o_m. This phenomenon exactly demonstrates the existence of astigmatism in the receiving optics. The distance between two virtual focal points is again called the astigmatic difference. Because the scattered light after passing trough the medium interface is no longer a beam bundle with unique focal point, it cannot be efficiently collected to the detector for further signal processing. Both the signal

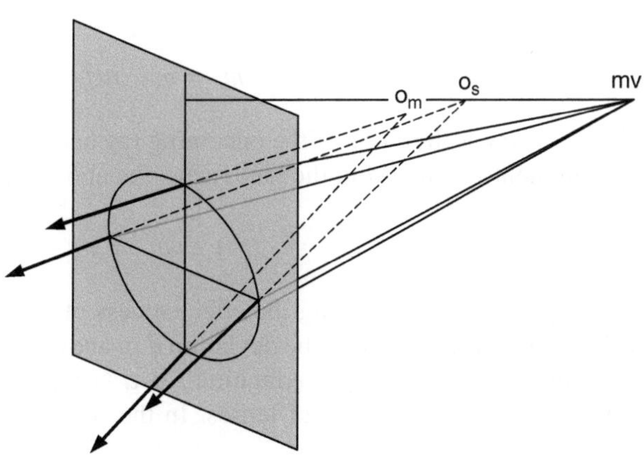

Fig. 14.16 Astigmatism in the receiving optics

quality and the strength will thus be considerably affected. As a result the reduction of the signal rate will come about.

14.8.2 Lens Dependence of Signal Qualities and Strengths

The existence of astigmatism in receiving optics deteriorates both the qualities and the strengths of light signals and hence leads to a reduction of the sampling rate in measurements. In a detailed investigation carried out by Zhang and Eisele (1996a, 1998b), it has been demonstrated that the reduction of the sampling rate in off-axis LDA measurements additionally and sensitively depends on the lens focal length specified at the LDA-head.

As known and explained in Sect. 4.2, the use of an LDA front lens with short focal length (f) will enhance the brightness of the measurement volume and consequently the strength of the scattered laser light. It is also known that this front lens as the receiving optics in turn has a large aperture to the measurement volume (proportional to $1/f^2$). For these two reasons the light signal received is then much stronger than that received by using a lens with a long focal length. A much high sampling rate in LDA measurements may usually be expected. This advantage of receiving optics with short focal length is lost, however, if the receiving optics is aligned off-axis, as in the case treated here. This can be demonstrated by employing the ray optics with respect to two lenses with different focal lengths ($f_1 < f_2$).

In order to model the astigmatism in LDA measurements, it is assumed, according to Fig. 14.7a, that the measurement volume (mv) is located at the distance s to the left of the lens and does not coincide with the focus of the lens. This modelling of astigmatism is based on the demonstration in the last section that hardly any elementary segments on the receiving lens have their focal points being coincident with the measurement volume. Because of the distance s the scattered laser light from the measurement volume will not be parallel to the lens axis after passing the lens. The slopes of the light rays before and after the lens are denoted by m and m', respectively. From the geometric optics the relationship between oncoming and departing light rays is given by

$$m' = m - h/f \tag{14.65}$$

With $m = h/s$ as the slope of the oncoming ray and $d = s - f$ as the deviation of the measurement volume from the lens focus, the above equation is rearranged to

$$f \cdot (f + d)\, m' = -h \cdot d \tag{14.66}$$

For the purpose of comparing the effectiveness of two changeable receiving lenses with different focal lengths, the deviation d in above equation should be considered. In the first instance, the approximation $f + d \approx f$ can be made because of $d << f_1$ and $d << f_2$ in both cases of lenses. In the second instance, the deviation d can be assumed to be constant i.e. $d_1 \approx d_2$. This assumption relies on the fact that the astigmatic difference is nearby independent of the light bundle thickness, as this can be verified by comparing Fig. 14.11 as being calculated for a thin light bundle with Fig. 14.9 as being calculated for a thick light bundle which comprises four laser beams. The nearby constant astigmatic difference points out that the same astigmatic difference will be given when using a lens with another focal length. With respect to these two aspects of the distance d Eq. (14.66) can be applied in comparison of optical effectiveness of two receiving lenses with different focal lengths. It yields then from Eq. (14.66) with $h = \text{const}$

$$f_1^2 m_1' = f_2^2 m_2' = -d \cdot h \tag{14.67}$$

This equation is available for comparison between corresponding segments (condition $h_1 = h_2 = h$) on both lenses. As known, the non-zero slope of the light ray after a lens segment implies that this light will not be efficiently collimated to the small aperture in the receiving optics and the corresponding segment on the lens behaves as blind to the measurement volume. Obviously the lens with short focal length ($f_1 < f_2$) behaves as more blind ($m_1' > m_2'$) to the measurement volume than the lens with long focal length, as shown in Fig. 14.17b. For achieving better optical conditions in off-axis LDA measurements, the applying of the lenses with long focal lengths is therefore suggested.

Analogous to Eq. (14.67) another comparison with regard to the slopes of both departing rays for equal oncoming rays ($m_1 = m_2 = m$) can be made. Combining

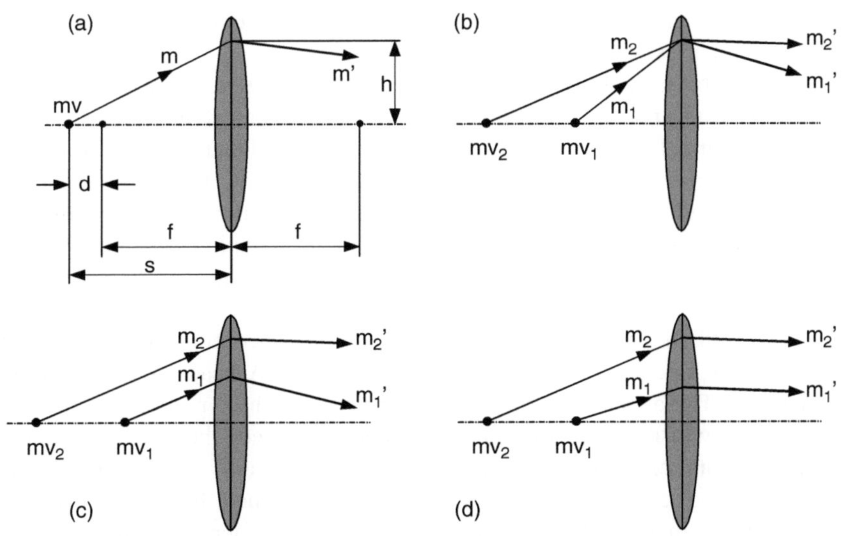

Fig. 14.17 Effects of the lens focal length on the signal transmission performance (quality and strength of signals) in the receiving optics

$m = h/s$ and $d = s - f$ yields $h = (d + f)\,m$ which is again inserted into Eq. (14.65). One obtains then

$$f_1 m_1' = f_2 m_2' = -md \tag{14.68}$$

Since $m_1 = m_2$ represents the same aperture of the lens segments to the measurement volume, Eq. (14.68) implies that for the same amount of incident light and thus of incident light energy (Fig. 14.17c) the inclination of departure rays is inversely proportional to the lens focal length. The lens with the short focal length ($f_1 < f_2$) is thus comparatively inefficient ($m_1' > m_2'$) to conduct the scattered laser light from the measurement volume to be received by the receiving optics. It again suggests applying the lenses with long focal length.

A third comparison between lenses with different focal lengths can be made from Eq. (14.65) if equal slopes of departure rays from the lenses ($m_1' = m_2' = m'$) are considered. Combining $m = h/s$ and $d = s - f$ yields $m = h/(d + f)$ which is again inserted into Eq. (14.65). One obtains then with $f + d \approx f$

$$\frac{h_1}{f_1^2} = \frac{h_2}{f_2^2} = -\frac{m'}{d} \tag{14.69}$$

This equation shows that for detecting the departure lights whose inclinations are below a given value $|m'|$ the available lens height (h) is proportional to the square of the lens focal length. This means that the lens with a longer focal length ($f_2 > f_1$) also has a large available lens surface ($h_2 > h_1$) and thus a large aperture to the measurement volume, as shown in Fig. 14.17d. This circumstance again suggests the use of the lenses with long focal length.

Based on above analyses it can be concluded that at a given off-axis angle of an LDA head and for the given position of the measurement volume in the test medium, an LDA receiving lens with longer focal length is preferred in use instead of the lenses with short focal length. This ensures the effective conduction of scattered laser light from the measurement volume to the receiving optics and consequently the obtaining of signals of better qualities. The rapid reduction of velocity sampling rate because of the occurrence of astigmatism can then be successfully suppressed.

Figure 14.18 shows an experimental verification concerning the velocity sampling rate as a function of the off-axis angle of an LDA head with two different focal lengths. The appearance of astigmatism has been here realized by means of a plane window of Plexiglas (40 mm of thickness) between the LDA head and the flow. The orientation of the plane window to the optical axis was changed, so that a great number of cases with astigmatism could be obtained. It can be clearly seen from Fig. 14.18 that the data sampling rate by using the lens with the long focal length ($f = 400$ mm) has hardly changed with the LDA off-axis angle up to 20°. In contrast to this, the data sampling rate by using the lens with the short focal length ($f = 160$ mm) rapidly decreases, as the LDA off-axis angle increases.

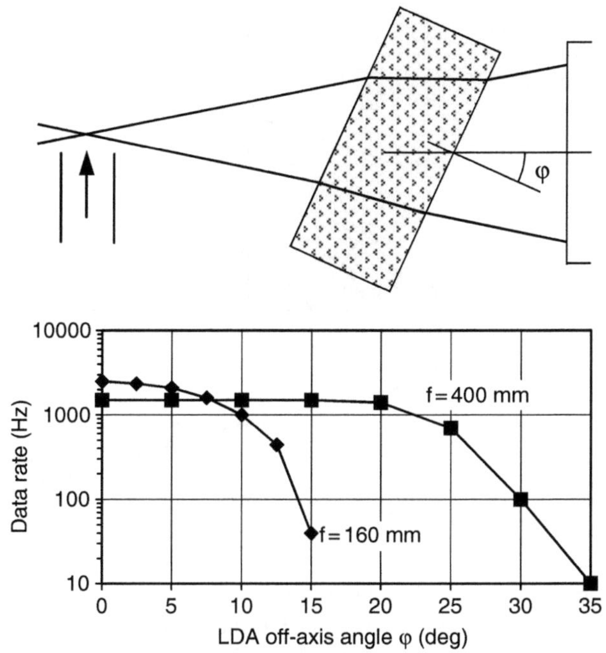

Fig. 14.18 Data rate reduction at off-axis LDA measurements and the dependence on the lens focal length

Another possibility to compensate for astigmatism and its influence on the signal strength and qualities is to use a water-filled prism if water flow is measured. Details in use of such a prism and its calculation will be given in Sect. 14.10.

14.9 Error Sensitivities in Forming the Measurement Volume

14.9.1 Beam Separation in the Test Medium

In the foregoing sections, the LDA off-axis alignment against the plane wall (i.e. the medium interface) is assumed to be within the optical plane containing two laser beams. This is the way that ensures the laser beam propagation, after being refracted, within the same plane and thus leads to perfect laser beam intersection. Any deviation in arranging the off-axis LDA from this requirement, as the consequence of inaccurate mechanical support of LDA for instance, will cause the imperfect laser beam intersection or laser beam separation. This undesired outcome comes about because two refracted laser beams after being refracted do not propagate in a unique plane (Fig. 14.19). For the practical use as well as for the reference purpose, the possible beam separations arising from errors and inaccuracy in LDA off-axis alignment will be presented based on investigations carried out by Zhang and Eisele (1998b).

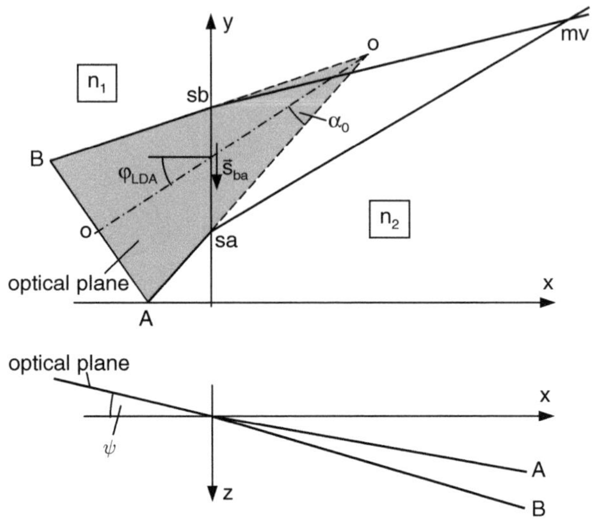

Fig. 14.19 Bias angle ψ and the separation of two laser beams after the refraction in the medium 2

The desired off-axis LDA alignment is given by arranging the LDA off-axis in the LDA optical plane. It can also be said that the normal of the medium interface lies in the optical plane. Any inaccurate LDA off-axis alignment can be considered as the combined outcome of the accurate off-axis alignment in the optical plane followed by

– tilting this plane by ψ (Fig. 14.19) and/or by
– rotating the optical plane about the optical axis by δ (Fig. 14.20).

These two bias angles are independent and represent two possible sources leading to separation between the normal of the medium interface and the LDA optical plane. As a result, no or an imperfect measurement volume can be formed. In fact,

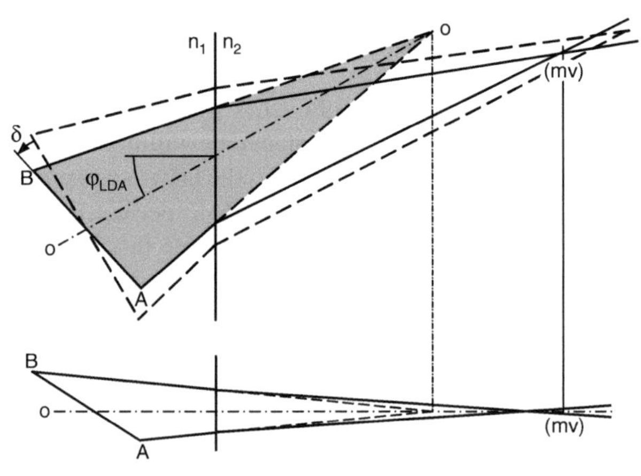

Fig. 14.20 Bias angle δ and the separation of two laser beams after the refraction in the medium 2

any improper LDA off-axis alignment can be expressed to be the additive effect of these two bias angles.

To simulate the improper off-axis LDA alignment for internal flow measurements, a one-component LDA head with two laser beams is assumed to generally have an arbitrary spatial orientation against the normal of the medium interface. In order to calculate the beam separation explicitly, both laser beams that are denoted by A and B, respectively, will be expressed by unit vectors $\vec{a}_1$ and $\vec{b}_1$ before and $\vec{a}_2$ and $\vec{b}_2$ after the refractions. As in the geometric optics, the unit vectors $\vec{a}_2$ and $\vec{b}_2$ can be determined from the unit vectors $\vec{a}_1$ and $\vec{b}_1$ by applying the law of refraction in the vector form according to Eqs. (3.8) and (3.9). For this reason, a coordinate system will be fixed at which the $y-z$ plane agrees with the medium interface (Fig. 14.19).

Two laser beams will be assumed to have their initial (i.e. virtual) focal point at $o\,(x_o, y_o, z_o)$. The intersection of laser beams with the medium interface ($y-z$ plane) at points *sa* and *sb*, respectively, is determined as follows:

$$x_{sa} = 0, \quad y_{sa} = y_o - \frac{a_{1y}}{a_{1x}} x_o, \quad z_{sa} = z_o - \frac{a_{1z}}{a_{1x}} x_o \tag{14.70}$$

$$x_{sb} = 0, \quad y_{sb} = y_o - \frac{b_{1y}}{b_{1x}} x_o, \quad z_{sb} = z_o - \frac{b_{1z}}{b_{1x}} x_o \tag{14.71}$$

Combining these two intersection points yields a new vector (from B to A) as

$$\vec{s}_{ba} = (0, y_{sa} - y_{sb}, z_{sa} - z_{sb}) \tag{14.72}$$

The arbitrary spatial orientation of the LDA head to the medium interface implies that the normal of the medium interface may not lie in the plane containing two laser beams (LDA optical plane). If this occurs, both refracted laser beams then no longer propagate in a common plane, so that no or imperfect intersection of two laser beams will be given. The separation between these two spatial laser beams is represented by the minimum distance in the direction perpendicular to both beam axes. The direction perpendicular to both laser beams is given by the unit vector $\left(\vec{a}_2 \times \vec{b}_2\right) / \left|\vec{a}_2 \times \vec{b}_2\right|$. The projection of the separation vector $\vec{s}_{ba}$ in this direction then gives the definite distance between both laser beams (plus and minus are available):

$$s = \frac{\vec{a}_2 \times \vec{b}_2}{\left|\vec{a}_2 \times \vec{b}_2\right|} \cdot \vec{s}_{ba} = \frac{n_1}{n_2} x_o \frac{a_{1y} b_{1z} - a_{1z} b_{1y}}{\left|\vec{a}_2 \times \vec{b}_2\right|} \cdot \left(\frac{a_{2x}}{a_{1x}} - \frac{b_{2x}}{b_{1x}} \right) \tag{14.73}$$

Because the expression $\left|\vec{a}_2 \times \vec{b}_2\right|$ represents the sine value of the angle between both laser beams with $\vec{a}_2$ and $\vec{b}_2$ as unit vectors, it can be found by

$$\left|\vec{a}_2 \times \vec{b}_2\right| = \sqrt{1 - \left(\vec{a}_2 \cdot \vec{b}_2\right)^2} = \sqrt{1 - \left(a_{2x} b_{2x} + a_{2y} b_{2y} + a_{2z} b_{2z}\right)^2} \tag{14.74}$$

The desired i.e. accurate off-axis LDA alignment is regarded to occur in the LDA optical plane. This means that, according to Fig. 14.19 with chosen coordinate system and $\psi = 0$, the LDA optical plane coincides with the $x - y$ plane. Because of $a_{1z} = 0$ and $b_{1z} = 0$ the distance s from Eq. (14.73) is zero. This corresponds to the case with perfect intersection of axes of two laser beams after being refracted in the medium 2. Deviations from this desired off-axis LDA alignment may be caused by inaccurate manipulation or the inaccurate mechanical support. As explained before, any deviation can be considered as the combined and additive effect of two bias angles ψ and δ. Because these two parameters are independent of each other, the beam separations respectively caused by these two bias angles can be separately calculated.

14.9.1.1 Beam Separation Due to the Bias Angle ψ

This case with the bias angle ψ as an error parameter corresponds to the case that has been shown in Fig. 14.19. Both incident laser beams can be represented by the unit vector $\vec{a}_1$ and $\vec{b}_1$, respectively, that are given by the following components:

$$a_{1x} = \cos(\varphi_{LDA} + \alpha_0) \cdot \cos\psi \tag{14.75}$$

$$a_{1y} = \sin(\varphi_{LDA} + \alpha_0) \tag{14.76}$$

$$a_{1z} = \cos(\varphi_{LDA} + \alpha_0) \cdot \sin\psi \tag{14.77}$$

$$b_{1x} = \cos(\varphi_{LDA} - \alpha_0) \cdot \cos\psi \tag{14.78}$$

$$b_{1y} = \sin(\varphi_{LDA} - \alpha_0) \tag{14.79}$$

$$b_{1z} = \cos(\varphi_{LDA} - \alpha_0) \cdot \sin\psi \tag{14.80}$$

In these equations, the constant α_0 denotes the half intersection angle between two laser beams.

With available incident beams given in above equations the spatial separation between two laser beams after refraction in the medium 2 can be calculated from Eq. (14.73). For the reason of compact writing form, only the expression $a_{1y}b_{1z} - a_{1z}b_{1y}$ in Eq. (14.73) is replaced by respective components listed above, leading to

$$\frac{s}{x_o} = \frac{n_1}{n_2} \frac{\sin\psi \sin 2\alpha_0}{\sqrt{1 - \left(a_{2x}b_{2x} + a_{2y}b_{2y} + a_{2z}b_{2z}\right)^2}} \left(\frac{a_{2x}}{a_{1x}} - \frac{b_{2x}}{b_{1x}}\right) \tag{14.81}$$

In this equation, all components of unit vectors representing the refracted laser beams in medium 2 can be calculated by applying the law of refraction that has already been presented by Eqs. (3.8) and (3.9). One then obtains

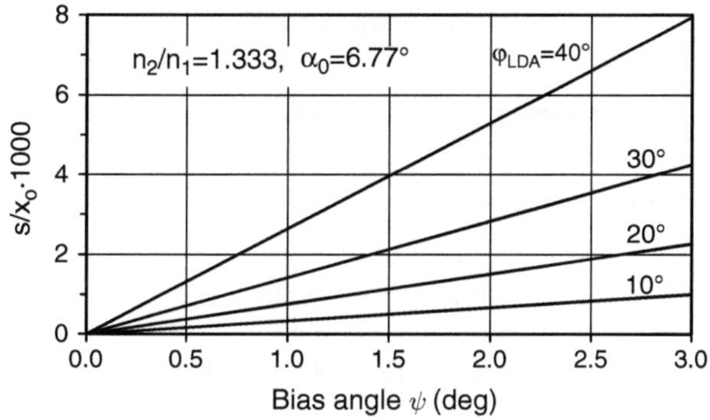

Fig. 14.21 Beam separation caused by the bias angle ψ

$$a_{2y} = \frac{n_1}{n_2} a_{1y}, \quad a_{2z} = \frac{n_1}{n_2} a_{1z}, \quad a_{2x} = \sqrt{1 - a_{2y}^2 - a_{2z}^2} \tag{14.82}$$

$$b_{2y} = \frac{n_1}{n_2} b_{1y}, \quad b_{2z} = \frac{n_1}{n_2} b_{1z}, \quad b_{2x} = \sqrt{1 - b_{2y}^2 - b_{2z}^2} \tag{14.83}$$

According to Eq. (14.81) the related separation between two refracted laser beams has been shown as a function of the off-axis angle of the LDA head φ_{LDA} and the bias angle ψ. For an example of calculations, Fig. 14.21 shows the beam separation in medium 2 (water) in an air-water system. This simplified system can be applied when the window thickness of an air-window-water system is negligible. The half intersection angle between two laser beams is taken as $\alpha_0 = 6.77°$. As it can be read out for instance with $\varphi_{\text{LDA}} = 30°$ and $\psi = 1°$, the resultant beam separation is given by $s/x_o = 1.4 \cdot 10^{-3}$. If the initial focal point is positioned by $x_o = 100$ mm, to which the measurement volume in the medium 2 is found at about $1.65x_o$ (Fig. 14.5), then the beam separation takes $s = 0.14$ mm. It should be mentioned that this beam separation may no longer be neglected in the practical LDA measurements because the thickness of the measurement volume itself is only in the order of 0.1 mm. The undesired outcome of the beam separation is the drop of velocity sampling rate and in the worst case the total loss of LDA signals.

14.9.1.2 Beam Separation Due to the Bias Angle δ

This case with the bias angle δ corresponds to the case that has been shown in Fig. 14.20. According to Appendix B, both incident laser beams can be again represented by the unit vector $\vec{a}_1$ and $\vec{b}_1$, respectively, that are given by the following components:

$$a_{1x} = \cos\varphi_{\text{LDA}} \cos\alpha_0 + \sin\varphi_{\text{LDA}} \sin\alpha_0 \cos\delta \tag{14.84}$$

$$a_{1y} = \sin\varphi_{\text{LDA}} \cos\alpha_0 - \cos\varphi_{\text{LDA}} \sin\alpha_0 \cos\delta \tag{14.85}$$

$$a_{1z} = \sin\alpha_0 \sin\delta \tag{14.86}$$

$$b_{1x} = \cos\varphi_{LDA}\cos\alpha_0 - \sin\varphi_{LDA}\sin\alpha_0\cos\delta \tag{14.87}$$

$$b_{1y} = \sin\varphi_{LDA}\cos\alpha_0 + \cos\varphi_{LDA}\sin\alpha_0\cos\delta \tag{14.88}$$

$$b_{1z} = -\sin\alpha_0\sin\delta \tag{14.89}$$

Substituting respective expressions in Eq. (14.73) yields

$$\frac{s}{x_o} = -\frac{n_1}{n_2}\frac{\sin\varphi_{LDA}\sin 2\alpha_0\sin\delta}{\sqrt{1-\left(a_{2x}b_{2x}+a_{2y}b_{2y}+a_{2z}b_{2z}\right)^2}}\left(\frac{a_{2x}}{a_{1x}}-\frac{b_{2x}}{b_{1x}}\right) \tag{14.90}$$

In this equation, all components of unit vectors representing the refracted laser beams in medium 2 can be calculated by Eqs. (14.82) and (14.83).

From the above equation the resultant separation between both refracted laser beams has been shown as a function of the off-axis angle of the LDA head φ_{LDA} and the bias angle δ. For an example of calculations, Fig. 14.22 shows the beam separation in medium 2 (water) in a simplified air-water system. The half intersection angle between two laser beams is again taken as $\alpha_0 = 6.77°$. As it can be read out for instance with $\varphi_{LDA} = 30°$ and $\delta = 1°$ typically, the resultant beam separation is $s/x_o = 0.7 \cdot 10^{-3}$. If the initial focal point is positioned by $x_o = 100$ mm, to which the measurement volume in the medium 2 is found at about $1.65x_o$, then the beam separation takes $s = 0.07$ mm. This value has also to be considered as sufficiently large if compared with the measurement volume whose thickness is only about 0.1 mm. As said before, the undesired outcome of the beam separation is the drop of velocity sampling rate or the total loss of LDA signals.

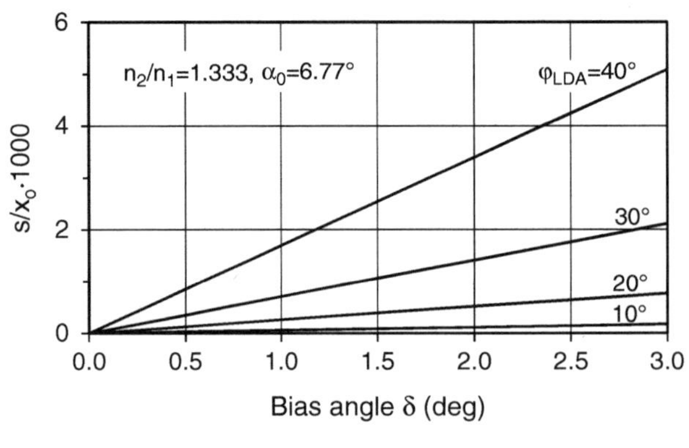

Fig. 14.22 Beam separation caused by the bias angle δ

14.9.1.3 Statement Regarding the Inaccurate Laser Beam Alignment

Calculations made above and shown in Figs. 14.21 and 14.22 clearly pointed out that the uncertainty in forming the measurement volume is very sensitive to the imperfect off-axis LDA alignment, especially at large off-axis angles. The resultant beam separation implies that either no measurement is possible (for large beam separation) or the geometrical size of the measurement volume decreases. The signal rate will thus be much affected. This knowledge undoubtedly suggests that the off-axis LDA alignment has to be accurately accomplished, so that the laser beams after refractions do propagate in the same optical plane containing two incident beams. It is, however, usually not necessary to exactly know the error i.e. the beam separation s while forming the measurement volume. In addition, both bias angles (ψ and δ) are usually not available.

If possible, LDA alignment at large off-axis angles should be avoided. This is recommended also in concerning the rapid reduction of the signal rate because of astigmatism, as already treated in Sect. 14.8.

In dealing with multiple refractions e.g. in an air-window-water system, calculations of the beam separation will be given in the next section.

14.9.2 Beam Separation After Multiple Refractions

Most LDA applications to internal flow measurements are characterized by at least two-time refraction of each laser beam, for instance at medium interface air-glass and glass-water. Analogous to the derivation of Eq. (14.73) the beam separation after passing through a plane layer, say a glass window of thickness d, can be calculated as (Zhang and Eisele 1998b)

$$\frac{s}{x_o} = \frac{n_0}{n_2}\frac{a_{0y}b_{0z} - a_{0z}b_{0y}}{\sqrt{1-\left(a_{2x}b_{2x}+a_{2y}b_{2y}+a_{2z}b_{2z}\right)^2}}\left[\left(\frac{a_{2x}}{a_{0x}} - \frac{b_{2x}}{b_{0x}}\right) - \frac{d}{x_o}\frac{n_0}{n_1}\left(\frac{a_{2x}}{a_{1x}} - \frac{b_{2x}}{b_{1x}}\right)\right] \tag{14.91}$$

Herein the indices 0, 1 and 2 are commonly rearranged to the medium air, glass and water. x_o indicates the location of the virtual focal point of laser beams. The origin of x-axis lies on the first refraction surface.

14.9.3 Possible Impact on PDA Measurements

As stated, LDA is a method for flow velocity measurements. It operates on the wave theory and makes use of the Doppler effect in the light that is scattered by a moving particle in the flow. Based on the LDA principle, the LDA method has also been extended to measure the particle size. The extended method makes use of the phase difference in the scattered laser lights that are detected from two spatially different directions. Because this phase difference is proportional to the size

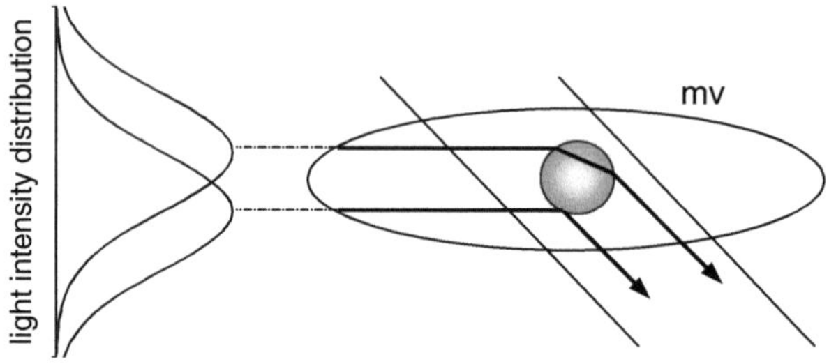

Fig. 14.23 Gaussian beam effect in PDA measurements as the consequence of the imperfect laser beam intersection

of the particle that scatters the laser light, the particle size can be calculated from the measured phase difference. It again deals with a method that does not need any calibration. This extended LDA for particle size measurements is known as the Phase Doppler Anemometry (PDA). It contains a transmitting optics that is the same as that in the LDA method and a separate receiving optics. Detailed descriptions of the functionality of the PDA method can be found for instance by Albrecht et al. (2003).

The analyses of the outcome of errors in LDA alignment, as described above with regard to the LDA off-axis angle φ_{LDA} and two bias angles ψ and δ, also apply to the transmitting optics of the PDA method in measuring the particle size in the flow. Because of the beam separation, the Gaussian distributions of the light intensity in both laser beams may no longer be coincident in the measurement volume, as illustrated in Fig. 14.23. It is then quite possible that the light of one laser beam is mainly reflected by a particle, whereas the light of another laser beam is mainly refracted by the same particle. The superposition and detection of two such different types of scattered laser light will lead to erroneous interpretation of the particle size. In addition, only large particles have a high probability of simultaneously scattering both laser beams. This again leads to mistaken interpretation of the size distribution of particles in the flow. Furthermore, because the effective flow area in the measurement volume could not be clearly specified, accurate measurements of both the concentration and the mass flux of particles in the flow could not be achieved. This will in turn largely limit the fundamental evaluations of related physical and engineering processes with particle flows (Zhang et al. 1998, Zhang and Eisele 1999, Zhang and Ziada 2000).

14.10 Method for Compensation of Astigmatism

The astigmatism in LDA measurements has been confirmed to greatly influence the measurement accuracy. In worst cases, no measurement can be carried out at all because of the separation between two laser beams. To compensate for the astigmatism, for instance in measuring the internal water flow, a water-filled prism according to Fig. 14.24, with or without the air gap, can be applied. Such an optical arrangement has already found its practical applications (Booij and Tukker 1994).

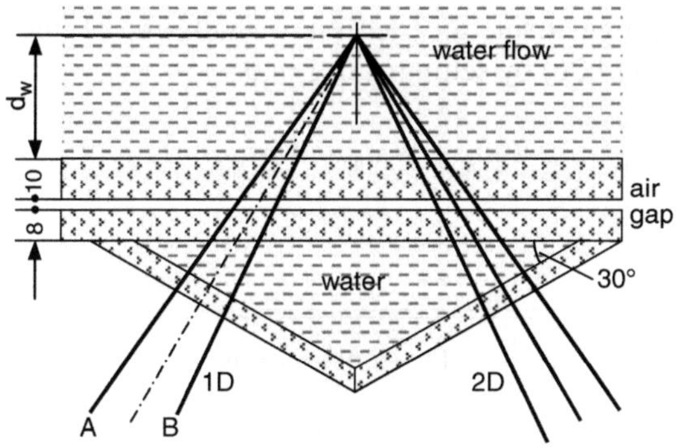

Fig. 14.24 Application example of using a water-filled prism to reduce and compensate for the optical aberration (astigmatism) in LDA measurements of water flows

The air gap serves to cause a "negative" astigmatism so that a complete compensation of astigmatism in both the transmitting and the receiving optics can be achieved. Because the perpendicular entry of laser beams into the prism does not cause any significant optical aberrations (see Chap. 13 and especially Sect. 14.7.3), the use of a water-filled prism acts as to submerge the LDA head into the same water. The occurrence of astigmatism on the side of transmitting optics thus changes from the case of air-glass-water to the case of water-glass-air-water. This is meaningful not only because of the possibility of completely suppressing the astigmatism, but also because of the simplicity of doing this, as will be shown below.

With the help of the water-filled prism the LDA head acts as to be submerged in water. This ensures that all four laser beams within the water of the water-filled prism are free of the optical aberrations and would intersect at a unique virtual focal point. The water in the prism should thus be considered as the reference medium for calculating the astigmatic difference in relying on calculation fundamentals in Sect. 14.5. The first non-perpendicular refraction of the optical axis occurs at $\varphi_0 = 30°$, while the optical axis intersects the next medium interface. The same optical property of transparent walls on both sides of the air gap is assumed. Hence for further calculations an equivalent wall thickness to the sum of both can be used.

The half intersection angle between two laser beams in the reference medium α_w should be used in place of α_0. Between α_w and α_0 it applies the law of refraction with $n_0 = 1$

$$\sin\alpha_w = \frac{n_0}{n_w}\sin\alpha_0 \tag{14.92}$$

The two laser beams in the meridian plane are designated by A and B. There are generally three distances that cause astigmatism and contribute to the astigmatic difference: the glass thickness (d_g), the air gap (d_a) and the depth of the measurement volume in the water flow (d_w). Against the reference medium the air gap causes a negative astigmatic difference. In order to achieve the complete compensation of astigmatism, the necessary thickness of the air gap can be calculated from Eq. (14.26) by setting $\Delta x_{\mathrm{m,s}} = 0$:

$$\Psi_{\mathrm{g}} d_{\mathrm{g}} + \Psi_{\mathrm{a}} d_{\mathrm{a}} + \Psi_{\mathrm{w}} d_{\mathrm{w}} = 0 \tag{14.93}$$

Herein

$$\Psi_{\mathrm{g}} = \frac{\cos\alpha_{\mathrm{w}} \cos\varphi_{\mathrm{o}}}{\sqrt{\frac{n_{\mathrm{g}}^2}{n_{\mathrm{w}}^2} - \left(1 - \cos^2\alpha_{\mathrm{w}} \cos^2\varphi_{\mathrm{o}}\right)}} - T_{\mathrm{g,w}} \tag{14.94}$$

$$\Psi_{\mathrm{a}} = \frac{\cos\alpha_{\mathrm{w}} \cos\varphi_{\mathrm{o}}}{\sqrt{\frac{n_{\mathrm{a}}^2}{n_{\mathrm{w}}^2} - \left(1 - \cos^2\alpha_{\mathrm{w}} \cos^2\varphi_{\mathrm{o}}\right)}} - T_{\mathrm{a,w}} \tag{14.95}$$

$$\Psi_{\mathrm{w}} = 0 \tag{14.96}$$

$$T_{\mathrm{g,w}} = \frac{\tan\varepsilon_{\mathrm{Ag}} - \tan\varepsilon_{\mathrm{Bg}}}{\tan\varepsilon_{\mathrm{Aw}} - \tan\varepsilon_{\mathrm{Bw}}} \tag{14.97}$$

$$T_{\mathrm{a,w}} = \frac{\tan\varepsilon_{\mathrm{Aa}} - \tan\varepsilon_{\mathrm{Ba}}}{\tan\varepsilon_{\mathrm{Aw}} - \tan\varepsilon_{\mathrm{Bw}}} \tag{14.98}$$

The result $\Psi_{\mathrm{w}} = 0$ has been automatically obtained as expected since the test medium is equal to the reference medium (see Sect. 14.5). This just represents the advantage of using a water-filled prism for water flow measurements, because the astigmatism and thus the optical aberration become independent of the depth of the measurement volume in the flow. From Eq. (14.93) the necessary thickness of the air gap is calculated as

$$d_{\mathrm{a}} = -\frac{\Psi_{\mathrm{g}}}{\Psi_{\mathrm{a}}} d_{\mathrm{g}} \tag{14.99}$$

A calculation example of using the water-filled prism should be demonstrated here. The LDA head is specified by the laser beams with a half intersection angle equal to $\alpha_0 = 5.5°$. The water-filled prism is designed with $\varphi_{\mathrm{o}} = 30°$. Table 14.1 shows the calculation results, from which the thickness of the air gap is calculated as 1.9 mm.

As shown in Table 14.1, the parameter Ψ_{a} is negative. This means that the used air gap leads to a negative astigmatic difference which enables the total astigmatism to be compensated for. If no air gap is used, then an astigmatic difference, caused by the glass wall of the thickness $d_{\mathrm{g}} = 18$ mm, is calculated from Eq. (14.29) again with $\Psi_{\mathrm{w}} = 0$ to

$$\Delta x_{\mathrm{m,s}} = \Psi_{\mathrm{g}} d_{\mathrm{g}} = 1.1 \text{ (mm)} \tag{14.100}$$

This astigmatic difference which in effect represents a displacement between two measurement volumes appears to be too large for the 2D LDA optics in Fig. 14.24. For two component measurements, therefore, it is indispensable to use an air gap

Table 14.1 Calculation example of using a water-filled prism and the air gap to completely compensate for the astigmatism

Parameter specifications:		
Refractive index of water	n_w	1.333
Refractive index of transparent glass window	n_g	1.52
Half intersection angle of two laser beams at LDA head	α_0	5.5
Geometrical angle of the water-filled prism	φ_o	30
Glass window thickness total (mm)	d_g	18
Calculations:		
Half intersection angle of two laser beams in water	α_w	4.1
Refraction angle of beam A in glass	ε_{Ag}	29.5
Refraction angle of beam B in glass	ε_{Bg}	22.5
Refraction angle of beam A in air	ε_{Aa}	48.4
Refraction angle of beam B in air	ε_{Ba}	35.6
Refraction angle of beam A in water	ε_{Aw}	34.1
Refraction angle of beam B in water	ε_{Bw}	25.9
Parameter (glass – water)	$T_{g,w}$	0.78
Parameter (air – water)	$T_{a,w}$	2.13
Parameter (glass)	Ψ_g	0.061
Parameter (air gap)	Ψ_a	−0.58
Air gap needed (mm)	d_a	1.9

to completely compensate for the astigmatism. For 1D LDA optics in Fig. 14.24, however, the astigmatic difference $\Delta x_{m,s} = 1.1$ mm is irrelevant. This means that measurements of using the 1D LDA head can be performed without any significant difficulties and inaccuracies.

For the purposes of comparison, the case of off-axis LDA alignment without using the prism should be considered. Because the reference medium in this case is air (index 0), the parameter Ψ_1 (for glass), Ψ_2 (for water) and T_{20} are recalculated to be $\Psi_1 = 0.098$, $\Psi_2 = 0.091$ and $T_{20} = 0.608$, respectively. The astigmatic difference that is caused by the glass wall of the thickness $d = 10$ (Fig. 14.24) is calculated to $\Delta x_{m,s(glass)} = 1.4$, by concerning the first part in Eq. (14.26). Also worth being mentioned is that in this case the depth of the measurement volume in the flow additionally contributes to the astigmatic difference. Because of this a water-filled prism should always be applied when the LDA optics has to be aligned off-axis.

It should be further mentioned that in the case of Fig. 14.24 with full compensation of astigmatism, the shift of each measurement volume is generally two-dimensional, even though the LDA head outside of the flow merely moves along the normal direction of the plane wall. This leads to undesirable separation between the measurement volumes of one- and two-component LDA units. This two-dimensional shift of each measurement volume is similar to that considered in Sect. 14.4. The flow field behind the wall could only be measured, provided that the optical realignment would take place for each new measurement point in the flow.

Chapter 15
Flow Measurements in Circular Pipes

Fluid flows in circular pipes belong to the most familiar and most broadly investigated flows. But the measurement of such flows by means of laser Doppler anemometry is not at all straightforward because of the surface curvatures on both the inside and the outside of the pipe. As is well known, simultaneous two-component velocity measurements, that are taken by using two pairs of laser beams, cannot be carried out, as the four laser beams do not intersect at a unique point in the flow because of optical aberrations. One of these optical aberrations is comparable with the astigmatism that occurs when the LDA optical axis is aligned off-axis against the normal of a plane wall, as treated in the last chapter. Nearly all the optical features and the effects of astigmatism in the case of plane windows also exist in a similar but often more complex form in the measurement of the circular pipe flow. Laser beam refractions on the curved interfaces of a circular pipe, for instance, complicate the calculation of beam propagations with regard to both the location of the measurement volume and the intersection angle between two laser beams (Boadway and Karahan 1981). In order to avoiding the difficulty of calculating complex laser beam refractions on the curved flow interface, the method of matching the refractive index of the test fluid to that of the pipe wall has often been applied in small-scaled laboratory measurements. The method, however, is not applicable to the flows in most industrial applications where the refractive index matching is impossible or the gaseous flows are in use.

The problems that are related with direct measurements of flows in circular pipes have been additionally recognized in the signal strengths and qualities. Experiences show that in traversing the velocity profile across a circular pipe, the available signal qualities and thus the signal rate could be achieved only within a depth of about 1/3 of the pipe diameter. Beyond this depth both the signal strengths and qualities briskly drop down so that no measurements can be further carried out. The reason for such a disturbance in the respective measurement is the optical aberration in the receiving optics. This means, in concrete terms, that only a few elementary segments on the LDA receiving lens can see the measurement volume, while majorities are blind. This circumstance is comparable with that in flow measurements behind a plane wall by means of an off-axis LDA head, as already treated in Sect. 14.8. It appears that the flow in the centre area of a circular pipe could hardly be directly measured without matching the refractive index of the flow to that of the pipe wall.

Zh. Zhang, *LDA Application Methods*, Experimental Fluid Mechanics,
DOI 10.1007/978-3-642-13514-9_15,

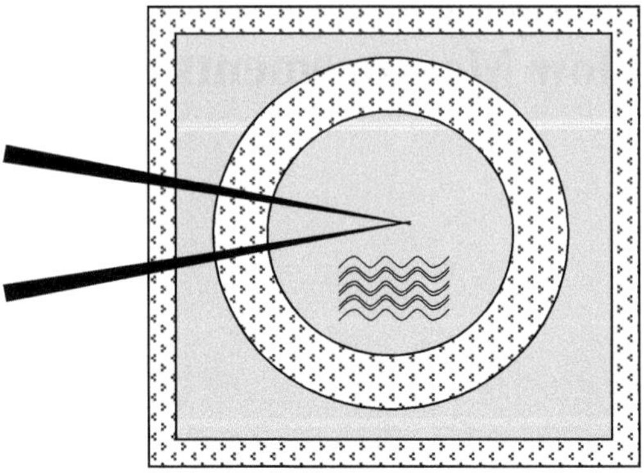

Fig. 15.1 Method to improve the optical condition in LDA measurements of flows in the circular pipe, effective but not quite convenient

Because of this, some LDA users applied the method of putting the circular pipe with flows into a rectangular water tank with plane walls (Fig. 15.1). In using such a measurement facility, the optical performance on the side of receiving optics can be significantly enhanced, so that better qualities of signals can be obtained. The problem of calculating the measurement volume, however, remains because each laser beam still suffers from two-time refraction on the curved interfaces.

In dealing with flow measurements in a circular pipe by means of the LDA method without matching the refractive index, the optical performance can be considerably improved if the outside of the pipe is cut off and made plane, as illustrated in Fig. 15.2 from an industrial application (Zhang and Casey 2007, Zhang and Parkinson 2002). This approach was initially considered to simplify calculations of laser beam transmissions, as each laser beam undergoes only one time refraction on the curved interface and the formation of the measurement volume is independent of the thickness of the pipe wall. The approach indeed additionally contributes to a lot of reductions of diverse optical aberrations, since the main optical aberration is then only restricted to the beam refraction on the internal surface of the circular pipe. Therefore high quality signals can be obtained even at a distance of about 2/3 of pipe diameters from the pipe wall. To get the complete distribution of each

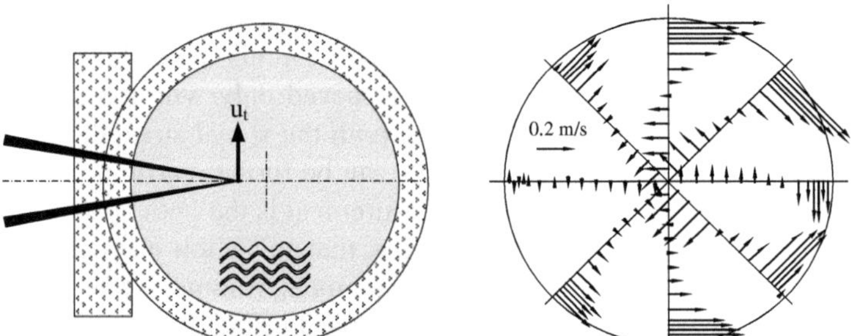

Fig. 15.2 Plane wall configuration on the outside of a circular pipe and measurements of the tangential velocity component (Zhang and Parkinson 2002)

velocity component across the pipe, additional measurements need to be performed from the opposite side by rotating the LDA head 180° around the pipe axis. The use of a moveable prism according to Fig. 15.2 creates a comparable plane cut-off on the outside of the pipe and enables the free alignment of the LDA head to the pipe and around the pipe axis. With this method both the main and the secondary flow distributions in the cross section of the respective pipe flow could be well measured. Clearly, measurements of all three velocity components in a pipe flow could only be achieved one after another.

As a matter of fact, a great number of flow measurements in circular pipes have been restricted to axial and tangential velocity components. The measurement of the tangential velocity component (secondary flow) does require some more calculations for tracking the laser beam in the flow. It actually does not represent a highly difficult task. In contrast, the measurement of the radial velocity component is obviously very restricted by complexities and intricacies of tracking the laser beams in the flow. It is in any case true that almost no such measurements have ever been carried out or can be referred to. Based on detailed analyses a fundamental guideline of measuring the radial velocity component of the flow in a circular pipe has been worked out by Zhang (2004a, b).

Another serious aspect associated with the laser beam refraction on the curved interface is the dislocation of all laser beam waists from the measurement volume. At large displacement between the measurement volume and the laser beam waists, the laser light intensity in the measurement volume decreases so that velocity signals will become too weak to be detected. This incident states an additional reason for the measurement limitation within a depth at about 2/3 of the pipe diameter. Furthermore, the dislocation of laser beam waists inevitably leads to the fringe distortion in the measurement volume and hence to measurement errors.

In the following sections, essential calculations of tracking the laser beams in the flow will be presented with respect to the shift and optical properties of the measurement volume. All three velocity components will be considered. In that case, possible influences on both the signal quality and the measurement accuracy will be revealed and quantified. Because the plane outside of the circular pipe, as shown in Fig. 15.2, has been confirmed to be very effective and helpful in improving the optical conditions, all calculations will be referred to in this optical configuration. Correspondingly, the LDA head is assumed to be perpendicularly aligned to this plane surface.

15.1 Measurements of Axial Velocities

The measurements of axial velocities require that the two laser beams lie in a plane parallel to the pipe axis. The plane containing two laser beams is called the optical plane. By arranging the optical plane to go through the pipe axis (Fig. 15.3b), the laser beam refraction on the inside of the circular pipe is comparable with that on a perpendicular plane surface. Both laser beams propagate in the same plane as the

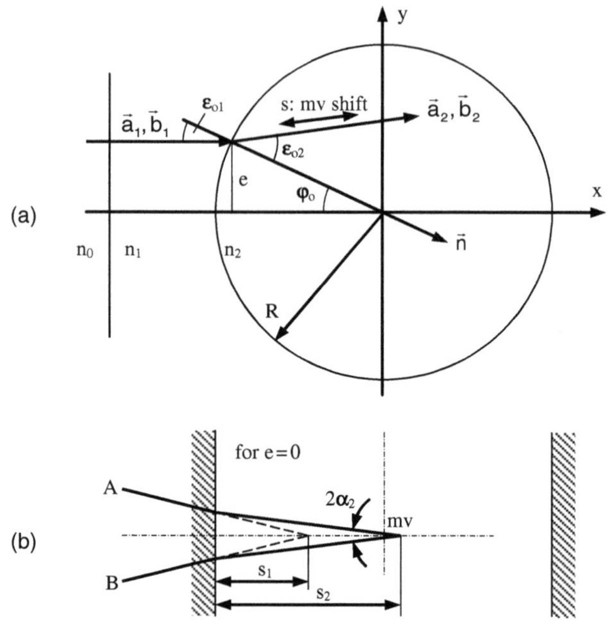

Fig. 15.3 Deviation of the optical plane from the pipe axis and its effect on the measurements of the axial velocity component

optical plane prior to refractions. The optical aberration in the transmitting optics remains at a minimum so that the best available optical condition is ensured.

The position of the measurement volume, measured from the pipe wall, can be directly written out from Eq. (13.6) in proportion to the position of the virtual measurement volume as

$$s_2 = \frac{n_2}{n_1} s_1 \tag{15.1}$$

with s_1 as the coordinate of the virtual intersection point of two laser beams.

A particular property of the optical layout for measurements of axial velocities should be highlighted here for future reference. As long as the optical plane is kept parallel to the pipe axis, without having to go through it, the intersection angle between the two refracted laser beams remains unchanged. This can be confirmed according to Fig. 15.3a which shows a deviation e of the optical plane from the pipe axis. With regard to the laser beam A for instance, the laser beam is denoted by the unit vector $\vec{a}_1$ in the medium 1 and $\vec{a}_2$ after the refraction in the medium 2. For the applied coordinate system the corresponding z-component of the unit vector $\vec{a}_2$ is given, according to Eq. (3.9), as $a_{2z} = n_1/n_2 \cdot a_{1z}$. The intersection angle $2\alpha_2$ between the two refracted laser beams A and B can then be calculated by

$$\cos 2\alpha_2 = \vec{a}_2 \cdot \vec{b}_2 \tag{15.2}$$

Because of the symmetry condition between two laser beams there are $a_{2x} = b_{2x}$, $a_{2y} = b_{2y}$ and $a_{2z} = -b_{2z}$ as well as $a_{2x}^2 + a_{2y}^2 = 1 - a_{2z}^2$ for the unit vector $\vec{a}_2$. From Eq. (15.2) it then follows

$$\cos 2\alpha_2 = 1 - 2a_{2z}^2 = 1 - 2\left(\frac{n_1}{n_2}a_{1z}\right)^2 \tag{15.3}$$

With respect to the trigonometric identity $\cos 2\alpha_2 = 1 - 2\sin^2\alpha_2$ and $a_{1z} = \sin\alpha_1$ for the incident laser beam A it further follows from the above equation

$$\sin\alpha_2 = \frac{n_1}{n_2}a_{1z} = \frac{n_1}{n_2}\sin\alpha_1 \tag{15.4}$$

The intersection angle of the two laser beams and thus the optical properties of the measurement volume in the test fluid remain the same as that on the pipe axis. This result indicates that the deviation in aligning the optical plane from going through the pipe axis does not lead to any error in velocity measurements. The analysis made here can be easily demonstrated by a table-top experiment when a scaled rule is inserted into a circular pipe with or without water (Fig. 15.4). Observation from the outside verifies that there is no longitudinal distortion in the rule scale along the pipe axis. The distortion only occurs in the radial direction.

Although a certain deviation of the optical plane from the pipe axis does not lead to any change in the beam intersection angle, there are, however, some other aspects that should be accounted for:

- With the displacement of the LDA head in the x-direction, the measurement volume in the test fluid gets moved along the bisector of two refracted laser beams i.e. along a path which is not parallel to the x-axis. Because of the comatic aberration, similar to that in Fig. 14.3a, the bisector of the two refracted laser

Fig. 15.4 A table-top experiment showing no any longitudinal distortion in the rule scale along the pipe axis

beams does not precisely coincide with the refracted optical axis. The difference between them can, however, be left to a first approximation. The refracted optical axis in the test fluid is calculated by using the law of refraction as

$$\sin \varepsilon_{o2} = \frac{n_1}{n_2} \sin \varepsilon_{o1} = \frac{n_1}{n_2} \frac{e}{R} \tag{15.5}$$

As an example, the beam refraction with $n_1/n_2 = 1.52/1.33$ is assumed. For $e/R = 0.5$ which means $\varepsilon_{o1} = 30°$ there is $\varepsilon_{o2} = 34.8°$. The inclination angle of the refracted optical axis in the flow against the x-axis is 4.8°.

- Because the intersection angle between two refracted laser beams in the flow is entirely a constant and thus independent of the position of the measurement volume, the shift of the measurement volume along the bisector (s) of two laser beams is proportional to the displacement of the LDA head along the x-axis. The corresponding shift ratio is for $\alpha_0 << 1$ and $\alpha_2 << 1$ simply

$$k_{mv} = \frac{\Delta s_{mv}}{\Delta x_{LDA}} = \frac{\tan\alpha_0}{\tan\alpha_2} \approx \frac{\sin \alpha_0}{\sin \alpha_2} = \frac{n_2}{n_0} \tag{15.6}$$

The LDA head is assumed to be positioned in air ($n_0 = 1$).

The x-component of the above shift ratio is calculated, according to Fig. 15.3, by

$$k_{mv,x} = k_{mv} \cos (\varepsilon_{o2} - \varphi_o) \tag{15.7}$$

Usually the inclination angle of the refracted optical axis in the flow is small. For instance for $e/R < 0.5$ there is almost $\varepsilon_{o2} - \varphi_o < 5°$, so that $k_{mv,x} \approx k_{mv}$ can be applied.

- At large deviations of the optical plane from the pipe axis ($e/R > 0.5$), which might be desired in some cases, the LDA optical axis tends to largely deviate from the normal $\vec{n}$ of the interface. This situation is comparable with large off-axis angles of the LDA head against the normal of a plane wall (see Chap. 14). As a result the optical aberration related to the LDA optics, or accurately speaking, the effect of astigmatism becomes significant and the quality of optical signals will be considerably deteriorated. This optical aberration can also be visualized with the table-top experiment mentioned above, at which the scale on the rule in the pipe becomes continuously more unclear along the pipe radius (Fig. 15.4).

Since the deviation of the LDA optical plane from the pipe axis causes undesirable features, large deviations should be avoided. For moderate deviations (say $e/R < 0.5$), as long as the signal rate is sufficiently high, no particular attention needs to be paid. The fringe distortion in the measurement volume, which results from the dislocation of laser beam waists, is not significant and can be neglected.

15.2 Measurements of Tangential Velocities

For measurements of tangential velocities the LDA head is aligned, so that the optical plane is perpendicular to the pipe axis (Fig. 15.2). With regard to the laser beam refractions on the circular surface, two main aspects need to be considered:

- The shift of the measurement volume in the flow is no longer proportional to the shift of the LDA head.
- The intersection angle between the two refracted laser beams and thus both the geometrical and optical properties of the measurement volume depend on the local position of the measurement volume in the flow.

Detailed calculations of tracking the laser beams in the flow should therefore be conducted to ensure correct LDA measurements.

15.2.1 Basic Geometrical Relationships

The laser beams are again considered as they are present in the transparent pipe wall (index 1). Owing to the symmetrical layout between two laser beams, only one beam will be considered. The measurement volume is then formed on the symmetrical axis i.e. the x-axis in Fig. 15.5. For clarity, the crossing angle between two laser beams has been shown exaggerated. The laser beam is assumed to have its start position 1, when the measurement volume is positioned on the inside of the circular pipe. The shift of the laser beam to the position 2 is given by a distance s_1 with which the virtual measurement volume is moved to m'. The distance s_1 is about n_1-times the movement of the LDA head in air. Because of the laser beam refraction the real measurement volume is found at m. The application of the law of sines to the triangles ocm' and ocm, respectively, yields

$$\frac{R - s_1}{\sin(\alpha_1 - \varphi)} = \frac{R}{\sin \alpha_1} \tag{15.8}$$

$$\frac{R - s_2}{\sin(\alpha_2 - \varphi)} = \frac{R}{\sin \alpha_2} \tag{15.9}$$

The law of refraction in such a case is given by

$$n_1 \sin(\alpha_1 - \varphi) = n_2 \sin(\alpha_2 - \varphi) \tag{15.10}$$

These three equations form the basic solutions for both the new position (s_2) and the geometrical parameter ($2\alpha_2$) of the measurement volume in the flow. For a given movement s_1 the following calculation process can be performed

$$s_1 \xrightarrow{(15.8)} \varphi \xrightarrow{(15.10)} \alpha_2 \xrightarrow{(15.9)} s_2 \tag{15.11}$$

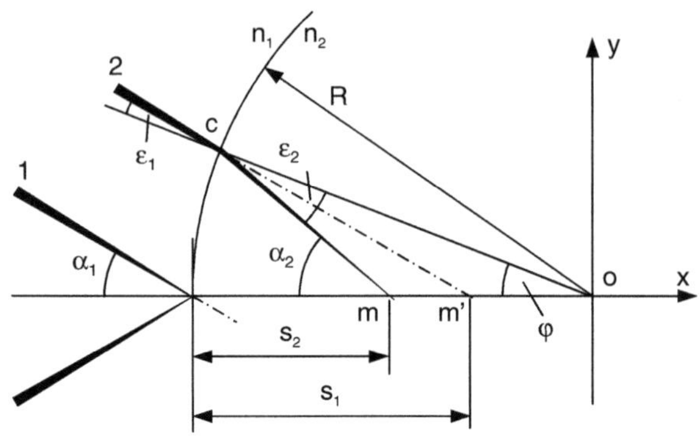

Fig. 15.5 Formation of the measurement volume in the flow for measurements of the tangential velocity component

An application example of the presented technique has already been shown in Fig. 15.2.

15.2.2 Simplifications of Calculations

In the above calculation, Eq. (15.11) represents the way of indirectly calculating the relevant parameters α_2 and s_2 that are related to the measurement volume. This process seems to be somewhat inconvenient. With respect to the configuration of common LDA systems, at which the half intersection angle of laser beams is usually not more than 10°, the half intersection angle of laser beams in the transparent pipe wall is then not more than 7° and the angle φ seldom exceeds 14°. For this reason all the angles under the symbol of sine functions ($\alpha_1, \alpha_2, \alpha_1 - \varphi$ and $\alpha_2 - \varphi$) in the basic equations from Eq. (15.8) to Eq. (15.10) can be considered as small angles. All three equations can then be simplified by using the approximation $\sin x = x$ for paraxial rays. The characteristic parameters for the measurement volume can then be directly interpreted as the function of the virtual position of the measurement volume as follows:

$$\frac{s_2}{R} = \frac{1}{1 + \frac{n_1}{n_2}\left(\frac{R}{s_1} - 1\right)} \tag{15.12}$$

$$\frac{\alpha_2}{\alpha_1} = \frac{n_1}{n_2} - \left(\frac{n_1}{n_2} - 1\right)\frac{s_1}{R} \tag{15.13}$$

$$\frac{\varphi}{\alpha_1} = \frac{s_1}{R} \tag{15.14}$$

At present, Eq. (15.14) is only written for completeness. It will not be applied to quantify the measurement volume.

The inaccuracy arising from the approximations made above should be estimated. An LDA system is assumed to have a half intersection angle $\alpha_1 = 4.45°$

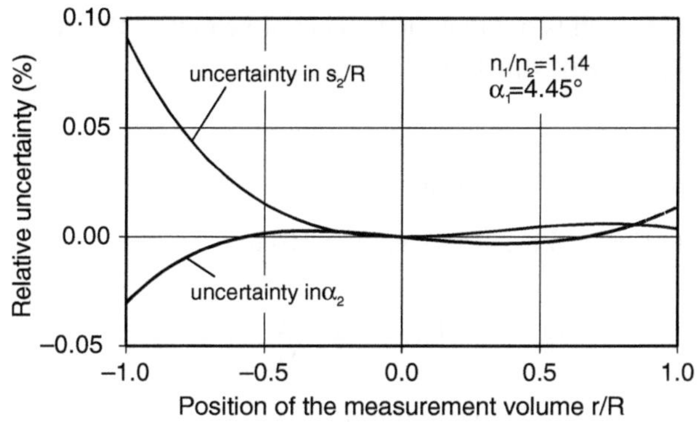

Fig. 15.6 Relative uncertainties in calculating the measurement volume in the water flow by using the simplified relations

between laser beams in the pipe wall ($n_1 = 1.52$). It is further assumed that the measurement volume traverses the pipe section for the case of water flows ($n_2 = 1.33$). From Eqs. (15.12) and (15.13), respectively, both the parameter s_2 and α_2 are calculated in the function of s_1. Their relative deviations from those as calculated from Eq. (15.11) can then be determined, as shown in Fig. 15.6. As can be seen, the maximum error arising from the approximations leading to Eqs. (15.12) to (15.14) is below 0.1% and really negligible. The same calculation can also be made for air flow ($n_2 = 1$). The maximum error embedded in Eq. (15.12) is 0.4%, while in Eq. (15.13) is only 0.1%.

15.2.3 Fringe Spacing and Velocity Corrections

By using the laser beam intersection angle in the fluid, the fringe spacing in the measurement volume is calculated as

$$\Delta x = \frac{\lambda_2}{2 \sin \alpha_2} \tag{15.15}$$

Because of $\lambda_1/\lambda_2 = n_2/n_1$ this equation is rewritten as

$$\Delta x = \frac{n_1}{n_2} \frac{\sin \alpha_1}{\sin \alpha_2} \frac{\lambda_1}{2 \sin \alpha_1} = \frac{n_1}{n_2} \frac{\sin \alpha_1}{\sin \alpha_2} \Delta x_0 \tag{15.16}$$

Herein Δx_0 represents the predefined fringe spacing in the LDA optics i.e. the reference fringe spacing, when the measurement volume is found in the open air or in the pipe wall.

Equation (15.16) points out that the tangential velocities that are measured have to be corrected by a factor equal to

$$k_{\mathrm{vel}} = \frac{u_{\mathrm{t}}}{u_{\mathrm{t,measured}}} = \frac{\Delta x}{\Delta x_0} = \frac{n_1}{n_2}\frac{\sin\alpha_1}{\sin\alpha_2} \tag{15.17}$$

or with respect to Eqs. (15.8), (15.9) and (15.10) to

$$k_{\mathrm{vel}} = \frac{R - s_2}{R - s_1} \tag{15.18}$$

In applying Eq. (15.12), this is again reformed as

$$k_{\mathrm{vel}} = 1 + \left(\frac{n_1}{n_2} - 1\right)\frac{s_2}{R} \tag{15.19}$$

It linearly depends on the depth of the measurement volume in the pipe flow. While in the near region to the pipe wall the correction factor is equal to unity, it takes n_1/n_2 on the pipe axis. For $n_1 = 1.52$ (glass) and $n_2 = 1.33$ (water) there is $k_{\mathrm{vel}} = 1.14$.

15.3 Measurements of Radial Velocities

The measurements of radial velocities in circular pipes represent a highly complex process, if no index matching method is applied. In positioning the LDA head for radial velocity measurements according to Fig. 15.7, the following problems have to be solved:

- How can the measurement volume (m) be accurately positioned?
- How can the beam intersection angle ($2\alpha_2$) be calculated?
- What about the orientation (τ) of the measurement volume?

The beam intersection angle and the measurement volume orientation have to be known in order to correct systematic measurement errors. In reality, because of $\tau \neq 0$ each measured velocity does not exactly represent the radial velocity component.

15.3.1 Accurate Positioning of the Measurement Volume

In principle, each LDA head position determines a position of the measurement volume in the flow. With the shift of the LDA head parallel to the y-axis, the measurement volume would generally travel along a two-dimensional path. This will considerably complicate the profile measurements of the radial velocity component. A relatively easy method is to calculate the necessary movement of the LDA

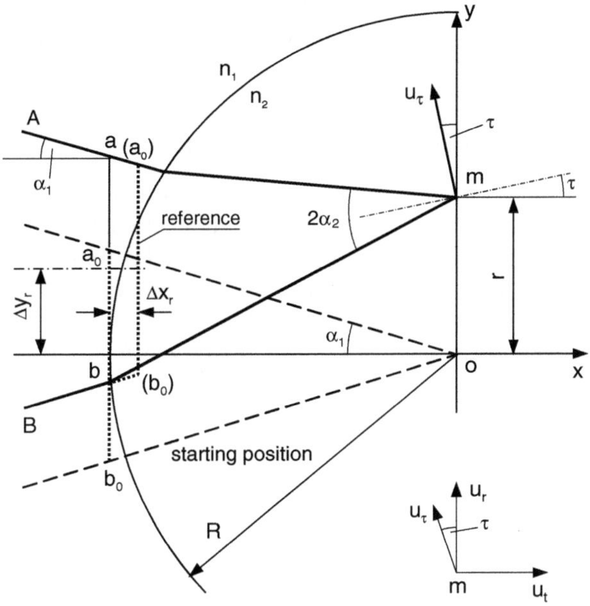

Fig. 15.7 Off-axis alignment of laser beams for measurements of the radial velocity component along the y-axis

head for the given path of the measurement volume along the y-axis, as shown in Fig. 15.7. The starting point is the LDA head position at which the measurement volume is positioned on the pipe axis (as the starting position). This can be achieved from the method described in the last section for measurement of the tangential velocity component. For each new position of the measurement volume (m) on the y-axis at r, the necessary movement of the beam pair in medium 1 from the starting position is assumed to be Δx_{r} and Δy_{r}. These are measured by shifting the reference line $a_0 b_0$ (denoted by a fine dotted line) from the starting position to the new position. Then the following relationships can be obtained:

$$\Delta x_{\mathrm{r}} = \frac{(y_{\mathrm{a}} - y_{\mathrm{b}}) - (y_{\mathrm{a}0} - y_{\mathrm{b}0})}{2 \tan \alpha_1} \tag{15.20}$$

$$\Delta y_{\mathrm{r}} = \frac{1}{2} (y_{\mathrm{a}} + y_{\mathrm{b}}) \tag{15.21}$$

In these two equations, a_0, b_0, a and b are intersections between laser beams and the tangent of the circular pipe at $x = -R$. The basic intersection coordinates $y_{\mathrm{a}0}$ and $y_{\mathrm{b}0}$ are given by

$$y_{\mathrm{a}0} = R \cdot \tan \alpha_1 \tag{15.22}$$

and

$$y_{\mathrm{b}0} = -R \cdot \tan \alpha_1 \tag{15.23}$$

The intersection points a and b represent positions which the laser beams A and B at the new position of the beam pair must pass through. Clearly detailed calculations of y_a and y_b for a given measurement volume position have to be completed.

15.3.1.1 Determination of Intersection Points y_a and y_b

The purpose of present calculations is to establish the functions of both $y_a = f(r)$ and $y_b = f(r)$ for the given measurement volume along the y-axis. Because it deals with the same calculations both for y_a and y_b, only detailed calculations of $y_a = f(r)$ for the laser beam A will be presented below.

According to Fig. 15.8 the function $y_a = f(r)$ can be obtained if the function $\varphi_a = f(r)$ is known. For this reason the function $\varphi_a = f(r)$ or equivalently $r = f(\varphi_a)$ will be established first. The position of the measurement volume on the y-axis can be expressed by

$$r = R\sin\varphi_a - R\cos\varphi_a\tan\alpha_{2a} \tag{15.24}$$

or related to R as

$$\frac{r}{R} = \sin\varphi_a - \cos\varphi_a\tan\alpha_{2a} \tag{15.25}$$

Here $\tan\alpha_{2a}$ needs to be expressed as a function of φ_a.

With respect to $\alpha_{2a} = \varphi_a - \varepsilon_2$ and the law of refraction in the form of $n_2 \sin\varepsilon_2 = n_1 \sin\varepsilon_1$ the following relationship can be obtained:

$$\tan\alpha_{2a} = \frac{\sqrt{T_a - 1}\tan\varphi_a - 1}{\sqrt{T_a - 1} + \tan\varphi_a} \tag{15.26}$$

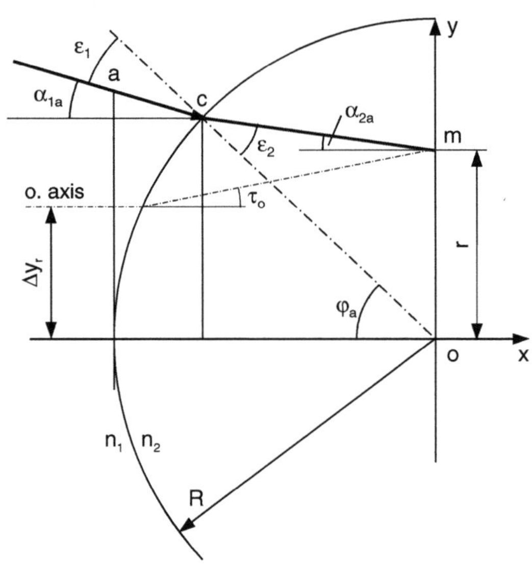

Fig. 15.8 Calculation of the laser beam transmission for the measurement volume located on the y-axis

Herein T_a is given by

$$T_a = \left(\frac{n_2}{n_1}\right)^2 \frac{1}{\sin^2 \varepsilon_1} = \left(\frac{n_2}{n_1}\right)^2 \frac{1}{\sin^2 (\varphi_a - \alpha_{1a})} \tag{15.27}$$

According to Eq. (15.26) there has to be $T_a \geq 1$. In fact, $T_a = 1$ just represents the beginning of the total reflection, as given in Eq. (15.27).

Together with Eq. (15.25), the position of the measurement volume on the y-axis is now expressed as a function of the angle φ_a:

$$\frac{r}{R} = \frac{1}{\sqrt{T_a - 1} + \tan \varphi_a} \frac{1}{\cos \varphi_a} = f(\varphi_a) \tag{15.28}$$

According to Figs. 15.7 and 15.8 the angle φ_a begins at $\varphi_a = \alpha_{1a}$, at which the measurement volume is positioned in the pipe centre.

In a similar way, the same function for the laser beam B could be found. As a matter of fact, the subscript a in Eq. (15.28) needs only to be replaced by the subscript b. However, attention should be paid to the fact that both the angle α_{1b} and α_{2b} are negative for the laser beam B. Correspondingly the angle φ_b begins at $\varphi_b = \alpha_{1b}$.

The intersection between laser beam A and the tangent at $x = -R$ is calculated according to Fig. 15.8 as

$$y_a = R \cdot \sin \varphi_a + R(1 - \cos \varphi_a) \tan \alpha_{1a} = f(r) \tag{15.29}$$

Similarly there is for the laser beam B

$$y_b = R \cdot \sin \varphi_b + R(1 - \cos \varphi_b) \tan \alpha_{1b} = f(r) \tag{15.30}$$

They are both functions of r because of Eq. (15.28) for the laser beam A and the corresponding equation for the laser beam B. Together with Eqs. (15.20) and (15.21) the necessary movements of the laser beam pair in the medium 1 for a given measurement volume at r can be determined.

15.3.1.2 Simplifications of Calculations

Both functions given in Eqs. (15.29) and (15.30) are not explicit, because according to Eq. (15.28) the polar angle φ_a cannot be expressed to be an explicit function of r/R. For a series of angles φ_a being assumed to be given, however, corresponding y_a/R and r/R can be calculated and related to each other, as these have been shown in Fig. 15.9 as an example. In this figure, the corresponding function for the laser beam B has also been shown. It can be concluded that below $r/R = 0.8$ there exist linear functions of both $y_a = f(r)$ and $y_b = f(r)$. For $r = 0$ there are $y_{a0} = R \tan \alpha_{1a}$ and $y_{b0} = R \tan \alpha_{1b}$ according to Fig. 15.7. To calculate these linear functions, one

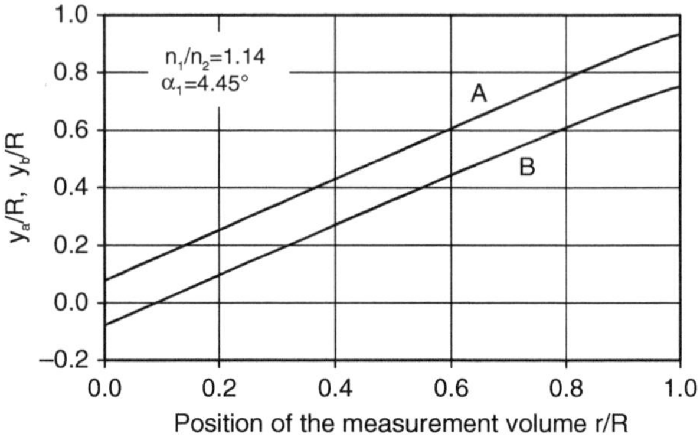

Fig. 15.9 Coordinates of intersections between the laser beams (A and B) and the circle tangent at $x = -R$

needs to further calculate the respective gradients at $r = 0$. From Eq. (15.29), for instance, for the laser beam A one obtains

$$\frac{\mathrm{d}y_\mathrm{a}}{\mathrm{d}r} = \frac{\mathrm{d}y_\mathrm{a}}{\mathrm{d}\varphi_\mathrm{a}}\frac{\mathrm{d}\varphi_\mathrm{a}}{\mathrm{d}r} = R\,(\cos\varphi_\mathrm{a} + \tan\alpha_{1\mathrm{a}}\sin\varphi_\mathrm{a})\,\frac{\mathrm{d}\varphi_\mathrm{a}}{\mathrm{d}r} \tag{15.31}$$

It is known that at $r = 0$ there is $\varphi_\mathrm{a} = \alpha_{1\mathrm{a}}$. Furthermore it results from Eq. (15.28) $T_\mathrm{a} \to \infty$ and then from Eq. (15.27) $\sqrt{T_\mathrm{a}}\sin(\varphi_\mathrm{a} - \alpha_{1\mathrm{a}}) = n_2/n_1$. With respect to these conditions it can be calculated from Eq. (15.28)

$$\left.\frac{1}{R}\frac{\mathrm{d}r}{\mathrm{d}\varphi_\mathrm{a}}\right|_{\mathrm{r}=0} = \frac{1}{\cos\alpha_{1\mathrm{a}}}\frac{n_1}{n_2} \tag{15.32}$$

Hence it follows from Eq. (15.31) for $r = 0$

$$\frac{\mathrm{d}y_\mathrm{a}}{\mathrm{d}r} = \frac{n_2}{n_1} \tag{15.33}$$

From similar calculations or immediately from Fig. 15.9 the same gradient is obtained for the laser beam B at $r = 0$.

The linear functions of both $y_\mathrm{a} = f(r)$ and $y_\mathrm{b} = f(r)$ are thus obtained as

$$\frac{y_\mathrm{a}}{R} = \tan\alpha_{1\mathrm{a}} + \frac{n_2}{n_1}\frac{r}{R} \tag{15.34}$$

$$\frac{y_\mathrm{b}}{R} = \tan\alpha_{1\mathrm{b}} + \frac{n_2}{n_1}\frac{r}{R} \tag{15.35}$$

To be mentioned again is $\alpha_{1\mathrm{b}} < 0$ and thus $\tan\alpha_{1\mathrm{b}} < 0$.

15.3.1.3 Necessary Movement Δx_r and Δy_r of Laser Beam Pair

The intersection coordinates shown in Eqs. (15.34) and (15.35) are now used to calculate the necessary movements of the laser beam pair in medium 1, when the measurement volume is shifted from the pipe centre ($r = 0$) to a given radius r on the y-axis. With respect to $\alpha_{1a} = -\alpha_{1b} = \alpha_1$ Eqs. (15.20) and (15.21) are calculated respectively to

$$\frac{\Delta x_r}{R} = 0 \tag{15.36}$$

$$\frac{\Delta y_r}{R} = \frac{n_2}{n_1}\frac{r}{R} \tag{15.37}$$

These beautiful results point out that for measurements along the y-axis the LDA head needs only to be moved in the parallel direction. The ratio between the LDA movement and the shift of the measurement volume is equal to the ratio of respective refractive indices. It deals with a simple way which enables the LDA measurement volume to be easily positioned in the flow. Measurements of radial velocities thus become possible.

15.3.2 Laser Beam Intersection Angle

It is generally the case that the results of velocity measurements must be corrected because of systematic errors arising from the change in the fringe spacing in the measurement volume. Related to this change is, in the first place, the intersection angle between two laser beams refracted in the flow. For this reason Eqs. (15.34) and (15.35) are again considered for further calculations. Combining these two equations with Eqs. (15.29) and (15.30), respectively, yields

$$\sin(\varphi_a - \alpha_{1a}) = \frac{n_2}{n_1}\frac{r}{R}\cos\alpha_{1a} \tag{15.38}$$

$$\sin(\varphi_b - \alpha_{1b}) = \frac{n_2}{n_1}\frac{r}{R}\cos\alpha_{1b} \tag{15.39}$$

Due to $\alpha_{1a} = -\alpha_{1b} = \alpha_1$ the r.h.s. of these two equations are equal to each other, so that it yields from equality of both terms on the l.h.s. of the two equations

$$\varphi_a - \varphi_b = \alpha_{1a} - \alpha_{1b} = 2\alpha_1 \tag{15.40}$$

With regard to Eq. (15.10) that represents the law of refraction Eqs. (15.38) and (15.39) are further written as

$$\sin(\varphi_a - \alpha_{2a}) = \frac{r}{R}\cos\alpha_1 \tag{15.41}$$

$$\sin(\varphi_b - \alpha_{2b}) = \frac{r}{R}\cos\alpha_1 \tag{15.42}$$

These two equations straightforwardly indicate

$$\varphi_a - \varphi_b = \alpha_{2a} - \alpha_{2b} \tag{15.43}$$

Because of Eq. (15.40) there is then

$$\alpha_{2a} - \alpha_{2b} = 2\alpha_1 \tag{15.44}$$

That is to say

$$\alpha_2 = \alpha_1 \tag{15.45}$$

This beautiful result indicates that the beam intersection angle remains constant and is the same as that in the medium 1 (pipe wall). This angle will be used to correct the measurement results which involve systematic errors.

15.3.3 Fringe Spacing and Velocity Corrections

In using the laser beam intersection angle in the fluid, the fringe spacing in the measurement volume is calculated to be:

$$\Delta x = \frac{\lambda_2}{2\sin\alpha_2} \tag{15.46}$$

Due to $\alpha_2 = \alpha_1$ and $\lambda_1/\lambda_2 = n_2/n_1$ this equation is rewritten as

$$\Delta x = \frac{n_1}{n_2}\frac{\lambda_1}{2\sin\alpha_1} = \frac{n_1}{n_2}\Delta x_0 \tag{15.47}$$

Herein Δx_0 represents the initial fringe spacing in the LDA optics i.e. the reference fringe spacing, when the measurement volume is found in the open air or in the pipe wall.

Equation (15.47) points out that the measured velocities must be corrected by a factor equal to $k_{vel} = n_1/n_2$:

$$u_\tau = \frac{n_1}{n_2}u_{\text{measured}} \tag{15.48}$$

It must be mentioned, however, that the measured velocity corresponds to the velocity component that is perpendicular to the bisector of two laser beams (Fig. 15.7). It should not be considered to be the radial velocity component.

15.3.4 Orientation of the Measurement Volume

The orientation of the measurement volume i.e. the bisector of two refracted laser beams determines the measured velocity component that generally differs from the radial velocity. The deviation angle τ of the bisector, as shown in Fig. 15.7, is considered positive and given by

$$\tau = \alpha_2 - \alpha_{2a} = \alpha_1 - \alpha_{2a} \tag{15.49}$$

For α_{2a} see Fig. 15.8.

In order to express the deviation angle of the measurement volume as a function of the radial position r of the measurement volume, Eqs. (15.38) and (15.41) are taken into account, respectively, so that

$$\tau = (\varphi_a - \alpha_{2a}) - (\varphi_a - \alpha_1) = \arcsin\left(\frac{r}{R}\cos\alpha_1\right) - \arcsin\left(\frac{n_2}{n_1}\frac{r}{R}\cos\alpha_1\right) \tag{15.50}$$

Clearly both the optical layout (α_1) and the medium properties (n_2/n_1) additionally influence the deviation angle of the measurement volume. Figure 15.10 for example shows the calculated deviation angle of the measurement volume in the function of the radial position r/R for a given optical configuration. For the measurement volume that is positioned within $r/R = 0.8$, the corresponding deviation angle slowly increases up to 10°.

The optical axis and its refraction should now be considered. In reality, the situation of the optical axis in the present case is equal to the situation that was shown in Fig. 15.3 concerning the measurement of the axial velocity component. The inclination angle of the refracted optical axis in the flow against the x-axis is thus

$$\tau_o = \varepsilon_{o2} - \varphi_o \tag{15.51}$$

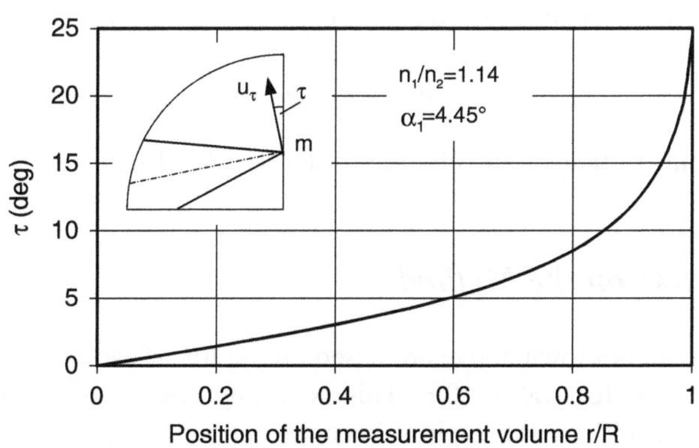

Fig. 15.10 Deviation angle of the bisector of the two refracted laser beams in the fluid

For the angle φ_o it applies that $\sin\varphi_o = \Delta y_r/R$. Because of Eq. (15.37) this is further written as

$$\sin\varphi_o = \frac{n_2}{n_1}\frac{r}{R} \tag{15.52}$$

The refraction angle of the optical axis in the present case is obtained directly from Eq. (15.5) as given by

$$\sin\varepsilon_{o2} = \frac{n_1}{n_2}\frac{\Delta y_r}{R} = \frac{r}{R} \tag{15.53}$$

Herein Δy_r, as the displacement of the optical axis from the x-axis, is applied according to Fig. 15.8 and Eq. (15.37).

Equation (15.51) then becomes

$$\tau_o = \arcsin\left(\frac{r}{R}\right) - \arcsin\left(\frac{n_2}{n_1}\frac{r}{R}\right) \tag{15.54}$$

This equation represents an approximation of Eq. (15.50) by assuming $\cos\alpha_1 \approx 1$ for small laser beam intersection angles. The difference between Eqs. (15.50) and (15.54) is less than 0.4% for the case considered in Fig. 15.10 for $r/R < 0.5$. Because of this Eq. (15.54) can be applied to specify the deviation angle of the measurement volume in the flow.

15.3.5 Determination of Radial Velocities

Because of $\tau \neq 0$, the measured velocity u_τ does not directly represent the radial velocity component. In principle, the radial velocity component can be obtained if the tangential velocity at the given measurement point on the y-axis is known. By assuming that the positive tangential velocity agrees with the x-direction, the radial velocity component can be resolved according to Fig. 15.7 from

$$u_\tau = u_r\cos\tau - u_t\sin\tau \tag{15.55}$$

For measurements of tangential velocities, refer to Sect. 15.2.

15.3.6 Remarks on the Method

The entire treatments shown above are based on the linearization according to Eqs. (15.34) and (15.35) for $r/R < 0.8$. This is in any case available because measurements beyond $r/R = 0.5$ rapidly become impossible, just as a result of the optical aberrations (e.g. astigmatism) and the growth of the beam thickness. The latter straightforwardly leads to the drop of laser light intensity in the measurement volume (total reflection occurs at $T = 1$). Calculations have shown that within $r/R = 0$ to $r/R = 0.5$ the maximum error in u_τ arising from the linearization is less than 1% (at $r/R = 0.5$). In most engineering flows, this error is well acceptable. The error in the deviation angle τ of the measurement volume, if Eq. (15.54) is applied, is less than 0.4%.

The optical aberration is related to the off-axis positioning of the measurement volume by r/R and can be represented, according to Eq. (15.52), by the off-axis angle φ_o of the optical axis. At large values of r/R, the off-axis angle becomes large. The entire optical properties then are comparable with those related with the off-axis LDA alignment to a plane wall, as thoroughly treated in Chap. 14. In the example of $n_1/n_2 = 1.14$ and $r/R = 0.6$ there is $\varphi_o = 32°$. This off-axis angle is associated with huge astigmatism effects, so that the effective aperture of the receiving lens and consequently the signal rate would be radically reduced, for reference see Fig. 14.9 as well as Fig. 14.18. Another serious problem at the large off-axis angle is the increased probability that due to a small inaccuracy in the optical layout both laser beams after refractions could no longer intersect at all. For this reason, it could be concluded that at positions of the measurement volume beyond $r/R = 0.5$, measurements would become impossible. This optical behaviour, in fact, also applies to the case for measurements of axial velocities with a deviation limit at $e/R < 0.5$, as already discussed in Sect. 15.1.

15.4 Optical Aberrations and Measurement Volume Distortion

Generally the optical condition of measurements continuously deteriorates as the depth of the measurement volume in the test fluid increases, for instance in measurements of both axial and tangential velocity distributions. From experience, as mentioned at the beginning of this chapter, flow measurements can be achieved at the most up to a depth of about 2/3 of the pipe diameter. The worsening of the optical condition, while positioning the measurement volume in the upper depth, is related to the increased optical aberration and the enlarged dislocation of laser beam waists. In measurements of radial velocities, the worsening of signal qualities becomes more and more significant with the shift of the measurement volume away from the pipe centre. This phenomenon as related to the optical aberration has already been discussed at the end of the last section.

The optical aberration and the beam waist dislocation will be quantified in the following sections. In particular, the analysis should comprehensibly reveal why the measurements of axial and tangential velocities will become impossible if the measurement volume is located in the pipe at a distance beyond 2/3 of the pipe diameter. To be mentioned is that analyses that will be made below are rather complex. They should mainly serve as the references and fundamentals for further extended investigations. For most LDA users it would be sufficient to correctly make measurements and afterwards to properly correct measurement results. They are therefore mostly referred to foregoing sections. In the case where the measurements do not run or both signal strengths and qualities are not satisfactory, the LDA users should know the reason that is often ascribed to the optical aberrations i.e. to the existence of astigmatism.

15.4.1 Optical Aberrations in Transmitting and Receiving Optics

The significance of using the plane cut-off on the outside of a circular pipe and aligning the LDA head perpendicular to this plane is to ensure laser beams in the pipe wall to be free of any optical aberrations. This is confirmed in that all four laser beams of a two-component LDA system intersect at an actual or a virtual unique point. The optical aberration then originates only at the internal surface of the circular pipe, leading to separations of two measurement volumes. In the case where the LDA optical axis goes through the pipe centre, these two measurement volumes correspond to those for measurements of axial and tangential velocities. To describe the optical aberration associated with the non-regular laser beam refractions on the curved pipe surface, the displacement between two measurement volumes for axial and tangential velocities, for the given LDA head position, can be applied. This is indeed the common case of using such a displacement as the astigmatic difference to represent the extent of the present optical aberration i.e. astigmatism, see Chap. 14.

The displacement between the two measurement volumes can be easily obtained from the calculation results that have already been obtained in Sects. 15.1 and 15.2. In using Eqs. (15.1) and (15.12), the displacement between two measurement volumes is obtained in function of the position of the virtual beam crossing point as

$$\Delta s_{\mathrm{R}} = \frac{s_{2,\mathrm{t}} - s_{2,\mathrm{a}}}{R} = \frac{1}{1 + \frac{n_1}{n_2}\left(\frac{R}{s_1} - 1\right)} - \frac{n_2}{n_1}\frac{s_1}{R} \tag{15.56}$$

This distance can also be shown in the function of the measurement volume position for tangent velocity component. The first term on the r.h.s. of the above equation is simply $s_{2,\mathrm{t}}/R$. The second term with s_1/R should be replaced by respective value from Eq. (15.12), so that

$$\Delta s_{\mathrm{R}} = \frac{s_{2,\mathrm{t}}}{R} - \frac{1}{\left(\frac{R}{s_{2,\mathrm{t}}} - 1\right) + \frac{n_1}{n_2}} \tag{15.57}$$

Figure 15.11 shows the calculated displacements between two measurement volumes in function of the position of the measurement volume for tangential velocity component. As can be seen, both measurement volumes get separated from each other, as the depth of the measurement volume in the flow increases. In other words, the optical aberration related to the LDA optics increases with the distance of the measurement volume from the pipe wall. This phenomenon significantly influences

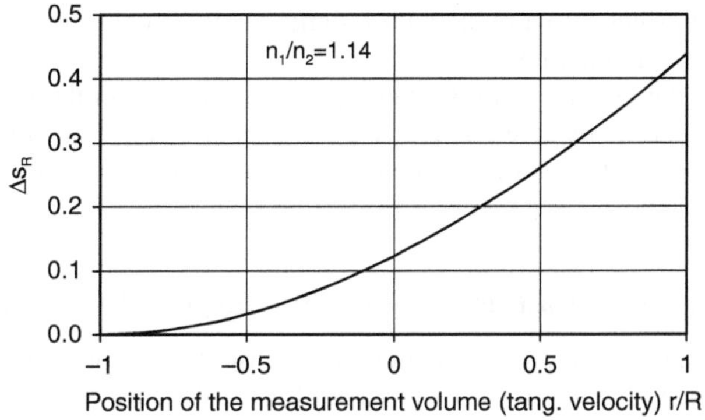

Fig. 15.11 Relative displacement between the tangential and the axial measurement volumes in function of the position of the measurement volume for the tangential velocity component

the measurements of both the axial and the tangential velocities. The influence mechanism is again found in the receiving optics.

Like in the case of astigmatism associated with the off-axis LDA against a plane medium interface (Chap. 14), the optical aberration in the current case implies that the larger the depth of the measurement volume in the flow, the fewer are the effective elementary segments on the receiving lens, that can see the measurement volume. Because the majority of elementary segments on the receiving lens behave as blind to the measurement volume, the signal strength diminishes and consequently the signal rate in data acquisition decreases. For this reason it is almost impossible to get optical signals of sufficiently high strengths and qualities in the flow area of a depth beyond 2/3 of the pipe diameter. Because of this the entire flow distribution through the pipe section could only be achieved, if an additional measurement is completed from the opposite side by rotating the LDA head for 180° around the pipe axis. This two-measurement feature applies to measurements of both the axial and tangential velocities. It provides, however, just an opportunity for utilizing the so-called Dual Measurement Method (DMM) with which the distribution of the very weak tangential velocity component can be exactly resolved from two measurements (see Chap. 9).

15.4.2 Dislocation of Laser Beam Waists from the Measurement Volume

Another undesirable outcome of the optical aberration related to measurements of tangential and radial velocities in a circular pipe is the dislocation i.e. the separation of laser beam waists from the measurement volume. This feature leads, on one hand, to the well-known fringe distortion in the measurement volume and, on the other hand, to the decrease of laser light intensity in the measurement volume and successively to the reduction of signal qualities. This type of signal disturbances additionally contributes to a reduction of signal qualities on account of the optical aberration just treated before.

To quantify the beam waist dislocation, the laser beams in the medium 1 prior to the test fluid are considered as focused light bundle and thus to be free of any optical aberration. Both cases of respectively measuring the tangential and the radial velocity components are separately considered.

15.4.2.1 Laser Beam Waists in Measuring the Tangential Velocities

With regard to the symmetrical refraction of two laser beams, only one laser beam is considered which is found in the $x-y$ plane (Fig. 15.12). Associated with the laser beam refraction on the internal surface of the pipe, the laser beam suffers from astigmatism effect. As a result there exist two particular focal points that are confirmed in the meridian and sagittal planes of the laser beam and denoted by p_{m} and p_{s}, respectively. The distance between these two focal points is called astigmatic difference. It is a measure of the extent of the associated optical aberration. To calculate the respective locations of these two focal points and to directly apply the analysis results in Sect. 14.6, a new coordinate ξ will be inserted that begins at the beam intersection point c and runs along the normal of the circular pipe i.e. goes through the pipe axis. The laser beam considered is represented here by the unit vector $\vec{r}$. Its corresponding ξ-coordinates prior to and after the refraction are represented by $r_{1\xi}=\cos\varepsilon_1$ and $r_{2\xi}=\cos\varepsilon_2$, respectively.

Because the laser beam is thin, the intersection area on the cylindrical surface is assumed as the plane surface. This assumption contributes to the simplification of beam refraction calculations. The respective meridian and sagittal focal points of the refracted laser beam in the test fluid have their ξ-coordinates which, according to Eqs. (14.41) and (14.42) for the plane interface, are given by

$$\xi_{\mathrm{m}}=\xi_{\mathrm{o}}\frac{n_2}{n_1}\frac{r_{2\xi}^3}{r_{1\xi}^3} \tag{15.58}$$

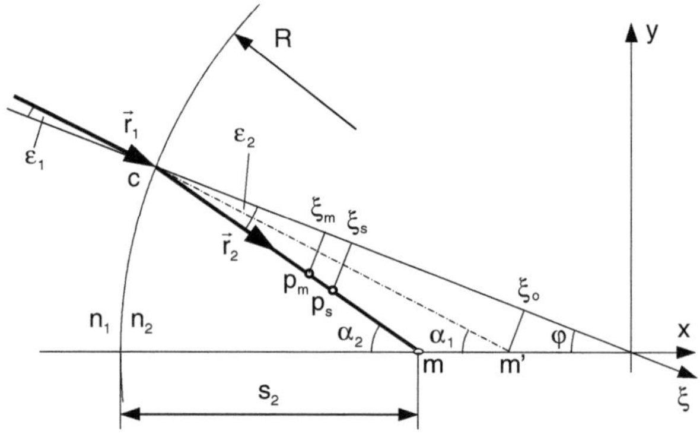

Fig. 15.12 Calculations of the laser beam waists and the measurement volume

and

$$\xi_s = \xi_o \frac{n_2}{n_1} \frac{r_{2\xi}}{r_{1\xi}} \tag{15.59}$$

Herein ξ_o denotes the ξ-coordinate of the virtual focal point m', as given by

$$\xi_o = cm' \cdot \cos\varepsilon_1 = \frac{R\sin\varphi}{\sin\alpha_1} r_{1\xi} \tag{15.60}$$

In the case of water flow in the circular pipe, it can be verified that $(r_{1\xi} - r_{2\xi})/r_{1\xi} < 0.1\%$ is nearly always satisfied, so that $r_{2\xi} \approx r_{1\xi}$ can be assumed. This shows that there is no need to differ from the meridian and sagittal focal points and the refracted laser beam can still be considered as a focused light bundle with a unique beam waist. For the sake of getting the simple form of the results, however, Eq. (15.59) is applied to represent the ξ-coordinate of the approximated unique laser beam waist:

$$\xi_w = \xi_s = \xi_o \frac{n_2}{n_1} \frac{r_{2\xi}}{r_{1\xi}} \tag{15.61}$$

On the refracted laser beam in the test fluid, this beam waist lies at a distance from the intersection point c:

$$\ell_w = \xi_w / \cos\varepsilon_2 = \xi_w / r_{2\xi} \tag{15.62}$$

Because of the symmetrical refraction at two laser beams the measurement volume lies on the x-axis at m and has a distance from the intersection point c:

$$\ell_{mv} = \frac{R\sin\varphi}{\sin\alpha_2} \tag{15.63}$$

The distance between the measurement volume and the waist of the laser beam is then given by $\ell_{mv} - \ell_w$ or in dimensionless form by

$$\ell_R = \frac{\ell_{mv} - \ell_w}{R} = \frac{\sin\varphi}{\sin\alpha_2} - \frac{n_2}{n_1} \frac{\sin\varphi}{\sin\alpha_1} \tag{15.64}$$

Clearly this distance depends on the local position of the measurement volume in the flow. Both the angle φ and α_2 in function of the position s_2 of the measurement volume can be obtained by Eq. (15.11). Figure 15.13 shows a calculation example, in which the position parameter s_2 is replaced by the radial position $r = R - s_2$. Evidently, there exists a largely extended dislocation of the beam waist from the measurement volume, when the measurement volume is located beyond the pipe centre. Such a dislocation of the beam waist doubtless implies the low brightness of the measurement volume and the undesirable fringe distortion in it. These two features related to the measurement volume on the side of LDA transmitting optics

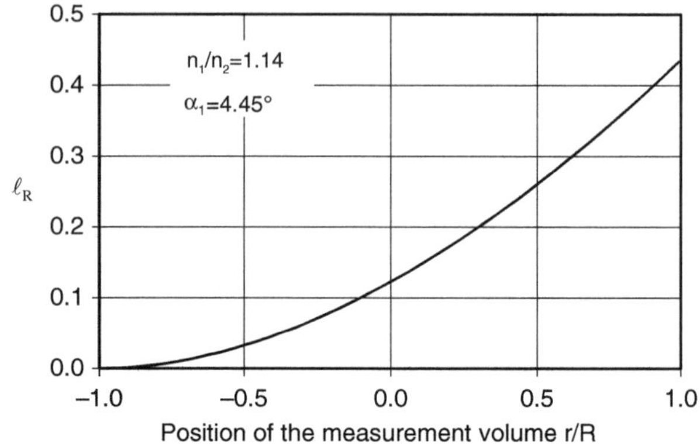

Fig. 15.13 Relative dislocation of the laser beam waists from the measurement volume at the measurements of the tangential velocity component

again signify that the measurement of tangential velocity components is more critical than that of axial velocities. In addition, it can be concluded from the values shown in Fig. 15.13 in dimensionless form that the absolute beam waist dislocation ($\ell_{mv} - \ell_w$) will be bigger, if circular pipes of large diameter are encountered in use. On the pipe axis for instance, the distance between the beam waist and the measurement volume is about $0.12R$, which will be 12 mm when $R = 100$ mm.

In comparing Fig. 15.13 with Fig. 15.11, it is evident that two curves are rather identical. This identity can be confirmed by applying the approximation for paraxial rays as in the form of $\sin\varphi \approx \varphi$, $\sin\alpha_1 \approx \alpha_1$ and $\sin\alpha_2 \approx \alpha_2$. Equation (15.64) can then be conducted to a form equal to Eq. (15.57).

The fringe distortion in the measurement volume in the current case is caused by the dislocation of beam waists in two symmetrical laser beams. It exactly represents the well-known type of fringe distortions at which the measurement volume is found prior to or after their respective waists located at equal distance from the beam intersection point. The optical occurrence in the present case is illustrated in Fig. 15.14. The fringe spacing in the measurement volume then linearly varies along the measurement volume. Corresponding errors involved in measurements of both the mean velocity and the flow turbulence have already been exactly analyzed by Zhang and Eisele (1997, 1998c). The apparent i.e. the affected mean velocity and its standard deviation are given as

$$\frac{\bar{u}_{app}}{\bar{u}} = 1 + \frac{1}{3}\gamma^2 \tag{15.65}$$

$$\frac{\sigma_{app}^2 - \sigma^2}{\bar{u}^2} = \gamma^2\left(\frac{\sigma^2}{\bar{u}^2} + \frac{1}{3}\right) \tag{15.66}$$

The exact derivation of these two equations will be presented in Chap. 16.

In these two equations, the fringe distortion number γ is a geometrical parameter that represents the relative change in the fringe spacing at the end of the measurement volume to the initial fringe spacing and is usually below 0.02. Because it deals

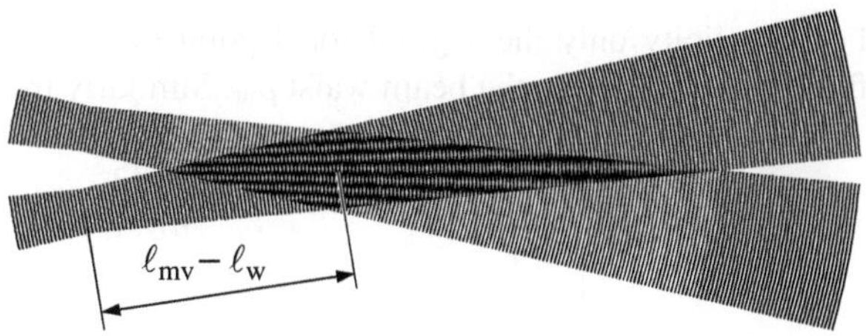

Fig. 15.14 Existence of the first type of the fringe distortion in the measurement volume at the measurement of the tangential velocity component in the circular pipe

with a quite small number, the fringe distortion in the LDA measurement volume usually does not appreciably influence the measurement accuracies.

15.4.2.2 Laser Beam Waists in Measuring the Radial Velocities

In Sect. 15.3, it has been indicated that for measurements of radial velocity components the measurement volume should be located on the y-axis, as shown in Fig. 15.7. It has also been indicated, relying on Eq. (15.36), that for shifting the measurement volume along the y-axis the laser beam pair in the medium 1 needs only to be shifted in the parallel direction. This implies that the virtual beam crossing point has also to lie on the y-axis, as it is so when the measurement volume is positioned at the pipe centre ($y = 0$). Corresponding locations of both the actual and virtual measurement volumes have been shown in Fig. 15.15, in which for simplicity only one laser beam has been illustrated. To be expected is that the dislocation of laser beam waists from the measurement volume also exists in this case. For calculations the laser beam is again represented by the unit vector $\vec{r}_1$ in the medium 1 and $\vec{r}_2$ in the medium 2.

Like in the above calculations, a new coordinate ξ will be inserted which begins at the beam intersection point c and runs along the normal of the circular pipe. Both unit vectors then have their components $r_{1\xi} = \cos\varepsilon_1$ and $r_{2\xi} = \cos\varepsilon_2$, respectively.

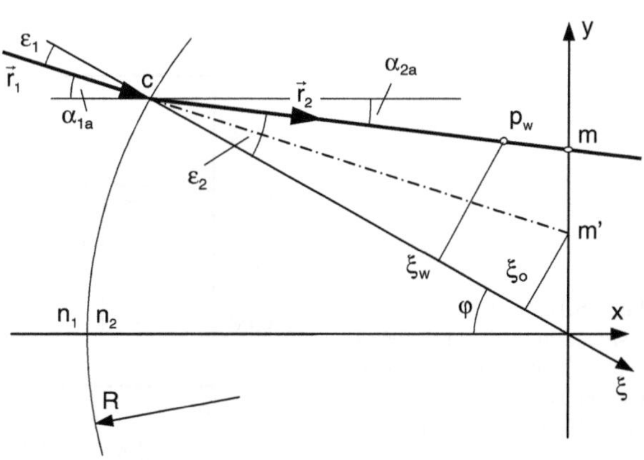

Fig. 15.15 Calculations of the laser beam waist (p_w) and the measurement volume (m)

For simplicity only the sagittal focal point of the refracted laser beam in the test fluid is considered as the beam waist p_w. Similarly there is

$$\xi_w = \xi_o \frac{n_2}{n_1} \frac{r_{2\xi}}{r_{1\xi}} \tag{15.67}$$

with

$$\xi_o = cm' \cos \varepsilon_1 = \frac{R \cos \varphi}{\cos \alpha_{1a}} r_{1\xi} \tag{15.68}$$

On the refracted laser beam in the test fluid, the beam waist lies at p_w with a distance from the intersection point c

$$\ell_w = \frac{\xi_w}{\cos \varepsilon_2} = \frac{\xi_w}{r_{2\xi}} \tag{15.69}$$

The measurement volume m has a distance from the same intersection point c

$$\ell_{mv} = \frac{R \cos \varphi}{\cos \alpha_{2a}} \tag{15.70}$$

The distance between the measurement volume (m) and the laser beam waist (p_w) is then given by $\ell_{mv} - \ell_w$. In dimensionless form and with subscripts a and b for laser beam A and B, respectively, there are

$$\ell_{R,a} = \frac{\cos \varphi_a}{\cos \alpha_{2a}} - \frac{n_2}{n_1} \frac{\cos \varphi_a}{\cos \alpha_{1a}} \tag{15.71}$$

$$\ell_{R,b} = \frac{\cos \varphi_b}{\cos \alpha_{2b}} - \frac{n_2}{n_1} \frac{\cos \varphi_b}{\cos \alpha_{1b}} \tag{15.72}$$

Clearly both of these distances depend on the local position r/R of the measurement volume in the flow. Angles φ_a, φ_b, α_{2a} and α_{2b} and their dependences on r/R

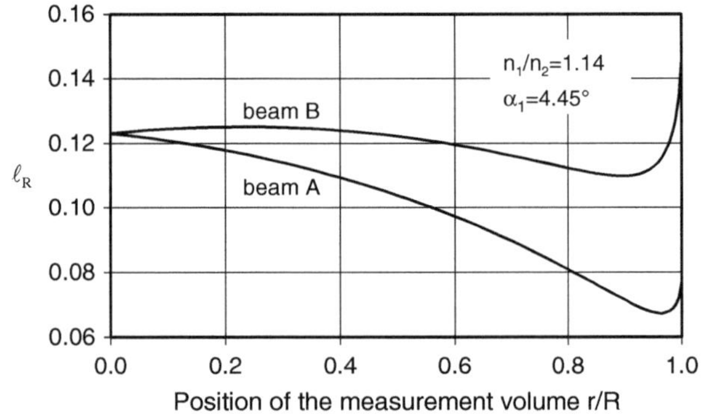

Fig. 15.16 Relative dislocation of the laser beam waists from the measurement volume at the measurement of the radial velocity component

have already been found to be given by Eqs. (15.38), (15.39), (15.40), (15.41), and (15.42).

Figure 15.16 shows the results calculated from Eqs. (15.71) and (15.72) for laser beam A and B, respectively. Because in the available area of $r/R < 0.6$ both beam waists are on the same side of the measurement volume and show no large differences in the distance to the measurement volume, it can be assumed that the fringe distortion in the manner according to Fig. 15.14 occurs. Like in the case of beam waist dislocation at measurements of tangential velocity components, the beam waist dislocation $(\ell_{\mathrm{mv}} - \ell_{\mathrm{w}})$ in the current case of measuring the radial velocities will also be large, if circular pipes of large diameter are encountered to use. On the pipe axis same values are obtained as in Fig. 15.13.

Chapter 16
Fringe Distortion Effects

From the LDA principle described in Chap. 3, the necessary condition for accurate LDA measurements is the uniformity of the fringe spacing in the measurement volume. The uniform fringe spacing can be achieved if the measurement volume that is formed by two laser beams coincides with two beam waists. Each deviation from this requirement will lead to fringe distortion in the measurement volume and hence to measurement errors. Because of the non-uniformity of the fringe spacing along the measurement volume a uniform constant laminar flow for instance will then be measured as a flow with velocity fluctuations. Measurements of both the mean velocity and the turbulence quantities thus suffer from systematic errors.

Fringe distortions in LDA measurement volumes have been historically considered as the consequence of improper optical layout. Two most well-known forms of the optical layout causing the fringe distortion have been shown in Fig. 16.1. The visible non-uniformity of the fringe spacing in these two cases is either along or across the measurement volume. Corresponding detailed investigations to characterize the non-uniformity of two such different fringe patterns have been performed for instance by Hanson (1973, 1975), Durst and Stevenson (1975) and Miles and Witze (1994, 1996). According to Hanson (1973, 1975) linear distributions of the fringe spacing exist in both longitudinally (Fig. 16.1a) and laterally (Fig. 16.1b) distorted measurement volumes. The influence of the fringe distortion on the measurement accuracy has been investigated by Zhang and Eisele (1997, 1998c) with respect to the longitudinal fringe distortion. As it has already been shown in Fig. 15.14, the first type of the fringe distortion in the measurement volume is confirmed as exactly taking place, when the tangential velocity of the flow in a circular pipe is measured without matching the refractive index of the fluid. It is obviously the most representative fringe distortion encountered in the practical applications. The second type of fringe distortion across the measurement volume (Fig. 16.1b), however, is still considered as merely a matter of the improper optical layout.

Another type or the third type of possible fringe distortions in the LDA measurement volume, as shown in Fig. 14.14b for a special case, is related to the astigmatism due to the laser beam refractions. Because of the irregular distribution of the beam waists around the LDA measurement volume and hence the complexity of the form of respective wave front of two laser beams, this type of fringe distortion may not yet

Zh. Zhang, *LDA Application Methods*, Experimental Fluid Mechanics,
DOI 10.1007/978-3-642-13514-9_16, © Springer-Verlag Berlin Heidelberg 2010

(a)

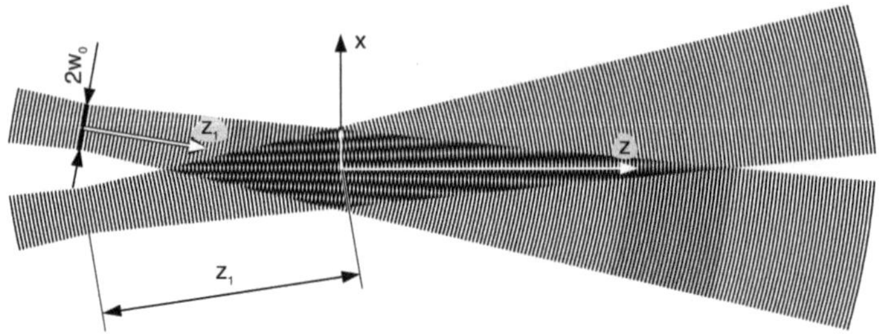

(b)

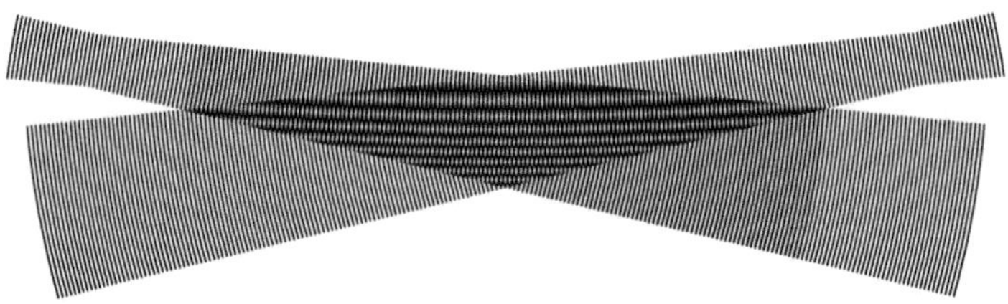

Fig. 16.1 Fringe distortions (first and second types) in the measurement volume; The third type of the possible fringe distortion is referred to Fig. 14.14b

be well characterized. A further type of fringe distortions in the measurement volume is known as the local fringe distortion which is caused by laser light diffraction through particles in the transmission path of the laser beams (Ruck 1991).

The outcome of the fringe distortion in the LDA measurement volume is the systematic measurement error in both the mean velocity and the turbulence quantities. In all cases of measuring the flow turbulence, the fringe distortion results in the broadening of the Doppler frequency and hence the overestimation of the turbulence intensity. This overestimation obviously depends on both the form and the scale of the complex fringe distortion. As an LDA user one is indeed interested in knowing the extent of respective measurement errors and the possibility of correcting them. For this purpose, the influence of the most representative fringe distortion in form of Fig. 16.1a on the flow measurement is considered here, in order to give a reference as well as to make a criterion for error estimations. The analysis assumes the linear distribution of the fringe spacing along the measurement volume length.

16.1 Linear Longitudinal Distribution of the Fringe Spacing

According to Fig. 16.1a the crossing of two Gaussian beams takes place after their respective waists located at equal distance from the beam intersection point. The same fringe distortion with the same consequence in LDA flow measurements will be given when the beam crossing is found prior to both beam waists. The assumption of linear distribution of the fringe spacing in the LDA measurement volume is based on earlier investigations of this type of fringe distortion and contributes to the simplification of calculations.

The uniform velocity distribution within the measurement volume length is assumed. From its measurement by $u = \Delta x \cdot \nu_D$ with non-uniform fringe spacing, one obtains the related non-uniform Doppler frequency as

$$\frac{1}{\nu_D}\frac{d(\nu_D)}{dz} = -\frac{1}{\Delta x}\frac{d(\Delta x)}{dz} \tag{16.1}$$

According to Hanson (1973, 1975) from accounting for the relative shift in Doppler frequency, the fringe spacing gradient is expressed as

$$\frac{1}{\Delta x}\frac{d(\Delta x)}{dz} = \frac{1}{R} \tag{16.2}$$

Herein R is the radius of curvature of two Gaussian beam wave fronts at the beam crossing point. It is calculated, according to Eq. (3.63), with the spot size (radius w_0) at the beam waist and the distance (z_1) of the beam waist from the beam crossing point as follows

$$R = z_1\left[1 + \left(\frac{\pi w_0^2}{\lambda z_1}\right)^2\right] = z_1\left[1 + \left(\frac{z_R}{z_1}\right)^2\right] \tag{16.3}$$

In this equation, z_R represents the Rayleigh length, as given in Eq. (3.66).

From Eq. (16.2) it can be shown that, by assuming $|z/R| << 1$ within the region of the measurement volume, the longitudinal fringe spacing varies linearly over the length of the measurement volume. That is with $k = \Delta x_0/R$

$$\Delta x = kz + \Delta x_0 \tag{16.4}$$

In this equation, Δx_0 is the fringe spacing at the centre of the measurement volume ($z = 0$). According to Hanson (1973), this value of fringe spacing is equal to that in the undistorted measurement volume ($\lambda/2\sin\alpha$).

16.2 Fringe Distortion Number and the Apparent Mean Velocity

It should be mentioned that the fringe distortion in the measurement volume also influences the measurement of mean velocities. Indeed, the error in the mean velocity will not disappear, even if linear fringe distortion according to Eq. (16.4) takes place. This can be easily demonstrated by assuming the uniform flow of velocity u_0 through the measurement volume. From the measured Doppler frequency and the specified constant fringe spacing in the software, the flow velocity is calculated as follows:

$$u = \Delta x_0 \nu_D = \Delta x_0 \frac{u_0}{\Delta x} \tag{16.5}$$

It is inversely related to the fringe spacing which linearly changes along the measurement volume. For this reason the ensemble average of velocities u from measurements is in no cases equal to the actual flow velocity u_0 and is therefore denoted as the apparent mean. For the general case a turbulent flow with random velocity fluctuations is considered to have a mean velocity equal to $\overline{u}$. The apparent mean velocity is calculated by the arithmetic average as

$$\overline{u}_{\text{app}} = \Delta x_0 \frac{1}{N} \sum_{i=1}^{N} \frac{u_i}{\Delta x_i} \tag{16.6}$$

On the side of the measurement volume, the measurement volume will longitudinally be divided into m partial volumes of equal distance. In each partial volume, the fringe spacing can be considered to be constant. On the side of the flow, the same and constant statistical flow properties are assumed to exist among all partial volumes. This also includes the assumption that particles have equal probability in passing through every partial volume. With respect to $N = m \cdot n$ and the mean velocity equal to $\overline{u}$ Eq. (16.6) is then written as

$$\overline{u}_{\text{app}} = \Delta x_0 \frac{1}{N} \sum_{j=1}^{m} \left(\frac{1}{\Delta x_j} \sum_{i=1}^{n} u_i \right) = \Delta x_0 \overline{u} \frac{1}{m} \sum_{j=1}^{m} \frac{1}{\Delta x_j} \tag{16.7}$$

By extending m to infinity and with substitution of Δx_j by Eq. (16.4) the summation in the above equation can be presented by the corresponding integral calculation, so that with $\Delta z/2$ as the half length of the measurement volume

$$\overline{u}_{\text{app}} = \overline{u} \frac{\Delta x_0}{\Delta z} \int_{-\Delta z/2}^{\Delta z/2} \frac{dz}{kz + \Delta x_0} \tag{16.8}$$

To simplify the calculation results, the fringe distortion number is introduced as defined by

$$\gamma = \frac{k\,\Delta z/2}{\Delta x_0} = \frac{\Delta z/2}{R} \tag{16.9}$$

It represents the relative change of the fringe spacing at the end of the measurement volume. Usually it is a small value because of $\Delta z << R$.

Equation (16.8) is then calculated as

$$\frac{\overline{u}_{\text{app}}}{\overline{u}} = \frac{1}{2\gamma} \ln \frac{1+\gamma}{1-\gamma} \approx 1 + \frac{1}{3}\gamma^2 \tag{16.10}$$

Herein the approximation has been made because of $\gamma << 1$.

This calculation result demonstrates that the measurement of the mean flow velocity will also be affected by the fringe distortion in the measurement volume, even though the linear fringe spacing distribution is assumed. The relative error, however, is mostly negligible because of $\gamma << 1$.

The fringe distortion number has been defined in Eq. (16.9) as a pure function of geometric parameters of the measurement volume. Since it specifies the relative change in the fringe spacing at the end of the measurement volume ($z = \Delta z/2$) against that at the measurement volume centre ($z = 0$), it also represents the frequency broadening $\Delta\nu_D/\nu_D$ at $z = \Delta z/2$.

According to Eq. (16.9) the fringe distortion number has also been shown to be a function of the radius of wave front curvature of two Gaussian beams at the beam crossing point. Substituting this curvature radius by Eq. (16.3) then yields

$$\gamma = \frac{\Delta z/2}{z_1\left[1 + (z_R/z_1)^2\right]} \tag{16.11}$$

For laser beams with given beam waist thickness $2w_0$ and hence given Raleigh length z_R the fringe distortion number has been shown as the function of the distance z_1 between the beam waist and the measurement volume centre. This functionality is illustrated in Fig. 16.2 for a given LDA optical set-up. The radius of the curvature of the Gaussian beam wave front has also been shown in function of the distance. At the Rayleigh length which is equal to $z_R = 30$ mm in this example, the fringe distortion number reaches its maximum.

Based on parameter quantifications in this example the fringe distortion number has been confirmed to be in the range of usually not exceeding 0.02. In applying this limit to Eq. (16.10), the error in the mean velocity is practically very small. By the way, the minimum radius of the curvature of the Gaussian beam wave front reads $R = 60$ mm. Compared with this value, the half length of the measurement volume ($\Delta z/2 = 1$ mm) is negligible. The assumption of $|z/R| << 1$ that leads to Eq. (16.4) is thus validated.

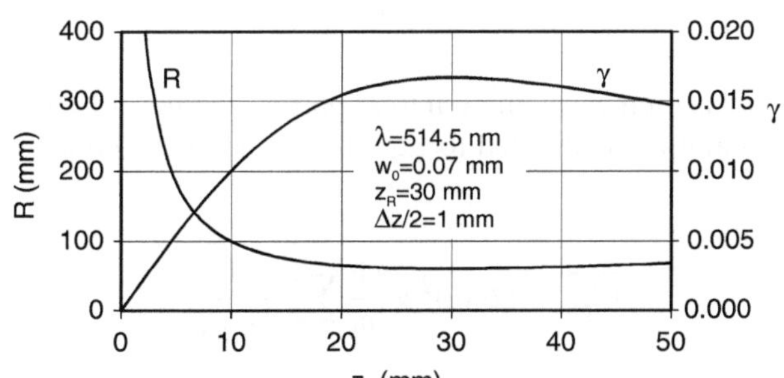

Fig. 16.2 Radius of the curvature of the Gaussian beam wave front and the fringe distortion number

16.3 Overestimation of the Flow Turbulence

As is well-known, the most significant outcome of the fringe distortion in LDA measurements is the Overestimation of the turbulence intensity. This is also called the broadening effect in turbulence measurements. In order to quantify this effect, a stationary turbulent flow is considered which is specified by a mean velocity $\bar{u}$ and a fluctuation velocity σ (standard deviation). From the viewpoint of statistics, the connection between the mean velocity and the standard deviation is given, according to Eq. (5.5), as

$$\sigma^2 = \overline{u^2} - \bar{u}^2 \tag{16.12}$$

with $\overline{u^2}$ as the mean square of velocity component u.

This relationship also applies in the case of velocity data from measurements undergoing the effect of fringe distortion in the measurement volume. Corresponding velocities are thus apparent, as given by

$$\sigma_{\mathrm{app}}^2 = \overline{u_{\mathrm{app}}^2} - \bar{u}_{\mathrm{app}}^2 \tag{16.13}$$

The overestimation of the turbulence intensity and the related quantities as a result of the fringe distortion in the measurement volume is then determined by

$$\sigma_{\mathrm{app}}^2 - \sigma^2 = \left(\overline{u_{\mathrm{app}}^2} - \overline{u^2}\right) - \left(\bar{u}_{\mathrm{app}}^2 - \bar{u}^2\right) \tag{16.14}$$

While the second term on the r.h.s. of this equation can be calculated by using Eq. (16.10) or simply set to zero, the first term requires similar calculations as those in Sect. 16.2. In response to each velocity event the Doppler frequency is again $u_{\mathrm{i}}/\Delta x_{\mathrm{i}}$. The apparent mean square of velocities in the above equation is basically calculated, in analogy to Eq. (16.6), by

$$\overline{u_{\mathrm{app}}^2} = (\Delta x_0)^2 \frac{1}{N} \sum_{\mathrm{i}=1}^{\mathrm{N}} \left(\frac{u_{\mathrm{i}}}{\Delta x_{\mathrm{i}}}\right)^2 \tag{16.15}$$

By dividing the measurement volume into m partial volumes of equal distance and based on same assumptions that led to Eq. (16.7), the above equation is converted into

$$\overline{u_{\mathrm{app}}^2} = (\Delta x_0)^2\, \overline{u^2} \frac{1}{m} \sum_{\mathrm{j}=1}^{\mathrm{m}} \frac{1}{\left(\Delta x_{\mathrm{j}}\right)^2} \tag{16.16}$$

The linear fringe spacing distribution according Eq. (16.4) is applied. By extending m to infinity, the summation in the above equation can be presented by the corresponding integral calculation, so that

$$\overline{u_{\text{app}}^2} = (\Delta x_0)^2 \overline{u^2} \frac{1}{\Delta z} \int_{-\Delta z/2}^{\Delta z/2} \frac{dz}{(kz + \Delta x_0)^2} \tag{16.17}$$

In using the fringe distortion number as defined by Eq. (16.9), one obtains

$$\overline{u_{\text{app}}^2} = \frac{1}{1-\gamma^2}\overline{u^2} \approx \left(1+\gamma^2\right)\overline{u^2} \tag{16.18}$$

Substitution of Eqs. (16.10) and (16.18) in Eq. (16.14) and with respect to $\overline{u^2} = \sigma^2 + \bar{u}^2$ yields

$$\frac{\sigma_{\text{app}}^2 - \sigma^2}{\bar{u}^2} = \gamma^2 \left(\frac{\sigma^2}{\bar{u}^2} + \frac{1}{3}\right) \tag{16.19}$$

Because of $\sigma_{\text{app}} + \sigma \approx 2\sigma$ (not for $\sigma = 0$) this equation can also be written as

$$\frac{\Delta\sigma}{\bar{u}} = \frac{1}{2}\gamma^2 \frac{\bar{u}}{\sigma}\left(\frac{\sigma^2}{\bar{u}^2} + \frac{1}{3}\right) \tag{16.20}$$

with $\Delta\sigma = \sigma_{\text{app}} - \sigma$ as the overestimation of the standard deviation of the mean flow velocity.

The overestimation of the turbulence intensity obviously depends on both the real flow turbulence to be measured and the extent of the fringe distortion that is specified by the fringe distortion number. In regards Eq. (16.20), the overestimation of the flow turbulence is illustrated in Fig. 16.3 in the function of the real turbulence intensity for different fringe distortion numbers. For typical fringe distortions ($\gamma < 0.02$) the overestimation of the flow turbulence has been found to be not significant, especially in the measurement of flows with high turbulence intensity.

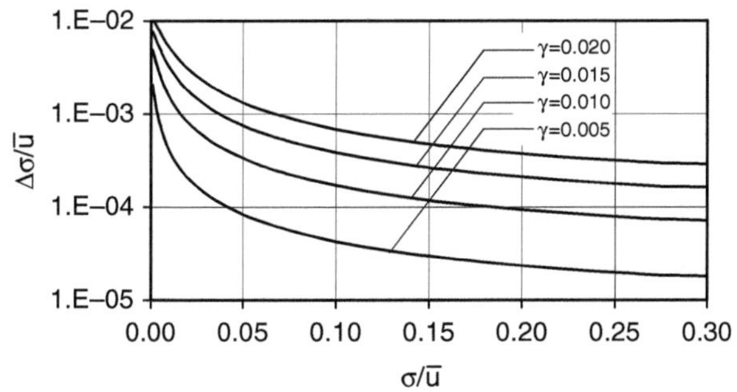

Fig. 16.3 Overestimations of the standard deviation in function of the real flow turbulence and the fringe distortion number γ

For a uniform laminar flow with $\sigma = 0$ the apparent turbulence intensity is calculated from Eq. (16.19) as

$$\frac{\sigma_{\text{app}}}{\overline{u}} = \sqrt{1/3} \cdot \gamma \tag{16.21}$$

Calculations made in this chapter basically refer to the fringe distortion with linear longitudinal variation in the fringe spacing (Fig. 16.1a). The calculation results can be used as the reference, if other types of fringe distortions and the respective influences on measurements should be accounted for. In general, the influence of the second type of fringe distortions (Fig. 16.1b) on turbulence measurements is rather smaller than that treated above, as this influence was already considered by Hanson (1975) in the earlier time to be negligible. The same might also be true for the case accounting for the third type of fringe distortions that is caused by the optical aberration i.e. astigmatism according to Fig. 14.14.

Chapter 17
Velocity Bias Effects

17.1 Velocity Bias as a Flow Phenomenon

Flow velocity measurements by means of LDA method are based on random sampling of velocity events which occur when particles pass through the measurement volume. The available velocity sampling rate depends on particle concentration, particle size, flow velocity and other flow and optical parameters. The dependence on the velocity magnitude in the turbulent flow becomes significant, if the sampling rate is considerably larger than the mean frequency of flow fluctuations. This type of dependence is generally characterized by the fact that velocities of large magnitudes will be more frequently sampled than velocities of small magnitudes. Accordingly, the time between two subsequent velocity samples is not equidistant. The sample mean of velocities according to Eq. (5.1) is then higher than that from velocities with equidistant time intervals. The shift of the sample mean of velocities towards the high value has been called the velocity bias and was first recognised by McLaughlin and Tiederman (1973). From the mechanism of velocity bias it is expected that not only the mean velocity but also other statistical parameters like the standard deviation show the difference to those from measurements with equidistant time intervals. Because of its dependence purely on velocity fluctuations, the velocity bias indeed represents a flow phenomenon. In the LDA terminology, the velocity bias i.e. the related difference in mean velocities has been categorized as being the measurement error. Corresponding investigations are generally restricted to two aspects: estimation and correction of effects of velocity bias.

In assuming homogeneous particle distribution in the flow, the particle arrival rate across the measurement volume and thus the velocity sampling rate are theoretically proportional to the velocity magnitude. From this viewpoint, considerable efforts have been made to estimate and correct the effect of velocity bias (Buchhave 1975, Buchhave et al. 1979, Lehmann 1986, Nobach 1998, Owen and Rogers 1975). An empirical estimation of the biased mean velocity has been given for instance by Edwards (1987) as a function of the turbulence intensity $\overline{u}_\mathrm{b}/\overline{u} = 1+\mathrm{Tu}^2$. In applying the numerical method, Nobach (1998) investigated bias effects which are included in both the mean velocity and the standard deviations in all three velocity components of a complex three-dimensional turbulent flow. Exact analytical quantifications of

Zh. Zhang, *LDA Application Methods*, Experimental Fluid Mechanics,
DOI 10.1007/978-3-642-13514-9_17, © Springer-Verlag Berlin Heidelberg 2010

the bias effects have been carried out by Zhang (2002), considering all three components of velocity fluctuations from zero to infinity. The analysis serves as a reference to estimate the extent of the bias effects.

One of the well-known effects of velocity bias on measurements of turbulent flows is the distortion of the symmetrical distribution of velocity fluctuations around the corresponding mean. Another effect is the misinterpretation of an isotropic turbulence as an apparent anisotropic one. In addition, the bias errors in all three velocity components obviously differ from each other.

Concerning the matter of correcting for the bias effect, different methods have been developed and applied in data processing of LDA measurements. The most common method uses the reciprocal velocity or the residence time of particles in the measurement volume as a weighting factor in the calculations for arithmetic averages, as these have been shown at Eqs. (5.13), (5.14), and (5.15). The method of using the reciprocal value of each velocity vector as the weighting factor is obviously very limited, because this value must be known from measurement. In addition, the result is only true if the flow fluctuation is one-dimensional. The method of using the residence time or the transit time as the weighting factor is based on the assumption of statistically uniform flow through the measurement volume (no velocity gradient within the measurement volume). This method indeed has been recognised to be able to account for both the three-dimensionality of flow fluctuations and the measurement volume shape (Buchhave et al. 1979). The transit time can be theoretically attained by measuring the burst lengths. Another correction method to be mentioned is the method of using controlled processors to externally impose a constant sampling frequency (Edwards 1987, Erdmann and Tropea 1981). Such processors are basically only recommended for measurements with high data densities. In reality, the constant sampling frequency that is externally imposed can also be obtained by the post data processing.

As it states, the velocity bias has been historically considered to affect the measurement accuracy. This viewpoint, however, has to be partially declined. According to Chap. 12, the velocity bias does represent a measurement error if the volumetric flow rate (or volumetric flux) is to be calculated without any correction. The velocity bias, however, precisely ensures correct velocity measurements and data processing if the momentum flux is interesting. This will be required when dealing with the momentum equations like Euler, Navier-Stokes or Reynolds equations. Concerning the matter of flow turbulence, the relevant physical turbulence quantities are generally turbulent Reynolds stresses according to Eqs. (2.7), (2.8), and (2.9). Because they all represent the momentum quantities (or in terms of the specific kinetic energy), the occurrence of velocity bias in LDA measurements does not mean any measurement errors. For this reason, it is generally necessary to basically clarify the physical significance of each flow parameter that needs to be obtained from measurements.

For the historical reason, nevertheless, the word "velocity bias" will still be applied to denote the related flow phenomenon.

For LDA users in engineering applications, it is rather more interesting to know about the extent of velocity bias and its influences than to correct it. In general, it is the responsibility of researchers and engineers to decide whether the associated measurement inaccuracy is acceptable or not, based on requirements of each related flow process. It is therefore reasonable to make distinctions between mean velocities and other flow parameters with and without the effect of velocity bias. The analysis made by Zhang (2002) will be referred to as the main content of this chapter.

It should be mentioned that in measurements of turbulent flows by the LDA method, another type of bias, the so-called angular bias, exists. This bias arises from the fact that the cross sectional area of the measurement volume and thus the particle arrival rate changes with the flow direction. It differs from the velocity bias physically in that the angular bias is not simply a flow phenomenon. Due to the three-dimensionality of flow fluctuations, it is usually impossible to distinguish between the velocity bias and the angular bias in the measurements.

17.2 Velocity Bias and the Momentum Flow Rate

It has been indicated that the velocity bias is in effect a flow phenomenon. This gives rise to basically clarifying its related flow dynamic properties. In Chap. 12 with regard to the non-uniform velocity distribution along the LDA measurement volume, it has been demonstrated that the related biased mean velocity precisely represents the mean velocity which is relevant for the momentum flow rate across the measurement volume. The ratio of the biased to the unbiased mean velocity simply stands for the so-called momentum flux correction factor. Because the velocity bias is traditionally referred to the turbulent flow with time-dependent velocity fluctuations, it is naturally of interest whether the same relation exists. The following analysis will demonstrate that this is true.

As known, the momentum flow rate is a vector quantity that is determined by the velocity vector ($\vec{u}$) and the normal vector ($\vec{n}_\mathrm{A}$) of the related cross-sectional area in the flow. The so-called momentum flux (momentum flow rate per unit area) is then given by $\dot{J} = \rho\vec{u} \cdot (\vec{n}_\mathrm{A} \cdot \vec{u})$. In this equation, the product in brackets represents the volumetric flux in unit of $\mathrm{m}^3/(\mathrm{m}^2\mathrm{s})$. For further discussion a one-dimensional stationary turbulent flow of velocity u_x is considered. The volumetric flux through the unit area perpendicular to x-axis is then simply equal to u_x. The mean momentum flux during a time interval Δt is calculated by

$$\bar{\dot{J}}_\mathrm{x} = \rho\frac{1}{\Delta t}\int_0^{\Delta t} u_\mathrm{x}^2 \cdot \mathrm{d}t \tag{17.1}$$

For large time intervals, the averaged momentum flux represents a statistical constant of the flow. Based on the ergodic hypothesis the average of such a process parameter over time is equal to the average over the statistical ensemble. In order to convert the above calculation into the ensemble average, the statistical distribution of fluctuation velocity, as the ensemble from measurements, is considered as being described by an existing probability density function (pdf) p. The most well-known pdf is obviously the Gaussian function (see Figs. 2.1 and 2.2). The ensemble averaged momentum flux can then be calculated by the corresponding integral

$$\bar{\dot{J}}_{\mathrm{x}} = \rho \int_{-\infty}^{\infty} p u_{\mathrm{x}}^2 \mathrm{d}u_{\mathrm{x}} \tag{17.2}$$

On the other side, the averaged momentum flux is usually represented by the product of an appropriate mean velocity and the mean of the volumetric flux $\overline{u}_{\mathrm{x}}$ as

$$\bar{\dot{J}}_{\mathrm{x}} = \rho \overline{u}_{\mathrm{x,J}} \overline{u}_{\mathrm{x}} \tag{17.3}$$

With respect to Eq. (17.2) the mean velocity $\overline{u}_{\mathrm{x,J}}$ that represents a significant parameter for the momentum flow rate or likewise for the momentum flux is calculated by

$$\overline{u}_{\mathrm{x,J}} = \frac{1}{\overline{u}_{\mathrm{x}}} \int_{-\infty}^{\infty} p u_{\mathrm{x}}^2 \mathrm{d}u_{\mathrm{x}} \tag{17.4}$$

In dealing with stationary laminar flows there is simply $\overline{u}_{\mathrm{x,J}} = \overline{u}_{\mathrm{x}}$.

For practical applications and in likeness to dealing with flows with non-uniform velocity distributions, the momentum flux correction factor β can be applied to the concerns of the effect of velocity fluctuations in each interested turbulent flow, as given in the form

$$\overline{u}_{\mathrm{x,J}} = \beta \overline{u}_{\mathrm{x}} \tag{17.5}$$

The momentum flux given in Eq. (17.3) is then expressed as

$$\bar{\dot{J}}_{\mathrm{x}} = \beta \cdot \rho \overline{u}_{\mathrm{x}}^2 \tag{17.6}$$

The velocity bias involved in LDA measurements will be again considered. As a result of velocity bias the probability density function p of the fluctuation velocity distribution will be distorted when calculated from measurements. In assuming the linear dependence of velocity sampling rate on the magnitude of velocities, the distorted i.e. biased probability density function then takes the form as given by

$$p_{\mathrm{b}} = k \left| u_{\mathrm{x}} \right| p \tag{17.7}$$

The constant k in this equation must fulfil the condition

$$\int_{-\infty}^{\infty} p_{\mathrm{b}} \mathrm{d}u_{\mathrm{x}} = 1 \tag{17.8}$$

from which one obtains

$$k = \frac{1}{\int\limits_{-\infty}^{\infty} p\,|u_x|\mathrm{d}u_x} \tag{17.9}$$

Like at Eq. (5.2), the sample mean of velocities based on Eq. (5.1) can be calculated equivalently by the following integral with respect to the biased probability density function

$$\overline{u}_{x,b} = \int\limits_{-\infty}^{\infty} p_b u_x \mathrm{d}u_x = k \int\limits_{-\infty}^{\infty} p u_x\,|u_x|\,\mathrm{d}u_x \tag{17.10}$$

In all low turbulent flows, the velocity fluctuations ($u'_x = u_x - \overline{u}_x$) are always small compared to the mean velocity so that the occurrence of the back flow is negligible and $u_x > 0$ at all times is satisfied. From Eq. (17.9) with respect to $u_x > 0$ it follows $k = 1/\overline{u}_x$ directly from the mean velocity definition. Consequently Eq. (17.10) becomes

$$\overline{u}_{x,b} = \frac{1}{\overline{u}_x} \int\limits_{0}^{\infty} p u_x^2 \mathrm{d}u_x \tag{17.11}$$

Because for $u_x < 0$ there is $p = 0$ this equation is equal to Eq. (17.4). It can then be pronounced that the biased mean velocity actually stands for the mean velocity that is applied to the momentum flow rate or similarly to the momentum flux at a local point in the flow. It can thus be directly applied in all momentum equations like the Euler, Navier-Stokes as well as Reynolds equations. On the other hand, both the use of the continuity equation and the calculation of the momentum flux according to Eq. (17.3) suppose the knowledge of the volumetric flux ($\overline{u}_x$). Thus the objective in dealing with the velocity bias is to estimate the difference between the mean for volumetric flux and its biased value for momentum flux. In other words, it is intended to determine the momentum flux correction factor according to Eq. (17.5).

17.3 Velocity Bias in One-Dimensional Flow Fluctuations

Although the flow fluctuation in a turbulent flow is always three-dimensional, the assumption of one-dimensional flow fluctuations contributes to the significant simplification of theoretical analyses of velocity bias. Since velocity fluctuations in the turbulent flow are of random nature, the fluctuations can then be approximated by the Gaussian probability distribution given by

$$p = \frac{1}{\sqrt{2\pi}\sigma} \mathrm{e}^{-\frac{(u-\overline{u})^2}{2\sigma^2}} \tag{17.12}$$

Herein $\overline{u}$ and σ are the volumetric mean velocity and its standard deviation, respectively. Clearly, the probability distribution of fluctuation velocity is symmetrical.

The symmetrical probability density function according to Eq. (17.12) will be distorted by the velocity bias in LDA measurements. As in the common case with the assumption of homogeneous particle distribution in the flow, the linear dependence of velocity sampling rate on the magnitude of velocities can be further assumed. The biased probability density function then must be written as

$$p_{\mathrm{b}} = k\,|u|\,p = \frac{k\,|u|}{\sqrt{2\pi}\,\sigma}\mathrm{e}^{-\frac{(u-\overline{u})^2}{2\sigma^2}} \tag{17.13}$$

As it states, the probability of negative velocity occurrence (back flow) because of velocity fluctuations has also been accounted for. From the same condition as given in Eq. (17.8) the constant k is determined as

$$\frac{1}{k\sigma} = \sqrt{\frac{2}{\pi}}\cdot \mathrm{e}^{-\frac{\overline{u}^2}{2\sigma^2}} + \frac{\overline{u}}{\sigma}\mathrm{erf}\left(\frac{\overline{u}}{\sqrt{2}\sigma}\right) \tag{17.14}$$

In this equation, the error function is defined by

$$\mathrm{erf}\,(x) = \frac{2}{\sqrt{\pi}}\int\limits_0^{\mathrm{x}} \mathrm{e}^{-u^2}\mathrm{d}u \tag{17.15}$$

The product of $k\sigma$ is given here as a function of relative flow fluctuations i.e. of the turbulence intensity $\sigma/\overline{u}$. Since the constant k appears always in the product form with σ, the product $k\sigma$ is designated here to be the bias product. It represents a measure of the distortion of the symmetrical probability distribution of the turbulent flow velocity. This has been demonstrated in Fig. 17.1 with two calculation examples.

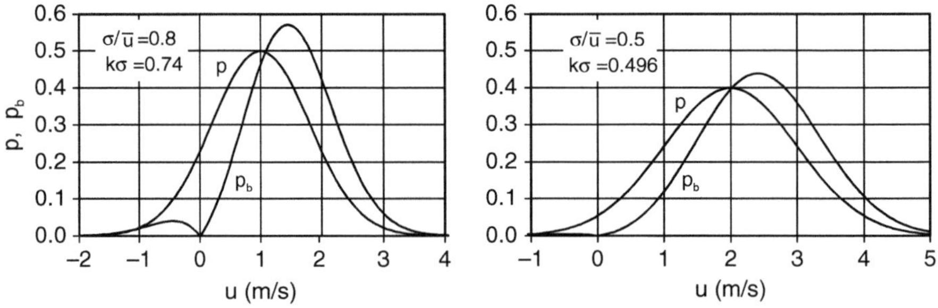

Fig. 17.1 Distortion of the distributions of the fluctuation velocity due to the effect of velocity bias in turbulent flows

For large turbulence i.e. $\overline{u}/\sigma << 1$ and due to the error function erf(0) = 0, the bias product is calculated from Eq. (17.14) as

$$k\sigma = \sqrt{\pi/2} \tag{17.16}$$

On the other hand, the bias product becomes zero if laminar flows ($\sigma/\overline{u} = 0$) are measured:

$$k\sigma = \sigma/\overline{u} = 0 \tag{17.17}$$

As shown in Fig. 17.1, there is $k\sigma \approx \sigma/\overline{u}$ i.e. $k \approx 1/\overline{u}$ already for $\sigma/\overline{u} \leq 0.5$. Actually, for low turbulent flow at which $u > 0$ can be assumed, $k = 1/\overline{u}$ is available as it has been already applied to Eq. (17.10), leading to Eq. (17.11).

In applying the biased probability density function and like in Eq. (17.10), the biased mean velocity is calculated by the integral

$$\overline{u}_\mathrm{b} = \int_{-\infty}^{\infty} p_\mathrm{b} u \mathrm{d}u \tag{17.18}$$

With respect to Eq. (17.13) for p_b the integral in above equation is calculated as

$$\frac{\overline{u}_\mathrm{b}}{\overline{u}} = 1 + k\sigma \frac{\sigma}{\overline{u}} \mathrm{erf}\left(\frac{\overline{u}}{\sqrt{2}\sigma}\right) \tag{17.19}$$

Like the bias product, the ratio of the biased to the initial mean velocity is again a function of the turbulence intensity $\sigma/\overline{u}$.

Similarly, the biased standard deviation because of the distortion of the probability distribution of velocities is calculated as

$$\begin{aligned}\frac{\sigma_\mathrm{b}^2}{\sigma^2} &= \frac{1}{\sigma^2}\int_{-\infty}^{\infty} p_\mathrm{b}\,(u-\overline{u}_\mathrm{b})^2\,\mathrm{d}u \\ &= 2 + \left(\frac{\overline{u}}{\sigma}\right)^2\left(1-\frac{\overline{u}_\mathrm{b}}{\overline{u}}\right)^2 + k\sigma\left(\frac{\overline{u}}{\sigma} - 2\frac{\overline{u}_\mathrm{b}}{\sigma}\right)\mathrm{erf}\left(\frac{\overline{u}}{\sqrt{2}\sigma}\right)\end{aligned} \tag{17.20}$$

or with Eq. (17.19) to replace $\overline{u}_\mathrm{b}$ as

$$\frac{\sigma_\mathrm{b}^2}{\sigma^2} = 2 - (k\sigma)^2\left[\mathrm{erf}\left(\frac{\overline{u}}{\sqrt{2}\sigma}\right)\right]^2 - k\sigma\frac{\overline{u}}{\sigma}\mathrm{erf}\left(\frac{\overline{u}}{\sqrt{2}\sigma}\right) \tag{17.21}$$

For flows with turbulence intensities $\sigma/\overline{u} < 0.5$, the error function tends to unity ($\mathrm{erf}(\overline{u}/\sqrt{2}\sigma) > 0.95$). Since this again leads to $k\sigma \approx \sigma/\overline{u}$ i.e. $k \approx 1/\overline{u}$, Eqs. (17.19) and (17.21) are then simplified to

$$\frac{\overline{u}_{\mathrm{b}}}{\overline{u}} \approx 1 + \frac{\sigma^2}{\overline{u}^2} \tag{17.22}$$

and

$$\frac{\sigma_{\mathrm{b}}^2}{\sigma^2} \approx 1 - \frac{\sigma^2}{\overline{u}^2} \tag{17.23}$$

respectively.

Equation (17.22) agrees exactly with the empirical estimation given by Edwards (1987), as mentioned in Sect. 17.1.

In Fig. 17.2, both the biased mean velocity and its standard deviation according to Eqs. (17.19), (17.21), (17.22) and (17.23), respectively, are shown in the function of turbulence intensity ($\mathrm{Tu} = \sigma/\overline{u}$). The turbulence intensities applied in Fig. 17.2 are restricted in the range between 0.01 and 10. This range, however, is sufficient to show the related effect of velocity bias for turbulence intensities varying from zero to infinity. As can be seen from the figure, the volumetric mean velocity is always overestimated because of the effect of velocity bias, while its standard deviation is underestimated at low and overestimated at high turbulence intensities. The approximations leading to Eqs. (17.22) and (17.23) are obviously valid for turbulence intensities below 50%. Again worth mentioning is that at low turbulent flows, at which $u_{\mathrm{i}} > 0$ applies to all velocity samples, the overestimated mean velocity exactly agrees with the mean velocity that is relevant for the momentum flux. This means that the biased mean velocity given in Eq. (17.22) can be directly applied to the momentum flux.

At flows with very high turbulence intensities, the biased mean velocity and its standard deviation become

$$\overline{u}_{\mathrm{b}} \,/\, \overline{u} = 2 \tag{17.24}$$

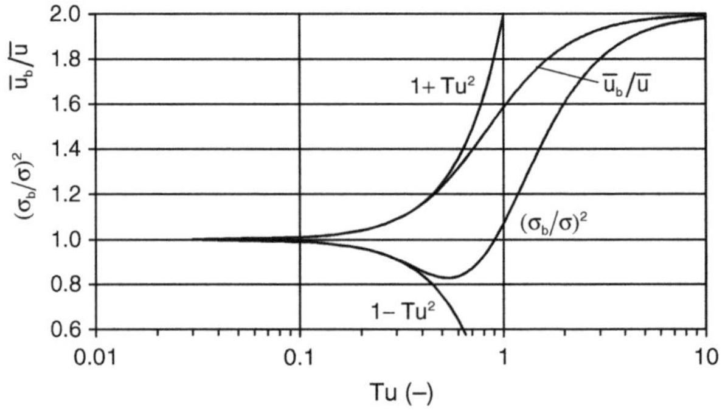

Fig. 17.2 Velocity bias in turbulent flow with one-dimensional flow fluctuations

and

$$\sigma_{\mathrm{b}}^2/\sigma^2 = 2 \tag{17.25}$$

respectively.

For measurements of flows with low turbulence intensities ($\sigma/\overline{u} \approx 0$), the velocity bias becomes negligible ($\overline{u}_{\mathrm{b}} \approx \overline{u}, \sigma_{\mathrm{b}} \approx \sigma$).

17.4 Velocity Bias in Two- and Three-Dimensional Flow Fluctuations

In reality, turbulent flow fluctuations are always three-dimensional and consist of fluctuations in both the velocity magnitude and the flow direction. LDA measurements of such turbulent flows undergo the velocity bias in each measured velocity component. It is therefore reasonable to establish the relationship between biased velocity components. In the case of isotropic or mostly anisotropic turbulence, the mean velocity vector can be considered to be the value which is surrounded symmetrically by velocities with fluctuations (Fig. 17.3). It can then be assumed that to each velocity event existing on one side of the symmetry line, there is always a conjugate one on the opposite side. For this type of flow turbulence, the velocity bias does not result in any change in the main flow direction. The ratio of the biased and the unbiased velocity component (for instance u_{x}) is then

$$\frac{\overline{u}_{\mathrm{x,b}}}{\overline{u}_{\mathrm{x}}} = \frac{\overline{u}_{\mathrm{b}} \cos \overline{\varphi}}{\overline{u} \cos \overline{\varphi}} = \frac{\overline{u}_{\mathrm{b}}}{\overline{u}} \tag{17.26}$$

The relative bias in the velocity component u_{x} is thus independent of the mean flow angle $\overline{\varphi}$. This signifies that the relative bias in the mean velocity is constant amongst all the velocity components. This is obviously also true if one-dimensional velocity fluctuations are assumed, as considered in Sect. 17.3.

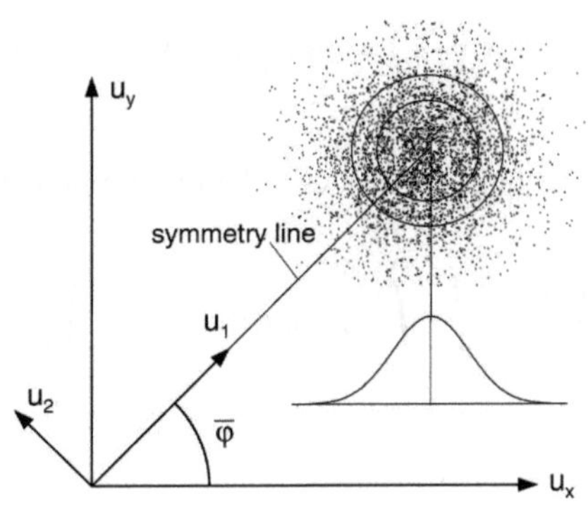

Fig. 17.3 Velocity fluctuations symmetrically around the mean velocity vector

17.4.1 Velocity Bias in Mean Velocities

With respect to Eq. (17.26) it is sufficient to estimate the velocity bias in the velocity component that corresponds to the main flow direction. For the sake of simplicity, an isotropic turbulent flow is considered as having a unique standard deviation σ of velocities. The coordinate system has been chosen, such that the first axis coincides with the main flow direction (Fig. 17.3).

17.4.1.1 Two-Dimensional Flow Fluctuations

In the case of two-dimensional flow fluctuations, the velocity components on the coordinates axis are denoted by u_1 and u_2, respectively. The Gaussian probability distribution of fluctuation velocity is given by

$$p = \frac{1}{2\pi\sigma^2} e^{-\frac{(u_1-\bar{u}_1)^2+u_2^2}{2\sigma^2}} \tag{17.27}$$

Because of the effect of velocity bias, this distribution is distorted, as this can be formulated by multiplying it with the factor $k\sqrt{u_1^2+u_2^2}$

$$p_\mathrm{b} = k\sqrt{u_1^2+u_2^2}\frac{1}{2\pi\sigma^2} e^{-\frac{(u_1-\bar{u}_1)^2+u_2^2}{2\sigma^2}} \tag{17.28}$$

The bias constant $k\sigma$ can be determined from the similar condition as that in Eq. (17.8), so that

$$\frac{1}{k\sigma} = \frac{1}{2\pi}\int_{-\infty}^{\infty} X\left(\frac{u_2}{\sigma}\right) \mathrm{d}\left(\frac{u_2}{\sigma}\right) \tag{17.29}$$

In this equation, the expression X is given by

$$X\left(\frac{u_2}{\sigma}\right) = \int_{-\infty}^{\infty} \sqrt{\left(\frac{u_1}{\sigma}\right)^2 + \left(\frac{u_2}{\sigma}\right)^2} \cdot e^{-\frac{1}{2}\left[\left(\frac{u_1}{\sigma}-\frac{\bar{u}_1}{\sigma}\right)^2+\left(\frac{u_2}{\sigma}\right)^2\right]} \mathrm{d}\left(\frac{u_1}{\sigma}\right) \tag{17.30}$$

The biased mean of velocity component u_1 is then calculated as

$$\frac{\bar{u}_{1,\mathrm{b}}}{\bar{u}_1} = \frac{1}{\bar{u}_1}\int_{-\infty}^{\infty}\int_{-\infty}^{\infty} p_\mathrm{b} u_1 \mathrm{d}u_1 \mathrm{d}u_2 = \frac{k\sigma}{2\pi}\frac{\sigma}{\bar{u}_1}\frac{1}{\sigma^4}\int_{-\infty}^{\infty}\int_{-\infty}^{\infty} u_1\sqrt{u_1^2+u_2^2}\, e^{-\frac{(u_1-\bar{u}_1)^2+u_2^2}{2\sigma^2}} \mathrm{d}u_1\mathrm{d}u_2 \tag{17.31}$$

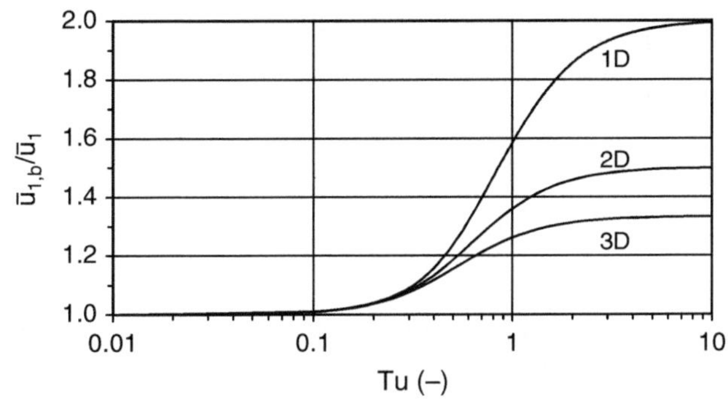

Fig. 17.4 Velocity bias in the mean velocity of the turbulent flow, respectively, with one-, two- and three-dimensional flow fluctuations

It is evident that the biased velocity component is again only a function of the turbulence intensity ($\mathrm{Tu} = \sigma/\bar{u}_1$). Because the integrations in both Eqs. (17.29) and (17.31) could not be analytically performed, the finite numerical method has to be applied. Figure 17.4 shows corresponding calculation results regarding the biased velocity component $\bar{u}_{1,\mathrm{b}} / \bar{u}_1$ in the function of the turbulence intensity (curve 2D). For the purposes of comparison, the biased mean velocity in the case of one-dimensional flow fluctuations, calculated from Eq. (17.19), has also been shown (curve 1D).

17.4.1.2 Three-Dimensional Flow Fluctuations

Similar calculation procedures can be applied to the flows with three-dimensional flow fluctuations. In place of Eq. (17.27), the Gaussian probability distribution of velocities takes the following form:

$$p = \frac{1}{\sqrt{8\pi}\pi\sigma^3} \mathrm{e}^{-\frac{(u_1-\bar{u}_1)^2+u_2^2+u_3^2}{2\sigma^2}} \tag{17.32}$$

Because of the effect of velocity bias this distribution is distorted, as this can be similarly formulated by multiplying it with the factor $k\sqrt{u_1^2+u_2^2+u_3^2}$, so that

$$p_\mathrm{b} = \frac{k\sqrt{u_1^2+u_2^2+u_3^2}}{\sqrt{8\pi}\pi\sigma^3} \mathrm{e}^{-\frac{(u_1-\bar{u}_1)^2+u_2^2+u_3^2}{2\sigma^2}} \tag{17.33}$$

The bias constant $k\sigma$ can be again determined from the similar condition as that in Eq. (17.8). The biased mean of velocity component u_1 is then calculated as

$$\frac{\overline{u}_{1,\mathrm{b}}}{\overline{u}_1} = \frac{1}{\overline{u}_1} \int_{-\infty}^{\infty}\int_{-\infty}^{\infty}\int_{-\infty}^{\infty} p_{\mathrm{b}} u_1 \mathrm{d}u_1 \mathrm{d}u_2 \mathrm{d}u_3$$

$$= \frac{k\sigma}{\sqrt{8\pi}\pi} \frac{\sigma}{\overline{u}_1} \frac{1}{\sigma^5} \int_{-\infty}^{\infty}\int_{-\infty}^{\infty}\int_{-\infty}^{\infty} u_1 \sqrt{u_1^2 + u_2^2 + u_3^2}\, \mathrm{e}^{-\frac{(u_1 - \overline{u}_1)^2 + u_2^2 + u_3^2}{2\sigma^2}} \mathrm{d}u_1 \mathrm{d}u_2 \mathrm{d}u_3 \tag{17.34}$$

The integration must be again performed by the finite numerical method. Corresponding calculation results have been shown in Fig. 17.4 (curve 3D) in the function of the turbulence intensity. Obviously the effect of velocity bias in the velocity component is lowest if the flow fluctuation is three-dimensional.

The above calculations for three cases of flow fluctuations (1D, 2D and 3D) help to understand the extent of influences of velocity bias in LDA measurements. The flow fluctuations in turbulent flows, in fact, are always three-dimensional. For this reason and by comparing Figs. 17.2 and 17.4 it can be concluded that the approximation $\overline{u}_{\mathrm{b}} / \overline{u} = 1 + \mathrm{Tu}^2$ is only applicable for real flows with turbulence intensities up to 30%.

17.4.2 Velocity Bias in Turbulent Normal Stresses

The influence of velocity bias on measurements of turbulence quantities is identified, for instance, in that the velocity bias will cause the actual isotropic turbulence to be measured and presented as an apparent anisotropic one. For simplicity, turbulent flows that show the symmetrical velocity fluctuations around the main flow direction are considered here. According to Chap. 8 and in considering a two-dimensional section in the flow field (Fig. 17.3), the turbulent normal stress in the velocity component u_{x} for instance is calculated by Eq. (8.7). Obviously, Eq. (8.7) also applies to the biased normal stress, so that

$$\sigma_{\mathrm{x,b}}^2 = \sigma_{1,\mathrm{b}}^2 \cos^2 \overline{\varphi} + \sigma_{2,\mathrm{b}}^2 \sin^2 \overline{\varphi} \tag{17.35}$$

For this reason, one needs only to determine the effects of velocity bias on the turbulence quantities $\sigma_{1,\mathrm{b}}^2$, $\sigma_{2,\mathrm{b}}^2$ and $\sigma_{3,\mathrm{b}}^2$ in a general turbulent flow. The simplest case would be again the turbulent flow with isotropic turbulence ($\sigma_1 = \sigma_2 = \sigma_3 = \sigma$) that is described by the turbulence intensity $\mathrm{Tu} = \sigma/\overline{u}_1$ and undergoes the velocity bias with $\sigma_{2,\mathrm{b}} \neq \sigma_{1,\mathrm{b}}$ and $\sigma_{2,\mathrm{b}} = \sigma_{3,\mathrm{b}}$. For this reason and because of Eq. (17.35) it is proposed that one only calculates the biased quantities $\sigma_{1,\mathrm{b}}$ and $\sigma_{2,\mathrm{b}}$, respectively. From the viewpoint of statistics leading to Eq. (5.5), the following relationship between the statistical quantities, as applied to the velocity component u_1 for instance, is available

$$\sigma_{1,\mathrm{b}}^2 = \overline{u_{1,\mathrm{b}}^2} - \overline{u}_{1,\mathrm{b}}^2 \tag{17.36}$$

As the biased mean velocity has been obtained in Sect. 17.4.1, only the first term on the r.h.s. of above equation needs to be calculated. For the purpose of again making a comparison between flows with two- and three-dimensional turbulence, respective flows will be separately considered, as presented below.

17.4.2.1 Two-Dimensional Turbulence

For an assumed flow with two-dimensional and isotropic velocity fluctuations, Eq. (17.36) is rewritten in detail

$$\frac{\sigma_{1,\mathrm{b}}^2}{\sigma^2} = \frac{1}{\sigma^2} \int_{-\infty}^{\infty} \int_{-\infty}^{\infty} u_1^2 p_\mathrm{b} \mathrm{d}u_1 \mathrm{d}u_2 - \frac{\overline{u}_{1,\mathrm{b}}^2}{\sigma^2} \tag{17.37}$$

The biased probability distribution p_b is the same as that which is given by Eq. (17.28) and has been used in Eq. (17.31). Equation (17.37) is expected to be a function of the turbulence intensity ($\mathrm{Tu} = \sigma/\overline{u}_1$) only.

Similarly, the biased standard deviation in the second velocity component i.e. the component perpendicular to the flow direction is given by

$$\frac{\sigma_{2,\mathrm{b}}^2}{\sigma^2} = \frac{1}{\sigma^2} \int_{-\infty}^{\infty} \int_{-\infty}^{\infty} u_2^2 p_\mathrm{b} \mathrm{d}u_1 \mathrm{d}u_2 \tag{17.38}$$

with $\overline{u}_2 = \overline{u}_{2,\mathrm{b}} = 0$.

Figure 17.5 shows corresponding calculation results in the function of the turbulence intensity. In the velocity component u_1, the related turbulent normal stress is underestimated for flows of the turbulence intensity below 65%. At the flows of the turbulence intensity $Tu < 30\%$, the over- and underestimation of the turbulent stress in the velocity component u_1 and u_2, respectively, are of the same order.

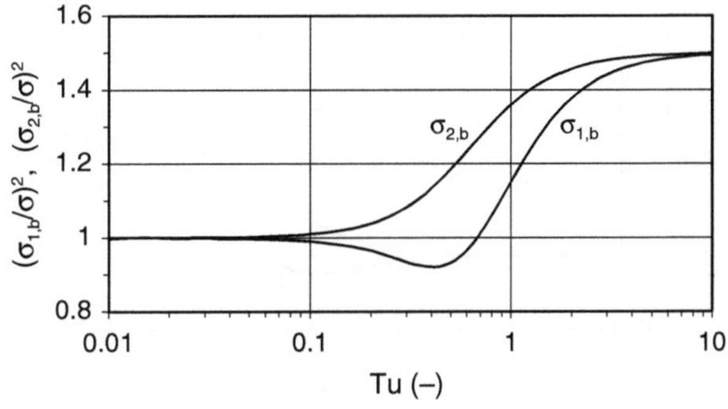

Fig. 17.5 Velocity bias in the turbulent stresses of the turbulent flow with two-dimensional flow fluctuations

17.4.2.2 Three-Dimensional Turbulence

Under the assumption of isotropic turbulence in a turbulent flow with three-dimensional velocity fluctuations, measurements of velocity components perpendicular to the main flow direction then undergo the same effect of velocity bias. This simply means $\sigma_{2,b}^2 = \sigma_{3,b}^2$. Because of this equality one needs to only estimate the biased quantities $\sigma_{1,b}^2$ and $\sigma_{2,b}^2$. The biased turbulence quantities in other velocity components can be generally calculated through the coordinate transformation of the related Reynolds stress matrix with $\sigma_{2,b}^2 = \sigma_{3,b}^2$. For the turbulent normal stress $\sigma_{x,b}^2$ for instance, it applies (Appendix C)

$$\sigma_{x,b}^2 = \sigma_{1,b}^2 \cos^2 \alpha_1 + \sigma_{2,b}^2 \sin^2 \alpha_1 \tag{17.39}$$

Herein α_1 is the angle between the main flow direction (axis with σ_1) and the x-axis (see also Fig. 6.1). The above equation is formally equal to Eq. (17.35). It applies, however, also to the flow in which the main flow direction does not lie in the $x-y$ plane.

Following the same calculation procedures, as presented above with respect to the two-dimensional flow fluctuations, the biased turbulence quantities $\sigma_{1,b}^2$ and $\sigma_{2,b}^2$ in the current case are calculated as

$$\frac{\sigma_{1,b}^2}{\sigma^2} = \frac{1}{\sigma^2} \int_{-\infty}^{\infty} \int_{-\infty}^{\infty} \int_{-\infty}^{\infty} u_1^2 p_b \mathrm{d}u_1 \mathrm{d}u_2 \mathrm{d}u_3 - \frac{\overline{u}_{1,b}^2}{\sigma^2} \tag{17.40}$$

$$\frac{\sigma_{2,b}^2}{\sigma^2} = \frac{1}{\sigma^2} \int_{-\infty}^{\infty} \int_{-\infty}^{\infty} \int_{-\infty}^{\infty} u_2^2 p_b \mathrm{d}u_1 \mathrm{d}u_2 \mathrm{d}u_3 \tag{17.41}$$

In these two equations, the biased probability density function p_b is the same as that given by Eq. (17.33) which has been applied in Eq. (17.34).

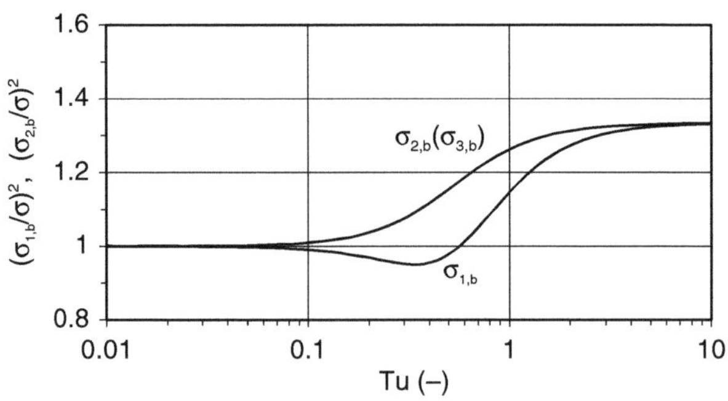

Fig. 17.6 Velocity bias in the turbulent stresses of the turbulent flow with three-dimensional flow fluctuations (The unbiased flow turbulence is isotropic)

Corresponding calculation results have been shown in Fig. 17.6. The effect of velocity bias in this case, if compared with Fig. 17.5 regarding two-dimensional flow fluctuations, has further diminished.

17.4.3 Velocity Bias in Turbulent Shear Stresses

Based on calculations of velocity bias in turbulent normal stresses in the last section, the velocity bias in the respective turbulent shear stress can be determined by means of the coordinate transformation of the related second order tensor i.e. the Reynolds stress matrix with $\sigma_{2,\mathrm{b}}^2 = \sigma_{3,\mathrm{b}}^2$. In interest of biased turbulent shear stress $\left(\overline{u'_{\mathrm{x}}u'_{\mathrm{y}}}\right)_{\mathrm{b}}$ for instance, it applies (Appendix C)

$$\left(\overline{u'_{\mathrm{x}}u'_{\mathrm{y}}}\right)_{\mathrm{b}} = \left(\sigma_{1,\mathrm{b}}^2 - \sigma_{2,\mathrm{b}}^2\right) \cos\alpha_1 \cos\beta_1 \tag{17.42}$$

In this equation, α_1 stands for the angle between the velocity components u_1 and u_{x}, as in Eq. (17.39). The angle β_1 conventionally denotes the angle between velocity components u_1 and u_{y}.

A special case is obtained when all four velocity components (u_1,u_2, u_{x} and u_{y}) are found in a plane. It is assumed that it deals with the $x - y$ plane, as shown in Fig. 17.3. Because of $\alpha_1 = \overline{\varphi}$ and $\beta_1 = \pi/2 - \alpha_1 = \pi / 2 - \overline{\varphi}$ it follows from Eq. (17.42)

$$\left(\overline{u'_{\mathrm{x}}u'_{\mathrm{y}}}\right)_{\mathrm{b}} = \frac{1}{2}\left(\sigma_{1,\mathrm{b}}^2 - \sigma_{2,\mathrm{b}}^2\right) \sin 2\overline{\varphi} \tag{17.43}$$

It completely agrees with Eq. (8.9) in Chap. 8.

Chapter 18
LDA Application Examples

18.1 High Speed Water Jet Flow in a Pelton Turbine

One of the most successful applications of the LDA method in flow measurements is the measurement of the high speed jet in a Pelton turbine. Such an application has been partly shown in Chap. 9 while describing the Dual Measurement Method (DMM). The high speed jet flow in a Pelton turbine represents a very important component which influences the entire hydraulic performance of the turbine system (Zhang 2009). For a long time, jet measurements were carried out by means of the pressure tubes and the photography with very limited accuracies. The first application of the LDA method to the high speed jet flow was conducted by Zhang et al. (2000a, b) in using a model injector of a Pelton turbine. In performing such measurements, the first step that must be completed is to bring the laser beams through the rough jet surface into the jet. This can be realized by using a transparent plane piece to slightly contact the jet, as illustrated in Fig. 18.1. The resulted disturbance on the jet flow is restricted within the boundary layer thickness on the rigid surface of the plane piece. Because this thickness is in the order of only about 0.1 mm, the flow disturbance can be neglected. In addition, the use of the plane piece as an optical window ensures the perfect optical conditions for LDA measurements.

It has been confirmed from measurements that the jet flow and the flow properties significantly depend on the flow condition ahead of the injector. The flow presented in Sect. 9.2 (see Fig. 9.2) corresponds to the condition of the inlet flow passing through a 90° bend. The secondary flow has been thus generated. Because it represents swirling flows, it persists while passing through the injector and thus also remains in the jet.

The simplest case of the jet flow is the jet flow without any secondary flow structure in it. This almost ideal flow can be generated by connecting the injector to a long straight pipe with uniform inlet flow. Corresponding measurements of flow distributions in such a high speed jet could be well carried out by means of the LDA method, as demonstrated in Fig. 18.2. At four different sections along the jet, the axial jet speeds are measured and normalized by the theoretical maximum jet speed $c_0 = \sqrt{2gh_0}$, with $h_0 = 30$ meters as the hydraulic head that is available at the

Zh. Zhang, *LDA Application Methods*, Experimental Fluid Mechanics,
DOI 10.1007/978-3-642-13514-9_18,

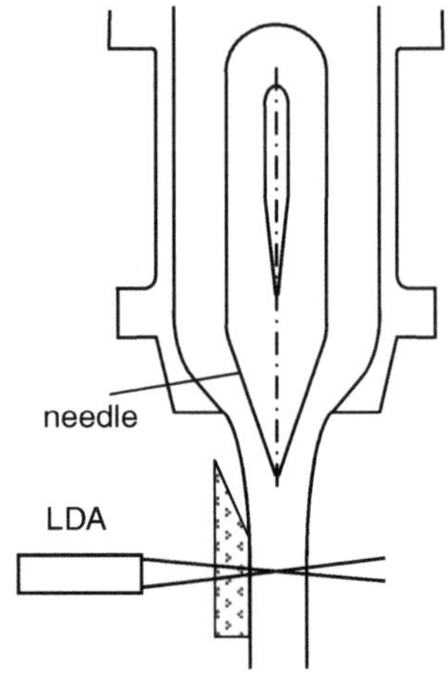

Fig. 18.1 High speed jet flow in a Pelton turbine and the LDA arrangement for flow measurements

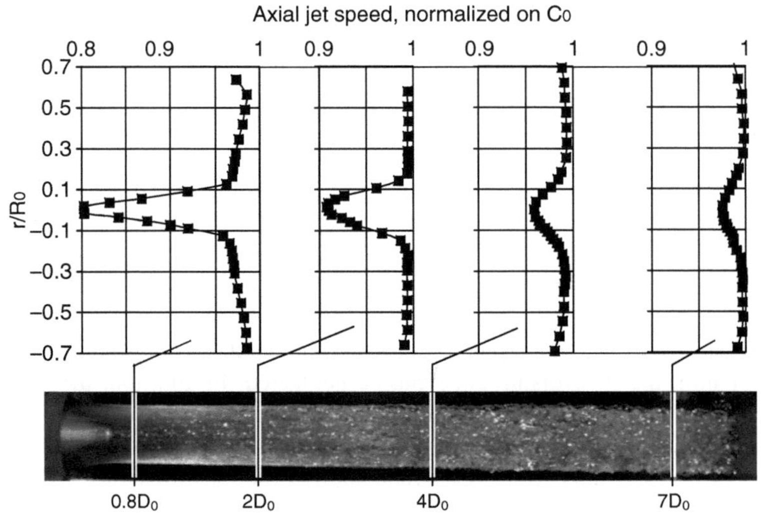

Fig. 18.2 Velocity distributions in the high speed jet flow measured by LDA

injector. Based on these accurate measurements, the following statements can be additionally made with regard to the flow dynamics.

A *Flow deficit in the jet core*

A significant flow deficit is confirmed in the jet core while the jet flow is formed at the nozzle exit. This is the wake arising from the viscous boundary layers on the needle surface. The boundary layer and its spatial limitation could thus be clearly identified.

B *Streamline curvature*

At the jet section $0.8D_0$ downstream of the nozzle, the flow velocity distribution across the jet and out of the jet core is non-uniform. The highest flow velocity is found near the edge of the jet. Such a velocity distribution indicates that the flow at this distance exhibits the streamline curvature in the length plane. This streamline curvature leads to the increase of the static pressure along the curvature radius towards the jet core and in a flow with constant total pressure, the velocity has to decrease. This appearance signifies that the slope of such a velocity distribution is simply a measure of the streamline curvature and thus of the pressure distribution, as this can be demonstrated here.

To calculate the streamline curvature in the jet near to the nozzle outlet, the jet flow is considered in the cylindrical coordinate system to have no circumferential velocity component. The streamline of the flow is then given by

$$r' = \frac{\mathrm{d}r}{\mathrm{d}z} = \frac{u_\mathrm{r}}{u_\mathrm{z}} \tag{18.1}$$

For the general case of $\mathrm{d}r/\mathrm{d}z = f(r, z)$, it is further differentiated as

$$r'' = \frac{\mathrm{d}^2 r}{\mathrm{d}z^2} = \frac{1}{u_\mathrm{z}^2}\left[u_\mathrm{z}\left(\frac{\partial u_\mathrm{r}}{\partial z} + \frac{\partial u_\mathrm{r}}{\partial r}\frac{\mathrm{d}r}{\mathrm{d}z}\right) - u_\mathrm{r}\left(\frac{\partial u_\mathrm{z}}{\partial z} + \frac{\partial u_\mathrm{z}}{\partial r}\frac{\mathrm{d}r}{\mathrm{d}z}\right)\right] \tag{18.2}$$

The jet flow out of the jet core can be considered to be the potential flow because a short distance downstream of the considered flow section, the flow with uniform and straight streamlines exhibits no vorticity. With respect to $\mathrm{d}r/\mathrm{d}z = u_\mathrm{r}/u_\mathrm{z}$ and $u_\mathrm{r}/u_\mathrm{z} << 1$ as well as $\partial u_\mathrm{r}/\partial z = \partial u_\mathrm{z}/\partial r$ for potential flow the above equation is simplified to

$$r'' = \frac{1}{u_\mathrm{z}}\frac{\partial u_\mathrm{z}}{\partial r} \tag{18.3}$$

The curvature radius of a streamline is calculated by

$$\frac{1}{R} = \frac{r''}{\left(1 + r'^2\right)^{3/2}} \approx r'' = \frac{1}{u_\mathrm{z}}\frac{\partial u_\mathrm{z}}{\partial r} \tag{18.4}$$

The streamline curvature is thus directly related to the velocity gradient in the area out of the jet core. In the first approximation, the velocity distribution out of the jet core can be considered to be a linear function of the radial coordinate, so that the velocity gradient $\partial u_\mathrm{z}/\partial r$ is a constant. Based on calculations in Eq. (18.4) an abstract of calculated radii of streamline curvatures from a series of measurements with different hydraulic heads is shown in Table 18.1.

As can be seen, the calculated radii of streamline curvatures are nearly independent of the available hydraulic head. This indicates that the jet flows operating at three different hydraulic heads are hydraulically similar.

Table 18.1 Streamline curvature radii in the near region to the nozzle exit for a constant needle stroke (Zhang and Casey 2007)

	Needle stroke s=16 mm	
Hydraulic head h_0	$\partial u_z/\partial r$	Curvature radius R (m)
10 m	37.7	0.37
20 m	54.0	0.37
30 m	61.1	0.39

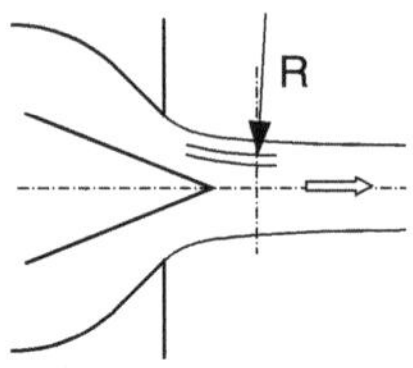

Streamline curvatures in the jet flow lead to pressure increase toward the jet core. This pressure increase can be calculated by applying the Euler momentum equation for the radial component as

$$-\frac{1}{\rho}\frac{\mathrm{d}p}{\mathrm{d}r} = u_\mathrm{r}\frac{\partial u_\mathrm{r}}{\partial r} + u_\mathrm{z}\frac{\partial u_\mathrm{r}}{\partial z} \tag{18.5}$$

As mentioned above the jet flow is considered to be the potential flow to which the irrotational flow condition $\partial u_\mathrm{r}/\partial z - \partial u_\mathrm{z}/\partial r = 0$ is satisfied. Combining this condition with $\partial u_\mathrm{z}/\partial r = u_\mathrm{z}/R$ from Eq. (18.4) leading to

$$\frac{\partial u_\mathrm{r}}{\partial z} = \frac{u_\mathrm{z}}{R} \tag{18.6}$$

With respect to $u_\mathrm{r} \approx 0$ the pressure gradient in the jet at the considered jet section is then calculated as

$$\frac{1}{\rho}\frac{\mathrm{d}p}{\mathrm{d}r} = -u_\mathrm{z}\frac{\partial u_\mathrm{r}}{\partial z} = -\frac{u_\mathrm{z}^2}{R} \tag{18.7}$$

The non-uniform flow distribution due to the streamline curvature clearly shows that the flow velocity in the considered section has not reached its maximum and the jet flow still undergoes acceleration. The mean velocity across this jet section is thus smaller than the mean velocity downstream at other sections. This first section was initially localized for detailed measurements as it was taken as the jet waist from photographic estimation. According to the presented LDA measurements, it is evident that the true jet waist is found further downstream. This also signifies that the LDA method is much more accurate than the photographic method even though for the purpose of localizing the jet waist.

It is worth mentioning here that because of the non-uniform static pressure distribution across the jet, the flow in the considered section could not be accurately measured by a conventional Pitot tube, which in reality only measures the total pressure. In assuming the constant static pressure, the Pitot tube measurement would then provide a uniform velocity distribution across the jet, which is clearly not true.

C *Limitation of measuring the jet diameter*
The use of the small wedge shaped Perspex window, as shown in Fig. 18.1, means that the jet surface position on the side of the jet adjacent to the wedge cannot be estimated. On the opposite side of the jet, the rough jet surface scatters so much laser light that the LDA optics rapidly reaches its limitation. For this reason it is impossible to accurately estimate the location of the edge of the jet, i.e. the jet thickness, from LDA measurements.

D *Negligible energy loss in the jet flow*
The mean velocity in the jet remains nearly constant along the jet, at least up to the section of a distance that is measured seven times of the nozzle exit diameter. This indicates that the jet is highly compact and the loss of the kinetic energy is negligible.

Measurements shown above were carried out at the hydraulic head up to 30 meters, as shown in Table 18.1. Successful LDA measurements of the jet flows at the hydraulic head of 90 meters have been carried out by Zhang et al. (2003). More about the measurements including those with a 90° bend ahead of the injector as well as at the hydraulic head of 90 meters can be found in Zhang and Casey (2007) that summarized the most relevant jet flow measurements by the LDA method.

18.2 Measurements of Warp Yarn Speed in a Weaving Machine

LDA method has been generally recognized to be a technique for flow measurements. In reality, it can also be applied to measurements of other object motions instead of small particle motions in the flow. One needs only to position the LDA measurement volume on the surface of the moving object of finite dimensions. The possible problem associated with the positioning of the measurement volume is obviously the light scattering that is very strong and will lead to rapid overloading of used photodetectors like the photomultiplier. In addition, the visibility of signals will considerably decrease, like in the case of large particles passing through the measurement volume. The high quality burst like that in Fig. 3.6 could not be achieved. Instead, continuous signals with highly diminished signal to noise ratio (SNR) will be obtained. Although the signal quality is largely deteriorated by the low SNR, signals can still be successfully evaluated by the powerful signal processor like the burst spectrum analyzer (BSA) that is based on the spectrum analysis using FFT and corresponds to the current state of the art LDA technologies. Because of this performance, the application area of the LDA method has been significantly extended, as it can be found for instance in measurements of mechanical vibrations at machines or machine components.

Figure 18.3 shows an example of extended LDA applications in the weaving machine. It represents the measurements of the time-dependent moving speed of a warp yarn with the purpose of checking and controlling the yarn tension in the

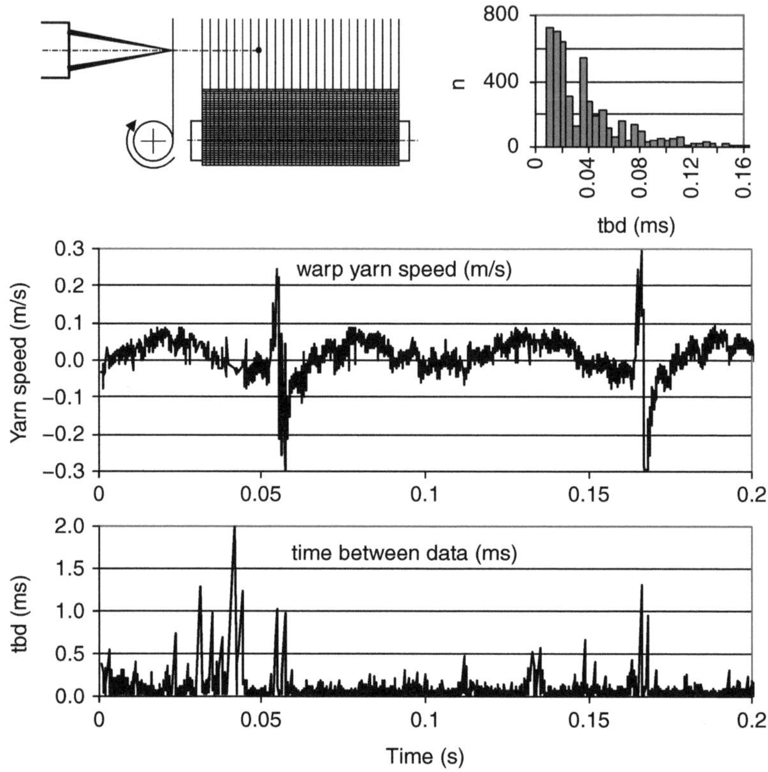

Fig. 18.3 LDA application to measurements of the warp yarn speed in a weaving machine

weaving machine. In the example given, the LDA measurement volume was directly positioned on the moving yarn. The measurement could be conducted with a very high data rate of about 30 kHz. This enables the highly dynamical motion of the yarn to be highly resolved.

In principle, the mean speed of the yarn movement can be calculated from the arithmetic average of measurement data. The calculation, however, has shown that this mean speed mostly does not agree with the actual mean speed of yarns while the machine is operating. The reason for this discrepancy is simply the non-constant data sampling rate. Worth remembering is that the non-constant data sampling rate as a result of the velocity bias in flow measurements (Chap. 17) usually does not significantly influence the determination of the mean velocity. In the present case, the non-constant data sampling rate arises from the non-constant light scattering on the moving yarn of about 0.1 mm in diameter. Such a non-constant data sampling rate can be shown by the time-dependent time-between-data ($tbd = t_{i+1} - t_i$), as also shown in Fig. 18.3 together with the corresponding probability distribution in form of the histogram. Obviously the measurements were carried out with a wide spectrum of the time-between-data i.e. data sampling rate. Because of the highly dynamical motion and comparably very low mean speed of yarns the non-constant data sampling rate will considerably sensitively influence the measurement of the mean speed of the moving yarn. In the example shown in Fig. 18.3, the relative fluctuation speed is measured at about $\sigma/\bar{u} \approx 100 = 10000\%$, with σ as the standard deviation of the mean speed. Worth mentioning is that in the measurement of a turbulent flow with comparable turbulence intensity the biased mean velocity is equal to two times of the unbiased mean, see Fig. 17.2.

It has been shown that in the present case the mean speed of the moving yarn could not be accurately determined. In reality, the most significant meaning of the current LDA measurements is in showing the dynamic behaviours of moving yarns in a weaving machine. The mean speed of the moving yarns could be accurately measured by simply using a meter and a stop watch.

The measurement shown in Fig. 18.3 clearly demonstrates the excellent applicability of the LDA method in areas other than the fluid flow mechanics.

18.3 Verification of the Shift Frequency in the Laser Beam

For purposes of resolving the flow direction in LDA measurements, the frequency of one laser beam in each laser beam pair is usually shifted, as described in Sect. 3.6. Sometimes it is necessary to verify the accuracy of the shift frequency, in order to estimate the system accuracy of the LDA unit and to identify the possible systematic error in LDA measurements. Direct measurements of the shift frequency as an optical quantity by using the interferometer, for instance, are not always available or usually very time-consuming. The simplest way to verify the shift frequency is probably to measure a non-moving object with the present LDA optics. For the correct setup and specification of shift frequency in LDA optics, zero velocity should be obtained. Any deviation of measured velocities from zero indicates the error in the shift frequency specifications.

The method of measuring the non-moving object seems to be somewhat inadequate, because it does not deal with the direct measurement of velocities. In addition, the object might likely be burned down at the place where the laser beams are focused. The strong laser light scattering at the object also leads to continuous overloading of the photodetectors such as the photomultiplier. For this reason it would be very helpful to use a rotating object such as a variable frequency optical chopper. Either the object surface, or better a thin wire on it, can be utilized to scatter the laser light in the configured LDA test.

For the purpose of merely verifying the shift frequency in the transmitting laser light, one doesn't need to compare the measured velocities with the actual velocity of the rotating object. The measurements can be quite easy and accurately performed, if the motor of the used optical chopper can be switched to rotate in both directions at equal rotational speed. The LDA measurement volume will be simply positioned on the rotating object without having to know where it is and which velocity component will be measured. One needs to only compare two velocities which are measured while the motor rotates, respectively, in two different directions at the same speed. Theoretically both velocities with different signs should exhibit the same absolute value when the specification of the shift frequency in the transmitting laser light is exact. Each difference in the absolute values of both velocities straightforwardly indicates that there is a difference between the shift frequency that is specified in the software and the actual shift frequency that is achieve by means of the Bragg cell, for instance. In reality, the error in the specification of the shift frequency can not only be confirmed, but also quantified. For this purpose, the error in the shift frequency is considered to result in an error in the shift speed of fringes in the measurement volume according to Eq. (3.60). This shift speed error is constantly involved in the verification measurements by means of the rotating object. As the motor rotates in two different directions, the measured velocity components are

$$u_+ = u_0 + \Delta u_{\mathrm{sh}} \tag{18.8}$$

and

$$u_- = -u_0 + \Delta u_{\mathrm{sh}} \tag{18.9}$$

respectively. Herein the actual velocity component is assumed to be u_0. By eliminating this actual velocity from above two equations, the error that is involved in the measurements is determined as

$$\Delta u_{\mathrm{sh}} = \frac{u_+ + u_-}{2} \tag{18.10}$$

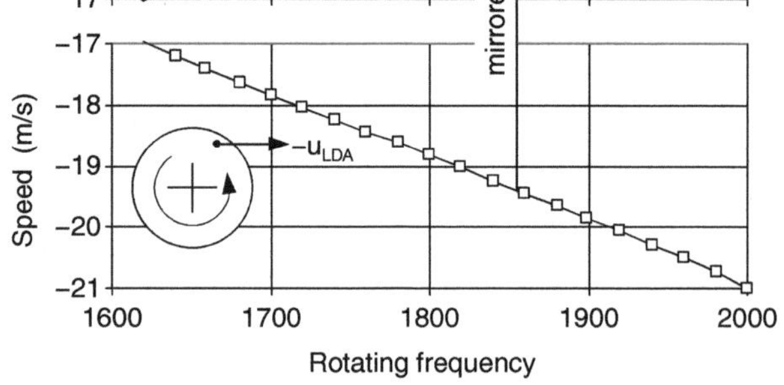

Fig. 18.4 Method of verifying the shift frequency accuracy in an LDA system

This error should remain constant, when the verification measurement is carried out at other rotational speeds of the motor. Figure 18.4 shows an example of the outlined verification measurement by using a variable frequency optical chopper that can be accurately set at each constant value of rotational speeds. As can be seen, the error in the outlined velocity measurements is highly constant. It indeed represents an error which is negligibly small. In the example shown in Fig. 18.4, the error in the shift speed takes $\Delta u_{\mathrm{sh}} = 0.047$. It corresponds to an additive amount to the velocity component that is positive in the LDA coordinate system. With respect to the fringe spacing $\Delta x = 2.18\,\mu\mathrm{m}$, the error in the shift frequency is calculated from Eq. (3.60) to be 22 kHz. If related to the applied shift frequency of 40 MHz, the error takes about 0.05%.

The method of rotating the object in two different directions represents an accurate method to examine any negligible difference between the actual and the specified shift frequency in an LDA unit. It is accurate because it is founded on the natural law and is occasionally called the two-step method. The most famous application example of this method might be the Michelson-Morley experiment (Hecht 1991) that was conducted to find out whether the light speed is different in different spatial directions. The same principle has also been applied in Chap. 9 to derive the Dual Measurement Method (DMM) which enables the very small secondary flow structures in the high speed jet flow to be accurately measured.

Appendix A
Off-axis LDA Alignment and Measurement Volume Displacement

For measurements of internal flows or flows behind a plane wall by means of the LDA method, it is sometimes necessary to align the LDA head off-axis against the normal of the plane wall. Such an arrangement of LDA optics requires a special concern that the LDA head should be off-axis aligned strictly in the plane containing two laser beams. This ensures the laser beams in both the window and the flow to propagate within the same plane and thus to intersect (Fig. A.1). One of the most significant outcomes of such an arrangement of a two-component LDA head with two pairs of laser beams is that the two measurement volumes (*m* and *s*) do not meet together. This phenomenon is known as astigmatism as a special form of optical aberrations. The distance between two intersection points, i.e. measurement volumes, is denoted as the astigmatic difference. Details about this form of optical aberrations have been described in Chap. 14, see Fig. 14.6.

To specify and evaluate the optical performance of an LDA system at the off-axis alignment, the distance between two separate measurement volumes has been considered to be a measure of the related optical aberration. It can be applied even in the case of using a one-component LDA with two laser beams. As a reference, this distance will be calculated in this appendix.

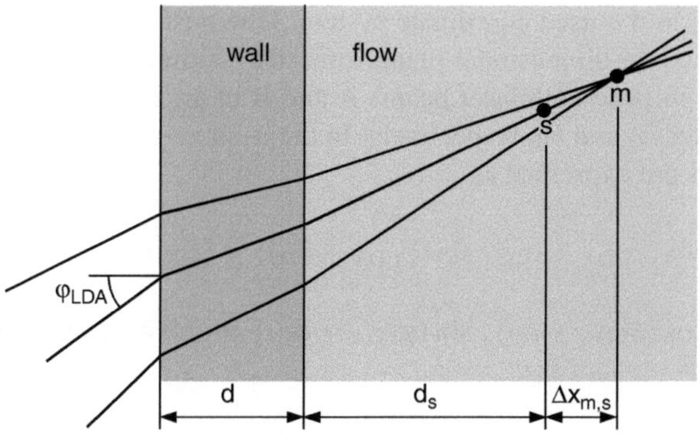

Fig. A.1 Off-axis alignment of a two-component LDA head and the separation of two measurement volumes (astigmatic difference)

Zh. Zhang, *LDA Application Methods*, Experimental Fluid Mechanics,
DOI 10.1007/978-3-642-13514-9, © Springer-Verlag Berlin Heidelberg 2010

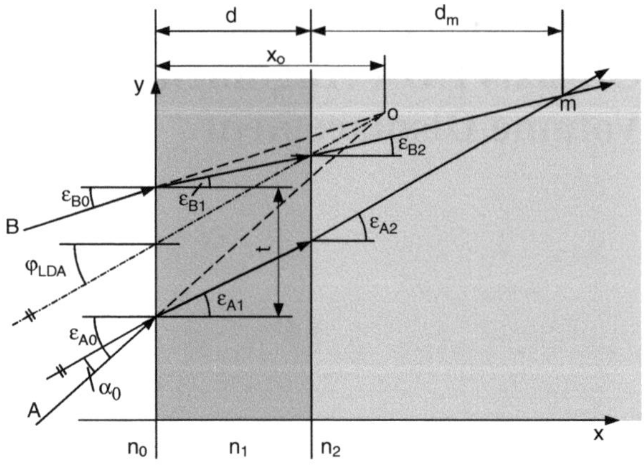

Fig. A.2 Propagation of two laser beams in the meridian plane

For the common purpose of regarding the astigmatism, the plane in which the off-axis alignment takes place is denoted as the meridian plane. The plane that is perpendicular to the meridian plane is denoted as the sagittal plane. For an LDA head with four laser beams then each plane contains two laser beams. The way to calculate the distance between two measurement volumes is to firstly calculate the intersection point of two laser beams in each laser beam pair.

The medium 0, in which all four laser beams intersect at a unique point, is denoted as the reference medium. Usually it refers to the air in which the LDA head is present.

A.1 Laser Beams in the Meridian Plane

The two laser beams in the meridian plane will be considered according to Fig. A.2 first. The initial i.e. virtual intersection point of two laser beams is fount at o that is measured by x_o in the used coordinate system. Obviously both laser beams propagate in the same two-dimensional plane from the medium 0 to the medium 2. For the calculation purpose both laser beams A and B in the medium 0 are represented by the unit vector $\vec{a}_0$ and $\vec{b}_0$, respectively. In the used x–y coordinate system, these two unit vectors are expressed as

$$\vec{a}_0 = [\cos(\varphi_{\text{LDA}} + \alpha_0), \sin(\varphi_{\text{LDA}} + \alpha_0)] = (\cos\varepsilon_{\text{A0}}, \sin\varepsilon_{\text{A0}}) \tag{A.1}$$

$$\vec{b}_0 = [\cos(\varphi_{\text{LDA}} - \alpha_0), \sin(\varphi_{\text{LDA}} - \alpha_0)] = (\cos\varepsilon_{\text{B0}}, \sin\varepsilon_{\text{B0}}) \tag{A.2}$$

In the medium 1 and 2, the corresponding laser beams are represented by respective unit vectors as

$$\vec{a}_1 = (\cos\varepsilon_{A1}, \sin\varepsilon_{A1}) \tag{A.3}$$

$$\vec{b}_1 = (\cos\varepsilon_{B1}, \sin\varepsilon_{B1}) \tag{A.4}$$

$$\vec{a}_2 = (\cos\varepsilon_{A2}, \sin\varepsilon_{A2}) \tag{A.5}$$

$$\vec{b}_2 = (\cos\varepsilon_{B2}, \sin\varepsilon_{B2}) \tag{A.6}$$

Both laser beams intersect with the y-axis. The distance between two intersection points, as shown by t, can be calculated from either the virtual intersection point o or the actual point m, as given by

$$t = (\tan\varepsilon_{A0} - \tan\varepsilon_{B0})\, x_o = (d\tan\varepsilon_{A1} + d_m\tan\varepsilon_{A2}) - (d\tan\varepsilon_{B1} + d_m\tan\varepsilon_{B2}) \tag{A.7}$$

from which one obtains

$$x_o = \frac{\tan\varepsilon_{A1} - \tan\varepsilon_{B1}}{\tan\varepsilon_{A0} - \tan\varepsilon_{B0}} d + \frac{\tan\varepsilon_{A2} - \tan\varepsilon_{B2}}{\tan\varepsilon_{A0} - \tan\varepsilon_{B0}} d_m \tag{A.8}$$

This equation relates the actual and virtual intersection points of two laser beams which lie in the meridian plane

A.2 Laser Beams in the Sagittal Plane

The two laser beams in the sagittal plane will be considered according to Fig. A.3. The virtual intersection point of two laser beams is again given by o and located at x_o. Obviously both laser beams propagate in the medium 1 and 2 in other two-dimensional planes than that in medium 0. The optical axis lies in the x–y plane and is inclined by φ_{LDA}. Two laser beams are denoted by C and D, respectively. Because of the symmetry condition only the beam C will be further considered.

The laser beam C in the medium 0 is represented by the unit vector $\vec{c}_0$. Because the projection of this unit vector on the optical axis in the x–y plane leads to $\cos\alpha_0$ and this in turn has its x- and y-components $\cos\alpha_0\cos\varphi_{LDA}$ and $\cos\alpha_0\sin\varphi_{LDA}$, respectively, the unit vector $\vec{c}_0$ is expressed by

$$\vec{c}_0 = (\cos\alpha_0\cos\varphi_{LDA}, \cos\alpha_0\sin\varphi_{LDA}, \sin\alpha_0) \tag{A.9}$$

The projection of this unit vector in the x–z plane (Fig. A.3b) has a slope that is given by

$$\frac{c_{0z}}{c_{0x}} = \frac{\sin\alpha_0}{\cos\alpha_0\cos\varphi_{LDA}} = \frac{\tan\alpha_0}{\cos\varphi_{LDA}} \tag{A.10}$$

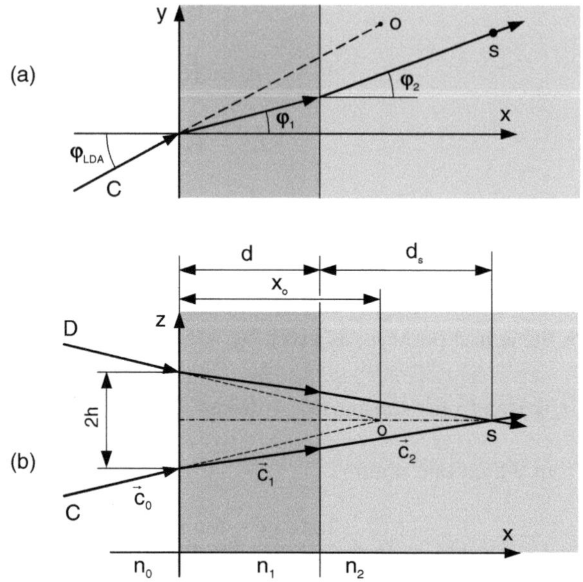

Fig. A.3 Propagation of two laser beams in the sagittal plane

According to Eqs. (3.8) and (3.9) the unit vector after the first refraction i.e. in the medium 1 is expressed as

$$\vec{c}_1 = \left(c_{1x}, \frac{n_0}{n_1} \cos\alpha_0 \sin\varphi_{LDA}, \frac{n_0}{n_1} \sin\alpha_0 \right) \tag{A.11}$$

Herein

$$c_{1x} = \sqrt{1 - c_{1y}^2 - c_{1z}^2} = \sqrt{1 - \frac{n_0^2}{n_1^2}\left(1 - \cos^2\alpha_0 \cos^2\varphi_{LDA}\right)} \tag{A.12}$$

The projection of the unit vector $\vec{c}_1$ in the x–z plane has a slope that is calculated as

$$\frac{c_{1z}}{c_{1x}} = \frac{n_0}{n_1} \frac{\sin\alpha_0}{c_{1x}} \tag{A.13}$$

In a similar way, the unit vector after the second refraction i.e. in the medium 2 is expressed as

$$\vec{c}_2 = \left(c_{2x}, \frac{n_0}{n_2} \cos\alpha_0 \sin\varphi_{LDA}, \frac{n_0}{n_2} \sin\alpha_0 \right) \tag{A.14}$$

with

$$c_{2x} = \sqrt{1 - c_{2y}^2 - c_{2z}^2} = \sqrt{1 - \frac{n_0^2}{n_2^2}\left(1 - \cos^2\alpha_0 \cos^2\varphi_{LDA}\right)} \tag{A.15}$$

The projection of the unit vector $\vec{c}_2$ in the x–z plane exhibits a slope confirmed by

$$\frac{c_{2z}}{c_{2x}} = \frac{n_0}{n_2}\frac{\sin\alpha_0}{c_{2x}} \tag{A.16}$$

Two laser beams in the medium 2 intersect at point s which is located at d_s. In order to find out the relation between x_o and d_s, two intersection points of two laser beams on the first medium interface ($y - z$ plane) are considered. Half of the distance $2\,h$ between two intersection points can be calculated from either the virtual intersection point o or the actual point s, as given by

$$h = \frac{c_{0z}}{c_{0x}}x_o = \frac{c_{1z}}{c_{1x}}d + \frac{c_{2z}}{c_{2x}}d_s \tag{A.17}$$

With respect to the slopes of respective unit vectors in the $x - z$ plane (Fig. A.3b), as calculated above, the following relationship can be found:

$$x_o = \left(\frac{n_0}{n_1}\frac{1}{c_{1x}}d + \frac{n_0}{n_2}\frac{1}{c_{2x}}d_s\right)\cos\alpha_0\cos\varphi_{\mathrm{LDA}} \tag{A.18}$$

This equation relates the actual and virtual intersection point of two laser beams which lie in the sagittal plane.

A.3 Combination

Combining Eqs. (A.8) and (A.18) to eliminate x_o yields

$$\begin{aligned}&\frac{\tan\varepsilon_{A1} - \tan\varepsilon_{B1}}{\tan\varepsilon_{A0} - \tan\varepsilon_{B0}}d + \frac{\tan\varepsilon_{A2} - \tan\varepsilon_{B2}}{\tan\varepsilon_{A0} - \tan\varepsilon_{B0}}d_m \\ &= \left(\frac{n_0}{n_1}\frac{1}{c_{1x}}d + \frac{n_0}{n_2}\frac{1}{c_{2x}}d_s\right)\cos\alpha_0\cos\varphi_{\mathrm{LDA}}\end{aligned} \tag{A.19}$$

With $d_m = d_s + \Delta x_{m,s}$ then the distance between the meridian and sagittal focal points $\Delta x_{m,s}$ (Fig. A.1) is resolved as

$$\Delta x_{m,s} = \frac{1}{T_{20}}\left(\Psi_1 d + \Psi_2 d_s\right) \tag{A.20}$$

Herein

$$\Psi_1 = \frac{n_0}{n_1}\frac{1}{c_{1x}}\cos\alpha_0\cos\varphi_{\mathrm{LDA}} - T_{10} \tag{A.21}$$

$$\Psi_2 = \frac{n_0}{n_2}\frac{1}{c_{2x}}\cos\alpha_0\cos\varphi_{\mathrm{LDA}} - T_{20} \tag{A.22}$$

$$T_{10} = \frac{\tan \varepsilon_{A1} - \tan \varepsilon_{B1}}{\tan \varepsilon_{A0} - \tan \varepsilon_{B0}} \tag{A.23}$$

$$T_{20} = \frac{\tan \varepsilon_{A2} - \tan \varepsilon_{B2}}{\tan \varepsilon_{A0} - \tan \varepsilon_{B0}} \tag{A.24}$$

In these equations, all refraction angles ($\varepsilon_{A1}, \varepsilon_{A2}, \varepsilon_{B1}, \varepsilon_{B2}$) can be calculated from incident angles ε_{A0} and ε_{B0} in the form of $\varepsilon_{A0} = \varphi_{LDA} + \alpha_0$ and $\varepsilon_{B0} = \varphi_{LDA} - \alpha_0$ by applying the law of refraction. Both c_{1x} and c_{2x} can be determined from Eqs. (A.12) and (A.15), respectively.

A special case will be considered in that the flow medium to be measured is equal to the reference medium. Then because of $n_2 = n_0$ it yields from Eqs. (A.15) and (A.24)

$$c_{2x} = \cos \alpha_0 \cos \varphi_{LDA} \tag{A.25}$$

$$T_{20} = 1 \tag{A.26}$$

These values are inserted into Eq. (A.22), yielding

$$\Psi_2 = 0 \tag{A.27}$$

The distance between two measurement volumes, as given in Eq. (A.20), is then reduced to

$$\Delta x_{m,s} = \Psi_1 d \tag{A.28}$$

It is independent of the location of the measurement volume in the flow. This also means that the depth of the measurement volume in the flow does not contribute to the astigmatic difference.

Appendix B
Laser Beam Orientation Under the Effect of the Bias Angle δ

This appendix refers to Sect. 14.9.1.

One reason that leads to inaccurate off-axis LDA alignment is confirmed by the off-axis alignment angle φ_{LDA} followed by the bias angle δ, at which the LDA head has turned around its own axis i.e. the optical axis. This type of imperfect off-axis LDA alignment indicates that the unit vector $\vec{n}$ that represents the normal of the plane containing two laser beams (known as the optical plane) deviates from z-axis by δ (Fig. B.1). Thus the normal vector $\vec{n}$ of the optical plane generally

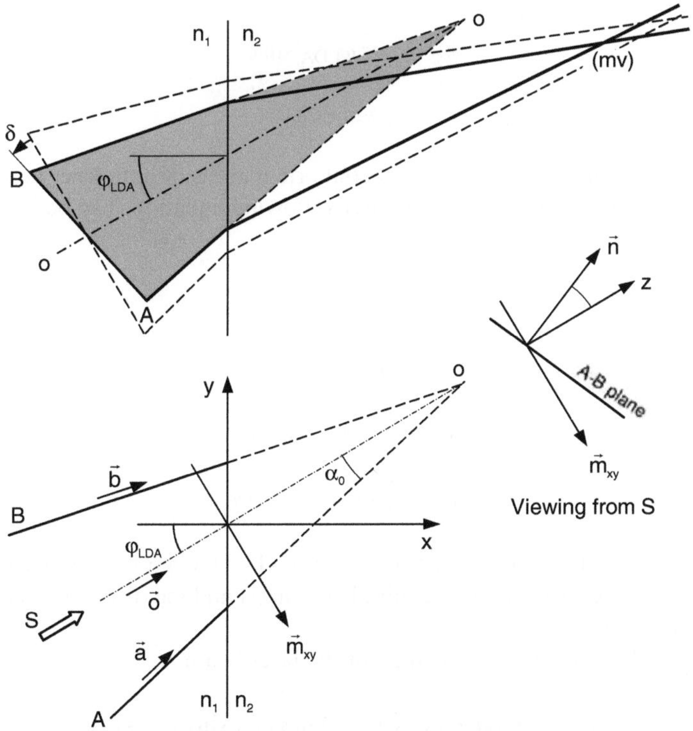

Fig. B.1 Laser beam orientation after rotating the LDA head by φ_{LDA} followed by the bias angle δ

has its spatial orientation. In order to determine the spatial direction of this normal vector, an auxiliary unit vector $\vec{m}_{xy}$ is introduced which lies in the $x-y$ plane and is perpendicular to the optical axis. Because the optical axis is represented by $\vec{o} = (\cos\varphi_{LDA}, \sin\varphi_{LDA}, 0)$, it yields for the unit vector $\vec{m}_{xy}$

$$\vec{m}_{xy} = (\sin\varphi_{LDA}, -\cos\varphi_{LDA}, 0) \tag{B.1}$$

Then, the following equations are used to derive the spatial direction of the normal vector $\vec{n}$:

$$\vec{n}\cdot\vec{z} = \cos\delta \tag{B.2}$$

$$\left(\vec{n}\times\vec{m}_{xy}\right)\cdot\vec{z} = 0 \tag{B.3}$$

In these equations, the z-axis has been considered as a unit vector $\vec{z} = (0, 0, 1)$. The triple product in Eq. (B.3) is zero because all three vectors lie in the same plane. From these two equations and with respect to $|\vec{n}| = 1$ the normal vector $\vec{n}$ can be solved as

$$n_x = -\sin\varphi_{LDA}\sin\delta \tag{B.4}$$

$$n_y = \cos\varphi_{LDA}\sin\delta \tag{B.5}$$

$$n_z = \cos\delta \tag{B.6}$$

In using these calculation results, the spatial orientations of both laser beams A and B can be calculated. According to Fig. B.1 following equations can be immediately obtained:

$$\vec{a}\cdot\vec{o} = \cos\alpha_0 \tag{B.7}$$

$$\vec{b}\cdot\vec{o} = \cos\alpha_0 \tag{B.8}$$

$$(\vec{a}\times\vec{o})\cdot\vec{n} = \sin\alpha_0 \tag{B.9}$$

$$\left(\vec{b}\times\vec{o}\right)\cdot\vec{n} = -\sin\alpha_0 \tag{B.10}$$

Equations (B.9) and (B.10) are derived from the knowledge that the vectors in the respective parentheses have the magnitude of $\sin\alpha_0$ and the directions parallel to $\vec{n}$ and $-\vec{n}$, respectively.

From Eqs. (B.7) and (B.9) it results for the laser beam A

$$a_x = \cos\varphi_{LDA}\cos\alpha_0 + \sin\varphi_{LDA}\sin\alpha_0\cos\delta \tag{B.11}$$

$$a_y = \sin\varphi_{LDA}\cos\alpha_0 - \cos\varphi_{LDA}\sin\alpha_0\cos\delta \tag{B.12}$$

$$a_z = \sin\alpha_0 \sin\delta \tag{B.13}$$

Analogous to the related calculations one obtains from Eqs. (B.8) and (B.10) for the laser beam B:

$$b_x = \cos\varphi_{LDA} \cos\alpha_0 - \sin\varphi_{LDA} \sin\alpha_0 \cos\delta \tag{B.14}$$

$$b_y = \sin\varphi_{LDA} \cos\alpha_0 + \cos\varphi_{LDA} \sin\alpha_0 \cos\delta \tag{B.15}$$

$$b_z = -\sin\alpha_0 \sin\delta \tag{B.16}$$

As can be seen, the bias angle $\delta \neq 0$ leads to $a_z \neq 0$ and $b_z \neq 0$. This means that both laser beams A and B, after refractions in the medium 2, do not lie in the same plane as in the medium 1. They will obviously propagate in different directions which indeed do not lie in a two-dimensional plane at all. Because of this, two laser beams will not intersect in the medium 2 to form the measurement volume. Corresponding calculations of using the above results have been performed in Sect. 14.9.1.

Appendix C
Coordinate Transformation of the Reynolds Stress Matrix

In representing the flow velocities and the velocity fluctuations, the method of using coordinate transformation has often been applied. In Chap. 6, transformations of both velocities and turbulence quantities in the two-dimensional $x-y$ plane have been presented. For reference purposes, as required for instance by Sect. 17.4.2, the velocity transformation in three-dimensional orthogonal coordinate systems should be shown in this appendix.

According to Fig. C.1, the Cartesian coordinate system is shown to represent the velocity vector $\vec{u}$ with components u_x, u_y and u_z. This coordinate system is usually directly related to the flow system. Another orthogonal coordinate system is also drawn to represent the same velocity vector, however in components of u_1, u_2 and u_3. The relative position between two coordinate systems is specified by angles α_i, β_i andγ_i. For instance, the coordinate axis for velocity component u_1 is given by angles α_1, β_1 and γ_1 in the $x-y-z$ coordinate system. The components of the velocity vector $\vec{u}$ in both coordinate systems are related by (see also Sect. 6.1.1)

$$\begin{bmatrix} u_1 \\ u_2 \\ u_3 \end{bmatrix} = \begin{bmatrix} \cos\alpha_1 & \cos\beta_1 & \cos\gamma_1 \\ \cos\alpha_2 & \cos\beta_2 & \cos\gamma_2 \\ \cos\alpha_3 & \cos\beta_3 & \cos\gamma_3 \end{bmatrix} \begin{bmatrix} u_x \\ u_y \\ u_z \end{bmatrix} = R \begin{bmatrix} u_x \\ u_y \\ u_z \end{bmatrix} \tag{C.1}$$

Herein R represents the orthogonal transformation matrix. As an orthogonal matrix its inverse is simply equal to its transpose, as given by $R^{-1} = R'$.

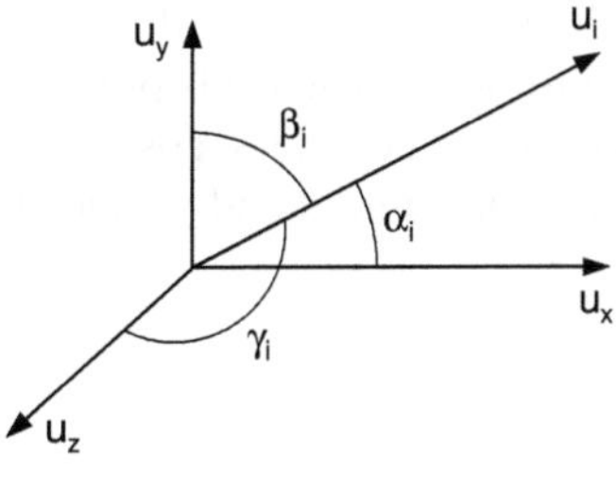

Fig. C.1 Relative position between two coordinate systems for velocity transformations

In order to transform the turbulent Reynolds stresses, the following matrix calculation is applied

$$\sigma_{\mathrm{mn}} = R' \sigma_{\mathrm{ij}} R \tag{C.2}$$

that is in concrete form

$$\begin{bmatrix} \sigma_{xx} & \tau_{xy} & \tau_{xz} \\ \tau_{yx} & \sigma_{yy} & \tau_{yz} \\ \tau_{zx} & \tau_{zy} & \sigma_{zz} \end{bmatrix} = R' \begin{bmatrix} \sigma_{11} & \tau_{12} & \tau_{13} \\ \tau_{21} & \sigma_{22} & \tau_{23} \\ \tau_{31} & \tau_{32} & \sigma_{33} \end{bmatrix} R \tag{C.3}$$

A special case will be considered in which the velocity component u_1 coincides with the main flow direction. Based on the zero correlation principle that has been presented in Chap. 8 for conducting the Zero Correlation Method (ZCM), the main flow direction also corresponds to the orientation of the principal normal stress. Because all turbulent shear stresses vanish in the related coordinate system it remains

$$\sigma_{\mathrm{ij}} = \begin{bmatrix} \sigma_{11} & 0 & 0 \\ 0 & \sigma_{22} & 0 \\ 0 & 0 & \sigma_{33} \end{bmatrix} \tag{C.4}$$

Starting from this state of turbulent stresses, all turbulent stresses in the $x - y - z$ coordinate system are then calculated from Eq. (C.2) as follows:

$$\sigma_{xx} = \sigma_{11} \cos^2 \alpha_1 + \sigma_{22} \cos^2 \alpha_2 + \sigma_{33} \cos^2 \alpha_3 \tag{C.5}$$

$$\sigma_{yy} = \sigma_{11} \cos^2 \beta_1 + \sigma_{22} \cos^2 \beta_2 + \sigma_{33} \cos^2 \beta_3 \tag{C.6}$$

$$\sigma_{zz} = \sigma_{11} \cos^2 \gamma_1 + \sigma_{22} \cos^2 \gamma_2 + \sigma_{33} \cos^2 \gamma_3 \tag{C.7}$$

$$\tau_{xy} = \sigma_{11} \cos \alpha_1 \cos \beta_1 + \sigma_{22} \cos \alpha_2 \cos \beta_2 + \sigma_{33} \cos \alpha_3 \cos \beta_3 \tag{C.8}$$

$$\tau_{zy} = \sigma_{11} \cos \beta_1 \cos \gamma_1 + \sigma_{22} \cos \beta_2 \cos \gamma_2 + \sigma_{33} \cos \beta_3 \cos \gamma_3 \tag{C.9}$$

$$\tau_{xz} = \sigma_{11} \cos \alpha_1 \cos \gamma_1 + \sigma_{22} \cos \alpha_2 \cos \gamma_2 + \sigma_{33} \cos \alpha_3 \cos \gamma_3 \tag{C.10}$$

A further special case will be considered. It is assumed that in the plane perpendicular to the main flow direction, i.e. in the plane of velocity components u_2 and u_3, isotropic turbulence is present. This simply means $\sigma_{22} = \sigma_{33}$. Then, with respect to $\cos^2 \alpha_1 + \cos^2 \alpha_2 + \cos^2 \alpha_3 = 1$ and the corresponding trigonometric identities for both the angle β and γ, one obtains

$$\sigma_{xx} = \sigma_{11} \cos^2 \alpha_1 + \sigma_{22} \sin^2 \alpha_1 \tag{C.11}$$

$$\sigma_{yy} = \sigma_{11} \cos^2 \beta_1 + \sigma_{22} \sin^2 \beta_1 \tag{C.12}$$

$$\sigma_{zz} = \sigma_{11} \cos^2 \gamma_1 + \sigma_{22} \sin^2 \gamma_1 \tag{C.13}$$

$$\tau_{xy} = \sigma_{11} \cos \alpha_1 \cos \beta_1 + \sigma_{22} (\cos \alpha_2 \cos \beta_2 + \cos \alpha_3 \cos \beta_3) \tag{C.14}$$

The last equation for τ_{xy} can be further simplified. Both the x- and the y-coordinate in the $x - y - z$ system will be represented by the respective unit vectors $\vec{x} = (\cos \alpha_1, \cos \alpha_2, \cos \alpha_3)$ and $\vec{y} = (\cos \beta_1, \cos \beta_2, \cos \beta_3)$ in the u_i-coordinate system. Because of $\vec{x} \perp \vec{y}$ there is $\vec{x} \cdot \vec{y} = 0$. This leads to

$$\cos \alpha_2 \cos \beta_2 + \cos \alpha_3 \cos \beta_3 = - \cos \alpha_1 \cos \beta_1 \tag{C.15}$$

By substituting this expression into Eq. (C.14) it then follows

$$\tau_{xy} = (\sigma_{11} - \sigma_{22}) \cos \alpha_1 \cos \beta_1 \tag{C.16}$$

This equation has been referred to in Sect. 17.4.2 by Eq. (17.42).

References

Albrecht H, Borys M, Damaschke N, Tropea C (2003) Laser Doppler and phase Doppler measurement techniques. Springer, Berlin

Boadway J, Karahan E (1981) Correction of laser Doppler anemometer readings for refraction at cylindrical interfaces. DISA Inf 26:4–6

Booij R, Tukker J (1994) 3-Dimensional laser Doppler measurements in a curved flume. 7th int. Symposium on Applications of Laser Techniques to Fluid Mechanics, Lisbon, Portugal, p 28.5

Bradshaw P (1978) Turbulence. Topics in applied physics, vol. 12. Springer, Berlin

Buchhave P (1975) Biasing errors in individual particle measurements with the LDA-counter signal processor. The Accuracy of Flow Measurements by Laser Doppler Methods, Proceedings of the LDA-Symposium, Copenhagen, Denmark, pp 258–278

Buchhave P, George WK, Lumley JL (1979) The measurement of turbulence with the laser-Doppler anemometer. Ann Rev Fluid Mech 11:443–503

Carter J, Martin K, Campbell W, Hall N, Ezekoye O (2001) Design of an oscillating flow apparatus for the study of low Reynolds number particle dynamics. J Exp Fluids 30:578–583

Durst F, Fischer M, Jovanovic J, Kikura H (1998) Methods to set up and investigate low Reynolds number, fully developed turbulent plane channel flows. ASME J Fluid Eng 120:496–503

Durst F, Kikura H, Lekakis I, Jovanovic J, Ye Q (1996) Wall shear stress determination from near-wall mean velocity data in turbulent pipe and channel flows. J Exp Fluids 20:417–428

Durst F, Martinuzzi R, Sender J, Thevenin D (1992) LDA-measurements of mean velocity, RMS-values and higher order moments of turbulence intensity fluctuations in flow fields with strong velocity gradients. 6th int. Symposium on Applications of Laser Techniques to Fluid Mechanics, Lisbon, Portugal, p 5.1

Durst F, Melling A, Whitelaw JH (1981) Principles and practice of laser-Doppler anemometry. 2nd edn, Academic Press, London

Durst F, Stevenson WH (1975): Moiré patterns to visually model laser-Doppler signals. The Accuracy of Flow Measurements by Laser Doppler Methods, Proceedings of the LDA-Symposium, Copenhagen, Denmark, pp 183–205

Edwards RV (1987) Report of the special panel on statistical particle bias problems in laser anemometry. Trans ASME, J Fluids Eng 109:89–93

Erdmann JC, Tropea C (1981) Turbulence induced statistical bias in laser anemometry. Proceedings of the 7th Symposium on Turbulence, University of Missouri-Rolla, USA

Hanson S (1973) Broadening of the measured frequency spectrum in a differential laser anemometer due to interference plane gradients. J Phys D Appl Phys 6:164–171

Hanson S (1975) Visualization of alignment errors and heterodyning constraints in laser Doppler velocimeters. The Accuracy of Flow Measurements by Laser Doppler Methods, Proceedings of the LDA-Symposium, Copenhagen, Denmark, pp 176–182

Hecht E (1990) Optics. 2nd edn, Addison-Wesley, Reading, MA

Hinze JO (1975) Turbulence. 2nd edn, McGraw-Hill, New York, NY

Hirt F, Jud E, Zhang Zh (1994) Investigation of the local flow topology in the vicinity of a prosthetic heart valve using particle image velocimetry. 7th. Int. Symposium on Applications of Laser Techniques to Fluid Mechanics, Lisbon, Portugal, p 37.3

Hüttmann F, Leder A, Michael M, Majohr D (2007) Wechselwirkungen runder Düsenfreistrahlen mit ebenen Wänden bei verschiedenen Auftreffwinkeln. 15. GALA-Fachtagung, Lasermethoden in der Strömungsmesstechnik, Rostock, Deutschland, pp 7.1–7.6

Jakoby R, Willmann M, Kim S, Dullenkopf K, Wittig S (1996) LDA-Messungen in rotierenden Bezugssystemen: Einfluss von Geschwindigkeitsgradienten auf die Bestimmung des Turbulenzgrads. 5. GALA-Fachtagung, Berlin, Deutschland, pp 41.1–41.10

Lehmann B (1986) Laser-Doppler-Messungen in einem turbulenten Freistrahl. DFVLR Forschungsbericht, DFVLR-FB 86–55

Li EB, Tieu AK (1998) Analysis of the three-dimensional fringe patterns formed by the interference of ideal and astigmatic Gaussian beams. 9th int. Symposium on Applications of Laser Techniques to Fluid Mechanics, Lisbon, Portugal, p 15.5

Lumley J, Acrivos A, Leal L, Leibovich S (1996) Research trends in fluid dynamics. American Institute of Physics, Woodbury, New York, NY

McLaughlin DK, Tiederman WG (1973) Biasing correction for individual realization of laser anemometer measurements in turbulent flows. Phys Fluids 16(12):2082–2088

Miles PC, Witze PO (1994): Fringe field quantification in an LDA probe volume by use of a magnified image. J Exp Fluids 16:330–335

Miles PC, Witze PO (1996): Evaluation of the Gaussian beam model for prediction of LDV fringe fields. 8th int. Symposium on Applications of Laser Techniques to Fluid Mechanics, Lisbon, Portugal, p 40.1

Nobach H (1998) Verarbeitung stochastisch abgetasteter Signale – Anwendung in der Laser-Doppler-Anemometrie. Diss., Univ. Rostock, Shaker Verlag, Aachen

Owen J, Rogers R (1975) Velocity biasing in laser Doppler anemometers. The Accuracy of Flow Measurements by Laser Doppler Methods, Proceedings of the LDA-Symposium, Copenhagen, Denmark, pp 89–114

Richter F, Leder A (2006) Wechselwirkungen runder Düsenfreistrahlen mit ebenen Wänden. 14. GALA-Fachtagung, Lasermethoden in der Strömungsmesstechnik. Braunschweig, Deutschland, pp 13.1–13.7

Ruck B. (1991) Distortion of LDA fringe pattern by tracer particles. J Exp Fluids 10:349–354

Thiele B, Eckelmann H (1994) Application of a partly submerged two component laser-Doppler anemometer in a turbulent flow. J Exp Fluids 17:390–396

Tropea C (1983) A note concerning the use of a one-component LDA to measure shear stress terms. Technical Notes, J Exp Fluids 1:209–210

Wittig S, Elsäßer A, Samenfink W, Ebner J, Dullenkopf K (1996) Velocity profiles in shear-driven liquid films: LDV-measurements. 8th Int. Symposium on Applications of Laser Techniques to Fluid Mechanics, Lisbon, Portugal, p 25.2

Yeh H, Cummins H (1964) Localized flow measurements with an He-Ne laser spectrometer. App Phys Lett 4:176

Zhang Zh (1995) Einfluss des Astigmatismus auf Laser Doppler Messungen. Sulzer Innotec Bericht, Nr. STT. TB95.022, Winterthur, Schweiz

Zhang Zh (1999) Null-Korrelationsmethode zur Bestimmung der anisotropen Strömungsturbulenz. Lasermethoden in der Strömungsmesstechnik. 7. GALA-Fachtagung, Lasermethoden in der Strömungsmesstechnik, St-Louis, Frankreich, pp 7.1–7.6

Zhang Zh (2000) Zur Bestimmung des "Biasing Error" in LDA-Messungen von komplexen turbulenten Strömungen. 8. GALA-Fachtagung, Lasermethoden in der Strömungsmesstechnik, Freising/München, Deutschland, pp 16.1–16.8

Zhang Zh (2002) Velocity bias in LDA measurements and its dependence on the flow turbulence. J Flow Meas Instrum 13:63–68

Zhang Zh (2004a) Optical guidelines and signal quality for LDA applications in circular pipes. J Exp Fluids 37:29–39

Zhang Zh (2004b) LDA-Methoden in Messungen aller drei Geschwindigkeitskomponenten in Rohrströmungen. 12. GALA-Fachtagung, Lasermethoden in der Strömungsmesstechnik, Karlsruhe, Deutschland, pp 8.1–8.8

Zhang Zh (2005) Dual-Measurement-Method and its extension for accurately resolving the secondary flows in LDA applications. J Flow Meas Instrum 16:57–62

Zhang Zh (2009) Freistrahlturbinen, Hydromechanik und Auslegung. Springer, Berlin

Zhang Zh, Bissel C, Parkinson E (2003) LDA-Anwendung zu Freistrahlmessungen bei einem Pelton-Turbine-Modell mit der Fallhöhe von 90 Metern. 11. GALA-Fachtagung, Lasermethoden in der Strömungsmesstechnik, Braunschweig, Deutschland, pp 13.1–13.6

Zhang Zh, Casey M (2007) Experimental studies of the jet of a Pelton turbine. Proc. IMechE Vol. 221 Part A: J Power Energy, 1181–1192

Zhang Zh, Eisele K (1995a) Off-axis alignment of an LDA-probe and the effect of astigmatism on the measurements. J Exp Fluids 19:89–94

Zhang Zh, Eisele K (1995b) Einfluss des Astigmatismus auf LDA-Messungen. 4. GALA-Fachtagung, Lasermethoden in der Strömungsmesstechnik, Rostock, Deutschland, pp 45.1–45.6

Zhang Zh, Eisele K (1996a) Neue Erkenntnisse über den Einfluss des Astigmatismus auf LDA-Messungen. 5. GALA-Fachtagung, Lasermethoden in der Strömungsmesstechnik, Berlin, Deutschland, pp 54.1–54.7

Zhang Zh, Eisele K (1996b) The effect of astigmatism due to beam refractions in the formation of the measurement volume in LDA measurements. J Exp Fluids 20:466–471

Zhang Zh, Eisele K (1997) On the broadening of the flow turbulence due to fringe distortion in LDA measurement volumes. Proceedings of the 7th Int. Conference Laser Anemometry, Advances and Applications, Karlsruhe, Germany, pp 351–357

Zhang Zh, Eisele K (1998a) On the directional dependence of turbulence properties in anisotropic turbulent flows. J Exp Fluids 24:77–82

Zhang Zh, Eisele K (1998b) Further considerations of the astigmatism error associated with off-axis alignment of an LDA-probe. J Exp Fluids 24:83–89

Zhang Zh, Eisele K (1998c) On the overestimation of the flow turbulence due to fringe distortion in LDA measurement volumes. J Exp Fluids 25:371–374

Zhang Zh, Eisele K (1998d) Zur Bestimmung des "Biasing Error" bei LDA-Messungen. 6. GALA-Fachtagung, Lasermethoden in der Strömungsmesstechnik, Essen, Deutschland, pp 37.1–37.37.7

Zhang Zh, Eisele K (1999) Mass flux measurements by PDA method. Proceedings of the 8th Int. Conference Laser Anemometry, Advances and Applications, Rome, Italy, pp 97–104

Zhang Zh, Eisele K, Geppert L (2000a) Untersuchungen am Freistrahl aus einer Modelldüse von Pelton-Turbinen mittels LDA. 8. GALA-Fachtagung, Lasermethoden in der Strömungsmesstechnik, Freising/München, Deutschland, pp 15.1–15.6

Zhang Zh, Eisele K, Hirt F (1996) Methode zur Bestimmung der Turbulenzgrößen in instationären Strömungen aus LDA-Messungen. 5. GALA-Fachtagung, Lasermethoden in der Strömungsmesstechnik, Berlin, Deutschland, pp 10.1–10.4

Zhang Zh, Eisele K, Hirt F (1997) The influence of phase-averaging window size on the determination of turbulence quantities in unsteady turbulent flows. J Exp Fluids 22: 265–267

Zhang Zh, Eisele K, Ziada S, Kälin R (1998) PDA measurements of water jet break-up in air cross-flow. Proceedings of the 7th European Symposium Particle Characterisation, Nürnberg, Germany, pp 131–140

Zhang Zh, Muggli F, Parkinson E, Schärer C (2000b) Experimental investigation of a low head jet flow at a model nozzle of a Pelton turbine. 11th int. Seminar on Hydropower Plants, Vienna, Austria, pp 181–188

Zhang Zh, Parkinson E (2001) Strömungsuntersuchungen am Freistrahl der Pelton-Turbine und Anpassen des LDA-Verfahrens. 9. GALA-Fachtagung, Lasermethoden in der Strömungsmesstechnik, Winterthur, Schweiz, pp 43.1–43.7

Zhang Zh, Parkinson E (2002) LDA application and the dual-measurement-method in experimental investigations of the free surface jet at a model nozzle of a Pelton turbine. 11th. Int. Symposium on Applications of Laser Techniques to Fluid Mechanics, Lisbon, Portugal, p 2

Zhang Zh, Zhang Ch (2002) Null-Korrelations-Methode und die Abweichungen in Turbulenzmessungen. 10. GALA-Fachtagung, Lasermethoden in der Strömungsmesstechnik, Rostock, Deutschland, pp 40.1–40.8

Zhang Zh, Ziada S (2000) PDA measurements of droplet size and mass flux in the three-dimensional atomisation region of water jet in air cross-flow. J Exp Fluids 28:29–35

Index

MIX
Papier aus verantwortungsvollen Quellen
Paper from responsible sources
FSC® C105338
FSC
www.fsc.org

If you have any concerns about our products,
you can contact us on
ProductSafety@springernature.com

In case Publisher is established outside the EU,
the EU authorized representative is:
Springer Nature Customer Service Center GmbH
Europaplatz 3, 69115 Heidelberg, Germany

Printed by Libri Plureos GmbH
in Hamburg, Germany